Athanase Papadopoulos

Editor

Surveys in Geometry II

 Springer

Editor
Athanase Papadopoulos
Institut de Recherche Mathématique
Avancée and Centre de
Recherche et Expérimentation sur l'Acte
Artistique
Strasbourg, France

ISBN 978-3-031-43509-6   ISBN 978-3-031-43510-2   (eBook)
https://doi.org/10.1007/978-3-031-43510-2

This Springer imprint is published by the registered company Springer Nature Switzerland AG
The registered company address is: Gewerbestrasse 11, 6330 Cham, Switzerland

Paper in this product is recyclable.

Preface

This is the second volume of a collection of surveys on topics that are at the forefront of current research in geometry. They are intended for graduate students and researchers. Some of these surveys are based on lectures given by their authors to middle-advance and graduate students, and all of them can be used as bases for courses on geometry. Each chapter concentrates on a topic which I consider particularly interesting and which is worth highlighting. The topics include Riemann surfaces, metric geometry, Finsler geometry, Riemannian geometry, projective geometry, symplectic geometry, Teichmüller spaces and combinatorial group theory.

I would like to thank Elena Griniari for her kind support and care for this project, and the reviewers of the various chapters for their valuable anonymous work. My warm thanks go to all the authors, for a fruitful and friendly collaboration.

Nisyros (Dodecanese), Greece   Athanase Papadopoulos
September 2023

Contents

Editor and Contributors

About the Editor

Athanase Papadopoulos (born 1957) is Directeur de Recherche at the French Centre National de la Recherche Scientifique. His main fields of interest are geometry and topology, the history and philosophy of mathematics, and mathematics and music. He has held visiting positions at the Institute for Advanced Study, Princeton (1984–1985 and 1993–1994), USC (1998–1999), CUNY (Ada Peluso Professor, 2014), Brown University (Distinguished Visiting Professor, 2017), Tsinghua University, Beijing (2018), Lamé Chair of the State University of Saint Petersburg (2019), and has had several month visits to the Max-Plank Institute for mathematics (Bonn), the Erwin Schrödinger Institute (Vienna), the Graduate Center of CUNY (New York), the Tata Institute (Bombay), Galatasaray University (Istanbul), the University of Florence (Italy), Fudan University (Shanghai), Gakushuin University (Tokyo) and Presidency University (Calcutta). He is the author of more than 220 published articles and 45 monographs and edited books.

Contributors

Norbert A'Campo Departement Mathematik und Informatik, Fachbereich Mathematik, Universität Basel Basel, Schweiz

Valerii Nikolaevich Berestovskiĭ Principal Scientific Researcher, Sobolev Institute of Mathematics of the SB RAS Novosibirsk, Russia

Bruno Luiz Santos Correia Institut de Mathématiques EPFL Lausanne, Switzerland

Boumediene Et-Taoui Université de Haute Alsace: IRIMAS Mulhouse, France

Peter Kristel Hausdorff Center for Mathematical Sciences Bonn, Germany

Árpád Kurusa Bolyai Institute, University of Szeged Szeged, Hungary

Gye-Seon Lee Department of Mathematics, Sungkyunkwan University Jangan-gu, South Korea

Ludovic Marquis Université Rennes, CNRS Rennes, France

Yuriĭ Gennadievich Nikonorov Principal Scientific Researcher, Southern Mathematical Institute of VSC RAS Vladikavkaz, Russia

Ken'ichi Ohshika Department of Mathematics, Faculty of Science, Gakushuin University Tokyo, Japan

Athanase Papadopoulos Institut de Recherche Mathématique Avancée and Centre de Recherche et Expérimentation sur l'Acte Artistique, Strasbourg, France

Eric Schippers University of Manitoba Winnipeg, MB, Canada

Graham Smith Departamento de Matemática, Pontifícia Universidade Católica do Rio de Janeiro (PUC-Rio) Rio de Janeiro, Brazil

Marc Troyanov Institut de Mathématiques EPFL Lausanne, Switzerland

Chapter 1
Introduction

Athanase Papadopoulos

Abstract This chapter contains a description of the various subjects covered in the book.

Keywords Conformal geometry · Metric geometry · Teichmüller spaces · Surfaces immersed in 3-manifolds · Symplectic geometry · Grassmann spaces · Finite homogeneous metric spaces · Polytopes · The Gauss–Bonnet formula · Isoperimetry · Coxeter groups

AMS Codes 12D10, 14H55, 20B25, 26C10, 30F10, 30F20, 20F55, 30F30, 30F60, 32G15, 51F20, 51F15, 51M35, 53A35, 53D30, 53C70, 57K10

This is the second volume of a collection of multi-authored surveys in geometry. The project of publishing these volumes arose from the conviction that the mathematical community is in need of good surveys of topics that are at the heart of current research. Thus I have asked some colleagues and friends to write an expository article on a subject they have been working on, which I find particularly important or interesting. I hope the result will be of use to mathematicians, both beginners and experienced.

The topics surveyed in the present volume include the conformal and the metric geometry of surfaces, Teichmüller spaces, surfaces immersed with prescribed extrinsic curvature in 3-dimensional manifolds, symplectic geometry, the metric theory of Grassmann spaces, finite homogeneous metric spaces, projective metric spaces, regular and semi-regular polytopes, the Gauss–Bonnet formula and its higher-dimensional versions, isoperimetry in finitely generated groups, and Coxeter groups. Let me review now each chapter in some detail.

A. Papadopoulos (✉)
Institut de Recherche Mathématique Avancée and Centre de Recherche et Expérimentation sur l'Acte Artistique, Strasbourg, France
e-mail: papadop@math.unistra.fr

© The Author(s), under exclusive license to Springer Nature Switzerland AG 2024
A. Papadopoulos (ed.), *Surveys in Geometry II*,
https://doi.org/10.1007/978-3-031-43510-2_1

Chapter 2, by Norbert A'Campo and me, is titled *Geometry on surfaces, a source for mathematical developments*. It is an overview of the theory of surfaces equipped with various structures: volume forms, almost complex structures, Riemannian metrics of constant curvature, quasiconformal structures, and others. Several topics discussed originate in the Riemann mapping theorem and its generalization, the uniformization theorem. The theory of Riemann surfaces is intertwined with topology and combinatorics. Higher-dimensional structures are also discussed. We also review several applications of a colored graph associated with a branched covering of the sphere, which we call a *net*, or a *Speiser net*. Nevanlinna, Teichmüller and others obtained criteria, using this graph, for the determination of the type of a simply connected surface, that is, to know whether the surface is conformally equivalent to the Euclidean plane or to the unit disc. The same graph appears in the theory of dessins d'enfants as well as in a realization theorem of Thurston concerning the characterization of some branched coverings of the sphere. Combinatorial characterizations of polynomials and rational maps among a class of branched covers of the sphere are discussed. We also survey the notion of rooted colored trees associated with slalom polynomials, and certain graphs that are used in the stratification of the space of monic polynomials. New models of the Riemann sphere and of hyperbolic 2- and 3-spaces appear at several places of the survey, some of them in an unexpected manner, using algebra (the ring of polynomials). By analogy, some constructions based on the field of complex numbers are extended to arbitrary fields.

Chapter 3, by Ken'ichi Ohshika, is titled *Teichmüller spaces and their various metrics*. This chapter is a survey of three different Finsler metrics on Teichmüller space: the Teichmüller metric, Thurston's asymmetric metric and the earthquake metric. In particular, the author presents some recent results he obtained with Y. Huang, H. Pan and A. Papadopoulos on the so-called earthquake metric, also introduced by Thurston. He reviews a duality established in that paper between the tangent space at an arbitrary point of Teichmüller space equipped with the earthquake norm and the cotangent space at the same point equipped with Thurston's co-norm. He also reports on a result, also obtained in that paper, saying that the earthquake metric is not complete, and he provides a description of its completion.

Chapter 4 by Marc Troyanov is titled *Double forms, curvature integrals and the Gauss-Bonnet formula*. We recall that the Gauss–Bonnet formula for surfaces is a major achievement of nineteenth century differential geometry, and is one of the very good examples of how topology is closely related to geometry. In its simplest form, the formula says that for a closed surface S equipped with a Riemannian metric, the integral of the Gaussian curvature is equal (up to a universal constant) to the Euler characteristic of the surface. There are much more evolved forms of the formula. Troyanov, in Chap. 4, recalls that the generalized Gauss–Bonnet formula is due to the effort of several mathematicians, namely, K. F. Gauss, P. Bonnet, J. Binet, and W. von Dyck, and that extensions of this formula to higher-dimensional Riemannian manifolds were obtained in the twentieth century by H. Hopf, W. Fenchel, C. B. Allendoerfer, A. Weil and S.S. Chern. The extended formula establishes relations between the Euler characteristic of a smooth, compact

Riemannian manifold with (possibly empty) boundary and a curvature integral over the manifold plus a boundary term that involves a combination of curvature and the second fundamental form over the boundary. In this chapter, the author revisits the higher-dimensional formula using the formalism of double forms, a tool introduced by G. de Rham and further developed by R. Kulkarni, J. Thorpe and A. Gray in the 1960s and 1970s. In particular, he surveys the history and the techniques around Chern's version of the formula, which uses É. Cartan's moving frame formalism. He explores the geometric nature of the boundary term and he provides examples and applications. This chapter is also an occasion for Troyanov to make a detailed account of the various tools that were used by various authors in the proof of the higher-dimensional Gauss–Bonnet formula, starting with H. Hopf's problem on the *Curvatura Intega*, including the theories of double forms with their geometric applications, the Pfaffian, the moving frame, the Gauss–Kronecker curvature of hypersurfaces, the Lipschitz–Killing curvature, and several other notions.

Chapter 5, by Graham Smith, is titled *Quaternions, Monge–Ampère structures and κ-surfaces*. This chapter combines Riemannian geometry, conformal geometry, symplectic geometry and quaternionic geometry (an analogue of complex geometry where the field of complex numbers is replaced by that of quaternions). The study is based on a work of F. Labourie. The latter developed, in his paper *Problèmes de Monge-Ampère, courbes pseudo-holomorphes et laminations*, published in 1997, a theory of surfaces immersed in three-manifolds with prescribed extrinsic curvature. This theory has found applications in hyperbolic geometry, general relativity, Teichmüller theory and other domains. Labourie's key insight is that if the second fundamental form of an immersed surface of prescribed extrinsic curvature is positive definite (the surface is then called infinitesimally strictly convex), then its Gauss lift is a pseudo-holomorphic curve for some suitable almost complex structure. This allows the application of Gromov's theory developed in his paper *Pseudo-holomorphic curves in symplectic manifolds* (1987), and in particular, his compactness results for families of immersed surfaces of prescribed extrinsic curvature in 3-dimensional Riemannian manifolds. Smith, in this chapter, builds on Labourie's work, of which he gives a quaternionic reformulation. This leads him to simpler proofs of Labourie's results, and at the same time, to generalisations to higher-dimensions. Two theorems of Labourie are in the background: a compactness result for sequences of quasicomplete pointed infinitesimally strictly convex immersed surfaces in a complete, oriented, 3-dimensional Riemannian manifold with prescribed curvature, and a description of the accumulation points of such a sequence. A key result of this chapter establishes a relation between the solutions of the 2-dimensional Monge–Ampère equation and pseudo-holomorphic curves.

Chapter 6, by Peter Kristel and Eric Schippers, is titled *Lagrangian Grassmannians of polarizations*. This chapter is an introduction to polarization theory in symplectic and orthogonal geometries. In this setting, one starts with a triple of structures on a real vector space, namely, an inner product, a symplectic form and a complex structure, the triple satisfying a compatibility condition that ensures that when we are provided with two out of these three structures, we can reconstruct the third one, if it exists (which is not always the case). The specification of such a compatible triple is equivalent to a decomposition of the complexified ambient vector space into the eigenspaces of the complex structure. A familiar example of such a triple of structures is that of a Kähler manifold, where the manifold is equipped with three compatible structures: a Riemannian metric, an integrable complex structure and a symplectic form. Kristel and Schippers adopt the natural point of view of fixing *one* of these structures, and studying the space of structures of a second type allowing the reconstruction of a compatible structure of the third type. For example, we can fix a symplectic manifold and study the space of integrable complex structures on the manifold such that this manifold admits a compatible Riemannian metric. In the case where either the symplectic form or the inner product is fixed, we get a Grassmannian of polarizations. Polarizations appear in several contexts of algebraic and complex geometry: in the study of moduli spaces, in the theory of metaplectic and spin representations, and in conformal field theory. In Chap. 6, Kristel and Schippers survey this circle of ideas, in which the underlying vector spaces are allowed to be infinite-dimensional, emphasizing the symmetry of the symplectic and orthogonal settings. The authors consider two particular situations: the Riemannian one, in which the polarization is an orthogonal decomposition, and the symplectic one, in which the polarization is a symplectic decomposition. The potential fields of applications of this theory of polarizations include representation theory, loop groups, complex geometry, moduli spaces, quantization, and conformal field theory.

The next three chapters are concerned with metric geometry.

Chapter 7, by Árpád Kurusa, is titled *Metric characterizations of projective-metric spaces*. The author starts by recalling Hilbert's Problem IV of the list of problems he proposed in 1900 at the Paris International Congress of Mathematicians. The problem asks for the construction of all the projective metrics on an open convex subset of projective space, that is, the metrics whose geodesics are the intersections of this open subset with the projective lines of the ambient space. Kurusa addresses the general question of characterizing such spaces under the effect of adding some additional conditions on the metric. On the same occasion, he surveys notions like Hilbert and Minkowski metrics, projective center, Ptolemaic metric, Erdös ratio, conics in metric spaces, general Finsler projective metrics, the Ceva and Menelaus properties, and other properties of triangles in a projective-metric setting. Notions such as plane and line perpendicularity, equidistance of lines, bisectors, medians, bounded curvature and others, that hold in a general metric space, are used. These properties were introduced in such a general setting by Herbert Busemann. The questions of characterizing Minkowski planes, of Hilbert geometries and of the three classical geometries, which were extensively investigated by Busemann, are addressed by Kurusa, who also mentions several open problems on this topic.

Chapter 8, by the same author, titled *Supplement to "Metric Characterization of projective-metric spaces"*, is a supplement to the previous chapter whose goal is to provide proofs of two theorems. The first one, due to B. B. Phadke, says that a projective-metric space is a Minkowski plane if all equidistants to geodesics are geodesics. The proof that the author provides is different from Phadke's original proof. The second theorem, due to the author, says that a projective-metric plane has the Ceva property (a generalization of the classical Ceva property) if and only if it is either a Minkowski plane or a model of hyperbolic or elliptic geometry. The proofs are long, which is the reason why these theorems are included in a separate chapter.

Chapter 9, by Boumediene Et-Taoui, is titled *Metric problems in projective and Grassmann spaces*. In this chapter, the author studies metric problems in real, complex and quaternionic spaces concerning equiangular lines and equi-isoclinic n-subspaces. Let us recall the setting. We take $\mathbb{F}$ to be the real, complex or quaternionic field. Let $p \geq 3$ and $r \geq 2$ be two integers. A collection of lines in $\mathbb{F}^r$ is said to be *equiangular* if any two lines in this collection make the same nonzero angle. A set of p n-subspaces in $\mathbb{F}^r$ is said to be *equi-isoclinic* if this set spans $\mathbb{F}^r$ and if any two lines in this set make the same non-zero angle. As the author notes, such structures appear, under various names, in fields such as discrete geometry, combinatorics, harmonic analysis, frame theory, coding theory and quantum information theory. The author addresses two natural questions, namely, (i) How many equiangular lines can be placed in $\mathbb{F}^r$? (ii) How many equiangular equi-isoclinic n-subspaces can be placed in $\mathbb{F}^r$? The chapter is a survey of the work done and the developments due to various authors in the last 70 years on these and related questions, interpreted in the setting of the metric geometry of projective and Grassmann spaces. This metric setting involves classical and fundamental works of Menger, Blumenthal, Lemmens, Seidel and others, as well as works of the author himself with co-authors.

Chapter 10, by Valeriĭ Berestovskiĭ and Yuriĭ Nikonorov, is titled *On the geometry of finite homogeneous subsets of Euclidean spaces*. The authors review recent results on finite homogeneous metric spaces, that is, spaces on which their isometry group acts transitively. They are especially interested in finite homogeneous metric subspaces of a Euclidean space that represent vertex sets (assumed to lie on a sphere) of compact convex polytopes whose isometry groups are transitive on the vertex set. In particular, the authors are led to the classification of regular and semiregular polytopes in Euclidean spaces, according to whether or not they satisfy the normal homogeneity property or the Clifford–Wolf homogeneity property on their vertex sets. The metric spaces that satisfy these two properties constitute a remarkable subclass of the class of homogeneous metric spaces. These properties are stronger than the usual homogeneity properties used for homogeneous metric spaces. The definitions of normal and Clifford–Wolf homogeneity involve a property satisfied by the isometry taking one point to the other, in the definition of homogeneity. The fact that such a study is closely related to the theory of convex polytopes in Euclidean spaces makes it natural to first check the presence of these properties for the vertex sets of regular and semiregular polytopes. Berestovskiĭ and Nikonorov are then led to the study of the m-point homogeneity property and to the notion of point homogeneity degree for finite metric spaces. They discuss several recent results, in particular, the classification of polyhedra with all edges of equal length and with 2-point homogeneous vertex sets, and they present results on the point homogeneity degree for some important classes of polytopes. While discussing these classification results, the authors explain in detail the main tools used for the study of the relevant objects, and they discuss prospects for future results, presenting several open problems.

Chapter 11, by Gue-Seon Lee and Ludovic Marquis, is titled *Discrete Coxeter groups*. It is an introduction to Coxeter groups, with a focus on how these groups can be used for the construction of discrete subgroups of Lie groups. Coxeter groups are geometrically defined groups. They were introduced in 1934 by H. S. M. Coxeter. They are generated by reflections in some spaces. They generalize the Euclidean reflection groups and the symmetry groups of regular polyhedra. They appear in several areas of mathematics, in particular in the theory of representations of discrete groups in Lie groups. The topics discussed in this chapter include, besides the general Coxeter groups, the theories of reflection groups in hyperbolic space, convex cocompact projective reflection groups, projective reflection groups, divisible and quasi-divisible domains in Hilbert geometry, and Anosov representations. The latter constitute a generalization of the class of discrete convex cocompact representations of hyperbolic groups into rank one Lie groups to the setting of representations of hyperbolic groups into semi-simple Lie groups. Anosov representations were introduced by Labourie in his paper *Anosov flows, surface groups and curves in projective space* (2006), in which he studies Hitchin representations.

Chapter 12, by Bruno Luiz Santos and Marc Troyanov, is titled *Isoperimetry in finitely generated groups*. The setting is that of infinite finitely generated groups equipped with word metrics associated with finite symmetric sets of generators. An isoperimetric inequality is an inequality between the size of an arbitrary finite set and the size of its boundary (for an appropriate definition of boundary). In this chapter, the authors revisit the work done in the 1980s–1990s by N. Varopoulos, T. Coulhon and L. Saloff-Coste on isoperimetric inequalities in finitely generated groups, adding new results and establishing relations with other topics. In particular, they obtain lower bounds for the isoperimetric quotient that appears in the isoperimetric inequality in terms of the $\mathcal{U}$-transform, which is a variant of the classical Legendre transform of a function, or its generalization, the Legendre–Fenchel transform, which is used in physics. The chapter also includes a review of some basic elements from geometric group theory (growth functions, amenability, the Cheeger constant, etc.) as well as some basic elements from the theory of $\mathcal{U}$-transform, including some computational techniques, and the relation between the $\mathcal{U}$-transform and the Legendre transform.

The first two chapters of this volume are based on lectures given by the authors at two thematic programs at Banaras Hindu University, in December 2019 and December 2022, which I organized with Bankteshwar Tiwari. The programs were funded by CIMPA (Centre International de Mathématiques Pures et Appliquées), IMU (International Mathematical Union), SERB (Science and Engineering Research Board of the Government of India), NBHM (National Board of Higher Mathematics of the Government of India), and SRICC (Sponsored Research and Industrial Consultancy Cell) and ISC-BHU (the Institute of Science of Banaras Hindu University).

Chapter 2
Geometry on Surfaces, a Source for Mathematical Developments

Norbert A'Campo and Athanase Papadopoulos

Abstract We present a variety of geometrical and combinatorial tools that are used in the study of geometric structures on surfaces: volume, contact, symplectic, complex and almost complex structures. We start with a series of local rigidity results for such structures. Higher-dimensional analogues are also discussed. Some constructions with Riemann surfaces lead, by analogy, to notions that hold for arbitrary fields, and not only the field of complex numbers. The Riemann sphere is also defined using surjective homomorphisms of real algebras from the ring of real univariate polynomials to (arbitrary) fields, in which the field with one element is interpreted as the point at infinity of the Gaussian plane of complex numbers. Several models of the hyperbolic plane and hyperbolic 3-space appear, defined in terms of complex structures on surfaces, and in particular also a rather elementary construction of the hyperbolic plane using real monic univariate polynomials of degree two without real roots. Several notions and problems connected with conformal structures in dimension 2 are discussed, including dessins d'enfants, the combinatorial characterization of polynomials and rational maps of the sphere, the type problem, uniformization, quasiconformal mappings, Thurston's characterization of Speiser nets, stratifications of spaces of monic polynomials, and others. Classical methods and new techniques complement each other.

Keywords Geometric structure · Conformal structure · Almost complex structure (J-field) · Riemann sphere · Uniformization · The type problem · Rigidity · Model for hyperbolic space · Cross ratio · Belyi's theorem · Riemann–Hurwitz formula · Chasles 3-point function · Branched covering · Type

N. A'Campo
Departement Mathematik und Informatik, Fachbereich Mathematik, Universität Basel, Basel, Switzerland
e-mail: Norbert.ACampo@unibas.ch

A. Papadopoulos (✉)
Institut de Recherche Mathématique Avancée and Centre de Recherche et Expérimentation sur l'Acte Artistique, Strasbourg, France
e-mail: athanase.papadopoulos@math.unistra.fr

© The Author(s), under exclusive license to Springer Nature Switzerland AG 2024
A. Papadopoulos (ed.), *Surveys in Geometry II*,
https://doi.org/10.1007/978-3-031-43510-2_2

problem · Dessin d'enfants · Slalom polynomial · Slalom curve · Space of monic polynomials · Stratification · Fibered link · Divide · Speiser curve · Speiser graph · Line complex · Quasiconformal map · Almost analytic function · Net · Speiser net

AMS Classification 12D10, 26C10, 14H55, 30F10, 30F20, 30F30, 53A35, 53D30, 57K10

2.1 Introduction

Given a differentiable surface, i.e., a 2-dimensional differentiable manifold, one can enrich it with various kinds of geometric structures. Our first aim in the present survey is to give an introduction to the study of surfaces equipped with locally rigid and homogeneous geometric structures.

Formally, a geometric structure on a surface S is given by a section of some bundle associated with its tangent bundle TS. We shall deal with specific examples, mostly, volume forms, almost complex structures (equivalently, conformal structures, since we are dealing with surfaces) and Riemannian metrics of constant Gaussian curvature. We shall also consider quasiconformal structures on surfaces. Foliations with singularities, Morse functions, meromorphic functions and differentials on almost complex surfaces induce geometric structures that are locally rigid and homogeneous only in the complement of a discrete set of points on the surface. Laminations, measured foliations and quadratic differentials are examples of less homogeneous geometric structures. They play important roles in the theory of surfaces, as explained by Thurston, but we shall not consider them here.

A theorem of Riemann gives a complete classification of non-empty simply connected open subsets of $\mathbb{R}^2$ that are equipped with almost complex structures. Only two classes remain! This takes care at the same time of the topological classification of such surfaces without extra geometric structure: they are all homeomorphic. The classical proof of this topological fact invokes the Riemann Mapping Theorem, that is, it assumes the existence of an almost complex structure on the surface. Likewise, only two classes remain in the classification of non-empty open connected and simply connected subsets of $\mathbb{R}^2$ that are equipped with a Riemannian metric of constant curvature: the Euclidean and the Bolyai–Lobachevsky plane. The latter is also called the non-Euclidean or hyperbolic plane.

No classification theorem similar to that of simply connected open subsets of $\mathbb{R}^2$ holds in $\mathbb{R}^3$, even if one restricts to contractible subsets. See [113] for the historical example, now called "Whitehead manifold", which, by a result of Gabai [39], is a manifold of small category, i.e., it is covered by two charts, both of which being copies of $\mathbb{R}^3$ that moreover intersect along a third copy of $\mathbb{R}^3$.

We shall be particularly concerned with almost complex structures, i.e., conformal structures, on surfaces. The theory of such structures is intertwined with topology. This is not surprising: Riemann's first works on functions of one complex variable gave rise at the same time to fundamental notions of topology. He conceived the notion of "n-extended multiplicity" (*Mannigfaltigkeit*), an early version of n-manifold, he introduced basic notions like connectedness and degree of connectivity for surfaces, which led him to the discovery of Betti numbers in the general setting (see Andé Weil's article [112] on the history of the topic), he classified closed surfaces according to their genus, he introduced branched coverings, and he was the first to notice the topological properties of functions of one complex variable (one may think of the construction of a Riemann surface associated with a multi-valued meromorphic function). At about the same time, Cauchy, in his work on the theory of functions of one complex variable, introduced path integrals and the notion of homotopy of paths. We shall see below many such instances of topology meeting complex geometry.

In several passages of the present survey, we shall encounter graphs that are used in the study of Riemann surfaces. They will appear in the form of:

1. Speiser nets associated with branched coverings of the sphere: these are used in Thurston's realization theorem for branched coverings (Sect. 2.6.3), in the type problem (Sect. 2.7.2), in the theory of dessins d'enfants (Sect. 2.8.1) and in a cell-decomposition of the space of rational maps (Sect. 2.8.4);
2. rooted colored trees associated with slalom polynomials (Sect. 2.8.2);
3. pictures of monic polynomials used for the stratification of the space of slalom polynomials (Sect. 2.8.3).

At several places, we shall see how familiar constructions using the field of complex numbers can be generalized to other fields. Conversely, algebraic considerations will lead to several models of the Riemann sphere and of 2- and 3-dimensional hyperbolic spaces. Relations with the theory of knots and links will also appear.

Let us give now a more detailed outline of the next sections:

In Sect. 2.2 we present a few classical examples of rigidity and local rigidity results in the setting of geometric structures on n-dimensional manifolds. A theorem due to Jürgen Moser, whose proof is sometimes called "Moser's Trick", deals with the classification up to isotopy of volume forms on compact connected oriented n-dimensional manifolds. We show how this proof can be adapted to the symplectic and contact settings. A local rigidity result (which we call a Darboux local rigidity theorem) gives a canonical form for volume, symplectic and contact forms on non-empty connected and simply connected open subsets (S, ω) of $\mathbb{R}^n$.

In dimension two, almost complex structures are also locally rigid, and we present a Darboux-like theorem for them. The question of the existence and integrability of J-structures on higher-dimensional spheres arises naturally. We survey a result due to Adrian Kirchhoff which says that an n-dimensional sphere admits a J-structure if and only if the $(n + 1)$-dimensional sphere admits a parallelism, that is, a global field of frames. This deals with the question of the existence of J-fields on higher-dimensional spheres, which we also discuss in the same section: only S^6 carries such a structure.

Section 2.3 is concerned with the first example of Riemann surface, namely, the Riemann sphere. We give several models of this surface. Its realization as the projective space $\mathbb{P}^1(\mathbb{C})$ leads to constructions that are valid for any field k and not only for $\mathbb{C}$. In the same section, we review a realization of $\mathbb{P}^1(\mathbb{C})$ with its round metric as a quotient space of the group SU(2) of linear transformations of $\mathbb{C}^2$ of determinant 1 preserving the standard Hermitian product. The intermediate quotient SU(2)/{$\pm$Id} is isometric to $P^3(\mathbb{R})$ and also to the space $T_{l=1}\mathbb{P}^1(\mathbb{C})$ of length 1 tangent vectors to S^2. In this description, oriented Möbius circles on $\mathbb{P}^1(\mathbb{C})$ (that is, the circles of the conformal geometry of $\mathbb{P}^1(\mathbb{C})$) lift naturally to oriented great circles on $S^3 =$ SU(2). Also, closed immersed curves without self-tangencies lift to classical links (that is, links in the 3-sphere).

In Sect. 2.4, we present another model of the Riemann sphere, together with models of the hyperbolic plane and of hyperbolic 3-space. A model of the Riemann sphere is obtained using algebra, namely, fields and ring homomorphisms. In this model, the point at infinity of the complex plane is represented by $\mathbb{F}_1$, the field with one element. The notion of shadow number is introduced, as a geometrical way of viewing the cross ratio. The hyperbolic plane appears as a space of ideals equipped with a geometry naturally given by a family of lines. In this way, the hyperbolic plane has a very simple description which arises from algebra. The cross ratio is used to prove a necessary and sufficient condition for a generic configuration of planes in a real 4-dimensional vector space to be a configuration of complex planes. We then introduce the notions of compatible (or J-conformal) Riemannian metric and we prove the existence and uniqueness of such metrics on homogeneous Riemann surfaces with commutative stabilizers. We describe several models of spherical geometry (surfaces of constant curvature +1), and of 2- and 3-dimensional hyperbolic spaces in terms of the complex geometry of surfaces. We then study the notion of J-compatible Riemannian metrics. An existence result of such metrics is the occasion to characterize homogeneous Riemann surfaces up to bi-holomorphic equivalence.

In Sect. 2.5, we reduce generality by assuming that the surface S is an open connected and path-connected non-empty subset of the real plane $\mathbb{R}^2$. Fundamental results appear. For instance, the theorem saying that two open connected and path-connected non-empty subsets of the real plane $\mathbb{R}^2$ are diffeomorphic, a consequence of the Riemann Mapping Theorem. This theorem says that any nonempty open subset of the complex plane which is not the entire plane is biholomorphically equivalent to the unit disc. The Riemann Mapping Theorem generalized to any simply connected Riemann surface (and not restricted to open subsets of the plane) is the famous Uniformization Theorem. It leads to the type problem, which we consider in Sect. 2.7.

The next section, Sect. 2.6, is concerned with some aspects of branched coverings between surfaces. A classical combinatorial formula associated with such an object is the Riemann–Hurwitz formula. It leads to some natural problems which are still unsolved. A combinatorial object associated with a branched covering of the sphere is a Jordan curve that passes through all the critical values and which we call a *Speiser curve*. Its lift by the covering map is a graph we call a *net*, or *Speiser net*, an object that will be used several times in the rest of the survey. A theorem of

Thurston which we recall in this section gives a characterization of oriented graphs on the sphere that are Speiser graphs of some branched covering of the sphere by itself. Thurston proved this theorem as part of his project of understanding what he called the "shapes" of rational functions of the Riemann sphere. In the same section, we introduce a graph on a surface which is dual to the net, often known in the classical literature under the name *line complex*, which we use in an essential way in Sect. 2.7. We reserve the name line complex to another graph.

The type problem, reviewed in Sect. 2.7, is the problem of finding a method for deciding whether a simply connected Riemann surface, defined in some specific manner (e.g., as a branched covering of the Riemann sphere, or as a surface equipped with some Riemannian metric, or obtained by gluing polygons, etc.) is conformally equivalent to the Riemann sphere, or to the complex plane, or to the open unit disc. We review several methods of dealing with this problem, mentioning works of Ahlfors, Nevanlinna, Teichmüller, Lavrentieff and Milnor. Besides the combinatorial tools introduced in the previous sections (namely, nets and line complexes), the works on the type problem that we review use the notions of almost analytic function and quasiconformal mapping.

In the last section, Sect. 2.8, combinatorial tools are used for other approaches to Riemann surfaces, in particular, in the theory of dessins d'enfants, in applications to knots and links and in the theory of slalom polynomials. Two different stratifications of the space of monic polynomials are presented.

2.2 Rigidity of Geometric Structures

In this section, we give several examples of locally rigid structures on surfaces. A classical example of a non-locally rigid structure is a Riemannian metric on any manifold of dimension ≥ 2.

2.2.1 Volume, Symplectic and Contact Forms

Moser's theorem says that only the total volume of a smooth volume form on a connected compact manifold matters, namely, two volume forms of equal total volume are isotopic. More precisely:

Theorem 2.2.1 (Moser [77]) *Let M be a compact connected oriented manifold of dimension n equipped with two smooth volume forms ω_0 and ω_1 of equal total volume. Then there exists an isotopy ϕ_t, $t \in [0, 1]$, satisfying $\phi_t^*(t\omega_1 + (1-t)\omega_0) = \omega_0$. In particular, we have $\omega_0 = \phi_1^*\omega_1$.*

Proof Clearly $\omega_1 = f\omega_0$ for some positive function f, since for any $p \in M$ and for any oriented frame $X_1, \cdots, X_n$ at p we have $\omega_0(X_1, \cdots, X_n) > 0$ and $\omega_1(X_1, \cdots, X_n) > 0$. It follows that $t \mapsto \omega_t = t\omega_1 + (1-t)\omega_0$ is a path of volume forms that connects the form ω_0 to the form ω_1 and we have

$$\frac{d}{dt} \int_{[M]} \omega_t = \int_{[M]} \frac{d}{dt}\omega_t$$

$$= \int_{[M]} \omega_1 - \omega_0 \text{ (differentiating the formula for } t \mapsto w_t)$$

$$= 0 \text{ (since the two forms have the same volume).}$$

Thus, the de Rham cohomology class $[\omega_1 - \omega_0]$ vanishes on the connected manifold M, therefore there exists a smooth $(n-1)$-form α with $d\alpha = \omega_1 - \omega_0$. Hence, $\frac{d}{dt}\omega_t = d\alpha$.

In order to construct the required isotopy ϕ_t satisfying $(\phi_t)^*\omega_t = \omega_0$, we need a time-dependent vector field X_t whose flow ϕ_t^X induces the isotopy ϕ_t and such that the equality $(\phi_t^X)^*\omega_t = \omega_0$ holds. Differentiating, using the Cartan formula and the fact that $d\omega_t = 0$, yields

$$0 = \frac{d}{dt}(\phi_t^X)^*\omega_t = (\phi_t^X)^*(d(i_{X_t}\omega_t) + d\alpha) = (\phi_t^X)^*(d(i_{X_t}\omega_t + \alpha)).$$

The family of vector fields $X = (X_t)_{t \in [0,1]}$ defined by $i_{X_t}\omega_t = -\alpha$ satisfies the above equation. The equation $i_{X_t}\omega_t = -\alpha$ has, for a given $(n-1)$-form α, has a unique solution, since for each $t \in [0, 1]$, ω_t is a non-degenerate volume form. Therefore the family of forms $(\phi_t^X)^*\omega_t$ is constant, hence $(\phi_1^X)^*\omega_1 = \omega_0$ as required. $\square$

The above result also holds for a symplectic form, that is, a closed nondegenerate differential 2-form, at the price of a stronger assumption. The proof works verbatim. Thus we get:

Theorem 2.2.2 (J. Moser) *Let M be a compact connected oriented manifold of dimension n equipped with two symplectic forms ω_0 and ω_1 of equal periods, i.e., with equal de Rham cohomology classes. Assume that the forms are connected by a smooth path ω_t of symplectic forms with constant periods, i.e., for all $t \in [0, 1]$, $[\omega_t] = [\omega_0]$ in $H^2_{\mathrm{dR}}(M)$. Then there exists an isotopy ϕ_t, $t \in [0, 1]$, with $\phi_t^*\omega_t = \omega_0$. In particular, $\omega_0 = \phi_1^*\omega_1$.*   ■

The so-called "Moser trick" works as a "simplification by d" in the equation $di_{X_t}\omega_t = -d\alpha$ and it amounts to noticing that for a volume form ω and for an $(n-1)$-form β the equation $i_X\omega = \beta$ has a unique solution X.

From symplectic structures, we pass to contact forms and contact structures.

A *contact form* α on an n-dimensional manifold M is a pointwise non-vanishing differential 1-form such that at each point p in M the restriction of $(d\alpha)_p$ to the kernel of α_p is non-degenerate.

A *contact structure* on M is a distribution of hyperplanes in the tangent bundle TM given locally as a field of kernels of a contact form.

Moser's method also works, even more simply, without "trick" and without extra stronger assumption, for families of contact structures and it gives another proof of the Gray stability theorem for contact forms [41]:

Theorem 2.2.3 (Gray [41]) *Let $(\alpha)_{t\in[0,1]}$ be a smooth family of contact forms on a compact manifold M. Then there exists a t-dependent vector field X_t on M with flow ϕ_t and* kernel$(\phi_t^*\alpha_t) = $ kernel(α_0). *In particular, there exists a family of positive functions $(f_t)_{t\in[0,1]}$ such that for all $t \in [0, 1]$, $\phi_t^*\alpha_t = f_t\alpha_0$.*

Proof A measure for the variation of the kernel $[\alpha_t]$ of α_t is the restriction $\dot{\alpha}_{t\,|[\alpha_t]}$ of $\dot{\alpha}_t = \frac{d}{dt}\alpha_t$ to the kernel $[\alpha_t]$. By the non-degeneration of the restriction $d\alpha_{t\,|[\alpha_t]}$, there exists a unique t-dependent vector field X_t in the distribution (that is, the family of subspaces) $[\alpha_t]$ with $\dot{\alpha}_{t\,|[\alpha_t]} + i_{X_t}d\alpha_{t\,|[\alpha_t]} = 0$. Hence the kernels of $\phi_t^*\alpha_t$ do not vary since $\frac{d}{dt}\phi_t^*[\alpha_t] = \phi_t^*(\dot{\alpha}_{t\,|[\alpha_t]} + i_{X_t}d\alpha_{t\,|[\alpha_t]}) = 0$. $\qquad\square$

For more applications, see [73]. The use of a proper exhaustion allows us to extend the above theorem to pairs of volume forms on connected non-compact manifolds of equal finite or infinite total volume.

Furthermore, the above proofs work also in a relative version: if the forms coincide on a closed subset A, then the time-dependent vector field X_t vanishes along the subset A and generates a flow that fixes the subset A. The Darboux type rigidity theorems for volume, symplectic and contact forms follow:

Theorem 2.2.4 (Local Darboux Rigidities) *Let ω be a volume or a symplectic form, and let α be a contact form on an n-, $2n$- or $(2n + 1)$-manifold M respectively. Then at each point of M there exists a coordinate chart $(x_1, \cdots, x_n)$ or $(x_1, \cdots, x_n, y_1, \cdots, y_n)$ or $(x_1, \cdots, x_n, y_1, \cdots, y_n, z)$ respectively such that the volume form is expressed by $\omega = dx_1 \wedge \cdots \wedge dx_n$, the symplectic form by $\omega = dx_1 \wedge dy_1 + \cdots + dx_n \wedge dy_n$ and the contact form by $\alpha = dz - y_1dx_1 - \cdots - y_ndx_n$.* $\qquad\blacksquare$

Remark The classical Darboux theorem holds in the setting of symplectic geometry, see [29]. This theorem says that any symplectic manifold of dimension $2n$ is locally isomorphic (in this setting, it is said to be symplectomorphic) to the linear symplectic space $\mathbb{C}^n$ equipped with its canonical symplectic form $\sum dx \wedge dy$. As a consequence, any two symplectic manifolds of the same dimension are locally symplectomorphic to each other.

2.2.2   Almost Complex Structures

An almost complex structure J on a differentiable surface S is an endomorphism of
the tangent bundle of S satisfying $J^2 = -\text{Id}$. More precisely, $J = \{J_p \mid p \in S\}$
is a smooth family of endomorphisms of tangent spaces $J_p : T_pS \to T_pS$ such
that at each point $p \in S$, we have $J_p^2 = -\text{Id}_{T_pS}$. The standard example is $(\mathbb{R}^2, J)$
where J is the constant family of endomorphisms given by the matrix $\left(\begin{smallmatrix} 0 & -1 \\ 1 & 0 \end{smallmatrix}\right)$. This
corresponds to the plane $\mathbb{C}$ equipped with multiplication by i.

The following proof is not based upon the above method.

Theorem 2.2.5 (Local J-Rigidity in Real Dimension 2) *Let J be an almost
complex structure on a surface S. Then at each point $p \in S$ there exists a coordinate
chart (x, y) such that $J(\frac{\partial}{\partial x}) = \frac{\partial}{\partial y}$ holds.*

Proof *(Sketch)* First construct, using a partition of unity, an almost complex
structure J_0 on the torus $T = \mathbb{R}^2/\mathbb{Z}^2$ such that the structures J and J_0 are
isomorphic when restricted to open neighborhoods U of p on S and U_0 of 0 on
T. Let ω be a volume form on T and let g_{ω,J_0} be the associated Riemannian metric
$g_{\omega,J_0}(u, v) = \omega(u, J_0(v))$. Let f be the real function on T satisfying $f(0) = 0$ and
solving the partial differential equation

$$d(df \circ J_0) = -k_{g_{\omega,J_0}}\omega,$$

where $k_{g_{\omega,J_0}}$ is the Gaussian curvature of the metric g_{ω,J_0}. By the Gauss–Bonnet
Theorem, $\int_T k_{g_{\omega,J_0}}\omega = 0$, therefore the equation admits a solution by Fourier theory.
Now use the Gauss curvature formula:

$$k_{g_{e^{2f}\omega,J_0}}\omega = k_{g_{\omega,J_0}}\omega + d(df \circ J_0) = 0.$$

The metric $g_{e^{2f}\omega,J_0}$ has constant curvature 0, therefore (T, J_0) is bi-holomorphic
to $\mathbb{C}/\Gamma$ for some lattice Γ (a 2-generator discrete subgroup), which shows the
statement for a local chart at $0 \in (T, J_0)$, and hence also for a local chart at any
$p \in (S, J)$.   □

For a detailed proof of Theorem 2.2.5, see [8, p. 114–117]. This theorem shows
that every almost complex structure on a differentiable surface S determines in a
unique way a holomorphic structure in the usual sense (that is, a structure defined
by an atlas of local charts with values in $\mathbb{C}$ and holomorphic local changes).

Exercise 2.2.1 Give a proof of Theorem 2.2.5 using Moser's trick.

Remark The first definition of an almost complex structure is due to Charles
Ehresmann who addressed the question of the existence of a complex analytic
structure on a topological (resp. differentiable) manifold of even dimension, from
the point of view of the theory of fiber spaces; cf. Ehresmann's talk at the 1950 ICM
[33]. Ehresmann mentions the fact that H. Hopf addressed the same question from a

different point of view. He notes in the same paper that by a method proper to even-dimensional spheres he showed that the 4-dimensional sphere does not admit any almost complex structure, a result which was also obtained by Hopf using different methods. See also McLane's review of Ehresmann's work [84]. Ehresmann and MacLane also refer to the work of Wen-Tsün Wu [114, 115], who was a student of Ehresmann in Strasbourg.

Remark The Nijenhuis tensor is an obstruction to local integrability of J-fields in higher dimensions, where Theorem 2.2.5 does not hold in the general case, see [8, p. 124–125]. Real dimension 2 is very special!

2.2.3 Almost Complex Structures on n-Spheres

The existence and integrability of J-structures in dimension 2 is very special. We mentioned that the 4-sphere S^4 does not admit any J-field (Ehresmann and Hopf), but the 6-sphere does.

Clearly only spheres of even dimension can carry J-fields. Adrian Kirchhoff, in his PhD thesis (ETH Zürich 1947) [57] established a relationship between two non-obviously related structures on spheres S^{2n} and S^{2n+1} of different dimensions; we report on this now.

Recall that a *parallelism* on a smooth n-manifold is a global field of frames, that is, a field of n tangent vectors which form a basis of the tangent space at each point.

Theorem 2.2.6 (Kirchhoff [58]) *The sphere S^n, $n \geq 0$, admits a J-field if and only if the sphere S^{n+1} admits a parallelism.*

Proof The case $n = 0$ is special: the tangent space TS^0 is of dimension 0, therefore $J = \mathrm{Id}_{TS^0}$ is a J-field and S^1 admits a parallelism.

"Only if" part for $n > 0$: In $V = \mathbb{R}^{n+2}$ with the standard basis $e_0, e_1, \cdots, e_{n+1}$, let S^n be the unit sphere in the span $[e_1, \cdots, e_{n+1}]$. Let S^{n+1} be the unit sphere of V. Assume that J is a J-field on S^n. Let $L : S^{n+1} \to \mathrm{GL}(V)$, $v \mapsto L_v$, be the continuous map satisfying $L_v(e_0) = v$, $v \in S^{n+1}$, defined as follows:

- First, for $v \in S^n$, seen as the equator of S^{n+1}, we set
  $L_v(v) = -e_0$, $L_v(e_0) = v$,
  $L_v(u) = v + J_v(u - v)$, $u \in [v, e_0]^\perp$.
- For $v \in S^{n+1}$, we can write $v = \sin(t)e_0 + \cos(t)v'$, $v' \in S^n$, $t \in]-\pi, \pi[$. We then set
  $L_v = \sin(t)\mathrm{Id}_V + \cos(t)L_{v'}$.

We have $L_v \circ L_v = -\mathrm{Id}_V$ for $v \in S^n$, hence $L_v = \sin(t)\mathrm{Id}_V + \cos(t)L_{v'} \in \mathrm{GL}(V)$ for $v \in S^{n+1}$, since the eigenvalues are $\sin(t) \pm \cos(t)i$. Observe that $T_{e_0}S^{n+1} = e_0 + [e_0]^\perp$ and $[e_0]^\perp = [e_1, \cdots, e_{n+1}]$. The differential $(DL_v)_{e_0} : T_{e_0}S^{n+1} \to V$ at e_0 of L_v maps the space $T_{e_0}S^{n+1}$ onto an affine space of dimension $n + 1$ in $T_{L_v(e_0)}V$ that intersects transversely the ray $[v]$. Then, for $v \in S^{n+1}$, the images

$(DL_v)_{e_0}(e_1), \cdots, (DL_v)_{e_0}(e_{n+1}) \in T_v V$ define a frame in $T_v S^{n+1} = v + [v]^{\perp}$ by the projection parallel to $[v]$ onto $v + [v]^{\perp}$.

"If" part: Work backwards.   □

Ehresmann in his ICM talk [33] mentions Kirchhoff's results [57].

In fact, by a celebrated result of Jeffrey Frank Adams [1], only the spheres S^1, S^3, S^7 admit a parallelism. This implies that S^4 does not admit any J-field and S^6 does. Adams' result was obtained several years after Kirchhoff's result.

The question of the existence of a complex structure on S^6 is still wide open. How the Nijenhuis integrability condition for a J-field on S^6 translates into a property of framings on S^7 is the subject of a recent paper [67].

We end this section on rigidity by a word on exotic spheres: Any two differentiable manifolds of the same dimension are locally diffeomorphic. But such manifolds may be homeomorphic without being diffeomorphic. The first examples of such a phenomenon are Milnor's exotic 7-spheres [75]. In later papers, Milnor constructed additional examples.

2.3 The First Compact Riemann Surface

A Riemann surface is a complex 1-dimensional real manifold, or a 2-dimensional manifold equipped with a complex 1-dimensional structure, that is, an atlas whose charts take values in the Gaussian plane $\mathbb{C}$, with holomorphic transition functions.

In this section, we shall deal with the simplest Riemann surface, the Riemann sphere.

2.3.1 The Riemann Sphere

The familiar round sphere in 3-space, together with its group of rigid motions, can be seen as a holomorphic object: its motions are angle-preserving. It is also a one-point compactification of the field of complex numbers. We shall see that this construction as a one-point compactification can be generalized to an arbitrary field.

First we ask the question:

Why do we need the Riemann sphere?

The statement: "Every sequence of complex numbers has a convergent subsequence" is very true, indeed true for bounded sequences. The statement is salvaged without this assumption if we introduce a wish object w with the property that every sequence of complex numbers, for which no subsequence converges to a complex number, converges to w. In this way, from the familiar Gaussian plane $\mathbb{C}$, we gain a new space, $\mathbb{C} \cup \{w\}$ in which the above statement improves from very true to true. This topological construction is the familiar one-point compactification of non compact but locally compact spaces.

Very true is also the statement: "The ratio $\frac{a}{b}$ is well defined as long as $(a, b) \neq$ $(0, 0)$". Again the statement becomes true if we introduce by wish a new object $w \notin k$ with $\frac{a}{0} = w, a \neq 0$. This algebraic construction applies to any field k, and not only $\mathbb{C}$.

In both constructions the object w appears as a newcomer, an immigrant, with a special restricted status.

It is Riemann who gave an interpretation of the new set $X = \mathbb{C} \cup \{w\}$ together with a very rich structure Σ on it, for which the new element gains unrestricted status. In short, the automorphism group of (X, Σ) acts transitively on this space, that is, the space X is homogeneous.

The above topological construction also shows that the newcomer w is above any bound, so from now on we use the symbol ∞ for w.

Here is another construction of an infinity, valid for any field.

Let k be a field. An element $\lambda \in k$ can be interpreted as a linear map $a \in k \mapsto \lambda a \in k$. Its *graph* $G_\lambda \subset k \times k$ is the vector subspace $\{(a, \lambda a) \mid a \in k\}$ of dimension 1 in $k \times k$. So we get an embedding $\iota : \lambda \in k \mapsto \mathbb{P}^1(k)$ of the field k in the projective space $\mathbb{P}^1(k)$ of all 1-dimensional vector subspaces in $k \times k$. The vector subspace $G = \{(0, b) \mid b \in k\}$ is the only one which is not in the image of the embedding ι.

The element λ can be retrieved from G_λ as a slope: indeed, for any $(a, b) \in G_\lambda$, if $(a, b) \neq (0, 0)$ then $a \neq 0$ and $\lambda = \frac{b}{a}$.

So the missing vector subspace G corresponds to the forbidden fraction $\frac{1}{0} = \infty$ and can be called G_∞.

In the case where $k = \mathbb{R}$, this is the well-known embedding of $\mathbb{R}$ in the circle of *directions up to sign*. Extending $\iota : k \cup \{\infty\} \to \mathbb{P}^1(k)$ by $\iota(\infty) = G_\infty$ gives the interpretation of $k \cup \{\infty\}$ as the projective space $\mathbb{P}^1(k)$. Each linear automorphism A of the k-vector space k^2 induces a self-bijection G_A of $k \cup \{\infty\} = \mathbb{P}^1(k)$. If the matrix of A is the 2×2-matrix $\begin{pmatrix} a & b \\ c & d \end{pmatrix} \in GL(2, k)$, then $G_A(G_\lambda) = G_{\lambda'}$ with $\lambda' = \frac{a\lambda+b}{c\lambda+d}$. The transformation $G \in \mathbb{P}^1(k) \mapsto G_A(G) \in \mathbb{P}^1(k)$ or $\lambda \mapsto \frac{a\lambda+b}{c\lambda+d}$ is called a fractional linear or Möbius transformation. Note that in particular $G_A(G_\infty) = \frac{a}{c}$.

Given a general field k, an important structure on $\mathbb{P}^1(k)$ is provided by a 4-point function which we shall study in Sect. 2.4.3.

At this stage, we restrict to the case $k = \mathbb{C}$. The above construction of $\mathbb{P}^1(k)$ for an arbitrary field k gives the familiar construction of $\mathbb{P}^1(\mathbb{C}) = (\mathbb{C}^2 - \{0\})/\mathbb{C}^*$, where $\mathbb{C}^*$ denotes the multiplicative group of nonzero complex numbers.

The set $\mathbb{C} \cup \{\infty\} = \mathbb{P}^1(\mathbb{C})$ carries many structures. First, there is the structure of a differentiable manifold given by the following atlas: We set $U_0 = \mathbb{P}^1(\mathbb{C}) \setminus \{G_\infty\}$ and $U_\infty = \mathbb{P}^1(\mathbb{C}) \setminus \{G_0\}$. Observe that $U_0 = \{G_\lambda \mid \lambda \in \mathbb{C}\}$ and that every $G \in U_\infty$ is of the type $G'_\sigma = \{(\sigma b, b) \mid b \in \mathbb{C}\}$ for $\sigma \in \mathbb{C}$.

Define maps $z_0 : U_0 \to \mathbb{C}$ by $z_0(G_\lambda) = \lambda$ and $z_\infty : U_\infty \to \mathbb{C}$ by $z_\infty(G'_\sigma) = \sigma$. Both maps are bijections. For $G \in U_0 \cap U_\infty$ the two maps are related; indeed, $z_0(G)z_\infty(G) = 1$. It follows that the system $((U_0, z_0), (U_\infty, z_\infty))$ is an atlas for a manifold structure with coordinates functions (z_0, z_∞). Its quality is hidden in the quality of the coordinate change. For $G \in U_0 \cap U_\infty$, from the above implicit relation it follows that $z_\infty(G) = 1/z_0(G)$, $z_0(G) = 1/z_\infty(G)$. This coordinate change is differentiable; therefore $\mathbb{C} \cup \{\infty\} = \mathbb{P}^1(\mathbb{C})$ is a smooth manifold with charts in the Gaussian plane $\mathbb{C}$.

The smooth 2-dimensional real manifold $\mathbb{C} \cup \{\infty\} = \mathbb{P}^1(\mathbb{C})$ is diffeomorphic to the unit sphere in the three-dimensional real vector space $\mathbb{R}^3$. More precisely, the coordinate change $\phi_{\infty,0} : \mathbb{C}^* = z_0(U_0) \to z_\infty(U_\infty) = \mathbb{C}^*$ is given in terms of the natural coordinate z on $\mathbb{C}^*$, by $\phi_{\infty,0}(z) = 1/z$. The smooth map $\phi_{\infty,0} : \mathbb{C}^* \to \mathbb{C}^*$ is moreover holomorphic, so the above atlas provides $\mathbb{C} \cup \{\infty\} = \mathbb{P}^1(\mathbb{C})$ with the structure of a Riemann surface. This Riemann Surface is the *Riemann Sphere*.

Let U be an open subset of the Gaussian plane $\mathbb{C}$. Riemann defined a map $\phi : U \to \mathbb{C}$ to be holomorphic without using an expression that evaluates the map at given points. The idea is the following. The real tangent bundles TU and $T\mathbb{C}$ come with a field m_i of endomorphisms. (The notation m_i stands for "multiplication by i".) The value $m_{i,p}$ of the field m_i at the point p is the linear map $m_{i,p} : T_pU \to T_pU$, $u \mapsto iu$. To be holomorphic by Riemann's definition is given by the following property of the differential:

$$(D\phi)_p(m_{i,p}(u)) = m_{i,\phi(p)}((D\phi)_p(u)).$$

In words, this means that the differential $D\phi$ is $\mathbb{C}$-linear.

Riemann's characterization of holomorphic maps together with the local J-Rigidity Theorem 2.2.5 allows us to define a Riemann surface (S, J) as a real 2-dimensional differentiable manifold S equipped with a smooth field of endomorphisms $J : TS \to TS$ of its tangent bundle satisfying $J \circ J = -\mathrm{Id}_{TS}$.

The Riemann Sphere is the first example of a compact Riemann surface. The most familiar non-compact Riemann surface is the Gaussian plane $\mathbb{C}$. Another most important Riemann surface is the unit disc in $\mathbb{C}$. This is also the image of the southern hemisphere by the stereographic projection from the North pole onto a plane passing through the equator. This projection is holomorphic. The importance of the unit disc stems from the fact that it is equipped with the Poincaré metric, which makes it a model for the hyperbolic plane.

2.3.2 The Group SU(2) and Its Action on the Riemann Sphere

Now that we are familiar with the Riemann sphere, we study a group action on it.

Let $< u, v >_{\mathrm{Herm}}$ be the usual Hermitian product on $\mathbb{C}^2$. This is the complex bilinear form on $\mathbb{C}$ defined by $< u, v >_{\mathrm{Herm}} = u_1\bar{v}_1 + u_2\bar{v}_2$. The Hermitian perpendicular $L^\perp$ to a complex vector subspace L is again a complex vector subspace.

The group of determinant 1 linear transformations of $\mathbb{C}^2$ that preserve $< u, v >_{\mathrm{Herm}}$ is the group SU(2) consisting of all matrices of the form $\left(\begin{smallmatrix} a & b \\ -\bar{b} & \bar{a} \end{smallmatrix}\right)$, $(a, b) \in \mathbb{C}^2$, $a\bar{a} + b\bar{b} = 1$. This group acts on the Riemann sphere by Möbius transformations, in fact, by rotations. The map $\left(\begin{smallmatrix} a & b \\ -\bar{b} & \bar{a} \end{smallmatrix}\right) \mapsto (a, b)$ defines a diffeomorphism SU(2) $\to S^3$ and induces a Lie group structure on the sphere S^3.

The group SU(2) acts transitively by conformal automorphisms on the Riemann sphere $\mathbb{P}^1(\mathbb{C})$. The stabilizer of $L = \{(\lambda, 0) \mid \lambda \in \mathbb{C}\}$ in SU(2) is the group $\left(\begin{smallmatrix} a & 0 \\ 0 & \bar{a} \end{smallmatrix}\right)$, $a \in \mathbb{C}, a\bar{a} = 1$, which is isomorphic to the group of complex numbers of norm 1. The quotient construction induces a Riemannian metric on

$$\mathbb{P}^1(\mathbb{C}) = \mathrm{SU}(2)/\mathrm{Stab}_{\mathrm{SU}(2)}(L).$$

A *marked element* in $\mathbb{P}^1(\mathbb{C})$ is a pair (L, u) where $u = (a, b) \in \mathbb{C}^2, a\bar{a} + b\bar{b} = 1$ and $L = [u] = \{\lambda u \mid \lambda \in \mathbb{C}\}$. Note that this representation is redundant since u determines $L = [u]$.

The group SU(2) acts simply transitively on marked elements in $\mathbb{P}^1(\mathbb{C})$.

The involution $L \mapsto L^\perp$ extends to marked elements: map $(L, u) = (L, (a, b))$ first to $u^\perp = (\bar{b}, -\bar{a})$ and next to $(L^\perp, u^\perp)$ with $L^\perp = [u^\perp]$.

A marked element (L, u) determines a path in $\mathbb{P}^1(\mathbb{C})$ by $L_u : t \in [0, \pi] \mapsto L_u(t) = [\cos(t)u + \sin(t)u^\perp]$, which in fact is a simple closed curve. Its velocity at $t = 0$ is a length 1 tangent vector $V_u \in T_{[u]}(\mathbb{P}^1(\mathbb{C}))$. Observe that $V_u = V_{-u}$ and $V_{iu} = -V_u$. The path L_u lifts to $H_u^\perp : t \in [0, \pi] \mapsto H_u^\perp(t) = \cos(t)u + \sin(t)u^\perp \in S^3$, which is a geodesic from u to $-u$ perpendicular to the foliation on S^3 by the Hopf circles $H_v = \{v' \in S^3 \mid v' = \lambda v\}$, $v \in S^3$. Hopf circles H_v map to points, and geodesics $H_u^\perp$ map to simple closed geodesics in $\mathbb{P}^1(\mathbb{C})$.

The map $\pm u \in S^3/\{\pm \mathrm{Id}\} = \mathbb{P}^3(\mathbb{R}) \mapsto V_u \in T(\mathbb{P}^1(\mathbb{C}))$ induces a bijection onto the length 1 vectors to $\mathbb{P}^1(\mathbb{C})$. Observe that SU(2) acts almost simply transitively on length 1 tangent vectors to $\mathbb{P}^1(\mathbb{C})$. The quotient group $\mathrm{PSU}(2) = \mathrm{SU}(2)/\{\pm \mathrm{Id}\}$ acts simply transitively on length 1 tangent vectors.

2.4 All Three Planar Geometries and Hyperbolic 3-Space Simultaneously

2.4.1 A Stratification of the Riemann Sphere Arising from Algebra

Bernhard Riemann was aware of the (Riemann) sphere being the complex plane union a point at infinity. His point of view on complex analysis was very geometric. In this section, we wish to describe an incarnation of the Riemann sphere which arises from algebra. For more details on this model, see [8, Chap. 3, §3.1] and [9, Chap. 1, §8.3].

The starting object is the set Σ of surjective ring homomorphisms from the ring $\mathbb{R}[X]$ of polynomials in one unknown X with real coefficients to a field F. On the set Σ we introduce two equivalence relations. The first relation, $\sim$, declares $f : \mathbb{R}[X] \rightarrow F$ and $f' : \mathbb{R}[X] \rightarrow F'$ to be equivalent if there exists a field isomorphism $\phi : F \rightarrow F'$ with $f' = \phi \circ f$.

The relation $f \sim f'$ holds if and only if the ideals kernel(f), kernel(f') in $\mathbb{R}[X]$ are equal.

The second relation, $\sim_X$, requires $f \sim f'$ and moreover $f(X) = f'(X)$ holds in $\mathbb{R}[X]/\text{kernel}(f) = \mathbb{R}[X]/\text{kernel}(f')$.

Up to field isomorphism, there are only three fields, F, that are hit by a surjective ring homomorphism $f : \mathbb{R}[X] \to F$, namely, the fields $\mathbb{C}$, $\mathbb{R}$ and $\mathbb{F}_1$ where $\mathbb{F}_1$ is the field with one element, that is, the field where $0 = 1$ holds. The field $\mathbb{F}_1$ corresponds to the ideal $\rho = \mathbb{R}[X]$ consisting of the whole ring, which is prime, maximal but not proper.

Exercise The two fields $\mathbb{R}$, $\mathbb{F}_1$ have only the identity as automorphism and the field $\mathbb{C} = \{a + bi \mid a, b \in \mathbb{R}\}$ only two automorphisms as $\mathbb{R}$-algebra, but as many field automorphisms the power set of the real numbers.

In the following we will describe the quotient sets $\Sigma/\sim$, $\Sigma/\sim_X$ together with natural structures on these sets.

All ideals in $\mathbb{R}[X]$ are principal, that is, any such ideal is generated by a single element (it is obtained by multiplication of such an element by an arbitrary element of the ring). Kernels of $f \in \Sigma$ are prime ideals, that is, the quotient of $\mathbb{R}[X]$ by such an ideal is an integral domain (the product of any two nonzero elements is nonzero). Thus, we have three kinds of kernels of f, namely, $\rho = (1) = \mathbb{R}[X]$, $(X - a)$, $a \in \mathbb{R}$, and $((X - a)^2 + b^2)$, $a, b \in \mathbb{R}$, $b > 0$. Therefore the set $\Sigma/\sim$ is identified with $\mathbb{C}_+ \cup \mathbb{R} \cup \{\rho\}$. Here we use the notation $\mathbb{C}_\pm = \{a + bi \mid a, b \in \mathbb{R}, \pm b > o\}$ for the upper/lower half planes.

The kernel of the ring homomorphism f is not sufficient in order to describe its class in $\Sigma/\sim_X$ if kernel(f) = $((X - a)^2 + b^2)$. One needs moreover to specify a root $a + bi \in \mathbb{C}_+$ or $a - bi \in \mathbb{C}_-$. Thus the set $\Sigma/\sim_X$ is a disjoint union of 4 strata $\Sigma/\sim_X = \mathbb{C}_+ \cup \mathbb{C}_- \cup \mathbb{R} \cup \{\rho\}$.

The fields $\mathbb{C}$, $\mathbb{R}$, $\mathbb{F}_1 = \{0\}$ are realized as sub-$\mathbb{R}$-algebras in $\mathbb{C}$, so an alternative description of the set $\Sigma/\sim_X$ is the set of $\mathbb{R}$-algebra homomorphism from $\mathbb{R}[X]$ to $\mathbb{C}$.

We shall see that the set $R = \Sigma/\sim_X$ and its strata carry a rich panoply of structures. The set $R = \mathbb{C}_+ \cup \mathbb{C}_- \cup \mathbb{R} \cup \{\rho\}$ is identified with the Riemann sphere $\mathbb{C} \cup \{\infty\}$. Structures, such as the Chasles three point function (defined below) on $\mathbb{R} \subset R$, or the hyperbolic geometry on $\mathbb{C}_+$, will appear naturally. Naturally means here that the construction that leads to the structure commutes with the $\mathbb{R}$-algebra automorphisms of $\mathbb{R}[X]$. For instance, it commutes with the substitutions that consist in translating X to $X - t$, $t \in \mathbb{R}$, or with stretching X to λX, $\lambda \in \mathbb{R}^*$. The ideal $(X - a)$ maps to the ideal $(X - a - t)$ by translation and to $(X - \frac{a}{\lambda})$ by stretching.

A first example is the Chasles 3-point function $\text{Ch}(A, B, C)$ on the stratum $\mathbb{R}$ consisting of the ideals $(X - a)$, $a \in \mathbb{R}$, defined as follows: Given three distinct such points, $A = (X - a)$, $B = (X - b)$, $C = (X - c)$, define $\text{Ch}(A, B, C) = \frac{b-a}{c-a}$.

In words, $\text{Ch}(A, B, C)$ is the ratio of the monic generators of B and C evaluated at the zero of the monic generator of A.

The next example is the 4-point function cross ratio $\mathrm{cr}(A, B, C, D)$: for 4 distinct points $A = (X - a), B = (X - b), C = (X - c), D = (X - d)$, define $\mathrm{cr}(A, B, C, D) = \mathrm{Ch}(A, B, C)\mathrm{Ch}(D, C, B) = \frac{b-a}{c-a} \cdot \frac{c-d}{b-d}$. In words, this is Chasles evaluated at the first three points times Chasles evaluated at the last three points in the reverse order. It is truly a remarkable fact that the cross ratio function extends to a 4-point real function on $\mathbb{P}^1(\mathbb{R})$ and to a 4-point complex function on $\mathbb{P}^1(\mathbb{C}) = \mathbb{C} \cup \{\rho\}$ if one transfers the above wishful calculus with $w = \infty$ to ρ.

The multiplicative monoid $\mathbb{R}^*[X]$ of polynomials which do not vanish at 0 has also automorphisms that do not directly fit with the interpretation as polynomials with unknown X. In particular, they are not ring automorphisms, but monoid automorphisms. A main example is the *twisted palindromic symmetry*: perform on a polynomial $P(X)$ the substitution $X \to \frac{-1}{X}$, followed by stretching with factor $(-X)^{\mathrm{degree}(P)}$. (The palindromic symmetry is said to be twisted, because of the minus signs.) Then the ideal $(X - a)$ maps to the ideal $(-X(\frac{-1}{X} - a)) = (1 + aX) = (X + \frac{1}{a})$. The Chasles function restricted to $\mathbb{R}^*$ does not commute with the symmetry $a \mapsto \frac{-1}{a}$, but the cross ratio commutes. (This property is among the ones that make the cross ratio more natural than the Chasles 3-point function.) This symmetry, which is an involution, extends to a fixed point free involution $\sigma_{\mathbb{P}}$ of $\mathbb{R} \cup \{\rho\} = \mathbb{P}^1(\mathbb{R})$. Remarkably, the symmetry $\sigma_{\mathbb{P}}$ commutes with the cross ratio cr. In this sense, cr is more natural than Ch.

The above operations of real translation and stretching, i.e., substituting $X - t$ for X or λX for X with $t \in \mathbb{R}$, $\lambda \in \mathbb{R}^*$, together with the twisted palindromic symmetry induce bijections of the set $\{(X - u)(X - \bar{u}) \mid u \in \mathbb{C}_+\}$ of monic polynomials of degree 2 without real roots. Composing these bijections generates a group G. It is a remarkable fact that this group is, as an abstract group, isomorphic to the group $\mathrm{PGL}(2, \mathbb{R})$. It is also a remarkable fact that the abstract group $\mathrm{PGL}(2, \mathbb{R})$ carries a unique structure of Lie group. So there is also a topology on G, which allows us to define the subgroup $G_0 \subset G$ as the connected component of the neutral element in the Lie group $G = \mathrm{PGL}(2, \mathbb{R})$. The group G_0 is isomorphic to the group $\mathrm{PSL}(2, \mathbb{R})$.

The fixed point free involution $\sigma_{\mathbb{P}}$ on $\mathbb{P}^1(\mathbb{R}) = \mathbb{R} \cup \{\rho\} = \partial \bar{C}_+$ extends to $\mathbb{C}_+$ by putting $\sigma_{\mathbb{P}}(u) = \frac{-1}{u}$ for an involution with i as unique fixed point.

The group G_0 acts transitively and faithfully on the above strata $\mathbb{C}_\pm$ and on $\mathbb{R} \cup \{\rho\}$. From this action one gets a topology on the strata and also, as we will explain, a geometry on $\mathbb{C}_+$. It is also remarkable that this geometry, in fact, the planar hyperbolic geometry, can also be explained in a more elementary way in term of the interpretation as ideals.

The action of G_0 on $\mathbb{C}_+$ is the so-called modular action of $\mathrm{PSL}(2, \mathbb{R})$ on $\mathbb{C}_+$. Thinking of an element $g \in G_0$ as a real 2×2 matrix of determinant 1 up to sign, $\pm(\begin{smallmatrix} a & b \\ c & d \end{smallmatrix})$, the action on $u \in \mathbb{C}_+$ is given by $(g, u) \mapsto \frac{au+b}{cu+d}$.

The modular action of G_0 on $\mathbb{C}_+$ extends to the projective action of $G = \mathrm{PGL}(2, \mathbb{R})$ on $\partial \bar{C}_+ = \mathbb{P}^1(\mathbb{R})$ and also to the projective action of the complex group $\mathrm{PGL}(2, \mathbb{C})$ on $\mathbb{P}^1(\mathbb{C})$.

The hyperbolic geometry on $\mathbb{C}_+$ can also be defined in terms of ideals in the following rather elementary way.

Points are the elements of $\mathbb{C}_+$. The monic generator of the corresponding ideal may be denote by $P_u(X) = (X - u)(X - \bar{u})$.

The notion of line is introduced using convex combinations. More precisely, given two distinct points u, v, the line $L_{u,v}$ through them is defined as follows:

Denote by $P_{t,u,v}(X) = t P_u(X) + (1 - t)P_v(X)$, $t \in [0, 1]$ the convex combination of the polynomials $P_u(X)$, $P_v(X)$. Note first that any polynomial obtained in this way is monic. Let $I_{u,v} =]m_{u,v}, M_{u,v}[$ be the maximal interval on which $P_{t,u,v}(X)$ has no real roots. Note that $[0, 1] \subset I_{u,v} \neq \mathbb{R}$ (one may start by checking that for $t = 0$ and $t = 1$ there are no real roots). This implies that the cross ratio $\delta(u, v) = \mathrm{cr}(m_{u,v}, 1, 0, M_{u,v})$ is a positive real number > 1. If $\mathrm{Re}(u) = \mathrm{Re}(v)$ define $L_{u,v} = \{w \in \mathbb{C}_+ \mid \mathrm{Re}(w) = \mathrm{Re}(u) = \mathrm{Re}(v)\}$. If $\mathrm{Re}(u) \neq \mathrm{Re}(v)$ let $c \in \mathbb{R}$ be the zero of $P_u(X) - P_v(X)$ and define $L_{u,v} = \{w \in \mathbb{C}_+ \mid |w - c|^2 = P_u(c)\}$.

With such a definition, the axiom of hyperbolic geometry saying that for any two distinct points there is a unique line passing through them is trivially satisfied.

Exercise For a better understanding of this construction of the hyperbolic plane, please check that the root of $P_{t,u,v}(X)$ travels along $L_{u,v}$ for $t \in I_{u,v}$ in case $\mathrm{Re}(u) \neq \mathrm{Re}(v)$. Note that $L_{u,v}$ is a half circle in $\mathbb{C}$ with centre in $\mathbb{R} \cup \{\rho\}$.

The hyperbolic Riemannian metric g_u is the Hessian at u of the function $v \in \mathbb{C}_+ \mapsto D(u, v)^2$.

In this model, we can define the hyperbolic distance between u and v as $D(u, v) = \frac{1}{2} \log(\delta(u, v))$. We can make a relation with the model of the hyperbolic plane that uses the Hilbert metric. In this way we have all the ingredients of hyperbolic geometry based on elementary algebra, that can be taught at high-school.

Remark The above twisted palindromic map $\sigma \colon \mathbb{R}[X] \to \mathbb{R}[X]$ induces an involution on $\mathbb{R}^*[X]$, the complement of the hypersurface of polynomials that vanish at 0. Therefore, σ is a birational involution and the above constructions would generate subgroups G, etc. in the Cremona group of birational transformations of $\mathbb{P}^1(\mathbb{C})$. Thus, the artefact of defining $\sigma_{\mathbb{P}}$ can be avoided. Even better, the geometry of Cremona groups is hyperbolic. For the Cremona group, see [21, 22, 26, 30, 95, 108, 116].

2.4.2   A Note on the Field with One Element

The importance of the field with one element $\mathbb{F}_1$, which was interpreted in the above construction of the Riemann sphere as the point at infinity of the complex plane, was first highlighted by Jacques Tits in his paper [105]. In this paper, Tits proposed the development of a geometry over the field $\mathbb{F}_1$ which would be the limit of geometries over the finite fields $\mathbb{F}_q$, where $\mathbb{F}_q$ denotes the field of cardinality $q = p^n$ where p is a positive prime. Note that if we make n tend to 0 in the notation $\mathbb{F}_q$, $q = p^n$, we obtain $\mathbb{F}_1$, which is an indication of the fact that the field $\mathbb{F}_q$ may be considered as a deformation of the field $\mathbb{F}_1$. One may also make an analogy with the fact that a limit $n \to 0$ is used in quantum topology for a geometric interpretation of the TQFT calculus of Turaev, Viro, Kauffman, etc. (The analogy is vague, but this is the nature of analogies.)

After Tits introduced his idea of studying a geometry over the field $\mathbb{F}_1$, many works were done on this theme. For instance, Manin, in his lectures on the zeta function [72], proposed the study of a "Tate motive over a one-element field". In the paper [28] titled *Fun with $\mathbb{F}_1$* by Connes, Consani and Marcolli, the field $\mathbb{F}_1$ becomes an actor in an approach to the Riemann hypothesis.

2.4.3 The Riemann Sphere and Shadow Numbers

Let V be a 2-dimensional vector space over a field k. If k has 3 or more elements, then the space $\mathbb{P}^1(k)$ of lines through the origin in V has at least 4 elements. We wish to define in a geometric way the cross ratio of a figure L_1, L_2, L_3, L_4 consisting of 4 distinct lines in V. This is done in the form of two exercises.

Exercise 2.4.1 Do the following: Identify $V = L_1 \oplus L_4$ with the Cartesian product of L_1 and L_4, and let ϕ_i be the linear map from L_1 to L_4 that has L_i as graph, $i = 2, 3$. Define $\mathrm{cr}(L_1, L_2, L_3, L_4)$ as the stretch factor of the linear map $\phi_3^{-1} \circ \phi_2 : L_1 \to L_1$. This is represented in Fig. 2.1, in which we start with the point X on L_1, with Y its image by ϕ_2 and Z the image of Y by ϕ_3^{-1}.

Please check now that if the lines are defined using a basis e, f and "numbers" $a_i \in k \cup \{\infty\}, i = 1, 2, 3, 4$ such that $e + a_i f \in L_i$, then the formula $\mathrm{cr}(L_1, L_2, L_3, L_4) = \frac{a_2 - a_1}{a_3 - a_1} \frac{a_4 - a_3}{a_4 - a_2}$ holds.

In this way, we recover the usual formula for the cross ratio.

In the above construction of the cross ratio, the use of a Cartesian product and graphs of maps suggest that parallel lines are essential. This is not the case. For instance, the construction also works inside a Euclidean triangle with vertices ABC, as in the following exercise, which asks for a construction of the cross ratio which is a projective version of the construction in Exercise 2.4.1, which is affine (it uses the notion of parallels):

Exercise 2.4.2 Let f_2, f_3 be interior points of the side BC of a triangle ABC, put $f_1 = B$, $f_4 = C$. Let L_i, $i = 1, 2, 3, 4$, be the segments connecting the vertex A with f_i.

Define $\phi_2 : L_1 \to L_4$ as follows: Connect $X \in L_1$ by a segment with C which insects L_2 in X'. The half-line $[B, X')$ intersects L_4 in $Y = \phi_2(X)$. Define in the same way ϕ_3. Now $\phi_3^{-1} \circ \phi_2 : L_1 \to L_1$ is not linear, but with fixed point A. Define the cross ratio $\mathrm{cr}(L_1, L_2, L_3, L_4)$ to be the stretch factor of the differential at A of the map $\phi_3^{-1} \circ \phi_2$. The construction is represented in Fig. 2.2, where, like in Fig. 2.1, X is a point in L_1, Y its image by ϕ_2 and Z the image of Y by ϕ_3^{-1}.

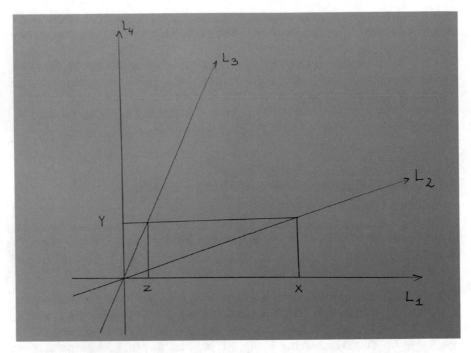

Fig. 2.1 Figure for Exercise 2.4.1

It is important to observe that $cr(L_1, L_2, L_3, L_4)$, in planar Euclidean geometry, does not depend upon the position of the side BC. This property does not hold in spherical or hyperbolic geometry. In spherical geometry there is a preferred choice for the side BC, namely, such that the triangle becomes bi-orthogonal. This was already done by Menelaus of Alexandria!

This construction also works for a configuration in general position of 4 linear subspaces E_1, E_2, E_3, E_4 of dimension 2 in a real vector space V of dimension 4. The invariant is now an element $\Lambda(E_1, E_2, E_3, E_4) \in GL(E_1)$. One may use this to prove the following:

Theorem 2.4.1 (The Four Complex Lines Theorem) *Let $A = (E_1, E_2, E_3, E_4)$ be a generic labelled configuration of planes in a real 4-dimensional vector space V. There exists a linear complex structure $J: V \to V$ with $J(E_j) = E_j$ (which means that each E_j is a complex line) for every $i = 1, 2, 3, 4$ if and only if*

$$\text{Trace}(\Lambda(A))^2 < 4\text{Det}(\Lambda(A))$$

or if $\Lambda(A)$ is a multiple λId_{E_1} of the identity Id_{E_1}, for some $\lambda \neq 1$.

The proof is contained in p. 73 of [8].

Theorem 2.4.1 should be seen in the setting of the following general question: Given a 4-dimensional real vector space V and a quadruple of 2-dimensional

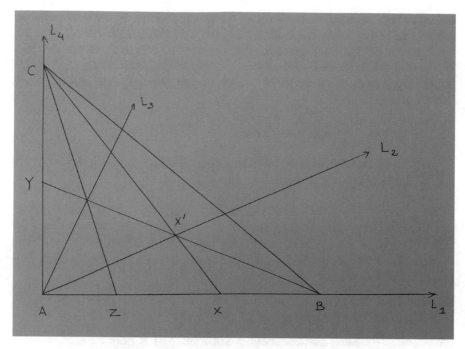

Fig. 2.2 Figure for Exercise 2.4.2

planes (E_1, E_2, E_3, E_4) in V satisfying some properties, does there exist an almost complex structure $J : V \to J$ making $(V, J) = \mathbb{C}^2$ such that (E_1, E_2, E_3, E_4) are complex lines?

The usual formula for the cross ratio with numbers is memo-technically speaking a headache. A more geometrically-rooted name would be more satisfying. We propose *shadow number*: the shadow issued from a light bulb of four concurrent lines lying in a plane on a plane shows the same number. The proof uses the above special property in Euclidean geometry and becomes simplified if one uses the above geometric definition. If the name should remember a person, then perhaps Leonardo da Vinci number, for Leonardo studied central projections and perspectivities about 500 years ago [69], or Menelaus number, for Menelaus studied shadows in spherical geometry about 2000 years ago. See [85] for the use by Menelaus of the invariance of the cross ratio in spherical geometry. Menelaus used this result in the proof of Proposition 71 of the *Spherics*,[1] and later medieval commentators, in an effort to provide full proofs of some of this and other difficult propositions in Menelaus'

---

[1] Menelaus was extremely concise in his *Spherics*, and the proofs of some of the propositions in this work are very difficult to follow. This is why several results of the *Spherics* [87] were later explained and commented on by Arab mathematicians of the Middle Ages, after the Greek mathematical schools were desintegrated. Regarding this particular proposition, see the two papers [85, 86].

Spherics, highlighted this invariance as a new proposition, in order to explain a proof in the *Spherics*; see Proposition 3.2 in [85] and [87, pp. 356–360].

The group $PGL(2, \mathbb{C})$ of linear transformations of $\mathbb{C}^2$ acts on the Riemann sphere R (see Sect. 2.4.1), since this group acts on complex lines through the origin of $\mathbb{C}^2$. From the geometric definition of the shadow number, it follows that this number is $PGL(2, \mathbb{C})$-invariant:

$$\lambda_{pqts} = \lambda_{Ap,Aq,At,As}, \quad A \in PGL(2, \mathbb{C}).$$

The shadow number is a 4-point function defined on the complement of the general diagonal in the Riemann sphere R:

$$\lambda : R^4 \setminus \text{Diag}(R) \to \mathbb{C}$$

where $\text{Diag}(R)$ is the subset of quadruples $(p, q, t, s) \in R^4$ of points in R with $\#\{p, q, t, s\} < 4$. Using the above formula one checks that the function λ is holomorphic with meromorphic extension to R^4.

Exercise 2.4.3 For p, q, t a triple of distinct points, study the partial function $f :$ $s \mapsto \lambda_{pqts}$. What are the level sets of $f, |f|, f \pm \bar{f}$?

Exercise 2.4.4 Reconstruct the complex structure J of R from the shadow function λ.

Exercise 2.4.5 For a 4 point function $f : R^4 \setminus \text{Diag}(R) \to \mathbb{C}$, define $\text{Aut}(R, f)$ to be the group of bijections $A : R \to R$ satisfying $f \circ A = f$. Consider the cases $f = \lambda$ or $f = |\lambda|$ or $f = \lambda \pm \bar{\lambda}$. Show that $PGL(2, \mathbb{C}) = \text{Aut}(R, \lambda)$ and $\text{Aut}(R, |\lambda|) = \text{Aut}(R, \lambda \pm \bar{\lambda})$.

2.4.4   J-Compatible Metrics

In this section, (S, J) is a differentiable compact connected surface equipped with a complex structure $J : TS \to TS$. A Riemannian metric g on S is called conformal with J if for every point $p \in S$ the map $J_p : T_pS \to T_pS$ is g_p-orthogonal. Thus, two metrics g, g' conformal with J differ pointwise by a positive factor: $g' = fg$ for some function f. A J-calibrated volume form ω on S, i.e., a differential 2-form satisfying $\omega_p(u, J_pu) > 0$, $p \in S$, $u \in T_pS$, $u \neq 0$, gives by $g_p(u, v) = \omega_p(u, J_pv)$ a Riemannian metric conformal with J whose associated volume form is precisely ω. All metrics g conformal with J are obtained in a unique way from such a construction by taking for ω the oriented volume form $\omega(u, Ju) = g(u, u)$ of the Riemannian metric g.

In conclusion, given J, the space of Riemannian metrics conformal with J is parametrized by the cone of J-calibrated volume forms. In this way, the uniqueness is not only up to multiplying by a constant factor.

We wish to strengthen the notion of being "conformal with J" for metrics and gain back uniqueness or controlled non-uniqueness up to multiplying by a constant factor. This will be achieved first for Riemann surfaces (S, J) that are homogeneous, i.e., whose group $\text{Aut}(S, J)$ acts transitively on S, such that moreover the stabilizers of points are compact.

The group $\text{Sim}(M, g)$ of a connected Riemannian manifold is the group of diffeomorphisms that multiply the metric by a constant factor. Such diffeomorphisms are called similarities.

Theorem 2.4.2 *Let (S, J) be a homogeneous Riemann surface with commutative stabilizers. Then there exists a Riemannian metric g conformal with J such that $\text{Sim}^+(S, g) = \text{Aut}(S, J)$. The Riemannian metric g is unique up to multiplication by a constant factor.*

Sketch of proof The list of homogeneous connected Riemann surfaces up to bi-holomorphic equivalence is short:

1. The Riemann sphere R;
2. $R \setminus \{\infty\}$;
3. $R \setminus \{0, \infty\}$;
4. the infinite family of elliptic curves $\mathbb{C}/\Gamma$;
5. $\mathbb{C}_+ = \{f_{a+bi} \mid a, b \in \mathbb{R}, b > 0\}$.

In this list, only the Riemann sphere R has non-compact and non-commutative stabilizers—the stabilizer of a point is the affine group of $\mathbb{C}$. Thus, we do not need to consider this surface.

The second surface in the list is bi-holomorphic to $\mathbb{C}$. The Euclidean metric $g_z(u, u) = u\bar{u}$ is up to multiplication by a constant factor the only metric whose bi-holomorphic equivalences are similarities. Note that its group of isometries is a strict subgroup of the group of similarities. It is not commutative. So we also do not consider this surface.

The third surface in the list is the image of the covering map $z \in \mathbb{C} \mapsto e^{2\pi i z} \in \mathbb{C}^*$ with deck transformations $z \mapsto z + k$, $k \in \mathbb{Z}$. These deck transformations are isometries of the Euclidean metric, therefore they define a metric g on $\mathbb{C}^*$ which is locally Euclidean. This metric g is defined by $g_z(u, u) = \frac{u\bar{u}}{z\bar{z}}$. Note the triple equality $\text{Sim}(\mathbb{C}^*, g) = \text{Iso}^+(\mathbb{C}^*, g) = \text{Aut}(\mathbb{C}^*, J)$.

A similar discussion holds for the family of elliptic curves. In this case, instead of the mapping $\exp : \mathbb{C} \to \mathbb{C}^* = \mathbb{C}/\mathbb{Z}$, one considers a mapping $\exp_\omega : \mathbb{C} \to \mathbb{C}/\text{Gr}(1, \omega)$ where $\text{Gr}(1, \omega)$ is the group of translations generated by the complex numbers 1 and ω.

The group $G = \text{PSL}(2, \mathbb{R})$ acts transitively on $\mathbb{C}_+$ by $z \in \mathbb{C}_+ \mapsto \frac{az+b}{cz+d} \in \mathbb{C}_+$. The stabilizer of i is the compact group $\text{Stab}(i) = \text{PSO}(2)$. The Killing form $k(H, K) = \text{Trace}(HK)$ on the Lie algebra $\text{Lie}(G)$ induces on $\mathbb{C}_+ = G/\text{Stab}(i)$ a Riemannian metric which is conformal with $J = m_i$ ("multiplication by i"). Again a triple equality of groups holds.

The interpretation of $z = a + bi \in \mathbb{C}_+$ as a ring homomorphism f_z and as an ideal $((X - a)^2 + b^2))$ (see Sect. 2.3.1) gives an elementary insight for the above explanation by Lie group theory.

Given $z \neq z'$, define the curve $L_{z,z'} = z_t$, $t \in]a^-(z, z'), a^+(z, z')[$ by the following: put $z = a + bi$, $z' = a' + b'i$, then z, z' are the roots in $\mathbb{C}_+$ of the polynomials $P_z = x^2 - 2ax + a^2 + b^2$, $P_{z'} = x^2 - 2a'x + a'^2 + b'^2$. Let $]a^-(z, z'), a^+(z, z')[$ be the maximal open interval containing the interval $[0, 1]$ such that for all $t \in]a^-(z, z'), a^+(z, z')[$ the convex combinations of polynomials $Q_t = t P_z + (1 - t) P_{z'}$ have no real roots. Define $L_{z,z'}$ as the path of roots in $\mathbb{C}_+$ of the polynomials Q_t.

If $a = a'$, then the curve $L_{z,z'}$ is the real half-line perpendicular to the real axis through z and z'. If $a \neq a'$, then the curve $L_{z,z'}$ is the semi-circle through z, z' with center on the real axis, hence perpendicular to the real axis.

The quantity $D(z, z') = \frac{1}{2} \log(\lambda_{a^-(z,z'),a^+(z,z'),0,1})$ defines a metric on $\mathbb{C}_+$, which is $\mathrm{PSL}(2, \mathbb{R})$-invariant. The curves $L_{z,z'}$ are length minimizing, hence geodesics. More precisely, the expansion of $D(z, z+u)^2$ at the point z gives the leading second order term $g_z(u, u) = \frac{u\bar{u}}{\mathrm{Im}(z)^2}$ and defines a Riemannian metric. From this, we can recover the hyperbolic metric on $\mathbb{C}_+$, see [9, §8.3].   □

In conclusion, from the formulae $\omega(u, v) = g(u, Jv)$ and $\omega(u, Jv) = g(u, v)$, each of the volume form ω and the metric g is determined from the other one and from the almost complex structure J. The same formulae determine J from ω and g because the 2-forms ω and g are non-degenerate.

In fact, more generally, the motto is that in the triple of structures on surfaces (volume form ω, Riemannian metric g, almost complex structure J) on $\mathbb{R}^{2n}$, two elements determine the third one. The result is that there are fruitful interactions between symplectic geometry, Riemannian geometry and complex geometry. This is expressed in a spectacular manner in Gromov's 1985 paper in which he introduced pseudo-holomorphic curves [42]. This introduced also notions of "positivity" and of "compactness" in the three geometrical settings. One result is that the stabilizers of w and g in $\mathrm{GL}(2n, \mathbb{R})$, if they are related by a J as above, coincide and form a maximal compact subgroup.

2.4.5 Spherical Geometry

The differential sphere is present as the Riemann sphere R. Spherical geometry is still missing. The group $\mathrm{PGL}(2, \mathbb{C})$ acts transitively on (R, J_R). The stabilizer $\mathrm{Stab}(\infty)$ is the group $z \mapsto \lambda z + t$, $\lambda \in \mathbb{C}^*$, $t \in \mathbb{C}$. Again by Lie theory, each maximal compact subgroup U of $\mathrm{Aut}(R, J_R)$ defines a spherical metric g_U on R that is conformal with J_R.

A perhaps more elementary approach is the following: Let I_R be the space of fixed-point free involutions that preserve the shadow function $\lambda : R^4 \setminus \mathrm{Diag} \to \mathbb{C}$. For two distinct involutions A, B, the composition $C = A \circ B$ has two fixed points

p, q. Let C_p, C_q be the determinants of the differentials $D_p C, D_q C$ at p, q. Define the distance $D(A, B)$ by

$$D(A, B) = \frac{1}{2} \log((C_p + C_q)/2).$$

The space (I_R, D) is a model for hyperbolic 3-space. The infinitesimal version g_{I_R} of the metric D is a Riemannian metric on I_R. For each $A \in I_R$, the half-rays from A define a diffeomorphism ϕ_A from the infinitesimal sphere S_A with center A to R. The map ϕ_A is conformal and carries the spherical metric of S_A to a metric g_A on R which is conformal with J_R. Think of S_A as the unit sphere in $(T_A I_R, g_{I_R, A})$. The group $\mathrm{Iso}(R, g_A)$ is a maximal compact subgroup in $\mathrm{Aut}(R, J_R) = \mathrm{PGL}(2, \mathbb{C})$. Moreover the equality $\mathrm{Sim}(R, g_A) = \mathrm{Iso}(R, g_A)$ holds.

We think of the Riemann sphere R as the space of 1-dimensional vector subspaces in $V = \mathbb{C}^2$ (Sect. 2.3.1). A positive Hermitian form on V is a real bilinear map $h : V \times V \to \mathbb{C}$ satisfying for $u, v \in V, \lambda \in \mathbb{C}$

- $h(\lambda u, v) = \lambda h(u, v)$,
- $h(u, v) = \overline{h(v, u)}$,
- $h(u, u) > 0$ for $u \neq 0$.

A positive Hermitian form on V defines by $L \in R \mapsto L^\perp \in R$ a fixed-point free involution A_h of the Riemann sphere R that preserves the shadow function λ. Here, associated with the complex line $L = tu, t \in \mathbb{C}, u \in V, u \neq 0$, is the complex line $L^\perp = \{v \in V \mid h(v, u) = 0\}$. Two positive Hermitian forms that differ by a positive factor give the same involution.

Moreover the stabilizer in $\mathrm{Aut}(R) = \mathrm{PGL}(2, \mathbb{C})$ is a maximal compact subgroup in $\mathrm{Aut}(R)$. Similar forms have the same stabilizers and all maximal compact subgroups correspond to a unique form.

If two lines L, L' in R are neither perpendicular nor equal, they give by $L, L', L^\perp, L'^\perp$ a quadruple of complex lines. The expression $D(L, L') = \lambda_{L, L', L'^\perp, L^\perp}$ defines a function on $R \times R$ with values in $\mathbb{R}_+ \cup \{+\infty\}$. The preimage of 0 is the diagonal, the set of pairs (L, L') with $L' = L$, and the preimage of $+\infty$ the set of pairs with $L' = L^\perp$. Its infinitesimal version along the diagonal, i.e., at $L \in R$ the Hessian of $L' \mapsto D(L, L')$ at $L' = L$, defines a Riemannian metric on R which is similar, even isometric, to the spherical geometry of Gaussian curvature $+1$ in dimension 2.

2.4.6   Models for Hyperbolic 3-Space and Metrics of Constant Positive Curvature on the Sphere

Even though the 3-dimensional hyperbolic space $\mathbb{H}^3$ does not admit a complex structure, its automorphism group is that of a complex manifold and is itself a complex Lie group, since we have

$$\mathrm{Iso}^+(\mathbb{H}^3) = \mathrm{Aut}_{\mathrm{hol}}(\mathbb{P}^1(\mathbb{C})) = \mathrm{PGL}(2, \mathbb{C}).$$

We have the following four models of the hyperbolic 3-space $\mathbb{H}^3$ that arise from the complex geometry of surfaces; the first two models are geometric (they are defined in terms of group actions) whereas the other two use algebra:

1. The space of fixed-point free shadow-preserving involutions of R.
2. The space of Riemannian metrics on R that are conformal with J_R and similar to the spherical metric.
3. The space of similarity classes of positive Hermitian forms on $\mathbb{C}^2$.
4. The space of maximal compact subgroups in $\mathrm{PGL}(2, \mathbb{C})$ (which is the automorphism group of the oriented hyperbolic 3-space).

The first model already appeared in Sect. 2.4.5, where we constructed the spherical geometry of the Riemann sphere. This is the sphere we called I_R there. To get another point of view on this model, consider first the well-known Poincaré model of the hyperbolic space $\mathbb{H}^3$ as a unit ball sitting in 3-space, with boundary the Riemann sphere R. For each point p in $\mathbb{H}^3$, an involution of R is defined by assigning to each point in R the intersection with the sphere of the line passing through this point and p. To see that we get a model of $\mathbb{H}^3$, use the Hilbert metric model of this space.

For the second model, we also consider the Poincaré unit ball model of the hyperbolic space, with boundary the Riemann sphere R. For each point of $\mathbb{H}^3$, we take the diffeomorphism which sends the infinitesimal round sphere (or a sphere in the tangent space) at that point to the boundary at infinity of the space, using the geodesic rays that start at this point. This is a conformal mapping. (One may prove this by trigonometry.) The mapping sends the conformal metric of the infinitesimal sphere to a metric on the sphere sitting at the boundary of the hyperbolic space. This construction commutes with the isometries of $\mathbb{H}^3$. It gives a 3-dimensional space of metrics in the same conformal class on $\mathbb{P}^1(\mathbb{C})$. Thus, the hyperbolic 3-space $\mathbb{H}^3$ appears as a space of metrics of curvature $+1$ on $\mathbb{P}^1(\mathbb{C})$. In other words, $\mathbb{H}^3$ appears as a space of special metrics, i.e., as a moduli space, namely, the space of Riemannian metrics g of Gaussian curvature $+1$ on R that are moreover in the conformal class of J_R. Alternatively, the hyperbolic 3-space $\mathbb{H}^3$ appears as the space of volume forms ω on R such that the Riemannian metric defined by $g(u, v) = \omega(u, J_R v)$ is a metric of Gaussian curvature $+1$.

For the third model, recall that a positive Hermitian form on $\mathbb{C}^2$ is a Riemannian metric on $\mathbb{R}^4$ such that one can measure the angle between two lines in this space. This model was explained in Sect. 2.3.2 above.

The fourth model is equivalent to the 3rd because the group $\mathrm{PGL}(2, \mathbb{C})$ acts on the space in question with stabilizer the automorphism group of the oriented hyperbolic 3-space, that is, the group $\mathbb{PGL}(2, \mathbb{C})$ quotiented by its maximal compact subgroup.

An explicit way of interconnecting the above models for the hyperbolic three-space $\mathbb{H}^3$ is as follows.

Think of S^2 as the unit sphere in $\mathbb{R}^3$ with its induced spherical metric together with the corresponding conformal structure, the space $\mathbb{R} \oplus \mathbb{C}$, for the Euclidean norm $||(t, z)||^2 = t^2 + z\bar{z}$.

The stereographic projection with pole $(0, -1)$ maps by

$$(0, z) \mapsto (\frac{1 - z\bar{z}}{1 + z\bar{z}}, \frac{2z}{1 + z\bar{z}})$$

the factor $\{0\} \times \mathbb{C}$ conformally to S^2. This projection extends to an (oriented) diffeomorphism

$$St : R = \mathbb{C} \cup \{\infty\} \to S^2$$

and equips by pullback the Riemann sphere with a Riemannian metric of Gaussian curvature $+1$.

Given an ordered triple of distinct points (a, b, c) in R, the shadow function gives by

$$p \in R \setminus \{c\} \mapsto Sh_{abc}(p) = (0, cr(a, p, b, c)) \in \{0\} \times \mathbb{C} \cup \{\infty\}$$

a mapping such that the composition $St_{abc} \circ Sh$ extends to a conformal diffeomorphism $\mu_{abc} : R \to S^2$.

By pullback we obtain a family g_{abc} of Riemannian metrics of curvature $+1$ on R. Two such metrics g_{abc} and $g_{a'b'c'}$ are related by an oriented isometry if and only if the composition $\mu_{abc} \circ \mu_{a'b'c'}^{-1}$ is an oriented isometry of S^2.

Now remember that the group of holomorphic automorphisms of R is isomorphic to $PGL(2, \mathbb{C})$ and acts simply transitively on triples of distinct points. So each triple (a, b, c) is obtained in a unique way as $(a, b, c) = M_{abc}(a_0, b_0, c_0)$ from a chosen triple (a_0, b_0, c_0) by applying a well-defined element $M_{abc} \in PGL(2, \mathbb{C})$. The above condition that the metrics g_{abc} and $g_{a'b'c'}$ are related by an oriented isometry translates into the fact that $M_{a',b',c'} \circ M_{a,b,c}^{-1} \in PU(2, \mathbb{C})$ and so it proves that the hyperbolic 3-space $\mathbb{H}^3$ is parametrized by the symmetric space $PGL(2, \mathbb{C})/PU(2, \mathbb{C})$.

From the above discussion, some basic ingredients of the geometry of $\mathbb{H}^3$ can be seen, such as the following:

- Ideal points are the points of R.
- Given $a \neq c$ in R, the points on the line L_{ac} from a to c are the metrics $\mu_{a,b,c}$ where b varies over $R \setminus \{a, c\}$.
- For $p \neq q \in H^3$, represented respectively by metrics μ_p, μ_q and volume forms ω_p, ω_q on R, these points lie on L_{ac} where the points a, c are the extrema of $\frac{\omega_p}{\omega_q}$.

2.4.7 Models for the Hyperbolic Plane

The hyperbolic plane $\mathbb{H}^2$, like the 3-dimensional hyperbolic space $\mathbb{H}^3$, has also several interpretations in terms of the complex geometry of surfaces. A standard interpretation of $\mathbb{H}^2$ is that it is the space of marked elliptic curves. We propose three other incarnations of this plane:

1. The hyperbolic plane $\mathbb{H}^2$ is the space of ideals $\mathbb{H}_I$ with its geometry naturally given by a family of lines $L_{((X-a)^2+b^2),((X-a')^2+b'^2)}$. Indeed, the plane $\mathbb{H}_I$ comes with a natural complex coordinate $z : \mathbb{H}_I \to \mathbb{C}_+ = \{a + ib, \ b > 0\}$, defined by

$$z((X - a)^2 + b^2)) = a + bi.$$

The coordinate z is a bijection with

$$Z : a + bi \mapsto ((X - a)^2 + b^2)$$

as inverse. Thus, $Z(a + bi)$ is the interpretation of the number $a + bi$ as an ideal. See the details in Sect. 2.4.4 above.

2. The hyperbolic plane is the space $\mathbb{H}_J$ of almost complex structures J on $\mathbb{R}^2$ (that is, endomorphisms of $\mathbb{R}^2$ satisfying $J^2 = -\text{Id}$) equipped with coordinates (x, y), and a volume form $\omega = dx \wedge dy$, and where J is calibrated by the inequality $\omega(u, Ju) > 0$ for all $u \in \mathbb{R}^2$.

   Another way of seeing this is the following. We start by defining a map: Fix : $H_J \to \mathbb{C}_+$, so that a point in the space H_J has a coordinate $z(J) = \text{Fix}(J) \in \mathbb{C}_+$ where $\text{Fix}(J)$ is the fixed point of J for its homographic action on $\mathbb{C}_+$.

   More explicitly, the matrix of J in the basis $(1, i)$ is of the form $\left(\begin{smallmatrix} a/b & * \\ 1/b & -a/b \end{smallmatrix}\right)$ where $h \in \mathbb{R}$ and $b > 0$ (use the fact that the trace is zero), and $*$ is obtained from the fact that the determinant of the matrix is equal to 1. Such a matrix acts on the upper half-plane with exactly one fixed point, and we assign to J this fixed point. It is possible to write explicitly the coordinates of this fixed point in terms of a and b.

   Consider the map

$$z\left(\left(\begin{smallmatrix} h & * \\ k & -h \end{smallmatrix}\right)\right) = \frac{h}{k} + \frac{i}{k}$$

where $k > 0$ and $* = \frac{1+h^2}{-k}$ is a complex coordinate. This is again a bijection, with inverse

$$Z : a + bi \mapsto \left(\begin{smallmatrix} a/b & * \\ 1/b & -a/b \end{smallmatrix}\right), \ b > 0,$$

and there is an interpretation of the number $a + bi$, $b > 0$ as a linear complex structure.

The incarnation $\mathbb{H}_J$ is very helpful for the construction of hyperbolic trigonometry. Only High-school/Freshman knowledge is required in this construction.

This incarnation $\mathbb{H}_J$ is also very helpful for the study of the space of complex structures $\mathbb{J}(TS)$ on a given surface. As a first example, the surface $\mathbb{H}_J$ itself has a complex structure. A tangent vector at $J \in \mathbb{H}_J$ is an endomorphism K of $\mathbb{R}^2$ such that the equation

$$(J + K)^2 = J^2 + J \circ K + K \circ J + K^2 = -\mathrm{Id}_{\mathbb{R}^2}$$

holds at first order in K. Therefore the tangent space $T_J \mathbb{H}_J$ at J of $\mathbb{H}_J$ is the vector space of endomorphisms K that anti-commute with J. The map $K \in T_J \mathbb{H}_J \mapsto J \circ K \in T_J \mathbb{H}_J$ is a canonical complex structure $J_{\mathbb{H}_J}$ on $\mathbb{H}_J$ which induces a complex structure on the infinite-dimensional manifold $\mathbb{J}(TS)$. This space $\mathbb{J}(TS)$ is canonically path-connected. Indeed two structures J_0, J_1 are connected by the path that for $p \in S$ is a geodesic in the space of linear complex structures on the oriented real vector space $T_p S$. The fact that the space $\mathbb{J}(TS)$ is contractible now follows, since these canonical paths depend continuously upon endpoints.

3. The bijective coordinates z on $\mathbb{H}_I$, $\mathbb{H}_J$ transport the geometries to a common geometry on $\mathbb{C}_+$. The coordinate $z : \mathbb{H}_J \to \mathbb{C}_+$ is $(J_{\mathbb{H}_J}, m_i)$-holomorphic. The model $\mathbb{C}_+$ with its conformal structure and hyperbolic metric, equipped with its action by the modular group $\mathrm{PSL}(2, \mathbb{R})$, is a very appreciated model of the hyperbolic plane.

2.5 Uniformisation

2.5.1 Riemann's Uniformisation of Simply Connected Domains

Theorem 2.5.1 (The Riemann Mapping Theorem [90]) *Let Ω be a non-empty open connected and simply connected subset of the complex plane which is not the entire plane. Let z_0 be a point in Ω and v a nonzero vector at z_0. Then, there is a unique holomorphic bijection f from Ω to the unit disc $\mathbb{D}$ such that $f(z_0) = 0$ and $df(v)$ is a complex number which is real and positive.*

Note that the inverse of a holomorphic bijection is also holomorphic (think of a holomorphic function as an angle-preserving map). Therefore the map f is biholomorphic.

Outline, After Riemann's Proof, See [52] This proof works in the case where Ω is an open subset of the plane bounded by a Jordan curve (a simple closed curve). We outline this proof.

We wish to find a biholomorphic mapping $f : \Omega \to \mathbb{D}$ which sends $\gamma = \partial\Omega$ to $\mathbb{S}^1 = \partial\mathbb{D}$.

Suppose that such a mapping f exists. Separating the real and imaginary parts, we set

$$f(z) = u(x, y) + iv(x, y).$$

Applying a translation, we may assume that $z_0 = 0$, therefore, $f(0) = 0$. Applying a rotation, we may also assume that the vector of coordinate $(0, 1)$ at 0 is sent by f to a vector pointing towards the positive real numbers.

Such a map f, it is exists, is unique, since any holomorphic automorphism of the disc which fixes the origin is a rotation, and a rotation that fixes a direction is the identity.

Since f is a bijection between Ω and the unit disc, the point $z = 0$ is the unique solution of the equation $f(z) = 0$. Furthermore, the differential (or the complex derivative) of f at 0 is nonzero, otherwise f would be non-injective in the neighborhood of zero. Thus, the function $f(z)/z$ is holomorphic on Ω. Since it does not take the value 0, a branch of its complex logarithm can be defined. Hence, there exists a holomorphic function $H(z)$ on Ω such that

$$\frac{f(z)}{z} = e^{H(z)}.$$

In polar coordinates we can write $z = r(x, y)e^{i\theta(x,y)}$, which gives

$$f(z) = ze^{H(z)} = r(x, y)e^{i\theta(x,y)}e^{H(z)}.$$

Separating the real and imaginary parts of H as $H(z) = P(x, y) + iQ(x, y)$, we get

$$\log f(z) = \log r(x, y) + i\theta(x, y) + P(x, y) + iQ(x, y),$$

which we can write as

$$\log f(z) = \log r(x, y) + P(x, y) + i(\theta(x, y) + Q(x, y)).$$

Notice that the function $\log|f|$ is identically zero on $\partial\Omega$, since points z on this curve satisfy $|f(z)| = 1$.

Furthermore, the function $\log|f|$ is the real part of the holomorphic function $\log f$, therefore it is harmonic at every point of Ω.

Now given a harmonic function defined on the simply connected subset Ω of the plane, there exists a unique holomorphic function which has this function as a real part. (The usual way to find this function is to use Cauchy's theory of path integrals.)

Reasoning backwards, the problem of mapping Ω to the unit disc can be solved if we can find a function P which is harmonic on Ω and which takes the values $-\log r$ on $\gamma = \partial\Omega$.

In this way, Riemann reduced the problem of finding a holomorphic mapping from Ω to $\mathbb{D}$ to a problem of finding a harmonic function on Ω with a prescribed value on $\partial\Omega$. To find such a function, he appealed to the so-called Dirichlet method, which consists of considering the functional I defined on the space of differentiable functions P on Ω by:

$$I(P) = \int_\Omega \left(\frac{\partial P}{\partial x}^2 + \frac{\partial P}{\partial y}^2\right) dx dy.$$

A function which realizes this infimum is harmonic. □

Remarks

1. Riemann also used the Dirichlet principle in his existence proof of meromorphic functions on general Riemann surfaces, obtaining them from their real parts (harmonic functions). It was realized later that the Riemann Mapping Theorem and the existence of meromorphic functions are equivalent.
2. A modern proof of the Riemann Mapping Theorem goes as follows (see Thurston's notes[104]): Given a simply connected open strict subset Ω of the complex plane and a point z_0 in Ω, consider the set of functions

$$\mathcal{F} = \{f : \Omega \to \mathbb{D}, \text{ holomorphic, injective, } f(z_0) = 0 \text{ and } f'(z_0) > 0\},$$

equipped with the topology of uniform convergence on compact sets. The desired conformal mapping is the mapping f in this family that maximizes the modulus of the derivative $|f'(z_0)|$. The existence of such a mapping uses the fact that for any family of holomorphic functions which is uniformly bounded on compact sets, every sequence has a subsequence which converges uniformly on compact sets to a holomorphic function (this is Montel's theorem).

We shall call a *Riemann mapping* a biholomorphic mapping that sends the unit disc to a simply-connected subset of the plane (which we sometimes call a domain). This is the inverse of the mapping $f : \Omega \to \mathbb{D}$ that we just considered.

For a general domain Ω, the Riemann mapping is not explicit. In some special cases, there are explicit formulae, e.g., when Ω is the upper half-plane, or a band in the plane (a region bounded by two parallel lines), or an angular sector (a connected component of the union of two intersecting Euclidean lines in the plane), or a domain bounded by two arcs of circle, or a sector of a disc, or a rectangle in the Euclidean plane, or a regular convex polygon, or a regular star polygon, or the interior of an ellipse or a parabola, and a few other cases. These examples and others are studied in the book by Julia [52].

There are also results that concern the boundary behavior of the Riemann mapping. The first name associated with such a boundary theory is Carathéodory.

He proved that if the boundary of the domain Ω is a Jordan curve, then the Riemann mapping extends continuously to a homeomorphism from the unit circle to the boundary of the domain. He also proved that in the general case the Riemann mapping extends continuously to the unit circle if and only if the boundary of the domain is locally connected (but this extension is generally not a homeomorphism). There are also measure-theoretic results and questions. For instance, in the case where the domain is bounded by a rectifiable curve γ, an extension of the Riemann mapping exists and sends any subset of measure zero (resp. of positive measure) of γ to a subset of measure zero (resp. of positive measure) of the circle. This was proved by Luzin and Privaloff [70] and Riesz [88]. One can also mention here the following result of Fatou [38]: If a function f defined on the unit disc is holomorphic and bounded, then, at almost every point t on the unit circle $|z| = 1$, the value of f at a sequence of points converging to that point on a path which is not tangential to the unit circle has a limit. For an exposition of related results, see Chapter II of Lavrentieff's book [65]. These results are useful in the modern developments of complex dynamics, in particular in the study of the conformal representation of the complement of the Julia set of a polynomial.

2.5.2 Uniformization of Multiply Connected Domains

It is natural to search for generalizations of the Riemann Mapping Theorem for multiply connected open subsets of the plane. The first natural question is whether there exist canonical domains onto which they can be mapped biholomorphically, in analogy with the fact that the disc and the plane are canonical domains for simply connected domains. It is known that two open subsets of the complex plane that have the same connectivity are not necessarily conformally equivalent. For instance, two circular annuli (that is, annuli bounded by two concentric circles) are conformally equivalent if and only if they have the same modulus, that is, if and only if the ratio of their outer radius to their inner radius is the same. Thus, there are no canonical domains for multiply connected open subsets of the plane. However, there are figures which may be considered as "standard" domains for such subsets, and we now discuss some of them.

One of the oldest works on this question was done by Koebe, who studied in his papers [62, 63] and others conformal mappings of multiply connected open subsets of the plane onto *circular domains*, that is, multiply connected domains whose boundary components are all circles (which may be reduced to points). He proved that every finitely connected domain in the plane is conformally equivalent to a circular domain. The question of whether there is an analogous result for open subsets of the plane with infinitely many boundary components is still open. This question is known under the name *Kreisnormierungsproblem*, or the *Koebe uniformization conjecture*. Koebe formulated it in his paper [61]. The reader interested in this question may refer to the paper [23] by P.L. Bowers in which the author surveys several developments of this conjecture, including related works on circle packings by Koebe, Thurston and others. See also [24].

Grötzsch, in his paper [44], works with two classes of domains which were considered by Koebe:

1. *Annuli with circular slits.* These are circular annuli from which a certain number (≥ 0) of circular arcs (which may be reduced to points) centered at the center of the annulus, have been deleted.
2. *Annuli with radial slits.* These are circular annuli from which a certain number (≥ 0) of radial arcs (which may be reduced to points) have been removed. Here, a radial arc is a Euclidean segment in the annulus which, when extended, passes through the center of the annulus.

The unit disc slit along an interval of the form $[0, r]$ ($r < 1$) was employed by Grötzsch in [44] as a standard domain and is known under the name *Grötzsch domain*. Teichmüller, in his paper [102], calls a *Grötzsch extremal region* the complement of the unit disc in the Riemann sphere $\mathbb{C} \cup \{\infty\}$ cut along a segment of the real axis joining a point $P > 0$ to the point ∞.

Another "standard model" for doubly connected domains is the Riemann sphere slit along two intervals of the form $[-r_1, 0]$ and $[r_2, \infty)$ where r_1 and r_2 are positive numbers. Such a domain is called a *Teichmüller extremal domain* by Lehto and Virtanen [68, p. 52]. In his paper [102], Teichmüller uses these domains in the study of extremal properties of conformal annuli. In the same paper, he works with other "standard" domains, e.g., the circular annulus $1 < |z| < P_2$ cut along the segment joining $z = P_1$ to $z = P_2$, where P_1 and P_2 are points on the real axis satisfying $1 < P_1 < P_2$ [102, §2.4]; see the discussion in Ahlfors' book [14, p. 76].

2.5.3   Uniformization of Simply Connected Riemann Surfaces

It is natural to consider conformal representations of surfaces that are not subsets of the complex plane. A wide generalization of the Riemann Mapping Theorem is the following, proved independently by Poincaré and Koebe:

Theorem 2.5.2 (Uniformization Theorem)   *Every simply connected Riemann surface can be mapped biholomorphically either to*

1. *the Riemann sphere (elliptic case);*
2. *the complex plane (parabolic case);*
3. *the unit disc (hyperbolic case).*

In each of these cases, the conformal biholomorphism from the model space onto the open Riemann surface is usually called a *uniformizing map*. Such a map associated to a Riemann surface is well defined up to composition by a conformal homeomorphism of the source. This implies that the mapping is determined in the first case by the images of three distinct points, in the second case by the image of two distinct points, and in the third case by the image of one point and a direction at that point.

If the Riemann surface is a simply connected open subset Ω of the complex plane which is not the entire plane, then it is necessarily of hyperbolic type (this is the Riemann Mapping Theorem). Uniformizing maps for such surfaces have been particularly studied.

For a proof of the uniformization theorem based on a historical point of view, see the collective book [92]. For another point of view on uniformization ("uniformization by energy"), see [8, Chapter 10].

There are several open questions on the behavior of the uniformizing maps of simply connected surfaces equipped with Riemannian metrics. Lavrentieff, in his paper [66], adresses the question of the Lipschitz behavior of a uniformizing map $f : \mathbb{D} \to \Omega$. More precisely, he asks for conditions on the surface Ω so that for any pair of points $x, x' \in \mathbb{D}$, the ratios $\frac{d_{\mathbb{D}}(x,x')}{d_S(f(x),f(x'))}$ and $\frac{d_S(f(x),f(x'))}{d_{\mathbb{D}}(x,x')}$ are bounded or unbounded (the distance in the disc being the Euclidean distance). To the best of our knowledge, this problem is still not settled.

2.6 Branched Coverings

2.6.1 The Riemann–Hurwitz Formula

The notion of branched covering between surfaces was conceived by Riemann. It will be used at several places in the rest of this chapter. We shall review in particular a theorem of Thurston on branched coverings of the sphere in relation with the question of the topological characterization of rational maps (Sect. 2.6.3). Branched coverings also appear in the study of the type problem (Sect. 2.7.2) and in the theory of dessins d'enfants (Sect. 2.8.1).

A map $f : S_1 \to S_2$ between two topological oriented surfaces is said to be a *branched covering* if it satisfies the following:[2]

1. for some discrete subset $\{a_i\}_{i \in I} \subset S_2$, the map f restricted to $f^{-1} \setminus \{a_i\}$ is a covering map in the usual sense;
2. for each point $x \in S_1$ which is the inverse image by f of a point $a_i \in S_2$ for some $i \in I$, there exists an orientation-preserving homeomorphism $\phi : U \to D$ between an open neighborhood U of x and an open neighborhood D of 0 in $\mathbb{C}$ satisfying $\phi(x) = 0$, and an orientation-preserving homeomorphism $\psi : V \to D$ between an open neighborhood V of a_i and D satisfying $\phi(a_i) = 0$, such that the mapping $\psi \circ f \circ \psi^{-1}$ defined on D coincides with the restriction of the map $z \mapsto z^k$ from the complex plane to itself, for some integer $k \geq 1$.

---

[2] In the older literature, this notion of *branched covering* is called *inner map in the sense of Stoilov*. Later, this was renamed *topologically holomorphic map*. For functions on con-compact surfaces, The term *branched covering* is rarely used. (We owe this remark to A. Eremenko)

The integer k in the above statement depends only on the point x and is called the *branching order* of f at x. If $k = 1$, then f is a local homeomorphism at x. If $k > 1$, then we say that x is a *critical point* of the covering and that f is *branched* at x. A point a_i which is the image of a critical point is called a *critical value*. The *degree* of the branched covering f is the usual degree of the covering induced by f on the surface $f^{-1} \setminus \{a_i\}$, that is, the cardinality of the pre-image by f of an arbitrary point in $S_2 \setminus \{a_i\}$. This degree is also equal to the sum of the branching orders of the points in the pre-image of an arbitrary point a_i. The degree of a branched covering is 1 if and only if this covering is a homeomorphism.

A meromorphic function on a Riemann surface defines a branched covering from this surface to the Riemann sphere. (Here, the adjective "meromorphic" can be replaced by "holomorphic" since the value ∞ is, from the complex structure, a point like any other point on the Riemann sphere.) Every connected compact Riemann surface S admits a branched covering $f : S \to R$ over the Riemann sphere R, with the conformal structure on S induced by lifting by f the conformal structure of the sphere. (This is one form of the so-called Riemann's Existence Theorem.) Special cases of branched coverings that we shall discuss below are the Belyi maps [20]: holomorphic maps from a compact Riemann surface to the Riemann sphere with at most 3 critical values (see Sect. 2.8.1).

The question of the topological characterization of analytic functions among the maps from a surface onto the Riemann sphere was known under the name *Brouwer problem* and it has many aspects, see the comments in [100, p. xv]. A theorem of Stoïlov says that if S is a topological surface, then a continuous mapping from S to the Riemann sphere is a branched covering if and only if f is open (i.e., images of open sets are open) and discrete (i.e., inverse images of points are discrete subsets of S, that is, each point is isolated in S). Furthermore, for each continuous mapping satisfying this property, there exists a unique complex structure on S such that this mapping is conformal [100, Chap. V].

René Thom was interested in such questions. In his paper [103], he solves the following problem:

Given n complex numbers (which are not necessarily distinct), does there exist a polynomial of degree n + 1 with these numbers as critical values?

Thom also considers the real analogue. At the end of his paper, he formulates the general problem:

Given a C^∞ function $f : \mathbb{R}^2 \to \mathbb{R}^2$ with only isolated critical points, can we transform it, by diffeomorphisms of the source and of the target, into a holomorphic function?

Thurston proved a fundamental result on the topological characterization of certain rational maps of the sphere, more precisely, he gave a characterization of postcritically finite branched coverings of the sphere by itself (branched coverings in which the forward orbits of the critical points are eventually periodic) that are topologically conjugate to rational maps. See [104] and the exposition in [25].

If S_1 and S_2 are compact surfaces, there is a classical combinatorial relation that is satisfied by any branched covering $S_1 \to S_2$, namely, the following:

Proposition 2.6.1 (Riemann–Hurwitz Formula) *For any branched covering $S_1 \to S_2$ between compact surfaces, we have*

$$\chi(S_1) = d \cdot \chi(S_2) - \sum (k(p) - 1),$$

where χ denotes Euler characteristic, and where the sum is taken over the critical points in S_1, $k(p)$ being the branching order at p, for each critical point p.

Sketch of Proof Consider a triangulation of S_2 whose set of vertices contains the set of critical values, and lift it to a triangulation of S_1. The formula follows from an Euler characteristic count.   □

A famous problem, called the Hurwitz problem, asks for a characterization of pairs of compact surfaces equipped with data satisfying the Riemann–Hurwitz formula that can be realized by branched coverings, see e.g., [82, 83].

A different kind of realization problem for branched coverings from the sphere to itself was formulated and solved by Thurston, see Sect. 2.6.3 below. It uses an object we call a *Speiser curve* which we introduce in the next subsection.

2.6.2   Speiser Curves, Nets and Line Complexes

In the setting of general holomorphic mappings $S \to R$ between a Riemann surface S and the Riemann sphere R, we must consider, besides the critical values, the asymptotic values. The union of the critical and asymptotic values is usually called the *singular values*.

Let S be a Riemann surface which is a branched covering $\psi : S \to R$ of the Riemann sphere R with finitely many singular values $a_1, \ldots, a_q \in R$, $q \geq 2$. It is understood in such a setting that the complex structure of S is induced from that of R, that is, the map ψ is holomorphic. We draw on R a Jordan curve γ passing through the points $a_1, \ldots, a_q$ in some cyclic order chosen arbitrarily and we rename these points accordingly. We equip this curve with the natural induced orientation. We call this curve a *Speiser curve*. (About the name, see the historical note below on Andreas Speiser.) The curve γ decomposes the sphere into two simply connected regions U_1 and U_2, each having a natural polygonal structure with vertices $a_1, \ldots, a_q$ and sides $(a_1 a_2), (a_2 a_3), \ldots (a_q a_1)$. The pre-image of γ by the covering map ψ is a graph Γ which we call a *net* or *Speiser net*.[3] It divides the surface S into cells that are also equipped with natural polygonal structures, each one projecting either to U_1 or to U_2. A point in the pre-image of some point $a_i \in R$, $i \in I$, may be unbranched; at such a point, there are only two polygons glued. At a

---

[3] The name *net* is due to È. Vinberg, see [109], and also [16]. (We owe this information to A. Eremenko.)

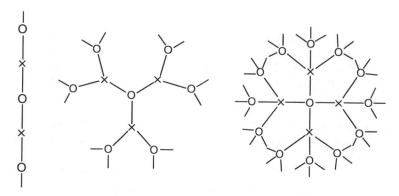

Fig. 2.3 Three examples of regular line complexes with their coloring. The associated branched coverings, restricted to the complement of the branch points, are universal coverings of the sphere with 2, 3 and 4 punctures respectively

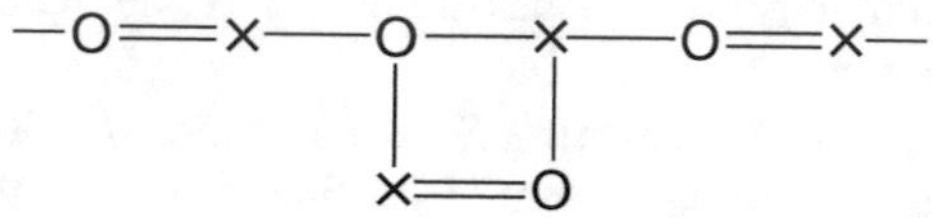

$$-\text{O}{=\!=}\text{x}-\text{O}-\text{x}-\text{O}{=\!=}\text{x}-$$
$$\quad\quad\quad\quad\ \big|\quad\quad\ \big|$$
$$\quad\quad\quad\quad\ \text{x}{=\!=}\text{O}$$

Fig. 2.4 The line complex of the Riemann surface of the function $f(z) = ze^{-z}$

branch point of order $m - 1$ in the pre-image of some a_i (with $m \geq 2$), there are $2m$ polygons glued.

Now, we construct a graph in R which is dual to the Speiser curve γ^4. We start by choosing two points P_1 and P_2 in the interior of U_1 and U_2 respectively. Then, for each edge $(a_i a_{i+1})$ of γ, we join P_1 to P_2 by a simple arc that crosses γ at a single point in the interior of this edge, in such a way that any two such simple arcs intersect only at their extremities. The union of these arcs constitutes a graph dual to γ. Its lift by ψ^{-1} is a graph G embedded in S, called the *line complex*[5] of the covering.

The line complex G is a homogeneous graph of degree q: each vertex is adjacent to q edges. Choosing a black/white coloring for the two points P_1 or P_2, every vertex of G is naturally colored using these two colors (depending on whether it is a lift of P_1 or of P_2). Thus, the line complex is bicolored, that is, its vertices can be colored using only two colors and no edge joins two vertices of the same color. A common notation for black and white vertices, for line complexes, is $\times$, $\circ$.

---

[4] In the present setting, when we talk about the *curve* γ, we usually mean this curve equipped with its marked points, $a_1, \ldots a_q$, that is, with its combinatorial structure which makes it a graph.

[5] Line complexes were popularized in Nevanlinna's book [80]. These objects penetrated holomorphic dynamics in the work of Eremenko and Lyubich, who called them Speiser graphs, see the two survey papers [34, 35] in which the author addresses line complexes and the type problem.

The connected components of the complement in S of the line complex G are called the *faces* of G. Around each vertex of the line complex G there is always the same number of faces, appearing with their indices, in the cyclic order, $a_1, \ldots, a_q$. The image of each face of G by the covering map ψ contains a unique branch value a_i in its interior. Such a face has a natural polygonal structure which it inherits from its image in R. This polygon may be of three types:

1. A polygon with an even number $2m$ of sides: such a polygon contains a critical point of finite order equal to $m - 1$ (the example in Fig. 2.4 contains polygons with 2 and 4 sides);
2. A polygon with an infinite number of sides: such a polygon is unbounded in S (the three examples in Fig. 2.3 contain such polygons);
3. A bigon: such a polygon contains an unbranch point over an a_i (the example in Fig. 2.4 contains bigons).

In Case 1, the face is said to be algebraic, and in Case 2 it is said to be logarithmic (the terminology comes from the theory of Riemann surfaces associated with holomorphic functions).

Topologically, the branched covering S of the sphere is uniquely determined by the points $a_1, \ldots, a_q$, the Jordan curve γ joining them and the line complex G. The line complex in the universal cover of the punctured surface encodes the way the various lifts of the polygons P_1 and P_2 (the complementary components of the Speiser curve) fit together.

The three line complexes represented in Fig. 2.3, with their coloring, correspond to the universal coverings of the sphere with 2, 3 and 4 punctures. In these cases, the surface S is simply connected and the associated line complex is regular. More generally, the line complex associated with the universal covering of the sphere punctured at $n \geq 2$ points is an n-valent regular graph. In terms of holomorphic functions, the line complex on the left hand side of the figure is associated with the Riemann surface of the exponential function. Figure 2.4 represents the line complex of the Riemann surface of the function $f(z) = ze^{-z}$, which is branched over three points of the sphere: the points 0 and ∞, each of infinite order, and the point $1/e$, of order one.

There are sections on line complexes in Nevanlinna's book [80], in Sario and Nakai's book [93] and in Volkovyskii's comprehensive survey on the type problem [110]. We shall return to line complexes in Sect. 2.7.2.

Historical Note Andreas Speiser (1885–1970) was a Swiss mathematician who studied in Göttingen, first with Minkowski and then with Hilbert. The latter became his doctoral advisor for a few months, after the death of Minkowski in 1909 (at age 44). Speiser defended his PhD thesis, in Göttingen, in 1909 and his habilitation in 1911, in Strasbourg. He worked in group theory, number theory and Riemann surfaces, and was the main editor of Euler's *Opera Omnia*. He edited 11 volumes and collaborated to 26 others of this huge collection. Ahlfors writes in his comments to his *Collected works* [16, vol. 1, p. 84], that the type problem, which we shall review in Sect. 2.7, was formulated for the first time by Andreas Speiser. The latter introduced line complexes (in a version slightly different from the one we use here) in his papers [98] and [99]. In his paper [15], Ahlfors writes: "Around 1930 Speiser had devised a scheme to describe some fairly simple Riemann surfaces by means of a graph and had written about it in his semiphilosophical style". Ahlfors' adjective is related to the fact that Speiser's articles do not contain formulae.

2.6.3 Thurston's Realization Theorem

A branched covering from the sphere to itself can be iterated (composed with itself). The *postcritical set* of such a branched covering is the union of the set of critical points with its forward images by the mapping. A branched covering is said to be *postcritically finite* if its postcritical set is finite. In 1983, Thurston obtained a characterization of branched coverings $f : S^2 \to S^2$ of a topological 2-sphere that are topologically equivalent to postcritically finite rational maps of the Riemann sphere, that is, quotients of two polynomials (see [104] and the exposition in [25]). In 2010, while he was considering again the question of understanding holomorphic mappings from the topological point of view, Thurston obtained a realization theorem for branched coverings of the sphere which we present now. The result involves the net associated with a branched covering of the sphere that we defined in Sect. 2.6.2. We now state this result.

Let $f : S^2 \to S^2$ be a generic degree d branched covering of the topological sphere S^2. Here, the word generic means that the cardinality of the set of critical values of f is $2d - 2$ (which is the largest possible). This implies that the cardinality of the set of critical points is also $2d - 2$. Every critical point of such a map is simple (of order 1).

Let γ be a Speiser curve associated with this covering, equipped with its orientation, and consider its lift $\Gamma = f^{-1}(\gamma)$. This is a graph we called net. Its vertices are all of valence 4 (we do not take into account lifts of the critical values that are not critical points).

The Speiser curve γ divides the sphere into two connected components which are naturally equipped with polygonal structures. We color these components in blue and white. Each connected component of the complement of Γ is sent by f to a connected component of γ, and therefore it carries naturally a color and a polygonal structure. We call such a component a *face* of $S^2 \setminus \Gamma$. Thurston addressed the following question:

To characterize the oriented 4-valent graphs in S^2 that can be obtained by the above construction, that is, as a Speiser graph associated with some branched covering $f : S^2 \to S^2$.

He proved the following:

Theorem 2.6.2 (Thurston, see [64]) *Let Γ be an oriented graph on S^2, with $2d - 2$ vertices, all of valence 4. Then Γ is a net of a degree-d branched covering of the* *sphere onto itself if and only if:*

1. *each face of $S^2 \setminus \Gamma$ is simply connected;*
2. *for any alternating blue-white coloring of the faces of $S^2 \setminus \Gamma$, there are d white faces and d blue faces;*
3. *for every oriented Jordan curve embedded in Γ and bordered by only blue faces on the left and only white faces on the right (excluding the corners), the Jordan disc bounded on the left by this curve contains strictly more blue faces than white faces.*

Conditions 2 and 3 are considered as *balance conditions.*

Thurston's proof of Theorem 2.6.2 given in [64] is based on the so-called "marriage theorem" (due to Philip Hall) from combinatorics, a theorem dealing with finite subsets of a given set; it gives a necessary and sufficient condition on such a family that guarantees the choice of one element from each set such that these elements are all distinct (monogamy is the rule). The authors in [64] ask for a better proof which would give an algorithm for constructing arbitrary rational functions whose critical values are on the unit circle [64, p. 255].

The Riemann–Hurwitz formula, which we recalled in Sect. 2.6.1, becomes in the case at hand, i.e., the case where f is a degree-d branched cover of the sphere by itself:

$$2d - 2 = \sum (k(p) - 1).$$

In this formula, the sum is taken as before over the critical points p, and $k(p)$ is the branching order at p. (This formula follows from the fact that the Euler characteristic of both surfaces is 1.) Written in this form, the Riemann–Hurwitz formula says that the branching information at the various points forms a partition of the integer $2d - 2$. The question of what are the partitions that are realized by a branched covering of the sphere to itself is still open. (Cases where the map does not exist are known to occur.)

Thurston's result is part of his project of understanding the shapes of rational functions, where by "shape" he means the evolution of the critical levels of such a function (see his comments on this problem in a MathOverflow thread he started in 2010, titled *What are the shapes of rational functions?*)[6].

The paper [6] which contains a stratification of spaces of monic polynomials of a give degree is inspired by Thurtson's question. See also [18].

---

[6] A. Eremenko pointed out to the authors that according to Koch and Tan Lei in [64], Thurston gave a proof of this theorem in the special case of graphs whose vertices have all degree 4, and J. Tomasini in [106] proved the general case.

2.7 The Type Problem

The Uniformization Theorem (Theorem 2.5.2) says that every simply connected Riemann surface S is either conformally equivalent to the Riemann sphere, or to the complex plane, or to the unit disc. In the first (respectively, second, third) case, S is said to be of elliptic (respectively, parabolic, hyperbolic) type. The names stem from the fact that the Riemann sphere, the complex plane and the unit disc are ground spaces for elliptic, parabolic and hyperbolic geometry respectively (elliptic and parabolic geometries are alternative names for spherical and Euclidean geometries). The first case is distinguished from the other two by topology, since in this case S is compact, whereas in the two other cases it is not.

The *type problem* asks for a practical way to decide whether a given simply connected non compact Riemann surface is of parabolic or elliptic type.

The answer depends on the context in which the surface is defined. For instance, the surface may be embedded in Euclidean 3-space and equipped with the induced metric, or it may be equipped with some abstract Riemannian metric, or it may be obtained by pasting smaller surfaces (Euclidean or non-Euclidean polygons, etc.). It may also be given as an infinite branched covering of the sphere with some data at each branching point, or it may be obtained by analytic continuation, like the universal cover of the Riemann surface of some holomorphic function. One can also imagine other situations. The multiplicity of these cases explains the variety of ways and techniques in which the type problem has been addressed; we shall see some examples below.

2.7.1 Ahlfors on the Type Problem

Ahlfors emphasized at several occasions the importance of the type problem, which was one of his main research topics during the period 1929–1941. In his paper [13], talking more precisely about the question of deducing the type of a Riemann surface associated with a univalent function, from the distribution of its singularities, he writes: "This problem is, or ought to be, the central problem in the theory of functions. It is evident that its complete solution would give us, at the same time, all the theorems which have a purely qualitative character on meromorphic functions." Nevanlinna, in the introduction of his book *Analytic functions* (1953), writes (p. 1 of the English edition [80]): "[...] Value distribution theory is thus integrated into the general theory of conformal mappings. From this point of view the central problem of the former theory is the *type problem*, an interesting and complicated question, left open by the classical uniformization theory." See also [11].

One of Ahlfors' earliest results on the type problem is a theorem contained in his paper [12]. It gives the following condition for a simply connected surface S which is a branched covering of the sphere whose branch values are all of finite degree, to be parabolic:

Theorem 2.7.1 (Ahlfors, [12]) *Let S be a simply connected surface which is a branched covering of the sphere with all branch values of finite degree, let p be an arbitrary point in S and let $n(t)$ $t \geq 0$ be a function defined on S which associates with t the number of branch points, counted with multiplicity, at distance $\leq t$ from p. If*

$$\int_a^\infty \frac{dt}{tn(t)} = \infty$$

for some (or equivalently for all) $a > 0$, then S is of parabolic type.

Thus, the theorem says that if the amount of branching is relatively small, the surface is parabolic. We shall see below other results on the type of branched coverings of the sphere that are also expressed in terms of the growth of the amount of branching. Heuristically, one can consider that the degree of branching measures a certain degree of negative curvature concentrated at the branch points, and consequently, the larger the amount of branching is, the more negative curvature is concentrated at that point; this explains the fact that a surface with a large amount of branching is hyperbolic.

In the setting of function theory, one also admits branch values of infinite degree, modeled on the ones that occur in the theory of Riemann surfaces of multi-valued analytic functions. The complex logarithm function is an example of a multivalued function which gives rise to a branch value of infinite degree.

In the next 3 subsections, we shall describe different approaches used by various authors to the study of the type problem. They involve combinatorial, analytical and geometrical methods.

2.7.2   Nevanlinna on the Type Problem

Nevanlinna, in the papers [78–80] considered in detail the type problem for simply connected surfaces that are branched coverings of the sphere with a finite number of critical values. Soon after, Teichmüller dealt with the same problem in his papers [101] and [102], disproving a conjecture made by Nevanlinna. The fundamental tool that is used in these works is the line complex that we recalled in Sect. 2.6.2.

Nevanlinna formulated a principle which amounts to formalizing the fact that the type of a simply connected Riemann surface S which is a branched covering of the Riemann sphere may be deduced from information on the associated line complex. The faces of this graph (i.e., the connected components of its complement) are in one-to-one correspondence with the branch points of the covering, if one discards bigons. A polygon with $2m$ sides is associated with a branch point of order $m - 1$. Roughly speaking, Nevanlinna's claim is that if the amount of branching of the covering is small, the surface is of parabolic type, and if the amount of branching is large, the surface is of hyperbolic type. He showed how the amount of branching can be deduced from the properties of the line complex. Nevanlinna writes, in [80,

p. 308]: "It is thus natural to imagine the existence of a critical degree of branching that separates the more weakly branched parabolic surfaces from the more strongly branched hyperbolic surfaces." To make this precise, he introduced the notions of mean branching and mean excess relative to the line complex. These are analogues of the notion of global mean curvature of a surface. We shall recall his definition.

To motivate the above principle, Nevanlinna first considered the case of a surface S which, instead of being an infinite-sheeted covering of the Riemann sphere, is finite-sheeted. In this case, the *mean branching* is simply the sum of the orders of the branch points divided by the degree of the covering.

In this case, the mean branching can also be computed using the line complex G. The number of vertices of the line complex is equal to twice the number of sheets. Indeed, if d is the number of sheets in the covering, then, using the notation in Sect. 2.6.2, there are d vertices which are lifts of the point P_1 and d others which are lifts of the point P_2. Each branch point of order $m - 1$ is in the interior of a polygonal component of $S \setminus G$ having $2m$ sides and $2m$ vertices. Considering that twice the order of this branch point, that is, $2m - 2$, is evenly distributed among these $2m$ vertices, each vertex of this polygon receives from the polygon itself a branching contribution equal to $\frac{2m-2}{2m} = 1 - \frac{1}{m}$. Taking the sum over the contributions coming from all the adjacent polygons, each vertex of the line complex receives a total branching contribution equal to $\sum(1 - \frac{1}{m})$, where the sum is over all the polygons adjacent to this vertex. Dividing by the total number of vertices, which is twice the number of sheets, gives the mean branching. Nevanlinna uses this second way of counting the mean branching for the definition of the mean branching in the case of an infinite covering.

Nevanlinna considered first the case of a *regularly* ramified surface. This is a surface whose group of covering transformations acts transitively on the set of branch points (therefore it acts transitively on the line complex). In this case, the mean ramification is equal to the ramification of an arbitrary polygonal component in the complement of the line complex. He proved:

Theorem 2.7.2 (Nevanlinna [80], p. 311) *For a regularly ramified surface, the following holds:*

1. *the surface is parabolic if the mean ramification is 2;*
2. *he surface is hyperbolic if the mean ramification is > 2;*
3. *The surface is elliptic if the mean ramification is < 2;*

Proof In his proof, Nevanlinna introduces a metric on each polygonal component of the complement of the line complex in which the measure of the angle at each vertex of such a polygon corresponding to a branch point of order $m - 1$ is equal to π/m.

The excess of a polygon is equal to $\sum \frac{1}{m} - q + 2$ where q is as before the cardinality of the set of critical values, which is also the number of sides of the polygon. Since we are in the case of a regularly ramified surface, the mean ramification is equal to the order of ramification at each branched point.

In case 1, the angle excess of each polygon is zero and we can realize this polygon as a regular polygon in the Euclidean plane. In case 2, the angle excess of each polygon is negative and we can realize it as a regular polygon in the hyperbolic plane. Case 3 corresponds to spherical geometry.   □

In the case of an infinitely-branched covering, the average of the branching over all the vertices of G is obtained by taking an exhaustion of this graph by an infinite number of finite sub-graphs, and taking the lower limit of a mean of the sum of the total branchings of the vertices [80, p. 309ff]. The conjecture is formulated by Nevanlinna as a question in [80, p. 312] in the following terms: *Is the surface parabolic or hyperbolic according as the mean excess is zero or negative?* After formulating this question, Nevanlinna notes that Teichmüller in his paper [102], disproved the conjecture by exhibiting a hyperbolic simply connected Riemann surface branched over the sphere with a line complex whose mean excess is zero.

2.7.3 Quasiconformal Mappings and Teichmüller's Work on the Type Problem

In the course of proving the above result, Teichmüller used quasiconformal mappings, and we next recall this notion.

To a differentiable complex-valued function $f(z) = u(x, y) + iv(x, y)$ defined on an open subset Ω of the complex plane (or on a Riemann surface), one asssociates its *complex dilatation* by taking first the partial derivatives $f_x = u_x + iv_x$ and $f_y = u_y + iv_y$, setting

$$f_z = \frac{1}{2}(f_x - if_y)$$

and

$$f_{\bar{z}} = \frac{1}{2}(f_x + if_y)$$

and defining the *complex dilatation* of f as

$$\mu(z) = \frac{f_{\bar{z}}(z)}{f_z(z)}.$$

The (real) *dilatation* of f is then given by the real function

$$K(z) = \frac{1 + |\mu(z)|}{1 - |\mu(z)|}$$

and f is said to be K-*quasiconformal*, for some $K > 0$, if this function satisfies

$$K(z) < K$$

for all z in Ω.

In practice, one allows f to be only absolutely continuous on almost every line which is parallel to the real or imaginary axis. This is sufficient for the partial derivatives $\frac{\partial f}{\partial x}$ and $\frac{\partial f}{\partial y}$ to exist almost everywhere.

The dilatation is then defined a.e., and the last inequality needs only to be satisfied a.e.

A computation shows that the *dilatation* of f at an arbitrary point is equal to the ratio of the major axis to the minor axis of the ellipses that are images by the differential of the map at that point, of circles in the tangent space centered at the origin. Therefore, a diffeomorphism f between two Riemann surfaces is K-quasiconformal if it takes an infinitesimal circle to an infinitesimal ellipse of eccentricity uniformly bounded by K.

The map f is quasiconformal if it is K-quasiconformal for some finite K.

This leads us to a notion of quasiconformally equivalent Riemann surfaces and *quasiconformal structure*, a weakening of the notion of conformal structure: A quasiconformal structure on a 2-dimensional manifold is a family of conformal structures such that the identity map between any two of them is quasiconformal. We say that two conformal structures in such a family are quasiconformally equivalent.

Teichmüller wrote a paper dedicated to applications of quasiconformal mappings the type problem, see [101], titled *An application of quasiconformal mappings to the type problem*. In the introduction to this paper, he writes: "Recently, the problem of determining the properties of the schlicht mapping of [a simply connected Riemann surface] $\mathfrak{M}$ from its line complex has been frequently studied. So far in the foreground has been the type problem: How can one determine, from the given line complex, if the corresponding surface $\mathfrak{M}$ can be mapped one-to-one and conformally onto the whole plane, the punctured plane, or the unit disk? One is still very far from finding sufficient and necessary conditions."

In this work on the type problem, Teichmüller started by showing that two branched coverings of the sphere whose line complexes are equal are quasiconformally equivalent. He gave a criterion for hyperbolicity and he proved that the type of a simply connected open Riemann surface is a quasiconformal invariant [101]. The latter result is a direct consequence of the fact that the complex plane and the unit disc cannot be quasiconformally mapped onto each other, a result which he proves in the same paper. In principle, this result reduces the type problem to a simpler problem, since quasiconformal mappings exist much more abundantly than conformal mappings. To show that a surface S is parabolic (respectively hyperbolic), it suffices to exhibit a quasiconformal homeomorphism between S and the complex plane (respectively the unit disc).

To end this subsection on Teichmüller's work on the type problem, let us mention the survey [17] titled *Teichmüller's work on the type problem*.

2.7.4 Lavrentieff on the Type Problem

Lavrentieff's work on the type problem is based on the notion of "almost analytic function" which he introduced, another weakening of the notion of conformal mapping which is close to (but slightly different from) the notion of quasiconformal mapping.

To state this property, let us first recall that there exist several equivalent conditions for a differentiable function f from the complex plane to itself (or, more generally, between surfaces) to be holomorphic, and weakening each of them gives a weakening of a function to be holomorphic. Among these conditions, we mention the following:

1. f is angle-preserving.
2. the real and imaginary parts of f satisfy the Cauchy–Riemann equations.
3. f satisfies $\frac{\partial f}{\partial \bar{z}} = 0$.
4. The Taylor series expansion of f around every point is convergent in some open disc around this point;
5. The following notion of holomorphicity is stated in terms of almost complex structures: Given a surface S equipped with a J-holomorphic function, a differentiable function $f : (S, J) \to \mathbb{C}$ is said to be J-*holomorphic* if for every point p on S and for every tangent vector u at p, we have

$$(df)_p(J_p u) = i(df)_p(u).$$

6. The function f possesses a complex derivative, that is, infinitesimally, it acts by multiplication by a complex number.

Lavrentieff's almost analyticity property is a weakening of Property 6, a property equivalent to the fact that at each point the Jacobian matrix of f acts on the tangent space as a rotation followed by a homothety. This is also expressed by the fact that the map sends infinitesimal circles to infinitesimal circles. (Thus, this property is close to the notion of quasiconformal mapping which we recalled in Sect. 2.7.2.)

In fact, in the most general case, the Jacobian matrix of a differentiable map between tangent spaces, since it acts linearly, takes circles centered at the origin to ellipses centered at the origin, and Lavrentieff's property is formulated in terms of the boundedness of the eccentricty and the directions of these ellipses. We now state this condition in precise terms.

Definition (Lavrentieff [66]) A function $f : \Omega \to \mathbb{C}$ defined on a domain $\Omega \subset \mathbb{C}$ is said to be *almost analytic* if the following three properties hold:

1. f is continuous.
2. f is orientation-preserving and a local homeomorphism on the complement of a countable closed subset of Ω.
3. There exist two real-valued functions $p : \Omega \to [1, \infty[$ and $\theta : \Omega \to [0, 2\pi]$, called the *characteristics of* f, defined as follows:

   • There is subset E of Ω which consists of finitely many analytic arcs such that p is continuous on $\Omega \setminus E$ and θ is continuous at each point z satisfying $p(z) \neq 1$. (E might be empty.)
   • On every domain $\Delta \subset \Omega \setminus E$ whose frontier is a simple analytic curve, p is uniformly continuous. Furthermore, if such a domain Δ and its frontier do not

contain any point satisfying $p(z) = 1$, then θ is also uniformly continuous on Δ.

- For an arbitrary point z_0 in $\Omega \setminus E$, let $\mathcal{E}$ be an ellipse centered at z_0, let $\theta(z)$ be the angle between its major axis and the real axis of the complex plane, and let $p(z_0) = \frac{a}{b} \geq 1$ be the ratio of the major axis a to its minor axis b. Let z_1 and z_2 be two points on the ellipse $\mathcal{E}$ at which the expression $|f(z) - f(z_0)|$ attains its maximum and minimum respectively. Then,

$$\lim_{a \to 0} \left| \frac{f(z_1) - f(z_0)}{f(z_2) - f(z_0)} \right| = 1.$$

One goal of Lavrentieff's article [66] is to show that almost analytic functions share several properties of analytic functions. This includes a compactness result for families of almost analytic functions with uniformly bounded characteristics, which is a generalization of a known compactness property that holds for families of analytic functions. It also includes a generalization of Picard's big Theorem [66, §3]. The generalization of the same theorem for quasiconformal mappings was already obtained by Grötzsch, see [45].

The question of finding an almost analytic function from its characteristics, which Lavrentieff solves in the same paper and which we state now, is a form of what was later called the "Measurable Riemann Mapping Theorem". This is a wide generalization of the Riemann Mapping Theorem:

Theorem 2.7.3 (Lavrentieff [66]) *Given any two functions $p(z) \geq 1$ and $\theta(z)$ defined on the closed unit disc $|z| \leq 1$ such that p and θ satisfy the properties in the above definition and such that $p(z) < M$ for some constant M, there exists an almost analytic function f realizing a self-homeomorphism of the closed disc $|z| \leq 1$ and having the characteristics p and θ.*

In the same paper, Lavrentieff proves the following result on the type problem. Like Ahlfors' theorem (Theorem 2.7.1) and like the theorem of Milnor that we shall quote below (Theorem 2.7.5), Lavrentieff's theorem reduces the question of proving parabolicity to that of proving that a certain integral diverges.

Theorem 2.7.4 (Lavrentieff [66]) *Let S be a surface in the 3-dimensional Euclidean space which is the graph of a differentiable function*

$$t = f(x, y), \quad x^2 + y^2 < \infty,$$

whose partial derivatives $\frac{\partial f}{\partial x}$ and $\frac{\partial f}{\partial y}$ are continuous. For each $r \geq 0$, let

$$M(r) = \max_{x^2+y^2=r^2} 1 + |\operatorname{grad} f(x, y)|.$$

If $\int_1^\infty \frac{dr}{rM(r)} = \infty$, then S is of parabolic type.

Lavrentieff, in his paper, reduced the type problem to Theorem 2.7.3, that is, to the problem of finding an almost analytic function having given characteristics p and θ, such that for $z = (x, y)$,

$$p(z) = \frac{1}{\cos \alpha(z)},$$

where $\alpha(z)$ is the angle formed by the (x, y)-plane and the tangent plane of S at (x, y, t) and where $\theta(z)$ is the angle formed by grad $f(x, y)$ and the x-axis.

2.7.5   Milnor on the Type Problem

Milnor, in his paper [76], studied the type problem for the case of a surface S embedded in Euclidean 3-space which is complete and whose Gaussian curvature depends only on the distance to some fixed point p. The Riemannian metric induced on this surface can be written in the polar coordinates (r, θ) as $dr^2 + g(r)^2 d\theta^2$ where $2\pi g(r)$ is a smooth positive function that denotes the length of the geodesic circle of radius r centered at p. The Gaussian curvature is then given by $K(r) = -(d^2 g/dr^2)/g$.

Milnor proves the following:

Theorem 2.7.5 *For a surface S as above we have:*

1. *S is parabolic if and only if for some (or equivalently for any) $a > 0$, the integral $\int_a^\infty dr/g(r) = \infty$.*
2. *If for an arbitrary $\epsilon > 0$ we have $K(r) \geq -(r^2 \log r)$ for r large enough, then S is parabolic.*
3. *If $K \leq -(1+\epsilon)/r^2 \log r)$ for large r, where ϵ can be any positive constant, and if the function $g(r)$ is unbounded, then S is hyperbolic.*

Items 2 and 3 are not surprising if we consider them again from the point of view that enough negative curvature corresponds to a hyperbolic type and enough positive curvature corresponds to a parabolic type. An interesting fact here is that the difference between the two cases is made by an ϵ which can be chosen to be arbitrarily small, and in this sense the result is the best possible in the given situation.

Milnor's Proof The proofs of Items 2 and 3 follow from Item 1 which, like the results of Ahlfors and Lavrentieff that we mentioned above, characterizes parabolicity in terms of the divergence of a certain integral.

For the proof of 1, Milnor introduces a new coordinate $\rho = \int_a^r ds/g(s)$. In the coordinates (ρ, θ), the metric becomes

$$g^2(d\rho^2 + d\theta^2),$$

and is thus conformally equivalent to the Euclidean metric. (Coordinates in which the metric is conformal to the Euclidean metric are called isothermal, another term which comes from the theory of heat transfer.) The exponential map $(\rho, \theta) \mapsto e^{\rho+i\theta}$ maps S conformally onto $\mathbb{C}$. (The mapping is a priori defined in the complement of the point p, but it extends to p because it is bounded, differentiable and conformal in a neighborhood of p, therefore this singularity is removable.)   □

Milnor, at the end of his paper, adresses the question of finding an effective criterion for deciding, for a simply connected open surface embedded in 3-space, whether it is parabolic or hyperbolic.

2.7.6   Probablistic Approaches

There are other approaches to the type problem that we shall not touch upon here. In particular, there is a probabilistic point of view. A classical result in this direction says that a surface equipped with a Riemannian metric is parabolic if and only if the associated Brownian motion is recurrent. There is a large literature on this question, see e.g., Kakutani's paper [56], which appeared the same year as Teichmüller's paper [101] on the type problem. Kakutani gives the following criterion: Let S be a simply-connected open Riemann surface which is an infinite cover of the sphere and let D be a simply connected subset of S bounded by a Jordan curve Γ. For a point ζ in $S \setminus D$, let $u(\zeta)$ denote the probability that the Brownian motion on S starting at ζ enters into Γ without getting out of $S \setminus D$ before. Then, one of the following two cases holds:

1. $u(\zeta) < 1$ everywhere in $S \setminus D$ and in this case S is hyperbolic;
2. $u(\zeta)$ is identically equal to 1 $S \setminus D$, and in this case S is parabolic.

Kakutani wrote later several papers related to the type problem, see, e.g., [53–55].

Let us also mention Z. Kobayashi's papers on Riemann surfaces in which he discusses the type problem; see in particular his paper [59] in which he gives sufficient conditions under which a surface is of parabolic type, and his paper [60] in which he presents a theory based on Kakutani's paper [56] and Teichmüller's paper [102], generalizing both works. Figure 2.5 is extracted from Kobayashi's paper [59]. It is reproduced here for the purpose of showing the nobility of this kind of drawing compared to computer-drawing everybody uses today in mathematical papers.

2.7.7   Electricity

P. Doyle and J. L. Snell expanded on the relation between the type problem, Brownian motion and the propagation of an electric current on the surface, see the

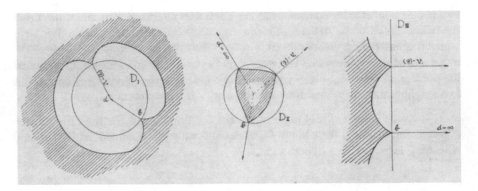

Fig. 2.5 From Kobayashi's paper [59] on the type problem

paper [32] where the main underlying idea is that the unit disc has finite electrical resistance while the plane has infinite resistance. See also Doyle's review in the Bulletin of the AMS, [31], in which the author reports on the type problem for a covering of the Riemann sphere with n punctures using the properties of the Speiser net of the covering, surveying the important work of Lyons–McKean–Sullivan [71] and of McKean–Sullivan [74] on this topic. Incidentally, the author seems not to be aware of the work of Teichmüller on these questions.

Introducing electricity in the type problem is completely in the tradition of Riemann who, at several places in his work on Riemann surfaces uses arguments from electricity. This occurs in particular in the two main papers where he deals with Riemann surfaces, namely, his doctoral dissertation [90] and his paper on Abelian functions [91]. For instance, in the Dirichlet principle that he used in his proof of the Riemann Mapping Theorem (see Sect. 2.5.3 above), Riemann considered the function f defined on the boundary of the domain Ω as a time-dependent electrical potential, and his conclusion in the solution to the Dirichlet problem consisted in letting the system evolve and reach an equilibrium state. The function obtained necessarily satisfies a mean value property and therefore is harmonic.

Historical Note In the summer of 1858, Riemann gave a course titled *The mathematical theory of gravitation, electricity and magnetism* [94]. Besides his mathematical papers in which he used electricity, Riemann also wrote papers on electricity. See his unfinished paper *Gleichgewicht der Electricität auf Cylindern mit kreisförmigen Querschnitt und parallelen Axen* (On the equilibrium of electricity on cylinders with circular transverse section and whose axes are parallel) [89] (1857) published posthumously in the second edition of his *Collected works* which concerns the distribution of electricity or temperature on infinite cylindrical conductors with parallel generatrices. The reader interested in more details may refer to the survey titled *Physics in Riemann's mathematical papers* [81].

Remark (Higher-Dimensional Generalizations) One advantage of introducing random walks and probabilistic methods in the theory of parabolicity/hyperbolicity of Riemann surfaces is that it allows this theory to be generalized to graphs, discrete groups and higher-dimensional manifolds. There are several works in this direction. I mention Marc Troyanov's paper [107] in which he discusses parabolicity for n-dimensional manifolds. More precisely, in this paper, Troyanov surveys an invariant of Riemannian manifolds which is related to the non-linear potential theory of the p-Laplacian and which determines a property which generalizes parabolicity or hyperbolicity of surfaces. The key notion that this leads to is that of p-parabolicity and p-hyperbolicity. In the generalized setting of the paper [107], the classical dichotomy parabolic/hyperbolic becomes 2-parabolic/2-hyperbolic, i.e., $p = 2$. The author focuses in particular on the relationship between the asymptotic geometry of a manifold and its parabolicity.

2.8 Uniformization: Geometry and Combinatorics

2.8.1 Dessins d'enfants

A *map* on a compact oriented surface S is a bicolored finite graph Γ embedded in S such that each connected component of $S \setminus \Gamma$ is contractible. The vertices of the graph are colored by 0, 1 and each edge joins vertices of different colors. A connected component of $S \setminus \Gamma$ is called a *cell* of the map. A map (S, Γ) can be upgraded to a triangulation (S, Δ_Γ) by choosing a point in each cell and connecting the chosen point by a star-shaped system of curves to the vertices of Γ that are on the boundary of the cell. The triangulation Δ_Γ is special in that it admits a checkerboard black-white coloring (that is, triangles meeting from opposite sides along an edge have opposite colors). To see this, we use the symbols $0, 1, \infty$ to color the vertices of the triangulation Δ_Γ in such a way that the chosen points in the cell gets the color ∞. Each triangle has vertices of different colors, hence becomes oriented by imposing the cyclic order $0, 1, \infty$. Finally we color white those triangles for which the induced orientation from S coincides with the cyclic orientation of the vertices.

The minimal example of a map is $M = (\mathbb{C} \cup \{\infty\}, [0, 1])$. Here, the graph is $\Gamma = [0, 1]$ with two vertices and one edge embedded in the Riemann sphere $\mathbb{C} \cup \{\infty\}$. This map has only one cell. Now choose the point ∞ connected by $[1, \infty]$ and $[\infty = -\infty_{\mathbb{R}}, 0]$ as an upgrading to a triangulation with two triangles, the white one being the so-called upper half-plane $\mathbb{C}_+$. We denote by M_∞ this minimal map together with the above upgrading to a triangulation.

Speiser nets discussed in Sect. 2.6.2 are maps in the above sense, except that the surfaces on which they are defined are not necessarily compact.

Let it be given the combinatorial data consisting of a triple $D = (S, \Gamma, \Delta_\Gamma)$ of a map Γ with its 0, 1 bicoloring on a compact surface S, together with the upgrading to a checkerboard colored triangulation Δ_Γ.

A smooth mapping $f_\Gamma : (S, \Gamma, \Delta_\Gamma) \rightarrow M_\infty$ is well defined up to isotopy. It sends vertices to vertices respecting colors, edges to edges, and its restriction to the complement of the vertex set of the triangulation is a submersion.

Let J_Γ be the complex structure on S obtained by pulling back the complex structure of $\mathbb{P}^1(\mathbb{C}) \setminus \{0, 1, \infty\}$ and by extending this structure using the removable singularity theorem.

The result is again an upgrading from the combinatorial data $\Gamma \subset S$ to a Riemann surface (S, J_Γ) together with a holomorphic function $f_D : (S, J_\Gamma) \rightarrow \mathbb{P}^1(\mathbb{C})$ having its critical values in the set $\{0, 1, \infty\}$.

The set of possible combinatorial data up to isotopy is countable and the set of Riemann surfaces up to bi-holomorphic equivalence is uncountable. Thus, a natural question arises, namely, what Riemann surfaces appear as an upgrading of a map. The answer is given by Belyi's Theorem:

Theorem 2.8.1 (Belyi [19]) *For a compact Riemann surface S the following are equivalent:*

1. *S carries a holomorphic map $f : S \rightarrow \mathbb{P}^1(\mathbb{C})$ with at most 3 critical values (which can be taken, without loss of generality, to be $\{0, 1, \infty\}$);*
2. *S is bi-holomorphically equivalent to an upgraded map surface;*
3. *S is bi-holomorphically equivalent to a complex curve, that is, a Riemann surface embedded in the projective space $P^n = \mathbb{P}^n(\mathbb{C})$ defined by a set of polynomials with coefficients in a number field, that is, a finite field extension of the field of rational numbers.*

The implication $3 \Rightarrow 1$ is the "easy part"; indeed, the inclusion $S \subset P^n$ as an algebraic curve defined by a set of polynomials with coefficients in a number field can be composed (with care) with projections to lower-dimensional spaces $P^n \rightarrow P^{n-1}$ until one gets a meromorphic function on S. Using the fact that this function is itself defined on a number field, there is a way of reducing the number of critical values to at most three.

The strength of Belyi's Theorem can already be appreciated in the case of genus 0 surfaces and maps having only one cell, or, equivalently, in the case where the graph Γ is a planar tree in the Gaussian plane $\mathbb{C}$. The above construction leads in this case of a map $\Gamma \subset \mathbb{C}$ to a polynomial $f_\Gamma : \mathbb{C} \rightarrow \mathbb{C}$ having at most two critical values. Such a polynomial $f_{\Gamma,e}$ becomes well defined if one marks an edge e of Γ and requires that $f_{\Gamma,e}$ sends its extremities to 0, 1.

The study of polynomial mappings $P : \mathbb{C} \rightarrow \mathbb{C}$ with at most two critical values was introduced by G. Shabat, see [96] and [97]. These polynomials generalize Chebyshev polynomials which have only two critical values and such that these critical points are all of Morse type. Equivalently, there is a maximal number of such critical points.

This fact leads to an action of the absolute Galois group $\mathrm{Gal}(\bar{\mathbb{Q}}, \mathbb{Q})$ on the set of isotopy classes of bicolored planar trees with marked edges. This action is faithful and a window for observing and perhaps understanding the absolute Galois group.

See Geothendieck's *Sketch of a program* [43]. We also refer to the review made in [10] of the relevant parts of the *Esquisse*.

The inverse image of the interval [0, 1] by $f_{\Gamma,e}$ may look like a drawing of a person by a child, hence the name "dessin d'enfants" given by Grothendieck to this mathematical object whose study he promoted in [43]. For an introduction to dessins d'enfants we refer to the original article [43] and to the expositions in [40, 46, 49]. The reader interested in the relation between dessins d'enfants and the deformation theory of Riemann surfaces may refer to [48].

Voevodsky and Shabat in [111] considered surfaces equipped with Euclidean structures with cone singularities obtained by gluing Euclidean equilateral triangles along their sides. They showed that the Riemann surface structure underlying such a metric space, as an algebraic curve, is defined over a number field. Cohen, Itzykson and Wolfart in [27] showed that the same construction works using congruent hyperbolic triangles (triangles in the Bolyai–Lobachevsky plane), instead of Euclidean.

2.8.2 Slalom Polynomials

We shall use the notions of slalom polynomial and slalom curve and we shall recall the definitions. For a more detailed exposition we refer to the original article [4] and the recent book [8] by the first author.

We start with the definition of the planar tree Γ_P associated with a generic monic polynomial P of degree $n + 1$. This is the closure of the union of the flowlines $\gamma : I \to \mathbb{C}$ of $-\text{grad}(\log |P|)$ that satisfy the following properties:

- P and its derivative P' do not vanish on the image of γ;
- the image of γ is maximal with respect to inclusion (in other words, γ is not the restriction of a flowline defined on an interval J strictly containing I);
- the image of γ is bounded in $\mathbb{C}$, except when I is the positive real line and where this image converges to ∞.

Consider now a rooted planar tree Γ with $n + 1$ edges in which one vertex is connected to $+\infty$ (considered as the root) by an unbounded edge. A *slalom curve* $Sl(\Gamma)$ of Γ is an immersed copy of the circle $\gamma : S^1 \to \mathbb{C}$ having n transversal double points at the midpoints of the bounded edges of Γ (Fig. 2.6). The curve $Sl(\Gamma)$ intersects Γ only and transversely twice at the midpoints of the bounded edges and once at the unbounded edge such that each bounded complementary region of the curve contains exactly one vertex.

A *slalom polynomial* $Sl P_\Gamma(z)$ *for* Γ is a generic monic polynomial P of degree $n + 1$ with $|P(s)| = 1$ for each zero s of P' such that the colored trees Γ_P and Γ are isotopic relative to $+\infty$. Generic means in this context that P has $n + 1$ distinct roots, and that P' has n distinct roots.

A *slalom polynomial* is a generic monic polynomial P that is a slalom polynomial for some tree Γ_P.

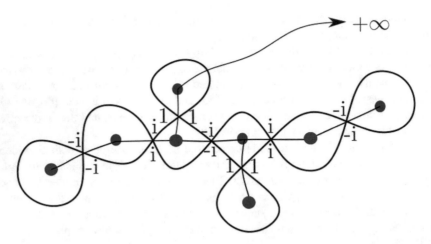

Fig. 2.6 A rooted colored tree having one edge asymptotic to $+\infty$ with a slalom curve. The double points of this curve are the critical points of the polynomial

Slalom polynomials P share the following properties with Shabat and Chebyshev polynomials: P' has $n = \mathrm{degree}(P) - 1$ distinct roots and all critical values of P have absolute value 1. Thus, the roots of P' are the double points of the curve $|P(z)| = 1$, which is a slalom curve for the tree Γ_P.

Theorem 2.8.2 (Existence of Slalom Polynomials) *Given a rooted planar tree Γ with $n + 1$ edges as above, there exists a slalom polynomial P for Γ.*

Proof There exists a continuous mapping $\phi : \mathbb{C} \to \mathbb{C}$ that is smooth except at the midpoints of Γ and such that

- the restriction of ϕ to the unbounded component of the complement of $Sl(\Gamma)$ is a degree $n + 1$ covering map of the complement of the closed unit disc in $\mathbb{C}$;
- the restriction of ϕ to a bounded region of the complement of $S(\Gamma)$ is a diffeomorphism to the open unit disc in $\mathbb{C}$ that sends the vertex to 0 and half edges to radial rays;
- ϕ is holomorphic near ∞.

Let J be the conformal structure on $\mathbb{C}$ such that $\phi : (\mathbb{C}, J) \to \mathbb{C}$ is holomorphic and let $U : \mathbb{C} \to (\mathbb{C}, J)$ be a uniformisation of the structure J. The composition $P = \phi \circ U : \mathbb{C} \to \mathbb{C}$ is a degree $n+1$ polynomial. The polynomial P has $n+1$ roots, the derivative has n roots at the points s with $U(s)$ being a midpoint of an edge. By adjusting the uniformisation map U one achieves that P is monic and generic.

The flowline from $+\infty$ ends at a root of P that we color red and which becomes the root of E_P. This red root can be changed by substituting λz to z. This substitution turns the picture and gives $2n$ possibilities for attaching the unbounded edge. Color the tree Γ with the two colors, red and blue. Choose a disc D_R that contains Γ up to a part of the real axis.

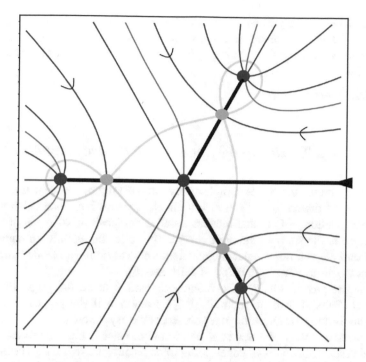

Fig. 2.7 Flowbox decomposition for $P = z^4 + z$. The green dots are the zeros of P', the saddle points of the function $|P|$ and bifurcation points of the vector field $-\mathrm{grad}(|P|)$. The zeros of P, which are the red and blue dots, span a planar tree of Dynkin diagram D_4. The flowline from $+\infty$ (the horizontal segment on the middle right part of the figure) ends at the triple point of D_4. The separating flowlines are marked with arrows. The flowlines $P^{-1}(i\mathbb{R})$ are colored blue

It follows that Γ and Γ_P are isotopic relative to $+\infty$, showing that all the $\mathrm{Cat}(n)$ combinatorial data consisting of a rooted planar tree as above are realized. Here $\mathrm{Cat}(n)$ designates the n-th *Catalan number*, that is, the number of distinct triangulations of a convex polygon with $n + 2$ sides obtained by adding diagonals.

The polynomial $P = P_\Gamma$ is a slalom polynomial for the tree Γ.   □

The slalom polynomial P_Γ is not unique up to holomorphic changes of coordinates in the range and in the target. Uniqueness can be forced for trees with maximum valence $\leq k$ if one imposes moreover that the critical values are roots of unity of order depending on k. See Fig. 2.6. For $k = 3$, order four works. In Fig. 2.6, the values $1, i, -i, -1$ are the values of the polynomial at the given points.

In the example of Fig. 2.7, the polynomial $z^4 + z$ is the slalom polynomial of the tree D_4. The Chebyshev polynomials are up to a real translation precisely the slalom polynomials with critical values among ± 1. See also [7].

As in [6] one obtains a cell-decomposition of the space of monic polynomials Pol_d of degree d and of the complement $\mathrm{Pol}_d \setminus \Delta$ of the discriminant Δ (the set of polynomials having at least one multiple root).

Theorem 2.8.3 *The equivalence relation* $\Gamma_P \sim \Gamma_Q$ *induces a stratification with* $\mathrm{Cat}(d - 1)$ *top-dimensional strata on the spaces* Pol_d *and* $\mathrm{Pol}_d \setminus \Delta$. *More-over, the top-dimensional strata are cells and have a representative by a slalom polynomial.* ∎

For this theorem, see [7].

2.8.3 A Stratification of the Space of Monic Polynomials

The space of monic univariate complex polynomials Pol_d of degree d is a complex affine space of dimension d. The discriminant Δ_d is an important sub-variety. A cell decomposition of Pol_d that induces a cell decomposition of Δ_d is of interest, for instance in the study of the braid group. The following defines an equivalence relation on Pol_d, whose associated equivalence classes define a cell decomposition such that the discriminant is a union of cells, see [6].

Define the *picture* $\mathrm{Pic}(P)$ of a monic polynomial to be the graph $\mathrm{Pic}(P) = P^{-1}(\mathbb{R} \cup i\mathbb{R})$. Call two polynomials P, Q equivalent if the graphs $\mathrm{Pic}(P)$ and $\mathrm{Pic}(Q)$ are isotopic by an isotopy that preserves the asymptotes.

The picture $\mathrm{Pic}(P)$ of a degree d monic polynomial is a planar forest without terminal vertices, but with $4d$ edges going to ∞ asymptotic to the rays corresponding to $4d$ roots of unity. The precise combinatorial characterization is given in [6]. The problem of showing that each possible equivalence class is realized was solved using Riemann's Uniformisation Theorem. The number of top-dimensional cells is the Fuss–Catalan number $\frac{1}{3d+1}\binom{4d}{d}$.

2.8.4 Rational Maps, Speiser Colored Cell Decompositions, Classical Knots and Links

Let $f : \mathbb{P}^1(\mathbb{C}) \to \mathbb{P}^1(\mathbb{C})$ be a holomorphic map of degree $d > 0$. Such a map is also called a rational map, since in the above coordinate z its expression $f(z) = \frac{P(z)}{Q(z)}$ is a ratio of two polynomials $P(z) \neq 0, Q(z) \neq 0$ of degrees $d_1, d_2 \geq 0$ with $d = \max\{d_1, d_2\}$.

Let $\Delta(f)$ be the finite set of critical values of f. The restriction of f to the complement of the set $f^{-1}(\Delta(f))$ is a covering map of degree d. The set $f^{-1}(\Delta(f))$ contains the set of critical points of f. For each $a \in \mathbb{P}^1(\mathbb{C})$, define the *defect* $\delta(a) \geq 0$ by $\delta(a) = d - \#f^{-1}(a)$. Observe that $\delta(a) = 0$ except for $a \in \Delta(f)$. The Riemann–Hurwitz formula becomes here

$$\sum_{a \in \Delta(f)} \delta(a) = 2d - 2,$$

as follows from computing the Euler characteristic of the domain of f,

$$2 = \chi(\mathbb{P}^1(\mathbb{C})) = d\chi(\mathbb{P}^1(\mathbb{C})) - \sum_{a \in \Delta(f)} \delta(a).$$

Let $\beta(f) = \#\Delta(f)$ be the number of critical values of f. Clearly, $\beta(f) \leq 2d-2$. Recall (Sect. 2.6.2) that a Speiser curve for f is an oriented simple closed smooth curve γ in $\mathbb{P}^1(\mathbb{C})$ that passes through all the critical values $\Delta(f)$. Observe that two such curves γ, γ' are smoothly isotopic, keeping $\Delta(f)$ fixed, if and only if the induced cyclic order on the elements of $\Delta(f)$ agree. So up to isotopy one has $(\#\Delta(f) - 1)!$ isotopy classes of curves γ through the points of $\Delta(f)$. A Shabat polynomial or a Belyi map with three real critical values admits up to isotopy two Speiser curves, one being $\mathbb{P}^1(\mathbb{R}) = \mathbb{R} \cup \{\infty\}$.

From a Speiser curve γ one constructs the Speiser net as the pair $\Gamma_{f,\gamma} = (f^{-1}(\gamma), f^{-1}(\Delta(f)))$ in $\mathbb{P}^1(\mathbb{C})$. The vertices are colored by the elements of the branching set $\Delta(f)$ and the edges are oriented by lifting the orientation of γ. The Speiser curve γ being smooth, the curve $f^{-1}(\gamma) \subset \mathbb{P}^1(\mathbb{C})$ is the image of the immersion $\iota : \dot{\cup}_k S^1 \to \mathbb{P}^1(\mathbb{C})$, $k \leq d$, of finitely many copies of S^1 ($\dot{\cup}$ denotes disjoint union). The immersion ι has no tangencies and equi-angular multiple points at the critical points of f (see Fig. 2.8). Each complementary region of $P_{f,\gamma}$ is a polygon with $\beta(f)$ vertices. We color such a polygon R by $\pm$: by $+$ only if the induced orientation on ∂R agrees with the orientation of $\Gamma_{f,\gamma}$, in which case the colors of the vertices of R appear in the cyclic order induced by γ. The $\pm$ coloring of the polygons is a checkerboard coloring.

The Speiser net $\Gamma_{f,\gamma}$ is in fact an immersion of circles without tangencies in $\mathbb{P}^1(\mathbb{C}) = S^2$. The immersion ι is not a generic immersion, since self-intersection of 3 or more local branches occur. But, at all self-intersections the local branches intersect pairwise transversely.

We recall that a *divide* in the sense of [3, 5] is the image of a relative generic immersion of a finite union of copies of the unit interval $(I, \partial I)$ in the unit disc $(D, \partial D)$. From a divide P, one defines a link $L(P)$ in S^3 by

$$L(P) = \{(x, u) \in T(P) \subset S^3 \text{ such that } \|(x, u)\| = 1\}$$

where the 3-sphere S^3 is seen here as the unit sphere in the tangent bundle $T(\mathbb{R}^2)$ of $\mathbb{R}^2$.

By a result of [3], if the image of the immersion defining a divide is connected, then the associated link is fibered. The theory of links associated with divides is closely related to singularity theory [2, 47]. In the paper [3], the monodromy of the fibered link associated with a connected divide is described in terms of the combinatorics of the divide.

Many constructions for divides still apply to Speiser nets.

Let $\dot{\Gamma}_{f,\gamma}$ be the set of length 1 tangent vectors to the Speiser net $\Gamma_{f,\gamma}$. Let $\dot{\Gamma}^+_{f,\gamma}$ be the set of oriented length 1 tangent vectors to $\Gamma_{f,\gamma}$.

Recall that the 3-sphere S^3 and the Lie group $SU(2)$ are diffeomorphic and cover by a $2 \to 1$ map c the space of length 1 tangent vectors to $\mathbb{P}^1(\mathbb{C})$. Hence $\dot{\Gamma}_{f,\gamma}$ and $\dot{\Gamma}^+_{f,\gamma}$ can be considered as subsets in $\mathbb{P}^3(\mathbb{R})$ and lift to the classical knots and links $c^{-1}(\dot{\Gamma}_{f,\gamma}), c^{-1}(\dot{\Gamma}^+_{f,\gamma})$ in S^3, which we call Speiser links. See Fig. 2.8 for an example of a Speiser net and Fig. 2.9 for its Speiser link.

Theorem 2.8.4 *The subsets $\dot{\Gamma}_{f,\gamma}$ and $\dot{\Gamma}^+_{f,\gamma}$ are links in $\mathbb{P}^3(\mathbb{R})$. The lifts to S^3 are classical links. Moreover, the link $c^{-1}(\dot{\Gamma}_{f,\gamma}) \subset S^3$ is fibered (Fig. 2.10).*

Proof Let $\theta : \gamma \to T\gamma \subset T_{l=1}\mathbb{P}^1(\mathbb{C})$ be a continuous oriented vector field of length 1 along γ. Each component of $\gamma \setminus \Delta(f)$ together with the restriction of the vector field θ lifts by c to $d = \mathrm{degree}(f)$ smooth arcs in $T_{l=1}\mathbb{P}^1(\mathbb{C})$. The $d\beta(f)$ lifted arcs are pairwise disjoint. At a critical point p of f, $2k$ arcs join. The local germ of $\dot{\Gamma}^+_{f,\gamma}$ at p is diffeomorphic to the germ at 0 of the set $\{a \in \mathbb{C} \mid \mathrm{Im}(a^k) = 0\}$. So, after a small C^1 perturbation of $\dot{\Gamma}_{f,\gamma}$ locally near the critical points of f, the graph $\dot{\Gamma}_{f,\gamma}$ can be deformed to a divide with checkerboard coloring $P_{f,\gamma}$ on $S^2 = \mathbb{P}^1(\mathbb{C})$. The connected components of the complement $P_{f,\gamma}$ are contractible and moreover $P_{f,\gamma}$ has no self-intersection other than double points. The construction in [3] associates the knot or link $c^{-1}(\dot{\Gamma}_{f,\gamma})$ in $T_{l=1}S^2 = \mathbb{P}^3(\mathbb{R})$. The fibration property as in [3] was extended by Masaharu Ishikawa and Hironobu Naoe [50, 51] and shows that $\dot{P}_{f,\gamma}$ is

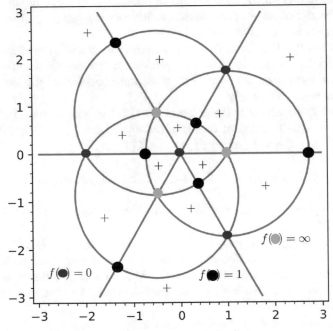

Fig. 2.8 The Speiser net $\Gamma_{f,\gamma}$ for the Belyi map $f = \frac{z^3(z^3+8)^3}{64(z^3-1)^3}$ of degree 12 and $\gamma = \mathbb{R} \cup \{\infty\}$. The covering has three critical values, $0, 1, \infty$, and γ is the Speiser curve passing through these points. The twelve regions marked with $+$ map to $\mathbb{C}_+$

a fibered knot or link in $\mathbb{P}^3(\mathbb{R}) = T_{l=1}S^2$. The same holds for its lift $c^{-1}(\dot{P}_{f,\gamma}) \subset S^3$ and also the link $c^{-1}(\dot{\Gamma}_{f,\gamma}) \subset S^3$.   □

The construction of the fiber surface and geometric monodromy for the link $c^{-1}(\dot{\Gamma}_{f,\gamma}) \subset S^3$ as in [3] is as follows. Let $P_{f,\gamma}$ be a small C^1-deformation of $\dot{\Gamma}_{f,\gamma}$ with only double points. The links $c^{-1}(\dot{P}_{f,\gamma})$ and $c^{-1}(\dot{\Gamma}_{f,\gamma})$ are isotopic, hence equivalent as links. The main step in the construction is the choice of a Morse function $h : S^2 \to \mathbb{R}$ having one maximum $+1$ in each $+$-region, one minimum -1 in each $-$-region, and $P_{f,\gamma} = h^{-1}(0)$ as critical level through all saddle points. Define $F_h = \{V_p \in T_p S^2 \mid h(p) \in]0, 1[, (Dh)_p(V_p) = 0, \|V_p\|_{\text{Eucl}} = 1\} \subset T_{l=1}S^2 = \mathbb{P}^3(\mathbb{R})$. The closure $\bar{F}_h$ of F_h in $\mathbb{P}^3(\mathbb{R})$ is a surface with boundary spanning the link $\dot{P}_{f,\gamma}$. The lift $c^{-1}(\bar{F}_h)$ is a spanning surface and its interior $F_{f,\gamma}$ a fiber surface for the link $c^{-1}(\dot{P}_{f,\gamma})$.

On the surface F appears a system of simple closed curves: for each maximum p of h the circle of length 1 tangent vectors in the kernel of $(Dh)_p$, each gradient line of h that connects a maximum to a minimum through a saddle point of h lifts to a simple closed curve on F, and to each minimum of h corresponds a simple closed curve built with pieces from preceding ones, see [3]. The monodromy is a composition of the positive Dehn twists along these curves.

The examples of Belyi maps are taken from the papers of K. Filom [36] and K. Filom and A. Kamalinejad [37] in which dessins on modular curves are computed, studied and used. In the first example the divide is the union of Moebius circles, which is not always the case. More complex components can appear as the second example shows.

The link $L_{f,\gamma}$ of a rational Belyi map $f : \mathbb{P}^1(\mathbb{C}) \to \mathbb{P}^1(\mathbb{C})$ where the Speiser curve $\gamma\mathbb{R} \cup \{\infty\}$ can be given as follows explicitly as the closure in $S^3 \subset \mathbb{C}^2$ of the set

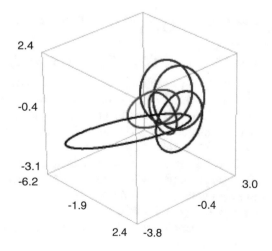

Fig. 2.9 The Speiser link $c^{-1}(\dot{\Gamma}_{f,\gamma})$ for the Belyi map $f = \frac{z^3(z^3+8)^3}{64(z^3-1)^3}$ of degree 12 consisting of 6 components. Each component is a great circle on the round S^3

2.4

-0.4

-3.1
-6.2

3.0

-1.9   -0.4

2.4   -3.8

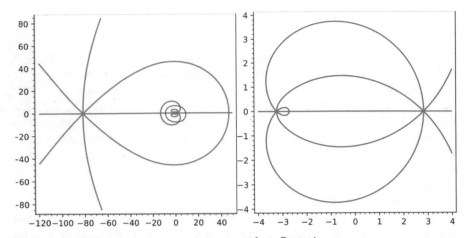

Fig. 2.10 The divide $\Gamma_{f,\gamma}$ for the Belyi map $f = \frac{(z^2+30\sqrt{8}z+264)^4}{216(z-\sqrt{8})^4(z+\sqrt{8})}$ of degree 8 and $\gamma = \mathbb{R} \cup \{\infty\}$. The black encircled region in the picture on the left is zoom enlarged and translated in the picture on the right

$$\{\frac{\lambda}{\sqrt{1+b\bar{b}}}(1,b) \mid b \in \mathbb{C}, f(b) \in \mathbb{R} \setminus \{0, 1, \infty\}, \lambda^2 = f'(b)/|f'(b)|\}$$

which is the union of $12d$ smooth arcs, $d = \mathrm{degree}(f)$.

The tangential lift of an oriented Moebius circle m on $\mathbb{P}^1(\mathbb{C})$ to $S^3 = \mathrm{SU}(2)$ is an oriented great circle on S^3. Indeed, by the isometric $\mathrm{SU}(2)$ action, move m to be the parametrized curve $s \in [0, 2\pi] \mapsto m(s) = r\cos(s) + ri\sin(s) \in \mathbb{C} \cup \{\infty\}$. The curve $M(s)$ on S^3 given by

$$s \in [0, 4\pi] \mapsto \frac{1}{r^2+1}(-\sin(s/2) + i\cos(s/2), -r\sin(s/2) + ri\cos(s/2)) \in \mathbb{C}^2$$

is its lift. From $M''(s) = \frac{-1/4}{r^2+1}M(s)$ follows the claim.

Let D_1, D_2 be closed disks in $\mathbb{P}^1(\mathbb{C})$ having as oriented boundary the Moebius circles m_1, m_2. If D_1, D_2 are complementary, the circles differ only by orientation as do their lifts. If the boundaries of D_1, D_2 touch tangentially in one point, the lifts M_1, M_2 intersect in two points. If the intersection $D_1 \cap D_2$ is open, then the lifts M_1, M_2 are disjoint great circles.

Orient $S^3 = \mathrm{SU}(2)$ as the boundary of the unit ball in $\mathbb{C}^2$. Assume that M_1, M_2 are disjoint, so that the boundaries of D_1, D_2 are disjoint or meet in two points. The linking number of the lifts M_1, M_2 of the oriented boundaries m_1, m_2 is given by $\mathrm{Lk}_{S^3}(M_1, M_2) = 1 - \#m_1 \cap m_2$.

In our example above with 6 great circles as lifts, all 15 pairwise linking numbers are equal to -1.

Acknowledgments Part of this survey is a revision of notes for lectures that were given by the first author at a CIMPA school in Varanasi (India), in December 2019. The authors would like to thank Bankteshwar Tiwari, the local organiser of this school, for his care and his excellent management. The second author is supported by the Interdisciplinary Thematic Institute CREAA, as part of the ITI 2021-2028 program of the University of Strasbourg, CNRS and Inserm (IdEx Unistra ANR-10-IDEX-0002), and the French Investments for the Future Program. The authors would like to thank Alexandre Eremenko for valuable suggestions and corrections after he read a first version of this chapter.

References

1. J.F. Adams, Vector fields on spheres. Ann. Math. **75**(2), 603–632 (1962)
2. N. A'Campo, Le groupe de monodromie du déploiement des singularités isolées de courbes planes. II, in *Proceedings of the International Congress of Mathematicians (Vancouver, B. C., 1974)*, vol. 1 (Canadian Mathematical Congress, Montreal, 1975), pp. 395–404
3. N. A'Campo, Generic immersions of curves, knots, monodromy and gordian number. Inst. Hautes Études Sci. Publ. Math. **88**, 151–169 (1998)
4. N. A'Campo, Planar trees, slalom curves and hyperbolic knots. Inst. Hautes Études Sci. Publ. Math. **88**, 171–180 (1998)
5. N. A'Campo, Real deformations and complex topology of plane curve singularities. Ann. Fac. Sc. Toulouse, Math. Série 6, t. **8**(1), 5–23 (1999)
6. N. A'Campo, Signatures of polynomials, in *In the Tradition of Thurston: Geometry and Topology*, ed. by K. Ohshika, A. Papadopoulos (Springer Verlag, Berlin, 2020), pp. 527–543
7. N. A'Campo, Flowbox decomposition for gradients of univariate polynomials, billiards, treelike configurations of vanishing cycles for A_n curve singularities and geometric cluster monodromy group. EMS Surv. Math. Sci. **9**(2), 389–414 (2022)
8. N. A'Campo, Topological, differential and conformal geometry of surfaces. *Universitext* (Springer, Berlin, 2021)
9. N. A'Campo, A. Papadopoulos, Notes on hyperbolic geometry, in *Strasbourg Master Class on Geometry*, ed. by A. Papadopoulos. IRMA Lectures in Mathematics and Theoretical Physics, vol. 18 (European Mathematical Society (EMS), Zürich, 2012), pp. 1–182
10. N. A'Campo, L. Ji, A. Papadopoulos, Actions of the absolute Galois group. *Handbook of Teichmüler Theory*, vol. IV, ed. by A. Papadopoulos (European Mathematical Society, Zürich, 2016), pp. 397–435
11. L.V. Ahlfors, Üntersuchungen zur Theorie der konformen Abbildung und der ganzen Funktionen. Acta Soc. Sci. Fenn., Nov. Ser. A1 **9**, 1–40 (1930). In *Collected Works*, vol. I, pp. 18–55
12. L. Ahlfors, Zur Bestimmung des Typus einer Riemannschen Fläche. Comment. Math. Helv. **3**(1), 173–177 (1931)
13. L.V. Ahlfors, Quelques propriétés des surfaces de Riemann correspondant aux fonctions méromorphes. Bull. de la S. M. F. **60**, 197–207 (1932)
14. L.V. Ahlfors, *Conformal Invariants: Topics in Geometric Function Theory* (AMS, Rhode Island, 1973)
15. L.V. Ahlfors, Riemann surfaces and small point sets. Annales Academiæ Scientiarum Fennicæ, Series A. I. Mathematica **7**, 49–57 (1982)
16. L.V. Ahlfors, Collected works, in *Series: Contemporary Mathematicians Series*, vol. 2 (Birkäuser, Boston-Basel-Stuttgart, 1982)
17. V. Alberge, M. Brakalova-Trevithick, A. Papadopoulos, Teichmüller's work on the type problem, in *Handbook of Teichmüller Theory*, vol. VII, ed. by A. Papadopoulos (European Mathematical Society, Berlin, 2020). IRMA Lect. Math. Theor. Phys. **30**, 543–560 (2020)

18. F. Bergeron, Combinatorial Cellular Decompositions for the space of Complex Coefficient Polynomials (2009). arXiv:0901.4020
19. G.V. Belyi, Galois extensions of a maximal cyclotomic field. Izv. Akad. Nauk SSSR Ser. Mat. **43**, 267–276 (1979). English transl. Math. USSR-Izv. **14**, 247–256 (1980)
20. G.A. Belyi, Galois extensions of a maximal cyclotomic field. Math. (Translated by Neal Koblitz) USSR Izv. **14**(2), 247–256 (1980)
21. J. Blanc, Groupes de Cremona, connexité et simplicité. Ann. Sc. Éc. Norm. Supér. (4) **43**(2), 357–364 (2010)
22. J. Blanc, S. Zimmermann, Susanna, Topological simplicity of the Cremona groups. Am. J. Math. **140**(5), 1297–1309 (2018)
23. P.L. Bowers, Combinatorics encoding geometry: the legacy of Bill Thurston in the story of one theorem, in *In the Tradition of Thurston*, ed. by V. Alberge, K. Ohshika, A. Papadopoulos (Springer Verlag, Berlin, 2020), pp. 173–239
24. W. Brägger, Kreispackungen und Triangulierungen. Enseign. Math. (2) **38**(3–4), 201–217 (1992)
25. X. Buff, G. Cui, L. Tan, Teichmüller spaces and holomorphic dynamics, in *Handbook of Teichmüler Theory*, vol. IV, ed. by A. Papadopoulos (European Mathematical Society, Zürich, 2014), pp. 717–756
26. S. Cantat, S. Lamy, Y. de Cornulier, Normal subgroups in the Cremona group. Acta Math. **210**(1), 31–94 (2013)
27. P. Cohen, C. Itzykson, J. Wolfart, Fuchsian triangle groups and Grothendieck dessins. Variations on a theme of Belyi. Commun. Math. Phys. **163**(3), 605–627 (1994)
28. A. Connes, C. Consani, M. Marcolli, Fun with $\mathbb{F}_1$. J. Number Theory **129**(6), 1532–1561 (2009)
29. G. Darboux, Sur le problème de Pfaff, Bulletin des Sciences Mathématiques et Astronomiques, Série 2, Tome **6**(1), 14–36 (1882)
30. J. Déserti, The Cremona group and its subgroups. *Mathematical Surveys and Monographs*, vol. 252 (American Mathematical Society, Providence, 2021)
31. P.G. Doyle, Random walk on the Speiser graph of a Riemann surface. Bull. Am. Math. Soc. New Ser. **11**, 371–377 (1984)
32. P. Doyle, J.L. Snell, *Random Walks and Electric Networks*. The Carus Mathematical Monographs, vol. 22 (American Mathematical Society, Providence, 1984)
33. C. Ehresmann, Sur les variétés presque complexes. Proc. Internat. Congr. Math. (Cambridge, Mass., Aug. 30-Sept. 6, 1950) **2**, 412–419 (1952)
34. A. Eremenko, Geometric theory of meromorphic functions, in *In the tradition of Ahlfors and Bers, III*, ed. by W. Abikoff, A. Haas. Proceedings of the 2nd Ahlfors-Bers colloquium, Storrs, CT, USA, October 18–21, 2001. Contemporary Mathematics, vol. 355 (American Mathematical Society (AMS), Providence, 2004), pp. 221–230
35. A. Eremenko, *Topics in geometric theory of meromorphic functions*. Preprint (2023)
36. K. Filom, The Belyi characterization of a class of modular curves. J. Théor. Nombres Bordeaux **30**(2), 409–429 (2018)
37. K. Filom, A. Kamalinejad, Dessins on Modular Curves (2016). arXiv:1603.01693
38. P. Fatou, Séries trigonométriques et séries de Taylor. Acta Math. **30**, 335–400 (1966)
39. D. Gabai, The Whitehead manifold is a union of two Euclidean spaces. J. Topol. **4**(3), 529–534 (2011)
40. G. Girondo, G. González-Diez, Introduction to compact Riemann surfaces and dessins d'enfants. *London Mathematical Society Student Texts*, vol. 79 (Cambridge University Press, Cambridge, 2012)
41. J.W. Gray, Some global properties of contact structures. Ann. Math. **69**(2), 421–450 (1959)
42. M. Gromov, Pseudo-holomorphic curves in symplectic manifolds. Invent. Math. **82**, 307–347 (1985)
43. A. Grothendieck, Esquisse d'un programme. Unpublished manuscript (1984). English translation: Sketch of a program, in *Geometric Galois Actions*, vol. 1. London Mathematical Society Lecture Note Series, vol. 242 (Cambridge University Press, Cambridge, 1997), pp. 5–48

44. H. Grötzsch, Über einige Extremalprobleme der konformen Abbildung. Ber. Verhandl. Sächs. Akad. Wiss. Leipzig Math.-Phys. Kl. **80**, 367–376 (1928). English translation by A. A'Campo-Neuen, On some extremal problems of conformal mappings, in *Handbook of Teichmüller Theory*, vol. VII , ed. by A. Papadopoulos (EMS Publishing House, Zürich, 2020), pp. 355–363
45. H. Grötzsch, Über die Verzerrung bei schlichten nichtkonformen Abbildungen and über eine damit zusammenhängende Erweiterung des Picardschen Satzes. Ber. Verhandl. Sächs. Akad. Wiss. Leipzig Math.-Phys. Kl. **80**, 503–507 (1928). English translation by M. Brakalova-Trevithick, On the distortion of schlicht non-conformal mappings and on a related extension of Picard's theorem, in *Handbook of Teichmüller Theory*, vol. VII, ed. by A. Papadopoulos (EMS Publishing House, Zürich, 2020), pp. 371–374
46. P. Guillot, A primer on dessins. *Handbook of Teichmüller Theory*, vol. VI, ed. by A. Papadopoulos (European Mathematical Society, Zürich, 2016), pp. 437–466
47. S.M. Gusein-Zade, Dynkin diagrams for singularities of functions of two variables. Funct. Anal. Appl. **8**, 10–13, 209–300 (1974)
48. W.J. Harvey, Teichmüller spaces, triangle groups and Grothendieck dessins. *Handbook of Teichmüller Theory*, vol. I. IRMA Lectures in Mathematics and Theoretical Physics, vol. 11 (European Mathematical Society, Zürich, 2007), pp. 249–292
49. F. Herrlich, G. Schmithüsen, Dessins d'enfants and origami curves. *Handbook of Teichmüller Theory*, vol. II, ed. by A. Papadopoulos (European Mathematical Society, Zürich, 2009), pp. 767–809
50. M. Ishikawa, Tangent circle bundles admit positive open book decompositions along arbitrary links. Topology **43**, 215–232 (2004)
51. M. Ishikawa, H. Naoe: Milnor fibration, A'Campo's divide and Turaev's shadow, Singularities, Kagoshima 2017, in *Proceedings of the 5th Franco-Japanese-Vietnamese Symposium on Singularities, World Scientific Publishing* (2020), pp. 71–93
52. G. Julia, *Leçons sur la représentation conforme des aires simplement connexes* (Gauthier-Villars, Paris, 1950)
53. S. Kakutani, Applications of the theory of pseudoregular functions to the type-problem of Riemann surface. Jpn. J. Math. **13**, 375–392 (1937)
54. S. Kakutani, Two-dimensional Brownian motion and harmonic functions. Proc. Imp. Acad. Tokyo **20**, 706–714 (1944)
55. S. Kakutani, Two-dimensional Brownian motion and the type problem of Riemann surfaces. Proc. Jpn. Acad. **21**, 138–140 (1945)
56. S. Kakutani, Random walk and the type problem of Riemann surfaces, in *Contributions to the Theory of Riemann Surfaces*. Annals of Mathematics Studies, vol. 30 (Princeton University Press, Princeton, 1953), pp. 95–101
57. A. Kirchhoff, Sur l'existence de certains champs tensoriels sur les sphères. Comptes Rendus Acad. Sc. Paris, t. **225**, 1258–1260 (1947)
58. A. Kirchhoff, Beiträge zur topologischen linearen Algebra. Compos. Math. **11**, 1–36 (1953)
59. Z. Kobayashi, Theorems on the conformal representation of Riemann surfaces. Sci. Rep. Tokyo Bunrika Daigaku, Sec. A **39**, 125–165 (1935)
60. Z. Kobayashi, On Kakutani's theory of Riemann surfaces. Sci. Rep. Tokyo Bunrika Daigaku, Sec. A **76**, 9–44 (1940)
61. P. Koebe, Über die Uniformisierung beliebiger analytischer Kurven (Dritte Mitteilung). Nachr. Ges. Wiss. Gött., Math.-Ph. Kl. 337–358 (1908)
62. P. Koebe, Abhandlungen zur Theorie der konformen Abbikdung. IV Abbildung mehrfach zusammenhängender schlichter Bereiche auf Schlitzbereiche. Acta Math. **41**, 305–344 (1918)
63. P. Koebe, Abhandlungen zur Theorie der konformen Abbikdung. V Abbildung mehrfach zusammenhängender schlichter Bereiche auf Schlitzbereiche (Fortsetzung). Math. Z. **2**, 198–236 (1918)
64. S. Koch, L. Tan, On balanced planar graphs, following W. Thurston, in *What's Next? The Mathematical Legacy of William P. Thurston*, ed. D. P. Thurston. Annals of Mathematics Studies, vol. 205 (Princeton University Press, Princeton, 2020), pp. 215–232

65. M.A. Lavrentieff, *Sur les fonctions d'une variable complexe représentables par des séries de Riemann* (Hermann, Paris, 1936)
66. M.A. Lavrentieff, Sur une classe de représentations continues. Mat. Sb. **42**, 407–423 (1935). On a class of continuous representations, English translation by V. Alberge, A. Papadopoulos, *Handbook of Teichmüller Theory*, vol. VII, ed. by A. Papadopoulos (EMS Publishing House, Berlin, 2020), pp. 417–439
67. O. Lázaro, D. Rodríguez, A note on Kirchhoff's theorem for almost complex spheres I. arXiv:1804.05794
68. O. Lehto, L. Virtanen, *Quasiconformal Mappings in the Plane* (Springer-Verlag, New York/Heidelberg/Berlin, 1973). English translation of the book *Quasikonforme Abhildungen* (Springer-Verlag, New York/Heidelberg/Berlin)
69. V. da Leonardo, *The Notebooks of Leonardo da Vinci*, vol. 1 (Dover, New York, 1970)
70. N. Lusin, J. Priwaloff, Sur l'unicité et la multiplicité des fonctions analytiques. Annales scientifiques de l'École Normale Supérieure, Série 3 Tome **42**, 143–191 (1925)
71. T.J. Lyons, H.P. McKean, Jr., Winding of the plane Brownian motion. Adv. Math. **5**(1), 212–225 (1984)
72. Y.I. Manin, Lectures on zeta functions and motives (according to Deninger and Kurokawa), in *Number Theory Seminar* (Columbia University, New York, 1992). Astérisque **228**(4), 121–163 (1995)
73. D. McDuff, D. Salamon, *Introduction to Symplectic Topology*. Oxford Mathematical Monographs, 2nd edn. (Oxford University Press, Oxford, 1998)
74. H.P. McKean, Jr., D. Sullivan, Brownian motion and harmonie functions on the class surface of the thrice-punctured sphere. Adv. Math. **5**(1), 203–211 (1984)
75. J.W. Milnor, On manifolds homeomorphic to the 7-sphere. Ann. Math. **64**(2), 399–405 (1956)
76. J.W. Milnor, On deciding whether a surface is parabolic or hyperbolic. Am. Math. Mon. **84**(1), 43–46 (1977)
77. J. Moser, On the volume elements on a manifold. Trans. AMS **120**(2), 286–294 (1965)
78. R. Nevanlinna, Über die Riemannsche Fläche einer analytischen Funktion, in *Proceedings of the International Congress of Mathematicians* (Zürich, 1932), pp. 221–239
79. R. Nevanlinna, Über Riemannsche Fläche mit endlich vielen Windunspunkten. Acta Math. **58**, 295–373 (1932)
80. R. Nevanlinna, *Eindeutige analytische Funktionen* (Springer Verlag, Berlin, 1953). English translation: Analytic functions tr. by P. Emig. Die Grundlehren der mathematischen Wissenschaften, vol. 162 (Springer Verlag, Berlin-Heidelberg-New York, 1970)
81. A. Papadopoulos, Physics in Riemann's mathematical papers, in *From Riemann to Differential Geometry and Relativity*, ed. by L. Ji, A. Papadopoulos, S. Yamada (Springer, Berlin, 2017), pp. 151–207
82. E. Pervova, C. Petronio, On the existence of branched coverings between surfaces with prescribed branch data, I. Algebr. Geom. Topol. **6**, 1957–1985 (2006)
83. E. Pervova, C. Petronio, On the existence of branched coverings between surfaces with prescribed branch data. II. J. Knot Theory Ramifications **17**(7), 787–816 (2008)
84. S. MacLane The genesis of mathematical structures, as exemplified in the work of Charles Ehresmann. Cahiers de topologie et géométrie différentielle catégoriques, tome **21**(4), 353–365 (1980)
85. R. Rashed, A. Papadopoulos, On Menelaus' *Spherics* III.5 in Arabic mathematics, I: Ibn 'Irāq. Arabic Sci. Philos. **24**, 1–68 (2014)
86. R. Rashed, A. Papadopoulos, On Menelaus' *Spherics* III.5 in Arabic mathematics, II: Naṣīr al-Dīn al-Ṭūsī and Ibn Abī Jarrāda. Arabic Sci. Philos. **25**, 1–32 (2015)
87. R. Rashed, A. Papadopoulos, Menelaus' Spherics. Early Translation and al-Māhānī/al-Harawī's Version. *Scientia Graeco-Arabica*, vol. 21 (De Gruyter, Berlin, 2018)
88. F.M. Riesz, Über die Randwerte einer analytischen Funktion Quatrième Congrès des Mathématiciens Scandinaves. *Stockholm 1916* (Uppsala, Almqvist, 1920), pp. 27–44
89. B. Riemann, Gleichgewicht der Electricität auf Cylindern mit kreisförmigen Querschnitt und parallelen Axen, Conforme Abbildung von durch Kreise begrenzten Figuren, Nachlass XXVI, 1857, in *Ges. math. Werke*, pp. 472–476

90. B. Riemann, Grundlagen für eine allgemeine Theorie der Functionen einer veränderlichen complexen Grösse (Göttingen, 1851), in *Gesammelte mathematische Werke, wissenschaftlicher Nachlass und Nachträge*. Nach der Ausgabe von Heinrich Weber und Richard Dedekind neu herausgegeben von Raghavan Narasimhan. Teubner Verlagsgesellschaft, Leipzig, 1862 (Springer-Verlag, Berlin, 1990), pp. 3–48

91. B. Riemann, Theorie der Abel'schen Functionen. J. Reine Angew. Math. **54**, 115–155 (1857). Gesammelte mathematische Werke, wissenschaftlicher Nachlass und Nachträge. Nach der Ausgabe von Heinrich Weber und Richard Dedekind neu herausgegeben von Raghavan Narasimhan. Teubner Verlagsgesellschaft, Leipzig, 1862 (Springer-Verlag, Berlin, 1990), pp. 88–144

92. H.P. de Saint-Gervais (pseudonym for a collective book), Uniformisation des surfaces de Riemann: Retour sur un théorème centenaire, ENS Editions (2010). English translation by R. G. Burns, Uniformization of Riemann surfaces. Revisiting a hundred-year-old theorem. *Heritage of European Mathematics* (European Mathematical Society, Zürich, 2016)

93. L. Sario, M. Nakai, *Classification Theory of Riemann Surfaces* (Springer-Verlag, Berlin, 1970)

94. E. Sellien, Notes on Riemann's courses *The mathematical theory of electricity and magnetism* and *Selected physical problems* of summer 1858. Nachlass Riemann. Handschriftenabteilung der Staats- und Universitätsbibliothek Göttingen. Riemann papers. Cod. Ms. B. Riemann Blatt **45**, 1–23 (1858)

95. J.-P. Serre, Le groupe de Cremona et ses sous-groupes finis, in *Séminaire Bourbaki*, vol. 2008/2009. Exposés 997–1011. Astérisque 332, 75–100, Exp. No. 1000 (Société Mathématique de France, Paris, 2010)

96. G.B. Shabat, V.A. Voevodsky, Drawing curves over number fields, in *The Grothendieck Festschrift*, vol. III. Progress in Mathematics, vol. 88 (Birkhäuser, Boston 1990), pp. 199–227

97. G. Shabat, A. Zvonkin, Plane trees and algebraic numbers, in *Jerusalem Combinatorics, '93*. Contemporary Mathematics, vol. 178 (American Mathematical Society, Providence, 1994), pp. 233–275

98. A. Speiser, Probleme aus dem Gebiet der ganzen transzendenten Funktionen. Comment. Math. Helv. **1**(1), 289–312 (1929)

99. A. Speiser, Über Riemannsche Flaechen. Commun. Math. Helvetici **2**, 284–293 (1930)

100. S. Stoïlov, *Leçons sur les principes topologiques de la théorie des fonctions analytiques* (Gauthier-Villars, Paris, 1956)

101. O. Teichmüller, Eine Anwendung quasikonformer Abbildungen auf das Typenproblem. Deutsche Math. **2**, 321–327 (1937). *Gesammelte Abhandlungen*, ed. by L. V. Ahlfors, F. W. Gehring (Springer-Verlag, Berlin-Heidelberg-New York 1982), pp. 171–178. English translation by M. Brakalova Trevithick, An application of quasiconformal mappings to the type problem, in *Handbook of Teichmüller Theory*, ed. by A. Papadopoulos, vol. VII (EMS Publishing House, Zürich/Berlin, 1982), pp. 453–461

102. O. Teichmüller, Untersuchungen über konforme und quasikonforme Abbildung. Deutsche Math. **3**, 621–678 (1938). *Gesammelte Abhandlungen*, ed. by L. V. Ahlfors, F. W. Gehring (Springer-Verlag, Berlin-Heidelberg-New York 1982), pp. 205–262. English translation by M. Brakalova, M. Weiss, Investigations on conformal and quasiconformal mappings, in *Handbook of Teichmüller Theory*, ed. by A. Papadopoulos, vol. VII (EMS Publishing House, Zürich/Berlin, 1982), pp. 463–529

103. R. Thom, L'équivalence d'une fonction différentiable et d'un polynôme. Topology **3**, suppl. 2, 297–307 (1965)

104. W. Thurston, *The Combinatorics of Iterated Rational Maps*. Preprint (Princeton University, Princeton, 1983). Published as: On the dynamics of iterated rational maps, in *Collected Works of William P. Thurston with Commentary*, vol. III. (American Mathematical Society, Providence, 2022), pp. 107–176; and the version, Complex dynamics. *Families and Friends*, ed. by D. Schleicher, N. Selinger, W. P. Thurston (A K Peters, Wellesley, 2009), pp. 3–109

105. J. Tits, Sur les analogues algébriques des groupes semi-simples complexes, in *Colloque d'algèbre supérieure* (Gauthier-Villars, Bruxelles/Paris, 1957), pp. 261–289
106. J. Tomasini, Realizations of branched self-coverings of the 2-sphere. Topology Appl. 196, Part A, 31–53 (2015)
107. M. Troyanov, Parabolicity of manifolds. Siberian Adv. Math. **9**(4), 125–150 (1999)
108. C. Urech, S. Zimmermann, A new presentation of the plane Cremona group. Proc. AMS **147**(7), 2741–2755 (2019)
109. É.V. Vinberg, Real entire functions with prescribed critical values (Russian). Vopr. Teor. Grupp Gomologicheskoi Algebry 9, 127–138 (1989). Yaroslav. Gos. Univ., Yaroslavl' (1989)
110. L. Volkovyskii, Investigation of the type problem for a simply connected Riemann surface. Trudy Mat. Inst. Steklov, vol. 34 (USSR Academy of Science, Moscow-Leningrad, 1950), pp. 3–171
111. V.A. Voevodski, G. B. Shabat, Equilateral triangulations of Riemann surfaces, and curves over algebraic number fields (Russian). Dokl. Akad. Nauk SSSR **304**(2), 265–268 (1989). Translation in Soviet Math. Dokl. **39**(1), 38–44 (1989)
112. A. Weil, Riemann, Betti and the birth of topology. Arch. Hist. Exact Sci. **20**(2), 91–96 (1979)
113. J.H.C. Whitehead, A certain open manifold whose group is unity. Q. J. Math. **6**(1), 268–279 (1935)
114. W.-T. Wu, Sur la structure presque complexe d'une variété differentiable réelle de dimensions 4. Comptes Rendus Acad. Sc. Paris, t. **227**, 1076–1078 (1948)
115. W.-T. Wu, Sur les classes caractéristiques des structures fibrées sphériques. Thèse, Strasbourg, 1949
116. S. Zimmermann, The abelianisation of the real Cremona group. Duke Math. J. **167**(2), 211–267 (2018)

Chapter 3
Teichmüller Spaces and Their Various Metrics

Ken'ichi Ohshika

Abstract This chapter is a survey on three different kinds of Finsler metrics on Teichmüller spaces. We start from the most classical one, which appeared indeed in the original definition of Teichmüller space, the Teichmüller metric. We give its definition and look at its relation with quadratic differentials and extremal lengths. Then we turn to a hyperbolic analogue of the Teichmüller metric, Thurston's asymmetric metric, and shall give descriptions of its Finsler structure and of cotangent spaces of Teichmüller space with respect to the norm induced by this metric. In the final part, we turn to the earthquake metric, which was first introduced by Thurston, but has not been seriously studied until quite recently. We shall look at the duality between tangent spaces to Teichmüller space equipped with the earthquake norm and cotangent spaces equipped with Thurston's norm. We also touch upon the incompleteness of this last metric and its completion.

Keywords Teichmüller space · Thurston's asymmetric metric · Earthquake

2020 Mathematics Subject Classification 30F60, 57K20

3.1 Introduction

Teichmüller space, which had appeared rather unsystematically in works of Fricke and Fenchel–Nielsen among others, was defined rigorously as a metric space by Teichmüller, and has been studied for a long time from various viewpoints. The metric introduced by Teichmüller is a Finsler metric measuring the distance between two marked Riemann surfaces making use of the best quasi-conformal homeomorphisms. Teichmüller completely characterised geodesics in this metric describing them in terms of holomorphic quadratic differentials. The Teichmüller metric

K. Ohshika (✉)
Department of Mathematics, Faculty of Science, Gakushuin University, Toshima-ku, Tokyo, Japan
e-mail: ohshika@math.gakushuin.ac.jp

relies on quasi-conformal homeomorphisms to compare two conformal/complex structures on a surface. This can be regarded as a generalisation of the viewpoint taken for the Teichmüller space of flat tori to surfaces of higher genera. Since we only look at conformal structures, not Riemannian metrics, to consider the Teichmüller metric, the lengths of closed curves using the ordinary definition do not make sense in this setting. Still, there is a notion of length, called extremal length, which depends only on conformal structures. The extremal length plays an important role to study quasi-conformal homeomorphisms and the Teichmüller metric, and indeed, we shall see that this gives a Finsler structure associated with the Teichmüller metric.

The second metric we are going to deal with is Thurston's asymmetric metric. Thurston introduced a metric based on hyperbolic structures on a surface instead of the conformal structures used in Teichmüller's work. Instead of quasi-conformal homeomorphisms, Lipschitz homeomorphisms (isotopic to the identity) are used to compare two hyperbolic structures on the same surface. It turns out that this metric is not symmetric, but that it is Finsler. Indeed, making use of comparison of the lengths of closed geodesics, one can express the infinitesimal form of this metric. The derivative of the log of the length function gives an embedding of the projective lamination space onto the unit sphere of the cotangent space at each point of Teichmüller space. This embedding relates Thurston's metric with another metric, the earthquake metric, which is the topic of the last section.

Earthquakes are deformations of hyperbolic structures on surfaces along measured laminations, which are generalisations of classical Fenchel-Nielsen twists. Thurston introduced this notion, and it was used essentially in Kerckhoff's work to solve the Nielsen realisation problem. The fundamental theorem shown by Thurston on earthquakes is that any two points in Teichmüller space can be joined by a unique earthquake (once the direction, right or left is fixed). This implies also its infinitesimal version: any tangent vector of Teichmüller space is represented as an infinitesimal earthquake. The earthquake norm is defined using this expression, by defining the norm of a tangent vector to be the length of a measured lamination along which the infinitesimal earthquake is performed to represent the given vector. The earthquake metric is the Finsler metric associated with this norm. It will turn out that this metric is incomplete and its completion is topologically the same as the Weil–Petersson completion, that is, the augmented Teichmüller space. We shall also show that the tangent space at each point of Teichmüller space with the earthquake metric is linearly isometric to the cotangent space with the norm associated with Thurston's metric, to which the projective lamination space was embedded in the preceding section.

This chapter grew out of the author's course given at the CIMPA school entitled "Geometric structures on surfaces, moduli spaces and dynamics" and held at Banaras Hindu University in Varanasi in December 2022. The author is very grateful to the organisers of the school, Bankteshwar Tiwari and Athanase Papadopoulos.

3.2   Teichmüller Space and the Teichmüller Metric

In this section, we give a geometric description of the original metric of Teichmüller space, which is now called the Teichmüller metric. We start with the definition of Teichmüller space, regarded as the space of marked complex structures. Throughout this chapter, we assume that S is a closed oriented surface of genus greater than 1. We shall also need to consider the case of genus 1 at some point, for this case constitutes a prototype of Teichmüller theory.

Definition 3.2.1 We consider the set of pairs $\mathcal{D}(S) = \{(\Sigma, f)\}$ where Σ is a Riemann surface, i.e. one-dimensinal complex manifold, and f is an orientation-preserving homeomorphism from S to Σ. We introduce an equivalence relation $\sim$ on $\mathcal{D}(S)$ by defining $(\Sigma_1, f_1) \sim (\Sigma_2, f_2)$ if and only if there is a bi-holomorphic map $g: \Sigma_1 \to \Sigma_2$ homotopic to $f_2 \circ f_1^{-1}$.

The Teichmüller metric measures how one complex structure is deformed from the other using the best possible quasi-conformal map. We now define quasi-conformal maps and their dilatations. First, we define these for maps between two regions in the complex plane.

Definition 3.2.2 Let U and V be open regions in the complex plane $\mathbb{C}$, and K a real number with $K \geq 1$. An orientation-preserving homeomorphism $f: U \to V$ is said to be K-quasi-conformal if the following two conditions are satisfied.

(1) f is ACL(absolutely continuous on lines). This means that both $\dfrac{\partial f}{\partial x}$ and $\dfrac{\partial f}{\partial y}$ exist for almost every x and y, and the equalities

$$f(x, y) - f(a, y) = \int_a^x \frac{\partial f}{\partial x}(t, y)dt,$$

and

$$f(x, y) - f(x, b) = \int_b^y \frac{\partial f}{\partial y}(x, s)ds$$

hold almost everywhere.

(2)

$$|f_{\bar{z}}(z)| \leq \frac{K - 1}{K + 1}|f_z(z)|,$$

where $f_{\bar{z}}(z) = \dfrac{1}{2}\left(\dfrac{\partial f}{\partial x} + i\dfrac{\partial f}{\partial y}\right)(z)$ and $f_z(z) = \dfrac{1}{2}\left(\dfrac{\partial f}{\partial x} - i\dfrac{\partial f}{\partial y}\right)(z)$ for $z = x + iy$.

We shall next look at the geometric meaning of the condition (2) under the assumption that f is differentiable. For $z = x + iy$, we express f by two

variable real valued functions u, v as $f(x + iy) = u(x, y) + iv(x, y)$. Since df
corresponds to a 2×2 matrix $\begin{pmatrix} \frac{\partial u}{\partial x} & \frac{\partial u}{\partial y} \\ \frac{\partial v}{\partial x} & \frac{\partial v}{\partial y} \end{pmatrix}$, it sends a unit tangent vector $\begin{pmatrix} \cos\theta \\ \sin\theta \end{pmatrix}$ to
$\begin{pmatrix} \frac{\partial u}{\partial x}\cos\theta + \frac{\partial u}{\partial y}\sin\theta \\ \frac{\partial v}{\partial x}\cos\theta + \frac{\partial v}{\partial y}\sin\theta \end{pmatrix}$. On the other hand, we can calculate

$$f_z e^{i\theta} + f_{\bar{z}} e^{-i\theta} = \frac{\partial u}{\partial x}\cos\theta + i\frac{\partial v}{\partial x}\cos\theta + \frac{\partial u}{\partial y}\sin\theta + i\frac{\partial v}{\partial y}\sin\theta,$$

which expresses the vector above as a complex number. Express f_z and $f_{\bar{z}}$ using the
polar coordinate as $f_z = |f_z|e^{i\alpha}$, $f_{\bar{z}} = |f_{\bar{z}}|e^{i\beta}$. Then we have

$$f_z e^{i\theta} + f_{\bar{z}} e^{-i\theta} = |f_z|e^{i(\alpha+\theta)} + |f_{\bar{z}}|e^{i(\beta-\theta)}.$$

Therefore, its absolute value attains the maximum $|f_z| + |f_{\bar{z}}|$ when $\alpha + \theta \equiv \beta - \theta$
mod 2π, and the minimum $|f_z| - |f_{\bar{z}}|$ when $\alpha + \theta \equiv \beta - \theta + \pi$ mod 2π. (We
note that $|f_z| > |f_{\bar{z}}|$ since f is assumed to be a local diffeomorphism preserving
the orientation.) Therefore, infinitesimally, the ratio of the major axis and the minor
axis is equal to $\dfrac{|f_z| + |f_{\bar{z}}|}{|f_z| - |f_{\bar{z}}|}$. The second condition for the K-quasi-conformality in
Definition 3.2.2 is equivalent to saying that this ratio is bounded by K.

We next define the dilatation of a homeomorphism.

Definition 3.2.3 Let $f: U \to V$ be a homeomorphism satisfying the condition (1)
of Definition 3.2.2. Then $\mu(z) := \dfrac{f_{\bar{z}}(z)}{f_z(z)}$ is called the *complex dilatation* of f at
$z \in U$, and $K_f(z) := \dfrac{1 + |\mu(z)|}{1 - |\mu(z)|}$ is called the *dilatation* of f at z.

The homeomorphism f is K-*quasi-conformal* if and only if $K_f(z) \leq K$ for almost
every $z \in U$. We say simply that f is quasi-conformal when there is K such that f
is K-quasi-conformal. This is equivalent to the condition that $\|\mu(z)\|_\infty < 1$. When
f is quasi-conformal ess. $\sup_{z\in U} K_f(z)$ is called the *maximal dilatation* of f and is
denoted by $K(f)$. We note that f is conformal if and only if $K(f) = 1$.

Lemma 3.2.4 *Let $f: U \to V$ and $g: V \to W$ be quasi-conformal homeomorphisms. Then we have $K(g \circ f) \leq K(g)K(f)$ and $K(f^{-1}) = K(f)$. In particular, if
g is conformal, then $K(g \circ f) = K(f)$ and if f is conformal, then $K(g \circ f) = K(g)$.*

Proof This follows from the geometric interpretation of maximal dilatation given
above. $\square$

Definition 3.2.5 Let Σ_1 and Σ_2 be homemorphic Riemann surfaces. For an
orientation-preserving homeomorphism $h: \Sigma_1 \to \Sigma_2$, its dilatation $K_h(x)$ at $x \in \Sigma_1$
is defined to be the dilatation of $\psi \circ h \circ \phi^{-1}|h^{-1}(V) \cap U$ at $\phi(x)$, where (U, ϕ)
and (V, ψ) are charts of Σ_1 and of Σ_2 at x and $h(x)$ respectively. The maximal
dilatation $K(h)$ of h is defined to be ess. $\sup_{x\in\Sigma_1} K_h(x)$. The homeomorphism h

is said to be K-quasi-conformal if $K(h) \leq K$. By Lemma 3.2.4, this definition is independent of choices of charts.

Now the Teichmüller distance d_T is defined as follows.

Definition 3.2.6 For two points $x_1 = (\Sigma_1, f_1), x_2 = (\Sigma_2, f_2)$ of $\mathcal{T}(S)$, the *Teichmüller distance* $d_T(x_1, x_2)$ is defined to be $\dfrac{1}{2} \inf\limits_{g \simeq f_2 \circ f_1^{-1}} \log K(g)$.

The function d_T satisfies the axioms of distance.

Lemma 3.2.7 *The function d_T is a well-defined distance function on $\mathcal{T}(S) \times \mathcal{T}(S)$.*

Proof To show that d_T is well defined, we need to prove that its value is finite. Since both f_1 and f_2 are homeomorphisms, there is a diffeomorphism $g \colon \Sigma_1 \to \Sigma_2$ homotopic to $f_2 \circ f_1^{-1}$. Therefore, g has finite dilatation at any point x on Σ_1. Since we assumed that S is compact, there is a constant K bounding the dilatations of g at all points of Σ_1. By definition, $d_T(x_1, x_2) \leq \frac{1}{2} \log K$, and hence it is always finite.

The symmetry follows from the fact that $K(g) = K(g^{-1})$ derived from Lemma 3.2.4. The triangle inequality also follows from Lemma 3.2.4.

It is evident that $d_T(x, x) = 0$. To show the converse, suppose that $d_T(x, y) = 0$ for $x = (\Sigma_1, f_1)$ and $y = (\Sigma_2, f_2)$. Then there is a sequence of quasi-conformal homeomorphism $g_i \colon \Sigma_1 \to \Sigma_2$ homotopic to $f_2 \circ f_1^{-1}$ such that $K(g_i) \longrightarrow 1$. Since this makes a normal family, passing to a subsequence, the sequence converges to a continuous map $g_\infty \colon \Sigma_1 \to \Sigma_2$ with $K(g_\infty) = 1$, which must be a conformal automorphism. This implies, by the definition of $\mathcal{T}(S)$, that $x = y$. □

Now we look at the special case when S is a torus. A torus with a structure of Riemann surface, which we call a Riemann torus, is uniformised by the complex plane. Considering a fundamental domain of the group of covering translations, it turns out that any Riemann torus is obtained from a parallelogram in the complex plane having its vertices at $0, 1, \alpha, 1 + \alpha$ with $\Im \alpha > 0$. By setting $\omega = \dfrac{-\alpha + i}{\alpha + i}$, the parallelogram is similar to the one having three vertices at $1 + \omega, 1 + i + \omega(1 - i)$ and $i(1 - \omega)$ for $\omega \in \mathbb{C}$ with $|\omega| < 1$. We consider a Riemann torus Σ_0 for $\omega = 0$ and another Riemann torus Σ_ω for a general ω, and the corresponding points in the Teichmüller space by x_0 and x_ω respectively.

Let $h \colon \Sigma_0 \to \Sigma_\omega$ be a homeomorphism induced from a homeomorphism $\tilde{h} \colon \mathbb{C} \to \mathbb{C}$ defined by $\tilde{h}(z) = z + \omega \bar{z}$. Then we have $h_{\bar{z}}/h_z \equiv \omega$, and hence $K(h) = \dfrac{1 + |\omega|}{1 - |\omega|}$. We can show that the map h realises the infimum of the dilatations, making use of extremal lengths which we shall deal with in the following section for instance. Therefore, we have

$$d_T(x_0, x_\omega) = \frac{1}{2} \log \frac{1 + |\omega|}{1 - |\omega|}.$$

This is exactly the same as the hyperbolic metric on the upper half plane.

We also note that this argument contains the fact that a natural affine map is a unique map realising the infimum dilatation between two rectangles, which is a prototype of Teichmüller maps.

3.3   Teichmüller Maps and Teichmüller Discs

We now turn to the fundamental work by Teichmüller.

Definition 3.3.1 Let Σ be a Riemann surface, which we assume to be closed as before. A *holomorphic quadratic differential* is a differential form on Σ expressed as $\phi(z)dz^2$ for a holomorphic function ϕ with respect to a holomorphic coordinate z.

Lifting this to the universal cover H of Σ, a holomorphic quadratic differential is regarded as a holmorphic function $\phi\colon H \to \mathbb{C}$ such that $\phi(\gamma z)(\gamma'(z))^2 = \phi(z)$ for all $\gamma \in \pi_1(\Sigma)$ and $z \in H$.

Now we shall show that there is an embedding of the Teichmüller space of a torus in a general Teichmüller space as follows.

Theorem 3.3.2 *For a holomorphic quadratic differential $\phi(z)dz^2$ on a Riemann surface Σ and a complex number ω with $|\omega| < 1$, we set $\mu(z) = \omega\dfrac{\overline{\phi(z)}}{\phi(z)}$. Then there is a quasi-conformal homeomorphism between Riemann surfaces $f_\omega\colon \Sigma \to \Sigma_\omega$ such that $\dfrac{(f_\omega)_{\bar{z}}}{(f_\omega)_z} = \mu$, almost everywhere. Its dilation is constantly $\dfrac{1}{2}\log\dfrac{1+|\omega|}{1-|\omega|}$ almost everywhere. Identifying S with Σ, the map taking ω to $(\Sigma_\omega, f_\omega)$ gives a totally geodesic embedding of the Poincaré disc $D = \{\omega \in \mathbb{C} \mid |\omega| < 1\}$ into $\mathcal{T}(S)$, which is called the Teichmüller disc directed by ϕ.*

We shall give some explanation of this theorem without proving it. Fixing the angle θ, we consider an arc ω in D expressed by $\omega(t) = e^{i\theta}\dfrac{e^{2t}-1}{e^{2t}+1}$. Then, for quasi-conformal homeomorphisms defined above, we can calculate $\dfrac{1}{2}\log K(f_{\omega(t)}) = t$. This shows that for $t \in [0, 1)$, the ray defined by $(\Sigma_{\omega(t)}, f_{\omega(t)})$ constitutes a geodesic ray with respect to the Teichmüller metric.

We now explain the geometric meaning of the definition above. A holomorphic quadratic differential ϕ on Σ defines two "measured foliations" on Σ, which are called horizontal and vertical trajectories, as follows. We set $z := \displaystyle\int \sqrt{\phi(\xi)}d\xi$, which is well defined on any simply connected chart away from the zeroes of ϕ, up to sign and a constant. Then $dz^2 = \phi(\xi)d\xi^2$ is well defined in such a chart. The differential form dz^2 defines a grid on Σ, defined by $d\Re z = 0$ and $d\Im z = 0$, which gives a Euclidean structure on the complement of the zeroes of ϕ. At a zero z_0 of ϕ of degree d, we can choose a coordinate ζ with respect to which the quadratic differential can be expressed as $\phi(\xi)d\xi^2 = \zeta^d d\zeta^2 = d(\zeta^{(d+2)/2})^2$. This gives a

Euclidean structure with a singularity of degree $-(d+2)/2$ at z_0 compatible with the Euclidean structure defined above .

The map f_ω given in Theorem 3.3.2 satisfies the equation $(f_\omega)_{\bar{z}} = \omega(f_\omega)_z$ with respect to the coordinate z as above. Therefore, f_ω should be expressed as $f_\omega(z) = z + \omega\bar{z}$. In particular when ω is a real positive number, by setting $z = x + iy$, we have $f_\omega(x+iy) = (1+\omega)x + i(1-\omega)y$. This means that f_ω corresponds to a map stretching the real direction by the factor $\dfrac{1+\omega}{1-\omega}$.

The following theorem is often referred to as Teichmüller's existence theorem. This was proved by Teichmüller [14] (see also [15] for its English translation).

Theorem 3.3.3 *For any two points* $x_1 = (\Sigma_1, f_2), x_2 = (\Sigma_2, f_2)$ *in* $\mathcal{T}(S)$, *there exists a holomorphic quadratic differential* ϕ *on* Σ_1, *which is unique up to real scalar multiplication, and* $\omega \in [0, 1)$ *such that* x_2 *coincides with* $(\Sigma_\omega, f_\omega)$ *given in Theorem 3.3.2. This map is the unique quasi-conformal map from* Σ_1 *to* Σ_2 *having the dilatation* $\dfrac{1}{2} \log \dfrac{1+\omega}{1-\omega}$.

This map $f_\omega \colon \Sigma_1 \to \Sigma_2$ with $\omega \in [0, \infty)$ is called the *Teichmüller map*. As we explained after Theorem 3.3.2, for real number $t \in [0, \infty)$, the arc $x_t = (\Sigma_{\omega(t)}, f_{\omega(t)})$ for $\omega(t) := \dfrac{e^{2t}-1}{e^{2t}+1}$ gives a geodesic ray with respect to the Teichmüller metric on $\mathcal{T}(S)$. Therefore, Theorem 3.3.3 shows that any two points in $\mathcal{T}(S)$ can be joined by a geodesic segment. We call such geodesics *Teichmüller geodesics*. Thus we have the following.

Corollary 3.3.4 *The Teichmüller space with the Teichmüller metric,* $(\mathcal{T}(S), d_T)$ *is a unique geodesic space.*

Furthermore, since the Teichmülle geodesic rays issued at $x = (\Sigma, f) \in \mathcal{T}(S)$ give a homeomorphism from the space of holomorphic quadratic differentials on Σ to the tangent space at x, we have the following.

Corollary 3.3.5 *Every non-zero tangent vector at* $x \in \mathcal{T}(S)$ *is represented by an infinitesimal Teichmüller geodesic.*

3.4 Extremal Lengths and Measured Foliations

As we explained up to the last section, the Teichmüller space S can be regarded as the space of complex/conformal structures on S. For a metric space, we can talk about lengths of curves, but as we have only a conformal structure, we cannot define lengths as in metric spaces. Extremal lengths correspond to lengths in the setting where a surface has only a conformal structure. Its definition dates back to the work of Beurling and Ahlfors, and its development can be found in Ahlfors–Beurling [1]. We adopt here a definition which can be found, for instance, in Jenkins [8].

Definition 3.4.1 Let Σ be a Riemann surface homeomorphic to S, and γ a non-contractible simple closed curve on Σ. The *extremal length* $\mathrm{Ext}_\Sigma(\gamma)$ of γ on Σ is defined to be $\sup_m \dfrac{\mathrm{length}_m(\gamma)^2}{\mathrm{Area}(\Sigma, m)}$, where m ranges over all Riemannian metrics compatible with the conformal structure on Σ.

This turns out to be equivalent to a more geometric definition as follows.

Let A be an annulus with a conformal structure. By the uniformisation theorem, A is conformally equivalent to a flat annulus A'. Then the modulus of A, denoted by modulus(A), is defined to be the height of A' divided by its circumference.

Lemma 3.4.2 *The extremal length of γ on Σ is equal to the infimum of $\dfrac{1}{\mathrm{modulus}(A)}$, where A ranges over all annuli embedded in Σ whose core curves are homotopic to γ.*

A proof can be found in [8] and [13].

The extremal length is closely related to the holomorphic quadratic differentials which we introduced in the previous section. To understand their relation, it is helpful to look at them in a broader perspective, through the space of measured foliations. We now define a measured foliation on a surface, first purely topologically.

Definition 3.4.3 Let S be a closed orientable surface. A *measured foliation* on S is a pair consisting of a codimension-one foliation $\mathcal{F}$ on S with singularities of negative degrees and a transverse measure μ with full support. Here a transverse measure is a system of positive measures equivalent to the Lebesgue measure given on all arcs transverse to the leaves of $\mathcal{F}$ which are invariant when an arc is moved along leaves.

For an isotopy class γ of non-contractible simple closed curves and a measured foliation $(\mathcal{F}, \mu)$, their intersection number $i(\gamma, (\mathcal{F}, \mu))$ is defined to be $\inf \int_{\gamma'} d\mu$, where γ' ranges over all simple closed curves in the class γ that are transverse to $\mathcal{F}$.

There are operations called *Whitehead moves* which deform measured foliations.

Definition 3.4.4 Two measured foliations $(\mathcal{F}_1, \mu_1)$ and $(\mathcal{F}_2, \mu_2)$ are said to be *Whitehead equivalent* if one is obtained from the other by performing the following operations finitely many times.

(1) Move a foliation together with its transverse measure by an ambient isotopy of S.
(2) Suppose that there is an arc a on a leaf which connects two singularities of degree d_1 and d_2. By collapsing a, we get a new measured foliation having a new singularity of degree $d_1 + d_2 + 1$ which was born out of merging the two singularities.
(3) The inverse of the preceding operation, i.e. splitting a singularity of degree d less than $-3/2$ into two singularities of degrees d_1, d_2 such that $d = d_1 + d_2 + 1$.

It is easy to see that if $(\mathcal{F}_1, \mu_1)$ and $(\mathcal{F}_2, \mu_2)$ are Whitehead equivalent, then for any simple closed curve γ, the intersection numbers $i(\gamma, (\mathcal{F}_1, \mu_1))$ and $i(\gamma, (\mathcal{F}_2, \mu_2))$ coincide.

The space of measured foliations on S is the set of Whitehead equivalence classes of measured foliations on S. We denote this set by $\mathcal{MF}(S)$, and put the weak topology with regard to the intersection numbers with non-contractible simple closed curves on S.

For a non-contractible simple closed curve c on S, we can construct a measured foliation all of whose leaves except for those containing singularities are isotopic to c, such that the total measure of an arc transverse to the non-singular leaves and intersecting each one at one point is equal to 1. In fact, such a foliation is unique up to Whitehead equivalence, and we denote it by $(\mathcal{F}(c), \mu_c)$. We can also identify a positively weighted simple closed curve wc with $(\mathcal{F}(c), w\mu_c)$. By this correspondence, we can regard the set of positively weighted simple closed curves $\mathbb{R}_+ \mathcal{S}$ as a subset of $\mathcal{MF}(S)$.

The following was proved by Thurston [18] (see Fathi–Laudenbach–Poénaru [4] for details).

Theorem 3.4.5 *The set of weighted simple closed curves $\mathbb{R}_+ \mathcal{S}$ is dense in $\mathcal{MF}(S)$. The space $\mathcal{MF}(S)$ is homeomorphic to $\mathbb{R}^{6g-6} \setminus \{\mathbf{0}\}$, where g denotes the genus of S*

Sometimes, it is more convenient to consider the projectivisation of this space, which is called the *projective foliation space*, and is denoted by $\mathcal{PMF}(S)$, obtained as a quotient of $\mathcal{MF}(S)$ by identifying scalar multiples with each other.

We now return to considering a point (Σ, f) in $\mathcal{T}(S)$. Let ϕ be a holomorphic quadratic differential on Σ. Then we can define as follows the vertical foliation and the horizonal foliation on Σ associated with ϕ, which corresponds to the horizontal and vertical directions of the singular Euclidean structure which we mentioned in the previous section. Recall from the previous section that for a simply connected chart disjoint from the zeroes of ϕ, there is a coordinate z such that $dz^2 = \phi(\xi)d\xi^2$. The measured foliation defined by $d\Im z = 0$ with its transverse $|d\Im z|$, is the horizontal foliation, and the one defined by $d\Re z = 0$ with its transverse $|d\Re z|$ is the vertical foliation. At a zero of ϕ of degree d, each foliation has a singularity of degree $-(d+2)/2$.

Conversely, for a Whitehead equivalence class of measured foliations on Σ, there is a holomorphic quadratic differential whose horizontal (resp. vertical) foliation is in that class. This fact was first proved by Jenkins [8] and Strebel [13] in the case when the foliation lies in $\mathbb{R}_+ \mathcal{S}$.

Proposition 3.4.6 (Jenkins, Strebel) *For any non-contractible simple closed curve c on Σ and a positive number w, there is a unique holomorphic quadratic differential ϕ_c on Σ whose horizontal foliation corresponds to $(\mathcal{F}_c, w\mu_c)$.*

This holomorphic quadratic differential realises the extremal length of c. To be more precise, the following holds.

Lemma 3.4.7 *We have the following equality:*

$$\mathrm{Ext}_\Sigma(c) = \mathrm{Area}(\Sigma, |\phi_c|) = \int_\Sigma |\phi_c| dx dy$$

In other words, an annulus realising the extremal length of c in Sect. 3.5 is attained by the singular Euclidean structure defined by $|\phi_c|$.

The extremal length function can be extended to the space of measured foliations as was shown by Hubbard–Masur [7] and Kerckhoff [9]. This can be done first by setting $\mathrm{Ext}_\Sigma(wc) = w^2 \mathrm{Ext}_\Sigma(c)$, and then making use of the density stated in Theorem 3.4.5. We can show the entire picture as follows.

Theorem 3.4.8 (Hubbard–Masur, Kerckhoff) *For any Whitehead equivalence class $[\mathcal{F}, \mu]$ of measured foliations on Σ, there is unique holomorphic quadratic differential $\phi_{[\mathcal{F},\mu]}$ whose horizontal measured foliation lies in the given class. The extremal length $\mathrm{Ext}_\Sigma([\mathcal{F}, \mu])$, which is a limit of $\{\mathrm{Ext}_\Sigma(w_i \gamma_i)\}$ for a sequence of weighted simple closed curves such that $[\mathcal{F}_{\gamma_i}, w_i \mu_{\gamma_i}]$ converges to $[\mathcal{F}, \mu]$ in $\mathcal{MF}(\Sigma)$, is well defined independently of the choice of sequences of weighted simple closed curves converging to $(\mathcal{F}, \mu)$. Furthermore, the following holds.*

$$\mathrm{Ext}_\Sigma([\mathcal{F}, \mu]) = \mathrm{Area}(\Sigma, |\phi_{[\mathcal{F},\mu]}|).$$

3.5   Finsler Structure of Teichmüller Space

A metric d on a manifold M is said to be a *Finsler metric* when a norm $\| \ \|_x$ is defined on the tangent space $T_x M$ at every $x \in M$, and $d(x, y)$ coincides with the infimum of $\int_\gamma \|\dot{\gamma}(t)\|_x dt$, where γ ranges over all differentiable paths connecting x and y. The norm $\| \ \|_x$ is often referred to as a *Finsler structure* associated with d.

The extremal length which we explained in the preceding section gives a Finsler structure for the Teichmüller metric. We shall here give a brief explanation of this structure. We refer the reader to Papadopoulos–Su [12] for a more complete exposition. Before showing this, we present the following important result due to Kerckhoff [9].

Theorem 3.5.1 (Kerckhoff) *Let $x_1 = (\Sigma_1, f_1)$ and $x_2 = (\Sigma_2, f_2)$ be two points in the Teichmüller space $\mathcal{T}(S)$. Then we have the following equality.*

$$d_T(x_1, x_2) = \frac{1}{2} \sup_{F \in \mathcal{MF}(S)} \log \frac{\mathrm{Ext}_{\Sigma_2}(f_2(F))}{\mathrm{Ext}_{\Sigma_1}(f_1(F))}.$$

We shall give a rough sketch of the proof following [9].

Sketch of Proof Let D be $\dfrac{1}{2} \sup\limits_{F \in \mathcal{MF}(S)} \log \dfrac{\mathrm{Ext}_{\Sigma_2}(f_2(F))}{\mathrm{Ext}_{\Sigma_1}(f_1(F))}$, and let d be $d_T(x_1, x_2)$.

Since $\dfrac{\mathrm{Ext}_{\Sigma_2}(f_2(F))}{\mathrm{Ext}_{\Sigma_1}(f_1(F))}$ is invariant under scalar multiplications on $\mathcal{F}$, we can take the sup in $\mathcal{PMF}(S)$. By Theorem 3.4.5, $\{[(\mathcal{F}_\gamma, \mu_\gamma)]\}_{\gamma \in S}$ is dense in $\mathcal{PMF}(S)$. Therefore we have $D = \dfrac{1}{2} \sup\limits_{\gamma \in S} \log \dfrac{\mathrm{Ext}_{\Sigma_2}(f_2(\gamma))}{\mathrm{Ext}_{\Sigma_1}(f_1(\gamma))}$.

We shall first show that $D \le d$. Consider an annulus A on Σ_1 with core curve isotopic to $f_1(\gamma)$. By Theorem 3.3.3, there is a quasi-conformal homeomorphism $g : \Sigma_1 \to \Sigma_2$ homotopic to $f_2 \circ f_1^{-1}$ with dilatation e^{2d}. Therefore, $g(f_1(A))$ is within the Teichmüller distance d from $f_1(A)$. It is known that the Teichmüller distance between annuli of moduli m_1, m_2 is $\dfrac{1}{2}|\log \dfrac{m_1}{m_2}|$. Take an annulus (not necessarily embedded at its boundary) A so that $1/\mathrm{modulus}(f_1(A))$ realises the extremal length of $f_1(\gamma)$ as in Lemma 3.4.7. Then $\mathrm{Ext}(f_2(\gamma)) \le 1/\mathrm{modulus}(gf_1(A))$, and hence $\dfrac{1}{2} \log \dfrac{\mathrm{Ext}_{\Sigma_2}(f_2(\gamma))}{\mathrm{Ext}_{\Sigma_1}(f_1(\gamma))} \le d$.

Next we shall show that $d \le D$. Take a Teichmuüller map from x_1 to x_2 as in Theorem 3.3.3, which corresponds to a Beltrami differential $w \dfrac{\bar\phi}{|\phi|}$. Then we have $d = \dfrac{1}{2} \log \dfrac{1 + w}{1 - w}$. Considering the horizontal foliation F_ϕ of ϕ, by Theorem 3.4.8, the right hand side is equal to $\dfrac{1}{2} \log \dfrac{\mathrm{Ext}_{\Sigma_2}(f_2(F_\phi))}{\mathrm{Ext}_{\Sigma_1}(f_1(F_\phi))}$. By Theorem 3.4.8 again, this right hand side can be expressed as a limit of the ratios of extremal lengths of simple closed curves. Thus we have $D \ge d$. $\qquad\square$

Making use of Theorem 3.5.1 a Finsler structure of the Teichmüller metric is given as follows.

Definition 3.5.2 Let x be a point in the Teichmüller space $\mathcal{T}(S)$, and v a tangent vector at $x = (\Sigma, f)$ represented by a Beltrami differential μ. Then we define the *Teichmüller norm* of v at x by

$$\|v\|_T = \sup_{\mathrm{Area}(\phi)=1} \int_\Sigma \mu\phi,$$

where ϕ ranges over the area-1 holomorphic quadratic differentials on Σ.

Theorem 3.5.3 *The Teichmüller norm gives a Finsler structure for $(\mathcal{T}(S), d_T)$.*

To prove this, we essentially use the following lemma due to Gardiner [5].

Lemma 3.5.4 *For $F \in \mathcal{MF}(S)$ and a point $x = (\Sigma, f) \in \mathcal{T}(S)$, let ϕ_F be a holomorphic quadratic differential on Σ whose horizontal foliation is isotopic to $f(F)$. We regard the extremal length of F as a function on $\mathcal{T}(S)$ and denote it by*

Ext_F. *Then for any tangent vector v at x represented by a Beltrami differential μ, we have*

$$(d\text{Ext}_F)_x(v) = -2\Re \int_\Sigma \mu\phi_F.$$

Admitting this lemma, we can prove Theorem 3.5.3 as follows.

Sketch of Proof of Theorem 3.5.3 By Corollary 3.3.5, for any non-zero tangent vector $v \in T_x\mathcal{T}(S)$, there is a Teichmüller geodesic ray $r(t) = (\Sigma_t, f_t)$ corresponding to a family of Beltrami differentials $\mu(t) = t\frac{\bar{\psi}}{|\psi|}$ such that $v = \dot{\mu}(0)$. We have only to show that $\|v\|_T = \dfrac{dd_T(x, r(t))}{dt} \Big|_{t=0}$. By Theorem 3.5.1 the right hand side can be computed as

$$\frac{d}{dt}\frac{1}{2}\log \sup_{F \in \mathcal{MF}(S)} \frac{\text{Ext}_{\Sigma_t}(F)}{\text{Ext}_{\Sigma_0}(F)}$$

$$= \frac{1}{2} \sup_{F \in \mathcal{MF}(S)} \frac{\frac{d}{dt}\text{Ext}_{\Sigma_t}(F)\,|_{t=0}}{\text{Ext}_{\Sigma_0}(F)}$$

$$= \sup_\phi \frac{\Re \int \mu\phi}{\int |\phi| dx dy}$$

$$= \sup_{\text{Area}(\phi)=1} \Re \int \mu\phi = \|v\|_T,$$

and we complete the proof.   □

In the process of the proof, we have also proved the following.

Corollary 3.5.5 *For any tangent vector $v = (\Sigma, f)$ at $x \in \mathcal{T}(S)$, we have*

$$\|v\|_T = \frac{1}{2} \sup_{F \in \mathcal{MF}(S)} \frac{d\text{Ext}_F(v)}{\text{Ext}_\Sigma(F)}.$$

3.6   Thurston's Asymmetric Metric

In this section, we shall deal with another Finsler metric which is called Thurston's asymmetric metric and was invented by Thurston in his very influential paper [16]. To define Thurston's metric, we need to regard Teichmüller space as the space of hyperbolic structures rather than conformal structures. To be more precise, the definition of Teichmüller space which we use in this section is: $\mathcal{T}(S) = \{(\Sigma, f)\}/\sim$, where Σ is an oriented hyperbolic surface, $f : S \to \Sigma$ an orientation-preserving homeomorphism, and $(\Sigma_1, f_1) \sim (\Sigma_2, f_2)$ if and only if there is an isometry from Σ_1 to Σ_2 homotopic to $f_2 \circ f_1^{-1}$.

Definition 3.6.1 Let (X, d_X) and (Y, d_Y) be metric spaces. For a homeomorphism $f \colon X \to Y$, we define

$$\mathrm{Lip}(f) = \sup_{x_1 \neq x_2} \frac{d_Y(f(x_1), f(x_2))}{d_X(x_1, x_2)},$$

allowing ∞ also as a value.

Now we can define *Thurston's (asymmetric) metric* as follows.

Definition 3.6.2 For $x_1 = (\Sigma_1, f_1)$, $x_2 = (\Sigma_2, f_2) \in \mathcal{T}(S)$, we define

$$d_{\mathrm{Th}}(x_1, x_2) = \log \inf_{g \simeq f_2 \circ f_1^{-1}} \mathrm{Lip}(g),$$

where g ranges over the homeomorphisms homotopic to $f_2 \circ f_1^{-1}$.

Lemma 3.6.3 *The function d_{Th} is an asymmetric metric on $\mathcal{T}(S)$.*

Sketch of Proof It is obvious that $d_{\mathrm{Th}}(x_1, x_1) = 0$. The triangle inequality follows from the inequality $\mathrm{Lip}(g \circ h) \leq \mathrm{Lip}(g)\mathrm{Lip}(h)$.

The proof of positivity, for which we follow [16], is a bit involved. Let L_0 be $\inf_{g \simeq f_2 \circ f_1^{-1}} \mathrm{Lip}(g)$, and suppose that $L_0 \leq 1$. Then there is a sequence of Lipschitz homeomorphisms $\{g_n \colon \Sigma_1 \to \Sigma_2\}$ homotopic to $f_2 \circ f_1^{-1}$ with $\mathrm{Lip}(g_n) \to L_0$. By Ascoli–Arzelá's theorem, passing to a subsequence, $\{g_n\}$ converges uniformly to a Lipschitz map $h \colon \Sigma_1 \to \Sigma_2$ homotopic to $f_2 \circ f_1^{-1}$ with $\mathrm{Lip}(h) = L_0$. Since it is homotopic to $f_2 \circ f_1^{-1}$ it must be surjective. We shall show that h must be an isometry.

For each point $x_0 \in \Sigma_1$, we can take $\epsilon > 0$ such that the ϵ-metric balls $D(x_0)$ and $D(h(x_0))$ centred at x_0 and $h(x_0)$ respectively are both topological discs. Since h is L_0-Lipschitz and $L_0 \leq 1$, $h(D(x_0))$ is contained in $D(h(x_0))$. If $D(h(x_0)) \setminus h(D(x_0))$ has positive measure, the area of $h(D(x_0))$ is strictly less than that of $D(h(x_0))$, which is the same as the area of $D(x_0)$. This implies that the image of h has area less than that of Σ_1, which contradicts the surjectivity of h. Therefore, $h|D(x_0)$ must be also surjective and the boundary of $D(x_0)$ is mapped to the boundary of $D(h(x_0))$. Making use of the L_0-Lipschitz property, we see that $h|\partial D(x_0)$ is a homeomorphism. Letting ϵ tend to 0 and considering polar coordinates and using the L_0-Lipschitz property again, we see that $h|D(x_0)$ is an isometry. Since x_0 is arbitrary, we see that h is an isometry, which implies that $L_0 = 1$ and $x_1 = x_2$. □

In the same way as Theorem 3.5.1 for the Teichmüller metric, we have another expression of Thurston's metric.

Theorem 3.6.4 *The following equality holds.*

$$d_{\mathrm{Th}}(x_1, x_2) = \log \sup_{s \in \mathcal{S}} \frac{\mathrm{length}_{\Sigma_2}(f_1(s))}{\mathrm{length}_{\Sigma_1}(f_2(s))},$$

where $\text{length}_\Sigma(s)$ *denotes the length of the closed geodesic (with respect to the hyperbolic metric on Σ) in the isotopy class s.*

We shall present a rough idea of how to prove Theorem 3.6.4. We set $d_K(x_1, x_2) := \log \sup_{s \in S} \dfrac{\text{length}_{\Sigma_2}(f_1(s))}{\text{length}_{\Sigma_1}(f_2(s))}$, and shall prove that $d_{\text{Th}} = d_K$. It is easy to see that $d_K(x_1, x_2) \leq d_{\text{Th}}(x_1, x_2)$, for any closed geodesic is mapped by $f_2 \circ f_1^{-1}$ to a closed curve whose geodesic length is multiplied at most by $d_{\text{Th}}(x_1, x_2)$.

To show the opposite inequality, we need to consider "ratio-maximising laminations". We first define geodesic laminations on a hyperbolic surface.

A *geodesic lamination* on a hyperbolic surface Σ is a closed subset of Σ which is a disjoint union of simple geodesics. Although a geodesic lamination depends on the hyperbolic structure on Σ, for a geodesic lamination λ on Σ_1 and a homeomorphism between two hyperbolic surfaces $f \colon \Sigma_1 \to \Sigma_2$, there is a unique geodesic lamination properly isotopic to $f(\lambda)$. In this situation, we refer to this geodesic lamination simply as $f(\lambda)$.

A disjoint union of closed geodesics is a geodesic lamination, which we call multi-curve. A geodesic lamination is called *chain recurrent* when it is a Hausdorff limit of a sequence of multi-curves.

Theorem 3.6.5 (Thurston [16]) *For any distinct two points $x_1 = (\Sigma_1, f_1)$ and $x_2 = (\Sigma_2, f_2)$ in $\mathcal{T}(S)$, there is a chain recurrent geodesic lamination $\mu(x_1, x_2)$ on Σ_1 which is stretched by the factor $e^{d_{\text{Th}}(x_1, x_2)}$. Furthermore, there is a unique chain recurrent geodesic lamination which is maximal among those with this property.*

This lamination is called the *maximal ratio-maximising lamination*. Although this is a very important result, the proof is complicated and uses notions which we have not given here. We give here just a rough idea of this proof. We refer the reader §8 of [16] for a genuine proof.

Set $L_0 = e^{d_{\text{Th}}(x_1, x_2)}$. Then there is a sequence of Lipschitz homeomorphisms $g_i \colon \Sigma_1 \to \Sigma_2$ homotopic to $f_2 \circ f_1^{-1}$ such that $\text{Lip}(g_i)$ tends to L_0. For any geodesic lamination λ on Σ_1, its image $g_i(\lambda)$ is homotopic to a geodesic lamination, and λ is stretched to $g_i(\lambda)$ at most by the factor $\text{Lip}(g_i)$. We can isotope g_i so that there is a chain recurrent geodesic lamination λ_i which is the locus where g_i stretches Σ_1 most, and the stretching factor is constant on λ_i. (This part is not straightforward. For instance, we need to show that at each point of Σ_1, there is only one direction to which the surface is stretched most.) Taking a limit of g_i, we get a Lipschitz map for which the Hausdorff limit of λ_i is the most stretched locus. The stretching factor must be equal to L_0 since the limit map must be L_0-Lipschitz. This shows that the limit lamination is ratio-maximising.

If there is more than one ratio-maximising geodesic laminations, they cannot intersect transversely. Indeed if there are two transversely intersecting laminations μ_1 and μ_2, we can show that there is a direction in which the ratio of geodesic length is greater than L_0. Therefore, there is a unique maximal ratio-maximising chain recurrent geodesic lamination.

Granting Theorem 3.6.5, we can prove Theorem 3.6.4 as follows.

Proof of Theorem 3.6.4 It remains to show that $d_{\text{Th}}(x_1, x_2) \leq d_K(x_1, x_2)$. Let $\mu(x_1, x_2)$ be the maximal ratio-maximising lamination. Since $\mu(x_1, x_2)$ is chain recurrent, there is a sequence of multi-curves μ_i which converges to $\mu(x_1, x_2)$ in the Hausdorff topology. Take a component γ_i of μ_i. Then we have $\log \lim_{i \to \infty} \dfrac{\text{length}_{\Sigma_2}(f_2(\gamma_i))}{\text{length}_{\Sigma_1}(f_1(\gamma_i))} = d_{\text{Th}}(x_1, x_2)$, and hence $d_K(x_1, x_2) = \log \sup_{s \in \mathcal{S}} \dfrac{\text{length}_{\Sigma_2}(f_2(s))}{\text{length}_{\Sigma_1}(f_1(s))} \geq d_{\text{Th}}(x_1, x_2)$. □

In contrast to the Teichmüller metric, Thurston's metric does not have uniqueness of geodesics. Still there is a special kind of geodesics which are called stretch lines. We shall explain it without formally defining it. For a point $(\Sigma, f) \in \mathcal{T}(S)$, let λ be a maximal geodesic lamination on Σ, that is, a geodesic lamination such that each component of $\Sigma \setminus \lambda$ is an ideal triangle. Then for each $K \geq 1$, we can define a stretch map $\phi_K : \Sigma \to \Sigma_K$ (with stretching locus λ) by stretching each leaf of λ by the factor K, and extend it to the entire Σ using horocyclic foliations on the ideal triangles which are the components of $\Sigma \setminus \lambda$. By defining the marking f_K to be $\phi_K \circ f$, we can regard (Σ_K, f_K) as a point in $\mathcal{T}(S)$. Now, for $r \in [0, \infty)$, the ray $\{(\Sigma_{e^r}, f_{e^r})\}$ constitutes a geodesic ray in $(\mathcal{T}(S), d_{\text{Th}})$, and is called the *stretch ray* with stretching locus λ issued at x. We call a subarc of a stretch ray a *stretch segment*, and use the term "stretching locus" also for stretch segments. Also, a stretch ray defines a tangent vector at x, which we call a *stretch vector*.

Thurston gave the following, which in particular implies that $(\mathcal{T}(S), d_{\text{Th}})$ is a geodesic space. The proof is rather long, and we refer the reader to [16].

Theorem 3.6.6 *Any two points x_1, x_2 in $\mathcal{T}(S)$ can be joined by a geodesic consisting of finitely many concatenation of stretch segments whose stretching loci contain the unique maximal ratio-maximising chain recurrent lamination $\mu(x_1, x_2)$.*

In a similar way to the Teichmüller metric, we can show that Thurston's metric is also Finsler. To describe the Finsler structure, we need to define measured laminations and the measured lamination space. A geodesic lamination which is equipped with a transverse measure (of full support) is called a *measured lamination*. For a measured lamination λ on Σ, its length $\text{length}_{\Sigma}(\lambda)$ is defined to be the integral of the product of the length element along leaves and the transverse measure over Σ. In the same way as the space of measured foliations, giving a weak topology with respect to the transverse arcs on the set of all measured laminations on S (with any fixed hyperbolic metric), we get the measured lamination space which is denoted by $\mathcal{ML}(S)$, and the set of weighted simple closed geodesics is dense in $\mathcal{ML}(S)$. We define the projective measured lamination space $\mathcal{PML}(S)$ to be the quotient space of $\mathcal{ML}(S) \setminus \{\phi\}$ obtained by identifying measured laminations obtained by multiplying positive scalars to transverse measures. For a measured lamination λ on S, we define the function $\ell_\lambda : \mathcal{T}(S) \to \mathbb{R}$ by $\ell_\lambda(\Sigma, f) = \text{length}_{\Sigma}(f(\lambda))$.

Theorem 3.6.7 *Let v be a tangent vector at a point $x = (\Sigma, f) \in \mathcal{T}(S)$. Then, the norm*

$$\|v\|_{\mathrm{Th}} := \sup_{\lambda \in \mathcal{ML}(S) \setminus \{\phi\}} d \log \ell_\lambda(v) = \sup_{\lambda \in \mathcal{ML}(S) \setminus \{\phi\}} \frac{d\ell_\lambda(v)}{\mathrm{length}_\Sigma(f(\lambda))}$$

gives a Finsler structure for Thurston's asymmetric metric.

Sketch of Proof A proof can be found in Papadopoulos–Su [12]. We shall explain its outline. What we need to show is that $d_{\mathrm{Th}}(x_1, x_2)$ is equal to $D_{\mathrm{Th}}(x_1, x_2) :=$ $\inf_\alpha \int_\alpha \|\frac{d\alpha(t)}{dt}\|_{\mathrm{Th}} dt$, where α ranges over all differentiable arcs connecting x_1 with x_2.

By the density of weighted simple closed geodesics in $\mathcal{ML}(S) \setminus \{\phi\}$, we see that

$$d_{\mathrm{Th}}(x_1, x_2) = \log \inf_{\gamma \in \mathcal{S}} \frac{\mathrm{length}_{\Sigma_2}(f_2(\gamma))}{\mathrm{length}_{\Sigma_1}(f_1(\gamma))} = \log \inf_{\lambda \in \mathcal{ML}(S) \setminus \{\phi\}} \frac{\mathrm{length}_{\Sigma_2}(f_2(\lambda))}{\mathrm{length}_{\Sigma_1}(f_1(\lambda))}.$$

Since the ratio $\dfrac{\mathrm{length}_{\Sigma_2}(f_2(\lambda))}{\mathrm{length}_{\Sigma_2}(f_2(\lambda))}$ is invariant under scalar multiplication on measured laminations, we can regard the domain as $\mathcal{PML}(S)$, which is compact. Therefore sup is attained by some measured lamination λ_{x_1, x_2}, that is, we have $d_{\mathrm{Th}}(x_1, x_2) = \log \dfrac{\mathrm{length}_{\Sigma_2}(f_2(\lambda_{x_1, x_2}))}{\mathrm{length}_{\Sigma_1}(f_1(\lambda_{x_1, x_2}))}$. It follows that for any differentiable path α connecting x_1 with x_2, we have $D_{\mathrm{Th}}(x_1, x_2) \geq \int_\alpha \|\frac{d\alpha(t)}{dt}\|_{\mathrm{Th}} dt \geq \int_\alpha d \log \ell_{\lambda_{x_1, x_2}} = \log \dfrac{\mathrm{length}_{\Sigma_2}(f_2(\lambda_{x_1, x_2}))}{\mathrm{length}_{\Sigma_1}(f_1(\lambda_{x_1, x_2}))} = d_{\mathrm{Th}}(x_1, x_2)$.

To show the opposite inequality, we need to use Theorem 3.6.6. Consider a geodesic arc α connecting x_1 with x_2. Let μ'_{x_1, x_2} be a minimal component of $\mu(x_1, x_2)$ equipped with a transverse measure. Then

$$d_{\mathrm{Th}}(x_1, x_2) = \log \frac{\mathrm{length}_{\Sigma_2}(f_2(\mu'_{x_1, x_2}))}{\mathrm{length}_{\Sigma_1}(f_1(\mu'_{x_1, x_2}))} \leq \int_\alpha d \log(\ell_{\mu'_{x_1, x_2}})$$

$$\leq \int_\alpha \|\frac{d\alpha(t)}{dt}\|_{\mathrm{Th}} dt \leq D_{\mathrm{Th}}(x_1, x_2).$$

□

The following theorem, which corresponds in a sense to Theorem 3.4.8 for the Teichmüller metric, is all the more important since it gives one of the motivations to consider the earthquake metric which we shall talk about in the following section. Recall that for $\lambda \in \mathcal{ML}(S)$, we have the function $\ell_\lambda : \mathcal{T}(S) \to \mathbb{R}$. Its derivative $(d\ell_\lambda)_x$ is a linear function defined on $T_x \mathcal{T}(S)$ for every $x \in \mathcal{T}(S)$, and hence an element in $T_x^* \mathcal{T}(S)$. Regarding λ as a variable, we have a continuous map

$(d\ell)_x \colon \mathcal{ML}(S) \to T_x^*\mathcal{T}(S)$ for every $x \in \mathcal{T}(S)$. Considering $d \log \ell$ instead of $d\ell$ the map descends to $\mathcal{PML}(S)$. In this setting, the following was proved by Thurston.

Theorem 3.6.8 ([16]) *For every* $x \in \mathcal{T}(S)$, *the map* $(d \log \ell)_x \colon \mathcal{PML}(S) \to T_x^*\mathcal{T}(S)$ *is an embedding onto the unit sphere with respect to the dual norm of* $\| \ \|_{\mathrm{Th}}$.

Sketch of Proof We shall first show that $(d \log \ell)_x(\lambda) = (d \log \ell_\lambda)_x$ has norm 1 with respect to the norm $\| \ \|_{\mathrm{Th}}^*$ dual to $\| \ \|_{\mathrm{Th}}$. Let v be a tangent vector at x with $\|v\|_{\mathrm{Th}} = 1$. By Theorem 3.6.7, we have $\sup_{\lambda \in \mathcal{PML}(S)}(d \log \ell_\lambda(v)) \le 1$, which implies that $(d \log \ell_\lambda)_x(v) \le 1$. By the definition of dual norm, this implies that $\|d \log \ell(\lambda)\|_{\mathrm{Th}}^* \le 1$.

On the other hand, let v be a unit stretch vector with stretching locus containing λ. Then $d \log \ell_\lambda(v) = 1$, which implies that $\|d \log \ell(\lambda)\|_{\mathrm{Th}}^* \ge 1$. Thus we have shown that the image of $(d \log \ell)_x$ lies on the unit sphere.

Next we shall show the injectivity of $(d \log \ell)_x$. Let λ and μ be measured laminations contained in distinct projective classes. We first consider the case when their supports $|\lambda|$ and $|\mu|$ differ. Then, by exchanging the roles of λ and μ if necessary, we can find a maximal geodesic lamination ν containing $|\lambda|$ which intersects μ transversely. Let v be a unit stretch vector with stretching locus ν. Then we have $d \log \ell_\lambda(v) = 1$ whereas $d \log \ell_\mu(v) < 1$, and hence $(d \log \ell)_x(\lambda) \ne (d \log \ell)_x(\mu)$.

Next suppose that $|\lambda| = |\mu|$, which is the case more difficult to deal with. Since λ and μ represent different projective classes in $\mathcal{PML}(S)$, there is a simple closed curve γ such that $i(\gamma, \lambda/\mathrm{length}_\Sigma(\lambda)) \ne i(\gamma, \mu/\mathrm{length}_\Sigma(\mu))$. (See [4] for the corresponding fact for measured foliations. To be more precise, we need the inequality $\int_\gamma \cos\theta d\lambda \ne \int_\gamma \cos\theta d\mu$, where θ denotes the angle formed by γ and $|\lambda| = |\mu|$ at their intersections.) Let v be a unit infinitesimal Fenchel–Nielsen twist around γ. Then by Kerckhoff's cosine formula, which we shall explain in the following section (Lemma 3.7.6), we have $\ell_\lambda(v) \ne \ell_\mu(v)$. $\qquad\square$

3.7 The Earthquake Metric

In this final section, we shall deal with another (asymmetric) Finsler metric on Teichmüller space which we call the *earthquake metric*. This notion first appeared in Thurston's paper [16] with a brief explanation, and has not been studied seriously until quite recently. Yi Huang, Athanase Papadopoulos, Huiping Pan and the present author started to study this metric comprehensively in [6]. Barbot–Fillastre also got interested in the same metric from the viewpoint of Anti-de-Sitter metrics [3]. Here we explain some of the results in [6] which shed light on the relation of this new metric and Thurston's asymmetric metric.

Before defining the metric, we explain what earthquake deformations are. The notion of earthquake is a generalisation of classical Fenchel–Nielsen deformation. Let $x = (\Sigma, f)$ be a pair consisting of a hyperbolic surface and a homeomorphism, regarded as a point in $\mathcal{T}(S)$ and γ a simple closed curve on S. The *Fenchel–Nielsen deformation* of x with respect to a weighted simple closed curve $w\gamma$ is an operation defined as follows: Cut Σ along the closed geodesic γ^* homotopic to $f(\gamma)$, and move one side at distance w relative to the other side to the left (viewed from the other side) and paste the two sides along γ^* again. This operation gives a new hyperbolic structure on Σ, the marking on Σ gives a marking on the new surface, and we can define a point $\tau_{w\gamma}(x)$ in $\mathfrak{T}(S)$, which is called the Fenchel–Nielsen deformation of x around $w\gamma$. It can be easily seen that fixing γ and w, the deformation $\tau_{w\gamma}$ is a homeomorphism from $\mathfrak{T}(S)$ to $\mathfrak{T}(S)$, which we call the Fenchel-Nielsen deformation around $w\gamma$. This operation can be extended to the measured lamination space $\mathcal{ML}(S)$.

Proposition 3.7.1 *Let λ be a measured lamination on S. By the density of weighted simple closed geodesics in $\mathcal{ML}(S)$, there is a sequence of weighted simple closed geodesics $\{w_i\gamma_i\}$ converging to λ. Then the sequence of Fenchel–Nielsen deformations $\{\tau_{w_i\gamma_i}: \mathfrak{T}(S) \to \mathfrak{T}(S)\}$ converges to a homeomorphism $E_\lambda: \mathfrak{T}(S) \to \mathfrak{T}(S)$, and the limit is independent of the choice of $\{w_i\gamma_i\}$ converging to λ.*

We call the limit map E_λ the *earthquake map* along λ. By convention, we define the earthquake map along the empty lamination to be the identity map. Concerning the earthquake maps, the following theorem by Thurston is essential.

Theorem 3.7.2 (Thurston [17], Kerckhoff [10]) *For any two points $x_1, x_2 \in \mathfrak{T}(S)$, there is a unique measured lamination λ such that $x_2 = E_\lambda(x_1)$.*

We shall next consider infinitesimal earthquake maps. For a measured lamination λ and $t \in [0, \infty)$, we consider a family of earthquake maps $\{E_{t\lambda}\}$. It was proved by Kerckhoff [11] that $E_{t\lambda}$ moves analytically (with respect to the coordinates of $\mathfrak{T}(S)$ given by the lengths of closed geodesics) with respect to t.

Definition 3.7.3 For a non-empty measured lamination λ and $x \in \mathfrak{T}(S)$, we define

$$\mathbf{e}_\lambda(x) := \frac{d}{dt}\Big|_{t=0} E_{t\lambda}(x).$$

The following is an infinitesimal version of Theorem 3.7.2.

Theorem 3.7.4 *For $x \in \mathfrak{T}(S)$ and for any tangent vector $\mathbf{v} \in T_x\mathfrak{T}(S)$, there is a unique measured lamination λ such that $\mathbf{e}_\lambda(x) = \mathbf{v}$.*

Sketch of Proof Regarding λ in $\mathbf{e}_\lambda(x)$ as a variable, we have a function $\mathbf{e}: \mathcal{ML}(S) \to T_x(\mathfrak{T}(S))$. Using the result of [11], it is easy to check this is continuous. Also the properness of this map is easy to see making use of Theorem 3.7.2. It then suffices to show the injectivity. This follows from a similar argument as the injectivity part of the proof of Theorem 3.6.8. □

Now, we can define the central notion of this section, the earthquake norm.

Definition 3.7.5 For $x = (\Sigma, f) \in \mathcal{T}(S)$ and a tangent vector $\mathbf{v} \in T_x\mathcal{T}(S)$, we define its *earthquake norm*

$$\|v\|_e := \text{length}_\Sigma(f(\lambda))$$

for $\lambda \in \mathcal{ML}(S)$ such that $\mathbf{e}_\lambda(x) = \mathbf{v}$ given in Theorem 3.7.4.

To see that $\|\cdot\|_e$ satisfies the axioms of (asymmetric) norm, we need to make use of the following due to Kerckhoff and Wolpert.

Lemma 3.7.6 ([10, 19]) *Let λ and μ be two measured laminations on S. Then*

$$(d\ell_\lambda)_x(\mathbf{e}_\mu(x)) = \int_\Sigma \cos(\theta_{\lambda,\mu})d\lambda d\mu = -(d\ell_\mu)_x(\mathbf{e}_\lambda(x)),$$

where $\theta_{\lambda,\mu}$ is the angle formed on Σ at the intersection points by the measured laminations $f(\lambda)$ and $f(\mu)$.

Now we shall show that $\|\cdot\|_e$ is a norm. The positivity is straightforward from the definition. We have only to verify the triangle inequality.

Lemma 3.7.7 *The triangle inequality holds for $\|\cdot\|_e$.*

Sketch of Proof For any non-empty measured lamination λ, $(d\log\ell)_x(\lambda)$ lies in $T_x^*\mathcal{T}(S)$. By Theorem 3.6.8, the image of $\mathcal{PML}(S)$ under $(d\log\ell)_x$ is a convex sphere, and hence there is a tangent vector $\mathbf{v} \in T_x\mathcal{T}(S)$ such that $(d\log\ell)_x(\cdot)(\mathbf{v})$ defined on $\mathcal{PML}(S)$ has a maximum at $[\lambda]$. Let μ be a measured lamination such that $\mathbf{v} = \mathbf{e}_\mu$, which is obtained from Theorem 3.7.4. By Lemma 3.7.6, we have

$$(d\log\ell)_x(\lambda)(\mathbf{e}_\mu) = \frac{(d\ell_\lambda)_x(\mathbf{e}_\mu)}{\text{length}_\Sigma(f(\lambda))} = \frac{-(d\ell_\mu)_x(\mathbf{e}_\lambda)}{\text{length}_\Sigma(f(\lambda))}$$

$$= -(d\ell)_x(\mu)\left(\frac{\mathbf{e}_\lambda}{\text{length}_\Sigma(f(\lambda))}\right).$$

Therefore, $\dfrac{\mathbf{e}_\lambda}{\text{length}_\Sigma(f(\lambda))}$ is a minimal point of $(d\ell)_x(\mu)$. It follows that the set $\{\mathbf{e}_\lambda/\text{length}_\Sigma(f(\lambda))\}_{\lambda\in\mathcal{ML}(S)\setminus\{\phi\}}$, which is the unit sphere of $\|\cdot\|_e$, is convex. Therefore, the triangle inequality holds for $\|\cdot\|_e$. $\qquad\square$

The Weil–Petersson paring gives a duality between Thurston's metric and the earthquake metric. We first explain briefly what is the Weil–Petersson pairing. Recall from Corollary 3.3.5 that for any $x = (\Sigma, f) \in \mathcal{T}(S)$, every tangent vector $\mathbf{v}$ at x corresponds to a Beltrami differential. For a Beltrami differential μ and a holomorphic quadratic differential ϕ, the integral $\int_\Sigma \mu\phi$ is well defined. Teichmüller theory which we described in this chapter from Sect. 3.2 to Sect. 3.5

implies that the space of holomorphic quadratic differentials $Q(\Sigma)$ on Σ can be identified with the cotangent space $T_x^*\mathcal{T}(S)$. An Hermitian form called the Petersson pairing in $Q(\Sigma)$ is defined to be $\langle \phi, \psi \rangle_P = \dfrac{i}{2} \displaystyle\int_\Sigma \frac{\phi\bar{\psi}}{\lambda^2} dz d\bar{z}$, where $\lambda^2 |dz|^2$ denotes the density of the hyperbolic metric on Σ. The Weil–Petersson metric is the real part of the pairing dual to this, and is defined on the tangent space $T_x\mathcal{T}(S)$. It is known that the Weil–Petersson metric is Kähler, and the Kähler form is denoted by $\omega\langle \cdot, \cdot \rangle$. Wolpert showed the following for Fenchel–Nielsen deformations. By continuity, we can show the same formula for measured laminations.

Lemma 3.7.8 ([20]) *For any point $x \in \mathcal{T}(S)$ and for any measured lamination λ, we have $\omega\langle \mathbf{e}_\lambda, \cdot \rangle = (d\ell)_x(\lambda)$.*

Then we have the following striking duality.

Theorem 3.7.9 ([6]) $(T_x\mathcal{T}(S), \| \cdot \|_e)$ *is linearly isometric to* $(T_x^*\mathcal{T}(S), \| \cdot \|_{Th})$.

Proof Consider the map $\Phi \colon T_x\mathcal{T}(S) \to T_x^*\mathcal{T}(S)$, taking $\mathbf{e}_\lambda$ to $\omega\langle \mathbf{e}_\lambda, \cdot \rangle$, which is evidently linear. By Lemma 3.7.8, the latter is equal to $(d\ell)_x(\lambda)$. Since $\|\mathbf{e}_\lambda\|_e =$ length$_\Sigma(f(\lambda)) = \|(d\ell)_x(\lambda)\|_{Th}$, the map Φ is also an isometry. $\qquad\square$

From the norm $\| \cdot \|_e$, we define the earthquake metric to be the associated Finsler metric.

Definition 3.7.10 For two points $x_1, x_2 \in \mathcal{T}(S)$, we define the *earthquake metric*

$$d_e(x_1, x_2) = \inf_\alpha \int_\alpha \left\| \frac{d\alpha(t)}{dt} \right\|_e dt,$$

where α ranges over all smooth arcs connecting x_1 with x_2.

There are several properties of the earthquake metric which we proved in [6]. As the last topic of this section, we mention one remarkable property of the earthquake metric.

Theorem 3.7.11 *The metric space* $(\mathcal{T}(S), d_e)$ *is incomplete.*

This theorem is proved by comparing the earthquake norm with the Weil–Petersson norm, which we denote by $\| \cdot \|_{WP}$. Indeed, we can prove the following:

Proposition 3.7.12 *There is a constant C depending only on S such that $\|\mathbf{v}\|_e \leq C\|\mathbf{v}\|_{WP}$ for any tangent vector $\mathbf{v}$ of $\mathcal{T}(S)$.*

Once we know that $(\mathcal{T}(S), d_e)$ is incomplete, it is natural to ask what is its completion. Since the metric is asymmetric, it is not straightforward to define its completion. Still, by an approach which is analogous to the work of Algom-Kfir [2], we can define its completion as a topological space, and it turns out that it is homeomorphic to the Weil–Petersson completion of $\mathcal{T}(S)$, that is, the augmented Teichmüller space. For further studies on the earthquake metric, we refer the reader to [6].

References

1. L. Ahlfors, A. Beurling, Conformal invariants and function-theoretic null-sets. Acta Math. **83**, 101–129 (1950)
2. Y. Algom-Kfir, The metric completion of outer space. Geom. Dedicata **204**, 191–230 (2020)
3. T. Barbot, F. Fillastre, Quasi-Fuchsian co-Minkowski manifolds, in *In the Tradition of Thurston—Geometry and Topology* (Springer, Cham, 2020), pp. 645–703
4. A. Fathi, F. Laudenbach, V. Poénaru, *Travaux de Thurston sur les surfaces*. Astérisque, vol. 66 (Société Mathématique de France, Paris, 1979). Séminaire d'Orsay
5. F.P. Gardiner, Measured foliations and the minimal norm property for quadratic differentials. Acta Math. **152**(1–2), 57–76 (1984)
6. Y. Huang, K. Ohshika, H. Pan, A. Papadopoulos, The earthquake metric on Teichmüller space. Preprint (2024)
7. J. Hubbard, H. Masur, Quadratic differentials and foliations. Acta Math. **142**(3–4), 221–274 (1979)
8. J.A. Jenkins, On the existence of certain general extremal metrics. Ann. Math. (2) **66**, 440–453 (1957)
9. S.P. Kerckhoff, The asymptotic geometry of Teichmüller space. Topology **19**(1), 23–41 (1980)
10. S.P. Kerckhoff, The Nielsen realization problem. Ann. Math. (2) **117**(2), 235–265 (1983)
11. S.P. Kerckhoff, Earthquakes are analytic. Comment. Math. Helv. **60**(1), 17–30 (1985)
12. A. Papadopoulos, W. Su, On the Finsler structure of Teichmüller's metric and Thurston's metric. Expo. Math. **33**(1), 30–47 (2015)
13. K. Strebel, *Quadratic Differentials*. Ergebnisse der Mathematik und ihrer Grenzgebiete (3), vol. 5 (Springer, Berlin, 1984)
14. O. Teichmüller, Extremale quasikonforme Abbildungen und quadratische Differentiale. Abh. Preuß. Akad. Wiss., Math.-Naturw. Kl. 1940 **22**, 197 s (1940)
15. O. Teichmüller, Extremal quasiconformal mappings and quadratic differentials (translated by Guillaume Théret), in *Handbook of Teichmüller Theory. Volume V* (European Mathematical Society, Zürich, 2016), pp. 321–483
16. W.P. Thurston, Minimal stretch maps between hyperbolic surfaces. https://arxiv.org/pdf/math/9801039.pdfV
17. W.P. Thurston, Earthquakes in two-dimensional hyperbolic geometry, in *Low-Dimensional Topology and Kleinian Groups (Coventry/Durham, 1984)*. London Math. Soc. Lecture Note Ser., vol. 112 (Cambridge University Press, Cambridge, 1986), pp. 91–112
18. W.P. Thurston, On the geometry and dynamics of diffeomorphisms of surfaces. Bull. Am. Math. Soc. (N.S.) **19**(2), 417–431 (1988)
19. S. Wolpert, On the symplectic geometry of deformations of a hyperbolic surface. Ann. Math. (2) **117**(2), 207–234 (1983)
20. S. Wolpert, On the Weil-Petersson geometry of the moduli space of curves. Am. J. Math. **107**(4), 969–997 (1985)

Chapter 4
Double Forms, Curvature Integrals and the Gauss–Bonnet Formula

Marc Troyanov

Abstract The Gauss–Bonnet Formula is a significant achievement in nineteenth century differential geometry for the case of surfaces and the twentieth century cumulative work of H. Hopf, W. Fenchel, C. B. Allendoerfer, A. Weil and S.S. Chern for higher-dimensional Riemannian manifolds. It relates the Euler characteristic of a Riemannian manifold to a curvature integral over the manifold plus a somewhat enigmatic boundary term. In this chapter, we revisit the formula using the formalism of double forms, a tool introduced by de Rham, and further developed by Kulkarni, Thorpe, and Gray. We explore the geometric nature of the boundary term and provide some examples and applications.

Keywords Gauss–Bonnet formula · Double forms · Curvature integrals

AMS Subject Classification 58A10, 53C20

4.1 Introduction

The Gauss–Bonnet Formula for Riemannian surfaces is a remarkable achievement of nineteenth century differential geometry, resulting from the cumulating work of K. F. Gauss, P. O. Bonnet, J.P. Binet, and W. von Dyck. The result is well-covered in a number of textbooks on Riemannian geometry.

Its extension to higher-dimensions is the result of the combined work of H. Hopf, W. Fenchel, C. B. Allendoerfer, A. Weil, and S. S. Chern, spanning the period from 1925 to 1945. The formula relates the Euler characteristic of a compact Riemannian manifold with a curvature integral over the manifold plus a boundary term involving the curvature of the manifold and the second fundamental form over the boundary.

M. Troyanov (✉)
Institut de Mathématiques EPFL, Lausanne, Switzerland
e-mail: marc.troyanov@epfl.ch

In his formulation of the Gauss–Bonnet Formula, Chern uses Élie Cartan's moving frame method. His proof is very direct but it does not readily reveal the geometric nature of the boundary term. In this chapter, we revisit and reframe the higher-dimensional Gauss–Bonnet Formula using double forms, a formalism introduced by G. de Rham and further developed by R. Kulkarni, J. A. Thorpe, and A. Gray in the 1960s and 1970s. Double forms offer a powerful tool for manipulating curvature identities and provide us with a different point of view on the Gaus-Bonnet Formula.

Our objective in this chapter is then to present the Gauss–Bonnet Formula within the framework of double forms. We provide details for Chern's original proof and give some examples and applications.

While we do provide some historical perspective, this chapter is not meant to be a historical essay. Further historical remarks and developments can be found in other sources such as [13, 41], as well as Volumes 3 and 5 of M. Spivak's treatise [36]. See also the interviews [4, 23].

Due to space and expertise limitations, we do not explore alternative proofs (such as based on heat kernel, probabilistic or PDEs methods) or broader topics such as the theory of characteristic classes, the Hirzebruch signature theorem and more generally the Atiyah–Singer index theorem. Applications to physics are also not discussed.

4.1.1 Organization of the Chapter

The rest of the chapter is organized as follows: In Sect. 4.3 we discuss the origin of the problem behind the Gauss–Bonnet Formula, beginning with Hopf's famous "Curvatura Integra" problem. Section 4.4 introduces the notion of double forms on a manifold. These are tensors that provide a convenient algebraic framework for manipulating various geometric quantities in Riemannian geometry.

Section 4.5 is dedicated to a fully detailed proof of the higher-dimensional Gauss–Bonnet Formula for compact, even-dimensional Riemannian manifolds with boundaries, following Chern's arguments. We also reinterpret the boundary term using double forms; the formula is stated in three different formulations in (4.4.16). Some examples and applications are discussed in Sect. 4.6.

Section 4.7 contains various remarks and comments, and a brief summary on the notion of double factorial that is often used throughout the paper. Some exercises are suggested in Sect. 4.7; the reader is advised to take a look at them at some point. The chapter concludes with an appendix that offers the necessary background in Riemannian geometry, including a detailed introduction to Cartan's moving frame methods.

As for prerequisites, we assume the reader is familiar with basic Differential and Riemannian geometry at the graduate level, including tensor fields and differential forms. Although we have tried to write a clean exposé, the reader should be aware that a number of algebraic calculations are quite intricate, and some of the details are left to the reader.

Let us conclude this introduction with a word about our notation. The manifolds we work with are assumed to be smooth and oriented. Typically, they are denoted as M or N, where $m = \dim(M)$ and $n = \dim(N)$. Throughout most of the discussion, m will be presumed to be an even number, while n may be either odd or even. The boundary ∂M of the manifold M, if non empty, will be given the orientation induced by the outer normal.

4.2 An Overview of Hopf's Problem on the Curvatura Integra

The 2-dimensional Gauss–Bonnet Formula, for a compact Riemannian surface (S, g) with piecewise smooth boundary, is conventionally stated as follows:

$$2\pi \chi(S) = \int_S K dA + \int_{\partial S} k_g ds + \sum_j \alpha_j,$$

where K represents the curvature, k_g denotes the geodesic curvature of its boundary, α_j refers to the exterior angles on the boundary and $\chi(S)$ is the Euler characteristic of the surface.

This formula has been extended to the case of higher-dimensional manifolds in a series of papers published between 1925 and 1945 by H. Hopf, W. Fenchel, C. Allendoerfer, A. Weil and S. S. Chern, as we now discuss. The first and fundamental breakthrough is due to Hopf, who proved in 1925 that for a closed hypersurface M in Euclidean space $\mathbb{R}^{m+1}$ the following holds if m is even:[1]

$$\frac{1}{2} \text{Vol}(S^m)\chi(M) = \int_M K \, d\text{vol}_M . \qquad (4.2.1)$$

Here K is the *Gauss–Kronecker curvature* of the hypersurface, that is, the product of its principal curvatures. We shall call Eq. (4.2.1) the *Gauss–Bonnet–Hopf* Formula; its right hand side is the *total curvature*, or *curvatura integra*, of the hypersurface.

To prove the Formula, Hopf generalized a result of H. Poincaré that relates the winding number of a vector field on the boundary of a planar domain to the indices of its singularities. We recall the definition: given a continuous vector field X on a smooth manifold M having an isolated zero at $p \in M$, and choosing a local coordinate system in the neighborhood U of p, one may identify X with a mapping

---

[1] No such formula can hold if m is odd, since in this case we always have $\chi(M) = 0$. See however Corollary 4.5.2.

$X : U \to \mathbb{R}^m$; for $\varepsilon > 0$ small enough, the following map is well defined:

$$\zeta : \partial B(p, \varepsilon) \to S^{m-1}, \quad \zeta(q) = \frac{X_q}{\|X_q\|},$$

where $B(p, \varepsilon) \subset M$ is the ball around p with respect to some Riemannian metric on M. The *index of X* at the point p is defined to be the degree of that map:

$$\text{Ind}(X, p) = \deg\left(\zeta|_{\partial B(p,\varepsilon)}\right).$$

The index depends only on the vector field and not on the chosen coordinate system or the auxiliary Riemannian metric. With this in mind, we state the Poincaré–Hopf Theorem:

Theorem 4.2.1 (Poincaré–Hopf) *Let M be a smooth compact manifold with (possibly empty) boundary. Let X be a vector field on M that is transverse to the boundary and outward pointing. If X has isolated zeroes at $x_1, \ldots, x_k \in M \setminus \partial M$, then the following equality holds:*

$$\sum_{i=1}^{k} \text{Ind}(X, x_i) = \chi(M),$$

where $\text{Ind}(X, x_i)$ represents the index of X at the zero x_i, and $\chi(M)$ denotes the Euler characteristic of M.

Hopf proved this result in [20, 22], mentioning Poincaré for the 2-dimensional case and earlier partial results by Brouwer and Hadamard in higher-dimensions. A very readable proof can be found in [30, page 35]. See also references [5, 17, 28] for the case of manifolds without boundary.

Corollary 4.2.2 *Let D be a smooth compact domain in $\mathbb{R}^m$. Denote by $\nu : \partial D \to S^{m-1}$ the exterior pointing Gauss map of the boundary. The degree of that map is equal to the Euler characteristic of the domain D, that is,*

$$\deg(\nu) = \chi(D).$$

We will give later a more general version of this result, see Corollary 4.5.2.

Proof Let us choose a vector field X on D with isolated zeroes at $x_1, \ldots, x_k \in D \setminus \partial D$ and such that X is outward pointing at the boundary. Let us set $D_\varepsilon = D \setminus \cup_{i=1}^{k} B(x_i, \varepsilon) \subset M$ for some $\varepsilon > 0$ small enough. We denote by ω the volume form on S^{m-1} and by $f : D_\varepsilon \to S^{m-1}$ the map defined by $f(q) = X_q / \|X_q\|$. Since ω is a closed form, we have

$$0 = \int_{D_\varepsilon} f^*(d\omega) = \int_{\partial D_\varepsilon} f^*(\omega) = \int_{\partial D} f^*(\omega) - \sum_{i=1}^{k} \int_{\partial B(x_i, \varepsilon)} f^*(\omega).$$

To complete the proof, observe that

$$\int_{\partial D} f^*(\omega) = \text{Vol}(S^{m-1}) \deg(f|_{\partial D}) \quad \text{and} \quad \sum_{i=1}^{k} \int_{\partial B(x_i, \varepsilon)} f^*(\omega)$$

$$= \text{Vol}(S^{m-1}) \sum_{i=1}^{k} \text{Ind}(X, x_i).$$

Moreover $\deg(f|_{\partial D}) = \deg(\nu)$ because $f|_{\partial D}$ is homotopic to the Gauss map. □

We are now in position to prove Hopf's formula (4.2.1). Recall first that, by the Jordan–Brouwer separation Theorem, any closed smooth hypersurface $M \subset \mathbb{R}^{m+1}$ is the boundary of a compact smooth domain with $D \subset \mathbb{R}^{m+1}$, see e.g. [17, page 89]. By definition, the Gauss–Kronecker curvature K of M is the Jacobian of the Gauss map $\nu : M \to S^m$, i.e. we have

$$K \, \text{dvol}_M = \nu^*(\omega),$$

where $\omega = \text{dvol}_{S^n}$ is the volume form of S^m. Using the previous corollary, we therefore have

$$\int_M K \, \text{dvol}_M = \deg(\nu) \int_{S^m} \omega = \chi(D) \text{Vol}(S^m),$$

and (4.2.1) follows now from the relation

$$\chi(M) = \chi(\partial D) = \frac{1}{2} \chi(D).$$

□

In a second paper [21], also published in 1925, Hopf established the validity of (4.2.1) for compact space forms, that is, Riemannian manifolds with constant sectional curvature. A proof will also be given later in Sect. 4.5.3. Motivated by these results, he asked for an intrinsic version of that result. Let us quote from Allendoerfer's paper [1]:

H. Hopf has repeatedly pointed to the problem of deciding whether one can define a curvature scalar in an arbitrary Riemannian manifold of even dimension with the following properties:

o In the special case of a hypersurface it becomes identical with the Gauss–Kronecker curvature mentioned above.
o It satisfies a formula of Gauss–Bonnet type, i.e.. its integral over a region of the manifold can be expressed by a curvature integral over the boundary.
o The integral of this curvature scalar is a topological invariant [in the case of a closed manifold].

In short, Hopf is asking for the definition of a locally defined, scalar function on the manifold, depending only on the metric tensor and whose integral is the Euler characteristic (plus a boundary term if $\partial M \neq \emptyset$). We will call this the *Hopf Problem on Curvature Integra*. A first observation is that, using Gauss' Equation relating the second fundamental form of a hypersurface M in some Euclidean space to its curvature tensor:

$$R_{ijkl} = h_{ik}h_{jl} - h_{il}h_{jk},$$

one can express the Gauss–Kronecker curvature $K(p)$ at any point p in M as follows

$$K = \frac{1}{m!!} \sum_{\sigma \in S_m} \sum_{\tau \in S_m} (-1)^\sigma (-1)^\tau R(e_{\sigma_1}, e_{\sigma_2}, e_{\tau_1}, e_{\tau_2}) \cdots R(e_{\sigma_{m-1}}, e_{\sigma_m}, e_{\tau_{m-1}}, e_{\tau_m}),$$

$$(4.2.2)$$

where $\{e_1, \ldots, e_m\}$ is an orthonormal basis of the tangent space of M at the point p. Here S_m is the permutation group on m symbols and $(-1)^\sigma$ is the signature of $\sigma \in S_m$. In the above formula, we have used the double factorial, which for an even number m is defined as

$$m!! = m \cdot (m-2) \cdots 2 = 2^{\frac{m}{2}} \left(\frac{m}{2}\right)!$$

we refer to Sect. 4.6.3 for more on this notion. A proof of (4.2.2) can be found on page 228 of the book [37] by Thorpe, but another proof will be given in Proposition 4.3.5 below. It is obvious from (4.2.2) that the Gauss–Kronecker curvature is intrinsic.

The Hopf problem stated above has been positively resolved in two papers, [1] and [12], published independently in 1940, by C. B. Allendoerfer in the U.S.A, and W. Fenchel in Denmark. They proved that (4.2.2) holds for any closed even-dimensional manifold M that can be isometrically embedded in some Euclidean space $\mathbb{R}^d$. Note that the embeddability condition is in fact not restrictive as was proved 16 years later by Nash in [31].

In both proofs, the authors consider the boundary of the ε-tubular neighborhood of M, with $\varepsilon > 0$ small enough; let us denote it by N, and observe that it is a hypersurface in $\mathbb{R}^d$. Moreover, the natural projection $\pi : N \to M$ is a fiber bundle with typical fiber the sphere S^{d-1-m}.

The scalar K_p, defined at any point p of M by integrating the Gauss–Kronecker curvature of the hypersurface N over the ε-sphere $\pi^{-1}(p) \subset N$, is the generalized curvature function answering Hopf's Problem. This is verified by applying (4.2.1) to the hypersurface N and computing. Fenchel mentions that R. Lipschitz and W. Killing already proved in the 1880s that K_p depends solely on the Riemannian metric of M and not on its embedding in Euclidean space. Alternatively, Allendoerfer used calculations from H. Weyl's 1938 paper on the tube formula [39] to establish that (4.2.2) still holds for the averaged Gauss–Kronecker curvature as defined.

In short, the Hopf Problem for a closed submanifold $M \subset \mathbb{R}^d$ is resolved by applying the Gauss–Bonnet–Hopf Formula (4.2.1) to a tubular neighborhood N of the manifold, then integrating the Gauss–Kronecker curvature over the fibers of the fibration $\pi : N \to M$, and proving that the resulting function is an intrinsic quantity attached to the Riemannian manifold.

The two aforementioned papers did not provide a complete solution to the Hopf Problem as stated earlier, since the second requirement explicitly called for a version of the Gauss–Bonnet formula applicable to manifolds with boundary. Also the condition that the manifold should be isometrically embedded in some Euclidean space was seen as a serious restriction (Nash's Theorem not being available at that time). The joint paper [2] published in 1943 by Carl B. Allendoerfer and André Weil addresses both issues with the following method: They consider a sufficiently fine triangulation of the manifold and use a result proved by E. Cartan in 1927 that guarantees the possibility to *locally* embed a Riemannian manifold isometrically into some $\mathbb{R}^d$, provided it is analytic. Using Cartan's Theorem, one can isometrically embed each cell of the triangulation in some Euclidean space $\mathbb{R}^d$, then consider a tubular neighborhood of that cell and repeat the previous construction. The result is a Gauss–Bonnet Formula, first for piecewise analytic Riemannian polyhedra, and then, invoking an approximability result by Whitney, for general Riemannian polyhedra of class C^2.

The result achieved by Allendoerfer and Weil indeed solved Hopf's problem, but it was not entirely satisfactory. The proof involved a high level of complexity and, furthermore, it relied on local embeddings in Euclidean space, which is clearly not the most natural approach for constructing an intrinsically defined generalized curvature function on a given Riemannian manifold.

According to the story, as told say in [41], Chern visited the Institute for Advanced Study in Princeton in August 1943. He was 31 years old. There, he met André Weil, who discussed with him the Hopf problem and the need for an intrinsic proof. Within 2 weeks, Chern solved the problem using the moving frame method that had been recently introduced by Élie Cartan, and some innovative ideas of his own. He published his proof in two papers in the Annals of Mathematics in 1944 and 1945. The first paper, titled *A Simple Intrinsic Proof of the Gauss–Bonnet Formula for Closed Riemannian Manifolds*, contains a short intrinsic proof of the Gauss–Bonnet Formula for closed even-dimensional Riemannian manifolds. The second paper, titled *On the Curvatura Integra in a Riemannian Manifold*, extends the investigation to manifolds of either odd or even dimension with boundary. The 1945 paper also gives additional formulas expressing some characteristic classes of sphere bundles using curvature integrals. Chern's work laid the foundation for the development of Chern–Weil theory, which relates characteristic classes to differential forms.

4.3   Double Forms and Their Geometric Applications

The notion of *double form* on a manifold N has been introduced by G. de Rham who defined them as differential forms on a manifold N with values in the Grassmann algebra of TN, see [11, §7]. Double forms have been used in the 1960s and 70s as a convenient tool in the computation of various curvature formulae in Riemannian geometry by R. Kulkarni, J. Thorpe and A. Gray. In this section we introduce the algebra of double forms following the book [14] by Gray and we refer to [26] for further developments.

4.3.1   Definition and Examples

(i) Let N be a smooth n-dimensional manifold. A *double-form* A of type (k, l) on N is a smooth covariant tensor field of degree $k + l$, which is alternating in the first k variables and also in the last l variables. This condition can be expressed as follows: If $X_1, \ldots, X_k, Y_1, \ldots, Y_l$ are $(k + l)$ vector fields on N and $\sigma, \tau \in S_k$, then

$$A(X_1, \ldots, X_k; Y_1, \ldots, Y_l) = (-1)^{\sigma}(-1)^{\tau} A(X_{\sigma_1}, \ldots, X_{\sigma_k}; Y_{\tau_1}, \ldots, Y_{\tau_l}).$$

We denote by $\mathcal{D}^{k,l}(N)$ the set of all (k, l)-double forms on N. This is a module over the algebra of smooth functions $C^{\infty}(N)$ that can be expressed as the tensor product

$$\mathcal{D}^{k,l}(N) = \mathcal{A}^k(N) \otimes_{C^{\infty}(N)} \mathcal{A}^l(N),$$

where $\mathcal{A}^k(N)$ is the usual space of differential k-forms on N. The direct sum of these spaces is denoted as

$$\mathcal{D}(N) = \bigoplus_{k,l=0}^{n} \mathcal{D}^{k,l}(N).$$

(ii) A natural product, denoted by $\oslash$, is defined on $\mathcal{D}(N)$ as follows: If $\psi_1 = \alpha_1 \otimes \beta_1 \in \mathcal{D}^{k,l}(N)$ and $\psi_2 = \alpha_2 \otimes \beta_2 \in \mathcal{D}^{p,q}(N)$ then $\psi_1 \oslash \psi_2$ is the following double-form of type $(k + p, l + q)$:

$$\psi_1 \oslash \psi_2 = (\alpha_1 \wedge \alpha_2) \otimes (\beta_1 \wedge \beta_2) \in \mathcal{D}^{k+p,l+q}(N).$$

The product is then extended to general elements of $\mathcal{D}(N)$ by linearity.; it can also be defined by the formula:

$$(\psi_1 \oslash \psi_2)(X_1, \ldots, X_{k+p}; Y_1, \ldots, Y_{l+q})$$

$$= \frac{1}{k!\, l!\, p!\, q!} \sum_{\sigma \in S_{k+p}} \sum_{\tau \in S_{l+q}} (-1)^\sigma (-1)^\tau \psi_1(X_{\sigma_1}, \ldots, X_{\sigma_k}; Y_{\tau_1}, \ldots, Y_{\tau_l})$$

$$\cdot \psi_2(X_{\sigma_{k+1}}, \ldots, X_{\sigma_{k+p}}; Y_{\tau_{l+1}}, \ldots, Y_{\tau_{l+q}})$$

We shall call this operation the *double wedge product* of ψ_1 and ψ_2.

(iii) The pair $(\mathcal{D}(N), \oslash)$ is the *algebra of double-forms* on the manifold N, it is a supercommutative bigraded algebra, meaning that if $\psi_1 \in \mathcal{D}^{p,q}(N)$ and $\psi_2 \in \mathcal{D}^{r,s}(N)$, then $\psi_1 \oslash \psi_2 \in \mathcal{D}^{k+r,l+s}(N)$ and

$$\psi_1 \oslash \psi_2 = (-1)^{pr+qs} \psi_2 \oslash \psi_1.$$

Remark 4.3.1 The product in the algebra $\mathcal{D}(N)$ is simply denoted as $\psi_1 \cdot \psi_2$ by various authors such as in [25] and [26], and it is denoted by $\psi_1 \wedge \psi_2$ in the book [14]. The notation $\oslash$ is used in [3] for symmetric double forms of type $(1, 1)$. We find it convenient to extend this notation to the whole algebra of double forms.

Examples

(a) The double forms of type $(0, 0)$ are the smooth functions on N, that is, $\mathcal{D}^{0,0}(N) = C^\infty(N)$.

(b) To any differential k-form $\alpha \in \mathcal{A}^k(N)$ we can associate the double forms $1 \oslash \alpha \in \mathcal{D}^{0,1}(N)$ and $\alpha \oslash 1 \in \mathcal{D}^{1,1}(N)$. This gives two natural embeddings of the exterior algebra into the algebra of double forms.

(c) The double forms of type $(1, 1)$ are the covariant tensor fields of order 2 on M. In particular the metric tensor g of a Riemannian manifold is an element in $\mathcal{D}^{1,1}(N)$.

(d) If N is a (cooriented) hypersurface in a Riemannian manifold (M, g), then the second fundamental form h of $N \subset M$ is another important double forms of type $(1, 1)$.

(e) The curvature tensor R of a Riemannian manifold is a double form of type $(2, 2)$. It is furthermore symmetric in the following sense:

$$R(X_1, X_2 : Y_1, Y_2) = R(Y_1, Y_2; X_1, X_2).$$

(f) The double wedge product of two symmetric $(0, 2)$ tensors fields $h', h'' \in \mathcal{D}^{1,1}(N)$ is usually called their *Kulkarni–Nomizu product*. It is explicitly given by

$$h_1 \otimes h_2(X_1, X_2; Y_1, Y_2) = h_1(X_1, Y_1)h_2(X_2, Y_2) - h_1(X_2, Y_1)h_2(X_1, Y_2)$$
$$+ h_1(X_2, Y_2)h_2(X_1, Y_1) - h_1(X_1, Y_2)h_2(X_2, Y_1).$$

(g) A covariant tensor field S of degree 3 can be seen as an element in $\mathcal{D}^{2,1}(N)$ if and only if it is skew symmetric in its first two variables: $S(X, Y; Z) + S(Y, X; Z) = 0$.

(h) Given $S \in \mathcal{D}^{2,1}(N)$ and a 1-form α seen as an element in $\mathcal{D}^{0,1}(N)$, the product $S \otimes \alpha \in \mathcal{D}^{2,2}(N)$ is given by

$$S \otimes \alpha(X, Y; Z, W) = S(X, Y; Z)\alpha(W) - S(X, Y; W)\alpha(Z).$$

(i) A smooth field $A \in \text{End}(\Lambda^k TN)$ of endomorphisms of the exterior power $\Lambda^k TN$ can be seen as the double form in $\mathcal{D}^{k,k}(N)$ defined by

$$(X_1, \ldots X_k; Y_1, \ldots Y_k) \mapsto \langle X_1 \wedge X_2, \wedge \cdots \wedge X_k, A(Y_1 \wedge Y_2, \wedge \cdots \wedge Y_k) \rangle,$$

where we have used the natural scalar product $\langle \, , \, \rangle$ on $\Lambda^k TN$ induced by g.

4.3.2 Transposition and Symmetric Double Forms

The *transpose* of a double form $\psi \in \mathcal{D}^{k,l}(N)$ is the double form $\psi^\top \in \mathcal{D}^{l,k}(N)$ defined by

$$\psi^\top(X_1, \ldots, X_l; Y_1, \ldots, Y_k) = \psi(Y_1, \ldots, Y_k; X_1, \ldots, X_l).$$

We trivially have $(\psi^\top)^\top = \psi$, hence transposition defines an involution of the algebra of double forms Furthermore we have

$$(\psi_1 \otimes \psi_2)^\top = \psi_1^\top \otimes \psi_2^\top.$$

A double form $\psi \in \mathcal{D}^{k,k}(N)$ is termed *symmetric* if $\psi^\top = \psi$. Note that the product of two symmetric double forms is itself a symmetric double form.

The symmetric elements of the double algebra $\mathcal{D}(N)$ form a commutative subalgebra, denoted by Kulkarni as $\mathcal{C}$. He refers to it as the *Ring of Curvature Structures*. Moreover, he considers sseveral subalgebras of $\mathcal{C}$, characterized by the Bianchi identities.

4.3.3   The Contraction Operator

Given a Riemannian metric g on the manifold N, and $1 \le k, l \le n = \dim(N)$, we inductively define the *contraction operator of order p*:

$$C^p : \mathcal{D}^{k,l}(N) \to \mathcal{D}^{k-p,l-p}(N)$$

for any integer p such that $0 \le p \le \min(k, l)$ as follows. For $\psi \in \mathcal{D}^{k,l}(N)$ we set $C^0(\psi) = \psi$ and at any point $x \in N$ we set

$$C^p(\psi)(X_1, \ldots, X_{k-p}; Y_1, \ldots, Y_{l-p})$$

$$= \sum_{i=1}^n C^{p-1}(\psi)(X_1, \ldots, X_{k-p}, e_i; Y_1, \ldots, Y_{l-p}, e_i),$$

for $1 \le p \le \min\{l, k\}$, where $e_1, \ldots, e_n \in T_x M$ is an orthonormal basis and $X_i, Y_j \in T_x M$ are arbitrary tangent vectors. This operation is independent of the chosen orthonormal basis.

As an example, the first and second contractions of the curvature tensor R of a Riemannian manifold are respectively its Ricci and scalar curvatures.

If $\psi \in \mathcal{D}^{p,p}(N)$, its *full contraction* is defined to be $C^p(\psi)$. This is a scalar quantity (i.e. an element of $C^\infty(M)$) and the following result is useful in computations:

Proposition 4.3.2 *Let $\{e_1, \ldots, e_n\}$ be a positively oriented orthonormal moving frame defined in a domain U of an oriented n-dimensional Riemannian manifold (N, g), and let $\{\theta^1, \ldots \theta^n\}$ be the dual coframe. Then the following holds: If $\mu_1, \ldots, \mu_p$ and $\nu_1, \ldots, \nu_p$ are two lists of pairwise distinct indices in $\{1, \ldots, m\}$, then*

$$\frac{1}{p!} \, C^p \left(\theta^{\mu_1} \wedge \cdots \wedge \theta^{\mu_p} \otimes \theta^{\nu_1} \wedge \cdots \wedge \theta^{\nu_p} \right) = \delta^{\nu_1, \ldots, \nu_p}_{\mu_1, \ldots, \mu_p}$$

is equal to $+1$ or -1, depending on whether $\nu_1, \ldots, \nu_p$ is an odd or even permutation of $\mu_1, \ldots, \mu_p$, and equal to zero otherwise.

Proof By definition we have

$$C^p \left(\theta^{\mu_1} \wedge \cdots \wedge \theta^{\mu_p} \otimes \theta^{\nu_1} \wedge \cdots \wedge \theta^{\nu_p} \right)$$

$$= \sum_{i_1, \ldots, i_m = 1}^m \left(\theta^{\mu_1} \wedge \cdots \wedge \theta^{\mu_p}(e_{i_1}, \ldots, e_{i_p}) \right) \cdot \left(\theta^{\nu_1} \wedge \cdots \wedge \theta^{\nu_p}(e_{i_1}, \ldots, e_{i_p}) \right)$$

$$= \sum_{i_1,\ldots,i_m=1}^{m} \det\left(a_k^j\right) \cdot \det\left(b_k^j\right)$$

$$= p!\, \delta_{\mu_1,\ldots,\mu_p}^{\nu_1,\ldots,\nu_p},$$

where we have used the notation

$$a_k^j = \left(\theta^{\mu_j}(e_{i_k})\right) \quad \text{and} \quad b_k^j = \left(\theta^{\nu_j}(e_{i_k})\right).$$

□

Corollary 4.3.3 *If $\alpha, \beta \in \mathcal{A}^p(N)$, then*

$$\langle \alpha, \beta \rangle = \frac{1}{p!}\, C^p\left(\alpha \otimes \beta\right),$$

where $\langle\ ,\ \rangle$ is the natural scalar product defined on the Grassmann algebra (see (A.1)).

Proof This follows immediately from the previous proposition and the fact that at any point x of the manifold, the family $\{\theta^{i_1} \wedge \cdots \wedge \theta^{i_p}\}_{i_1 < \cdots < i_p}$ is orthonormal in $\Lambda^p T_x^* N$.

□

In the next result we gather several useful contraction formulas:

Proposition 4.3.4 *With the same notation as above, we have*

(1) Any $\alpha \in \mathcal{A}^p(N)$ can be written as

$$\alpha = \frac{1}{p!} \sum_{i_1 < \cdots < i_p = 1}^{n} C^n\left(\alpha \otimes \theta^{i_1} \wedge \cdots \wedge \theta^{i_p}\right) \theta^{i_1} \wedge \cdots \wedge \theta^{i_p}.$$

(2) If $\alpha \in \mathcal{A}^n(N)$ is a differential form of degree n, then for any permutation $\sigma \in S_n$ we have

$$n!\, \alpha = (-1)^\sigma\, C^n\left(\alpha \otimes \theta^{\sigma_1} \wedge \cdots \wedge \theta^{\sigma_m}\right) \mathrm{dvol}_{N,g}.$$

(3) For any $\psi \in \mathcal{D}^{p,p}(N)$ we have

$$C^{p+q}\left(\psi \oslash g^q\right) = \prod_{j=1}^{q}(p+j)(n-p+1-j)\, C^p(\psi)$$

$$= (q!)^2 \binom{p+q}{q}\binom{n-p}{q} C^p(\psi)$$

for any integer $1 \le q \le (n-p)$.

In particular, we have

$$C^q(g^q) = \frac{n!q!}{(n-q)!} = \binom{n}{q}(q!)^2. \tag{4.3.1}$$

Proof The first formula follows immediately from the corollary and the expansion

$$\alpha = \sum_{i_1 < \cdots < i_p = 1}^{n} \left\langle \alpha, \theta^{i_1} \wedge \cdots \wedge \theta^{i_p} \right\rangle \theta^{i_1} \wedge \cdots \wedge \theta^{i_p},$$

and the second identity is a special case of the first.

We now prove the third formula for $q = 1$. Using the first identity, it suffices to consider the tensor $\psi = \theta^{\mu_1} \wedge \cdots \wedge \theta^{\mu_p} \otimes \theta^{\mu_1} \wedge \cdots \wedge \theta^{\mu_p}$ where the μ_is are pairwise distinct. We then have

$$C^{p+1}(\psi \oslash g) = \sum_{\nu=1}^{n} C^{p+1}\left(\theta^{\mu_1} \wedge \cdots \wedge \theta^{\mu_p} \wedge \theta^{\nu} \otimes \theta^{\mu_1} \wedge \cdots \wedge \theta^{\mu_p} \wedge \theta^{\nu}\right)$$

$$= (n-p)(p+1)!$$

because there are exactly $(n - p)$ indices $\nu \in \{1, \ldots, n\} \setminus \{\mu_1, \ldots, \mu_p\}$. Since $C^p(\psi) = p!$, we have proved that

$$C^{p+1}(\psi \oslash g) = (n-p)(p+1)C^p(\psi).$$

The general case $q \geq 2$ easily follows by induction.

$\square$

We can now prove that the Gauss–Kronecker Curvature of a hypersurface in $\mathbb{R}^{m+1}$ is an intrinsic invariant of the underlying Riemannian manifold if m is even. This observation can be seen as a generalization of Gauss' Theorema Egregium.

Proposition 4.3.5 *The Gauss–Kronecker curvature of a hypersurface M of even dimension m in Euclidean space $\mathbb{R}^{m+1}$ can be written in terms of its Riemann curvature tensor as follows:*

$$K = \frac{2^{\frac{m}{2}}}{(m!)^2} C^m(R^{\frac{m}{2}}) = \frac{1}{m!!} \sum_{\sigma \in S_m} \sum_{\tau \in S_m} (-1)^\sigma (-1)^\tau R_{\sigma_1 \sigma_2 \tau_1 \tau_2} \cdots R_{\sigma_{m-1} \sigma_m \tau_{m-1} \tau_m}.$$

$$(4.3.2)$$

In this formula, $R_{ijkl} = R(e_i, e_j; e_k, e_l)$, where $e_1, \ldots, e_m$ is a given orthonormal frame. If the basis is not orthonormal, the right hand side must be divided by $\det(g_{ij})$.

Proof From Lemma A.2 we know that

$$R = \frac{1}{4} \sum_{i,j,k,l=1}^{m} R_{ijkl}\, \theta^i \wedge \theta^j \otimes \theta^k \wedge \theta^l,$$

therefore

$$R^{\frac{m}{2}} = \frac{1}{4^{\frac{m}{2}}} \sum_{\mu_1,\ldots,\mu_n=1}^{m} \sum_{\nu_1,\ldots,\nu_n=1}^{m} R_{\mu_1\mu_2\nu_1\nu_2} \cdots R_{\mu_{m-1}\mu_m\nu_{m-1}\nu_m}\, \theta^{\mu_1} \wedge$$

$$\cdots \wedge \theta^{\mu_m} \otimes \theta^{\nu_1} \wedge \cdots \wedge \theta^{\nu_m},$$

and using Proposition 4.3.2, we obtain

$$\frac{1}{m!} C^m(R^{\frac{m}{2}}) = \frac{1}{4^{\frac{m}{2}}} \sum_{\sigma \in S_m} \sum_{\tau \in S_m} (-1)^{\sigma} (-1)^{\tau} R_{\sigma_1\sigma_2\tau_1\tau_2} \cdots R_{\sigma_{m-1}\sigma_m\tau_{m-1}\tau_m}.$$

The second fundamental form h of $M \subset \mathbb{R}^{m+1}$ is a double form $h \in \mathcal{D}^{1,1}(M)$, and by Gauss' Equation we have

$$R = \frac{1}{2} h \otimes h = \frac{1}{2} \sum_{i,j,k,l=1}^{m} h_{ik} h_{jl}\, \theta^i \wedge \theta^j \otimes \theta^k \wedge \theta^l.$$

Repeating the previous argument we have

$$\frac{1}{m!} C^m(R^{\frac{m}{2}}) = \frac{1}{m!} \frac{1}{2^{\frac{m}{2}}} C^m(h^m)$$

$$= \frac{1}{2^{\frac{m}{2}}} \sum_{\sigma \in S_m} \sum_{\tau \in S_m} (-1)^{\sigma} (-1)^{\tau} h_{\sigma_1\tau_1} h_{\sigma_2\tau_2} \cdots h_{\sigma_m\tau_m}$$

$$= \frac{m!}{2^{\frac{m}{2}}} \det\left(h_{ij}\right),$$

and we conclude that

$$K = \det\left(h_{ij}\right) = \frac{2^{\frac{m}{2}}}{(m!)^2} C^m(R^{\frac{m}{2}})$$

$$= \frac{1}{2^{\frac{m}{2}} m!} \sum_{\sigma \in S_m} \sum_{\tau \in S_m} (-1)^{\sigma} (-1)^{\tau} R_{\sigma_1\sigma_2\tau_1\tau_2} \cdots R_{\sigma_{m-1}\sigma_m\tau_{m-1}\tau_m}.$$

$\square$

4.3.4   Integrating Double Forms

Some interesting geometric quantities can be obtained by considering a geometrically defined double form $\psi \in \mathcal{D}^{p,p}(M)$ and integrating its full contraction $C^p(\psi)$ over the manifold. In particular, the following functional, which is defined on $\mathcal{D}^{2,2}(N) \times \mathcal{D}^{1,1}(N)$, will play an important role in the sequel:

Definition 4.3.6 Given a Riemannian manifold (N, g) of dimension n, and double forms $A \in \mathcal{D}^{2,2}(N)$ and $b \in \mathcal{D}^{1,1}(N)$, we define for any integer $0 \le k \le \frac{n}{2}$

$$\mathcal{Q}_k(A, b \mid N, g) = \frac{1}{n!\, k!\, (n - 2k)!} \int_N C^n(A^k \oslash b^{n-2k})\, \mathrm{dvol}_g, \qquad (4.3.3)$$

provided $C^n(A^k \oslash b^{n-2k})$ is integrable on N.

The special cases where $k = 0$ or $2k = n$ will also be denoted as

$$\mathcal{Q}_0(b \mid N, g) = \frac{1}{(n!)^2} \int_N C^n(b^n)\, \mathrm{dvol}_g = \int_N \det\left(b_{ij}\right) \mathrm{dvol}_g,$$

and, when n is even,

$$\mathcal{Q}_{\frac{n}{2}}(A \mid N, g) = \frac{1}{(n!)^2} \int_N C^n(A^{\frac{n}{2}})\, \mathrm{dvol}_g.$$

The notation $\mathcal{Q}_k(A, b \mid N, g)$ is somewhat heavy but it precisely indicates what is being integrated on which manifold. When there is no risk of ambiguity, we will simplify it to $\mathcal{Q}_k(A, b \mid N)$, or even $\mathcal{Q}_k(A, b)$.

If either A or b (or both) are curvature invariants of the manifold, the integral (4.3.3) will be referred to as a *curvature integral*. We will need the special cases where b is either the metric tensor g or the second fundamental form h in the case of a hypersurface, and A is a linear combination of the curvature tensor R and the double forms $g \oslash g$ or $h \oslash h$.

Example 4.3.7 As a first important example, the index of a smooth vector field ξ on an m-dimensional Riemannian manifold (M, g) with an isolated singularity at $p \in M$ can be expressed as

$$\mathrm{Ind}(X, p) = \lim_{\varepsilon \to 0} \frac{1}{\mathrm{Vol}(S^{m-1})} \int_{S_\varepsilon} \det(L)\, \mathrm{dvol}_{S_\varepsilon},$$

where $L : TM \to TM$ is defined as $L(Y) = \nabla_Y \zeta$ with $\zeta = \xi/\|\xi\|$. Introducing the double form $\psi \in \mathcal{D}^{1,1}$ defined by $\psi(X, Y) = \langle X, L(Y) \rangle$, one can write the index as

$$\mathrm{Ind}(X, p) = \lim_{\varepsilon \to 0} \frac{1}{\mathrm{Vol}(S^{m-1})} \mathcal{Q}_0(\psi \mid S_\varepsilon, g). \qquad (4.3.4)$$

4.3.5   The Lipschitz–Killing Curvatures

An important class of examples of curvature integrals is given by the *Lipschitz–Killing Curvatures*, which we define now:

Definition 4.3.8 The *total Lipschitz–Killing curvatures* of order j of an n-dimensional Riemannian manifold (N, g) are defined as $\mathcal{K}_j(N, g) = 0$ if j is odd and

$$\mathcal{K}_{2k}(N, g) = \mathcal{Q}_k(R, g \mid N), \tag{4.3.5}$$

if $j = 2k \leq n$ is even.

Remark Using Proposition 4.3.2, we have

$$C^n\left(R^k \oslash g^{n-2k}\right) = \frac{(n-2k)!\, n!}{(2k)!}\, C^{2k}\left(R^k\right),$$

therefore the Lipschitz–Killing curvatures can also be written as

$$\mathcal{K}_{2k}(N, g) = \frac{1}{k!\,(2k)!} \int_N C^{2k}(R^k)\, \mathrm{dvol}_N, \tag{4.3.6}$$

which is the definition given in [14].

Example 4.3.9

(a) The Lipschitz–Killing curvature of order 0 is integrable if and only if (N, g) has finite volume. In this case we have

$$\mathcal{K}_0(N, g) = \mathrm{Vol}_g(N).$$

(b) The total Lipschitz–Killing curvature of order 2 on a compact manifold (N, g) is half the integral of its scalar curvature:

$$\mathcal{K}_2(N, g) = \frac{1}{2} \int_N S_g\, \mathrm{dvol}_N.$$

(c) If (N, g) is a compact space form with constant sectional curvature a, then using $R = \frac{a}{2} g \oslash g$ we obtain

$$\mathcal{K}_{2k}(N, g) = \frac{n!\, a^k}{2^k k!(n-2k)!} \cdot \mathrm{Vol}(N, g). \tag{4.3.7}$$

A notable application of the Lipschitz–Killing curvatures is illustrated by the celebrated H. Weyl *Tube Formula* proved in 1938:

Theorem 4.3.10 *If N is a smooth closed submanifold of dimension n in $\mathbb{R}^d$, then the volume of a (small enough) ε-tubular neighborhood $U_\varepsilon(N) = \{x \in \mathbb{R}^d \mid \text{dist}(x, N) \leq \varepsilon\}$ is given by*

$$\text{Vol}\,(U_\varepsilon(N)) = \sum_{k=0}^{\lfloor \frac{n}{2} \rfloor} \alpha_{d,n,k}\, \mathcal{K}_{2k}(N)\, \varepsilon^{d-n+2k},$$

where

$$\alpha_{d,n,k} = \frac{\pi^{\frac{d-n}{2}}}{2^{k+1}\Gamma\left(\frac{d-n}{2} + 1 + k\right)}.$$

Here $\lfloor \frac{n}{2} \rfloor$ denotes the nearest integer that is less than or equal to $\frac{n}{2}$.

We refer to the original paper [39] by Weyl or the book [14] for a proof of the Tube Formula.

The other important result is the Gauss–Bonnet Formula itself, which, for a closed Riemannian manifold (M, g) of even dimension m, can be written as

$$\mathcal{K}_m(M) = (2\pi)^{\frac{m}{2}} \chi(M). \tag{4.3.8}$$

As mentioned in Sect. 4.2 and explained in detail in [14, §5.5]), Allendoerfer used some arguments and calculations from Weyl's paper in a crucial step of his proof.

We will now give a new proof of the Gauss–Bonnet–Hopf Formula (4.2.1), this time an intrinsic one, as a special case of the Gauss–Bonnet Formula (4.3.8). To this aim, it will be convenient to introduce the following definition:

Definition 4.3.11 The *Gauss–Bonnet density* of the Riemannian manifold (M, g) of even dimension m is the function $\mathcal{R} : M \to \mathbb{R}$ defined as

$$\mathcal{R} = \frac{1}{m!\left(\frac{m}{2}\right)!}\, C^m(R^{\frac{m}{2}}). \tag{4.3.9}$$

Observe that the top Lipschitz–Killing curvature is the integral of the Gauss–Bonnet density:

$$\mathcal{K}_m(M) = \int_M \mathcal{R} \, \text{dvol}_g. \tag{4.3.10}$$

Note also that the proof of Proposition 4.3.5 shows that the Gauss–Bonnet density of an oriented even-dimensional hypersurface in Euclidean space is equal, up to a multiplicative constant, to the Gauss–Kronecker curvature. More precisely we have

$$K = \frac{2^{\frac{m}{2}}}{(m!)^2}\, C^m(R^{\frac{m}{2}}) = \frac{m!!}{m!}\, \mathcal{R}. \tag{4.3.11}$$

Assuming the Gauss–Bonnet Formula (4.3.8) has been established, we now easily deduce (4.2.1):

$$\int_M K\,d\mathrm{vol}_N = \frac{2^{\frac{m}{2}} \left(\frac{m}{2}\right)!}{m!} \int_M \mathcal{R}\,d\mathrm{vol}_M = \frac{2^{\frac{m}{2}} \left(\frac{m}{2}\right)!}{m!} (2\pi)^{\frac{m}{2}} \chi(M)$$

$$= \frac{1}{2} \mathrm{Vol}(S^m)\chi(M), \tag{4.3.12}$$

because the volume of an even-dimensional sphere is

$$\mathrm{Vol}(S^m) = 2^{m+1}\pi^{\frac{m}{2}} \frac{\left(\frac{m}{2}\right)!}{m!}.$$

The proof of (4.3.8) will be given in next section.

Remark 4.3.12 The Gauss–Bonnet density $\mathcal{R}$ that we have just defined is also referred to as the *Gauss–Bonnet Integrand*. Fenchel proposed calling it the *Lipschitz–Killing curvature*, and some authors do adopt that terminology. However, we find it problematic considering our Definition 4.3.8. Note that $\mathcal{R}$ is precisely the intrinsically defined curvature function that was asked for in Hopf's Curvatura Integra problem (up to a possible multiplicative constant).

Remark 4.3.13 The Gauss–Bonnet density, as well as its integral, also made an appearance in the notable 1938 paper [27] by Lanczos. His work's context was the formulation of the fundamental laws of nature based on the principle of least action. Lanczos observed that the integral $\int \mathcal{R}$ remains invariant under variations of the metric g, establishing it as a differential invariant of the manifold. He concluded that, within the framework of field theory, the Gauss–Bonnet density doesn't bring any additional information. Lanczos' work was carried out independently of the concurrent efforts by mathematicians addressing the Hopf problem and the Gauss–Bonnet formula.

4.4 Chern's Insight and the Tao of Gauss–Bonnet

For a compact even-dimensional manifold with boundary (M, g), one needs to add a boundary term to (4.3.8). This additional term is a sum of curvature integrals involving the curvature tensor R of (M, g) and the second fundamental form h of the boundary ∂M. The Gauss–Bonnet Formula takes the following form:

$$(2\pi)^{\frac{m}{2}} \chi(M) = \mathcal{K}_m(M) + \sum_{k=0}^{\frac{m}{2}-1} b_{m,k} \, \mathcal{Q}_k(R, h \mid \partial M, g),$$

where the coefficient $b_{m,k}$ will be identified in due time. We will also obtain a similar formula, where the tensor R is being replaced by the intrinsic curvature $\bar{R}$ of the boundary.

Our goal in this section is to present a comprehensive and intrinsic proof of the above formula. Our approach closely follows the works of Chern [9, 10], with a few minor adjustments, which are discussed in Sect. 4.6.2. In what follows, the formalism of moving frames is extensively used; we refer the reader to the appendix, which contains all the necessary definitions and formulas. Also, a number of expressions below involve double factorials, see Sect. 4.6.3 for the definition ans some basic properties.

4.4.1 The Emergence of the Pfaffian

The ingredients in Chern's proof are differential forms and specifically E. Cartan's moving frame method. Referring to (4.3.10), we see that one needs to closely consider the differential form

$$\mathcal{R}\, \mathrm{dvol}_g,$$

where $\mathcal{R}$ is the Gauss–Bonnet density defined in (4.3.11). Invoking Lemma A.2, we see that the double form $R^m \in \mathcal{D}^{m,m}(M)$ can be written as

$$R^{\frac{m}{2}} = \frac{1}{2^{\frac{m}{2}}} \sum_{\mu_1,\dots,\mu_m=1}^{m} \Omega_{\mu_1,\mu_2} \wedge \dots \wedge \Omega_{\mu_{m-1},\mu_m} \otimes \theta^{\mu_1} \wedge \dots \wedge \theta^{\mu_m},$$

and Proposition 4.3.4 implies

$$\frac{1}{m!} \mathbf{C}^m (R^{\frac{m}{2}}) \theta^1 \wedge \dots \wedge \theta^m = \frac{1}{2^{\frac{m}{2}}} \sum_{\sigma \in S_m} (-1)^\sigma \Omega_{\sigma_1,\sigma_2} \wedge \dots \wedge \Omega_{\sigma_{m-1},\sigma_m}.$$

This relation can be written as

$$\mathcal{R}\, \mathrm{dvol}_g = \mathrm{Pf}(\Omega), \tag{4.4.1}$$

where the m-form

$$\mathrm{Pf}(\Omega) = \frac{1}{m!!} \sum_{\sigma \in S_m} (-1)^\sigma \Omega_{\sigma_1,\sigma_2} \wedge \dots \wedge \Omega_{\sigma_{m-1},\sigma_m}$$

is the *Pfaffian*[2] of the curvature form. Observe that (4.4.1) implies that the Pfaffian $\text{Pf}(\Omega)$ does not depend on the chosen moving frame.

Examples Recall that in dimension 2, there is only one non-zero curvature form (up to sign), namely $\Omega_{12} = -\Omega_{21}$. Therefore, for a Riemannian surface (S, g), the Pfaffian of its curvature form reduces to:

$$\text{Pf}(\Omega) = \frac{1}{2}(\Omega_{12} - \Omega_{21}) = \Omega_{12} = \frac{1}{2}R_{12ij}\theta^i \wedge \theta^j = R_{1212}\,\text{dvol}_S = K\,\text{dvol}_S,$$

where K is the Gauss curvature of the surface S.

If M is a 4-dimensional manifold, then its Pfaffian is given by:

$$\text{Pf}(\Omega) = \frac{1}{8}\sum_{\sigma \in S_4}(-1)^\sigma \Omega_{\sigma_1,\sigma_2} \wedge \Omega_{\sigma_3,\sigma_4}$$

is a sum of 24 terms. But using the symmetries $\Omega_{ab} = -\Omega_{bc}$ and $\Omega_{ab} \wedge \Omega_{cd} = \Omega_{cd} \wedge \Omega_{ab}$, it can be reduced to

$$\text{Pf}(\Omega) = \Omega_{1,2} \wedge \Omega_{3,4} + \Omega_{2,3} \wedge \Omega_{1,4} + \Omega_{3,1} \wedge \Omega_{2,4}.$$

See Exercise 2 in Sect. 4.7 for a generalization in higher dimension.

4.4.2 A Tale of Exactness, Featuring a Vector Field

The next result lies at the core of Chern's proof of the Gauss–Bonnet Formula. It states that when the manifold carries a non-vanishing vector field, the Pfaffian of the curvature form is exact.

Proposition 4.4.1 (Chern) *Let (M, g) be an oriented Riemannian manifold of even dimension m. If M admits an everywhere non vanishing vector field ξ, then $\text{Pf}(\Omega)$ is an exact differential form.*

We can rephrase this Proposition by expressing that the Gauss–Bonnet density $\mathcal{R}$ is a divergence. An immediate consequence is the following corollary, which is a special case of the Gauss–Bonnet Formula:

Corollary 4.4.2 *Let M be a compact oriented manifold without boundary of even dimension m. If M admits an everywhere non vanishing vector field ξ, then the top global Lipshitz–Killing curvature of M vanishes.*

---

[2] The *Pfaffian* of a skew-symmetric matrix $A = (a_{ij}) \in M_m(\mathbb{R})$ is defined for m even to be $\text{Pf}(A) = \frac{1}{m!!}\sum_{\sigma \in S_m}(-1)^\sigma \prod_{i=1}^{\frac{m}{2}} a_{\sigma_{2i-1},\sigma_i}$. The notion was introduced by A. Cayley.

Indeed, by Stoke's Theorem we have $\mathcal{K}_m(M, g) = \int_M \mathcal{R} \, \mathrm{dvol}_g = 0$.

The rest of his section is devoted to proving the above Proposition.

Proof Following Chern's ideas, we use the non-vanishing vector field ξ to construct an $(m-1)$-form Φ that satisfies $\mathrm{Pf}(\Omega) = d\Phi$. To achieve this, we first introduce the double form $h = h_\zeta \in \mathcal{D}^{1,1}(M)$ of type $(1, 1)$ on M, defined as

$$h(X, Y) = \langle \nabla_X \zeta, Y \rangle, \tag{4.4.2}$$

where ζ is the unit vector field $\zeta = \xi/|\xi|$. Next we define a collection of $(m-1)$-forms Φ_k, for $0 \le k < \frac{m}{2}$, as follows:

$$\Phi_k = -\frac{2^k}{(m-1)!} \cdot \mathbf{C}^{m-1} \left(R^k \oslash h^{m-1-2k} \right) \varpi, \tag{4.4.3}$$

where $\varpi = \varpi_\zeta$ is defined in (A.2). Additionally, we set $\Phi_{-1} = 0$ for convenience.

We will now build a primitive of $\mathrm{Pf}(\Omega)$ as a linear combination of the Φ_ks. To perform the necessary computations, we choose a positively oriented moving frame $e_1, \ldots, e_m$ such that $e_m = \zeta$. This is always possible in a neighborhood $U \subset M$ of any point. Referring to Definition A.1, we then have for any $1 \le i, j \le m-1$

$$h(e_i, e_j) = g(\nabla_{e_i} e_m, e_j) = \omega_{jm}(e_i),$$

that is,

$$h = \sum_{j=1}^{m-1} \omega_{jm} \otimes \theta^j. \tag{4.4.4}$$

We will now use some formulas established in the appendix. The first one is (A.3), which tells us that

$$\varpi = -\theta^1 \wedge \cdots \wedge \theta^{m-1},$$

because $m = \dim(M)$ is even. The second formula is stated in Lemma A.2:

$$R = \frac{1}{2} \sum_{i,j=1}^{m} \Omega_{ij} \otimes \theta^i \wedge \theta^j.$$

We therefore have

$$R^k \oslash h^{m-1-2k}$$

$$= \frac{1}{2^k} \sum_{\mu_1,\dots,\mu_{m-1}=1}^{m-1} \Omega_{\mu_1,\mu_2} \wedge \dots \wedge \Omega_{\mu_{2k-1}\mu_{2k}} \wedge \omega_{\mu_{2k+1},m} \wedge$$

$$\dots \wedge \omega_{\mu_{m-1},m} \otimes \theta^{\mu_1} \wedge \dots \wedge \theta^{\mu_{m-1}}.$$

Applying now Proposition 4.3.4, we obtain the following local expression for the $(m-1)$-form Φ_k:

$$\Phi_k = \sum_{\sigma \in S_{m-1}} (-1)^\sigma \Omega_{\sigma_1,\sigma_2} \wedge \dots \wedge \Omega_{\sigma_{2k-1},\sigma_{2k}} \wedge \omega_{\sigma_{2k+1},m} \wedge \dots \wedge \omega_{\sigma_{m-1},m}. \quad (4.4.5)$$

This is how the form Φ_k appears in Chern's paper [9, 10], and we will now closely follow Chern's computation of its exterior derivative. Because a 2-form commutes with any other differential form, we can write the exterior derivative of Φ_k as

$$d\Phi_k = k \sum_{\sigma \in S_{m-1}} (-1)^\sigma d\Omega_{\sigma_1,\sigma_2} \wedge \Omega_{\sigma_3,\sigma_4} \wedge \dots \wedge \Omega_{\sigma_{2k-1},\sigma_{2k}}$$

$$\wedge \omega_{\sigma_{2k+1},m} \wedge \dots \wedge \omega_{\sigma_{m-1},m}$$

$$+ (m - 2k - 1) \sum_{\sigma \in S_{m-1}} (-1)^\sigma \Omega_{\sigma_1,\sigma_2} \wedge \dots \wedge \Omega_{\sigma_{2k-1},\sigma_{2k}} \wedge d\omega_{\sigma_{2k+1},m}$$

$$\wedge \omega_{\sigma_{2k+2},m} \wedge \dots \wedge \omega_{\sigma_{m-1},m}.$$

We next use the second Structure Eq. (A.4) and the Bianchi identity (A.5). These identities allow us to rewrite $d\Omega_{\sigma_1,\sigma_2}$ and $d\omega_{\sigma_{k+1},m}$ as

$$d\omega_{\sigma_{k+1},m} = \Omega_{\sigma_{k+1},m} - \sum_{s=1}^m \omega_{\sigma_{k+1},s} \wedge \omega_{s,m},$$

and

$$d\Omega_{\sigma_1,\sigma_2} = \sum_{s=1}^m \left(\Omega_{\sigma_1,s} \wedge \omega_{s,\sigma_2} - \omega_{\sigma_1,s} \wedge \Omega_{s,\sigma_2} \right)$$

$$= \sum_{s=1}^m \left(\Omega_{\sigma_2,s} \wedge \omega_{\sigma_1,s} - \Omega_{\sigma_1,s} \wedge \omega_{\sigma_2,s} \right).$$

Carefully computing,[3] one then finds that

$$d\Phi_k = \frac{m - 2k - 1}{2k + 2}\Psi_k + \Psi_{k-1} + \Xi_k,$$

where Ψ_k is the m-form

$$\Psi_k = (-1)^{k+1}(2k + 2) \cdot \sum_{\sigma \in S_{m-1}} (-1)^\sigma \Omega_{\sigma_1,\sigma_2} \wedge \ldots \wedge \Omega_{\sigma_{2k-1},\sigma_{2k}}$$

$$\wedge \Omega_{\sigma_{2k+1},m} \wedge \omega_{\sigma_{2k+2},m}, \wedge \ldots \wedge \omega_{\sigma_{m-1},m}$$

and Ξ_k is a linear combinations of products of curvature and connection forms containing some ω_{ij} with $1 \leq i, j \leq (m - 1)$. The forms Ψ_k are defined for $0 \leq k \leq \frac{m}{2} - 1$, but it is convenient to also define $\Psi_{k-1} = 0$.

A key fact about the previous formula, observed by Chern, is that we must have $\Xi_k = 0$. The reason is that the forms Φ_k, and hence $d\Phi_k$, do not depend on the chosen moving frame $e_1, \ldots, e_m$, except for the constraint $e_m = \zeta$. Lemma A.5 implies therefore that all terms in $d\Phi_k$ containing some ω_{ij} with $1 \leq i, j \leq (m - 1)$ must cancel each other. We have thus established the following relation:

$$d\Phi_k = \frac{m - 2k - 1}{2(k + 1)}\Psi_k + \Psi_{k-1}. \tag{4.4.6}$$

This identity gives us Ψ_k as a linear combination of the $d\Phi_r$s. Let us indeed write

$$\Psi_k = \sum_{r=0}^{k} \lambda_{m,k,r}\, d\Phi_r, \tag{4.4.7}$$

where the unknown coefficients $\lambda_{m,k,r}$ are to be determined. From (4.4.6) we have

$$d\Phi_k = \sum_{r=0}^{k} \left(\frac{m - 2k - 1}{2(k + 1)}\lambda_{m,k,r} + \lambda_{m,k-1,r} \right) d\Phi_r.$$

The condition on $\lambda_{m,k,r}$ is then

$$\frac{m - 2k - 1}{2(k + 1)}\lambda_{m,k,r} + \lambda_{m,k-1,r} = \delta_{k,r}, \quad \text{with } \lambda_{m,k,r} = 0 \text{ for } r > k,$$

---

[3] See [29, Lemma 3.14] for a detailed explanation.

which is equivalent to

$$\lambda_{m,k,r} = (-1)^{k-r} \prod_{j=r}^{k} \frac{2j+2}{m-2j-1}. \tag{4.4.8}$$

We have established that each Ψ_k is an exact form. Considering the particular case $k = \frac{m}{2} - 1$, we have

$$\Psi_{\frac{m}{2}-1} = m \cdot \sum_{\sigma \in S_{m-1}} (-1)^\sigma \Omega_{\sigma_1,\sigma_2} \wedge \ldots \wedge \Omega_{\sigma_{m-1},\sigma_m}$$

$$= \sum_{\sigma \in S_m} (-1)^\sigma \Omega_{\sigma_1,\sigma_2} \wedge \ldots \wedge \Omega_{\sigma_{m-1},\sigma_m}$$

$$= m!! \, \mathrm{Pf}(\Omega),$$

and we conclude that $\mathrm{Pf}(\Omega)$ is an exact form. $\qquad \square$

Remarks

1. The differential forms Ψ_k have been defined using the connection and curvature forms of a particular moving frame, however Formula (4.4.7) implies that the Ψ_k do not depend on the chosen frame, provided $e_m = \zeta$.
2. The proof gives us an explicit formula for the primitive of the Pfaffian. We have

$$\mathrm{Pf}(\Omega) = d\Phi, \quad \text{with} \quad \Phi = \sum_{r=0}^{\frac{m}{2}-1} a_{m,r} \Phi_r, \tag{4.4.9}$$

where the $a_{m,r}$ can be computed from (4.4.8):

$$a_{m,r} = \frac{1}{m!!} \prod_{j=r}^{\frac{m}{2}-1} \frac{2j+2}{m-2j-1} = \frac{1}{(2r)!! \, (m-2r-1)!!}. \tag{4.4.10}$$

The coefficients $a_{m,r}$ can be written in several ways, which all appear in the literature,. For instance[4]

$$a_{m,r} = \frac{1}{2^r r! \, (m-2r-1)!!} = \frac{2^{\frac{m}{2}-2r} \left(\frac{m}{2}-r\right)!}{(m-2r)! \, r!} = \frac{\sqrt{\pi}}{2^{\frac{m}{2}} r! \, \Gamma\left(\frac{m-2r+1}{2}\right)}. \tag{4.4.11}$$

---

[4] These identities hold for m even and $r < \frac{m}{2}$.

4.4.3   Harvesting the Ripe Gauss–Bonnet Fruit

With the hard work behind us, we are now in a position to state and prove the Gauss–Bonnet Formula, as originally formulated by Chern (see Equation (19) in [10]).

Theorem 4.4.3 *The Euler characteristic of a compact Riemannian manifold* (M, g) *of even dimension* m *with (possibly empty) boundary has the following integral representation:*

$$(2\pi)^{\frac{m}{2}} \chi(M) = \int_M \mathrm{Pf}(\Omega) - \int_{\partial M} \Phi.$$

Proof For the proof, we choose a smooth vector field ξ on M with the following properties:

(i) ξ has only isolated singularities at $x_1, \dots, x_q \in M \setminus \partial M$.
(ii) At any boundary point, ξ is non vanishing, outward pointing and orthogonal to ∂M.

On the manifold $M' = M \setminus \{x_1, \dots, x_q\}$ we define the normalized vector field $\zeta = \xi/\|\xi\|$ and the double form $h(X, Y) = \langle X, \nabla_Y \zeta \rangle$.

Let us then denote by $B_i(\varepsilon)$ the open ball of radius ε centered at x_i. We assume $\varepsilon > 0$ small enough so that the balls $B_i(\varepsilon)$ are pairwise disjoint and have empty intersection with ∂M, and we set $M_\varepsilon = M \setminus \cup_{i=1}^q B_i(\varepsilon)$. The unit vector field $\zeta = \xi/\|\xi\|$ is well defined on M_ε and we have by Stokes Theorem:

$$\int_{M_\varepsilon} \mathrm{Pf}(\Omega) = \int_{\partial M} \Phi - \sum_{i=1}^q \int_{\partial B_i(\varepsilon)} \Phi,$$

where the form Φ has been defined in (4.4.9), see also (4.4.3). Chern observes then that

$$\lim_{\varepsilon \to 0} \int_{\partial B_i(\varepsilon)} \Phi_k = 0$$

for $k \geq 1$. Furthermore, from the definition (4.4.3) of Φ_0 and the formula (4.3.4) representing the index, we see that

$$\lim_{\varepsilon \to 0} \left(- \int_{\partial B_i(\varepsilon)} \Phi_0 \right) = \lim_{\varepsilon \to 0} \frac{1}{(m-1)!} \cdot \mathcal{Q}_0(h \mid \partial B_\varepsilon, g)$$

$$= (m-1)! \, \mathrm{Vol}(S^{m-1}) \, \mathrm{Ind}(\xi, x_i).$$

We have thus established that

$$\int_M \mathrm{Pf}(\Omega) - \int_{\partial M} \Phi = \lim_{\varepsilon \to 0} \sum_{i=1}^{q} \left(-a_{m,0}\right) \int_{\partial B_i(\varepsilon)} \Phi_0$$

$$= \lim_{\varepsilon \to 0} \sum_{i=1}^{q} \frac{a_{m,0}}{(m-1)!} \cdot \mathcal{Q}_0(h \mid \partial B_\varepsilon, g)$$

$$= a_{m,0}(m-1)! \, \mathrm{Vol}(S^{m-1}) \sum_{i=1}^{q} \mathrm{Ind}(\xi, x_i).$$

Recall that $a_{m,0} = \dfrac{1}{(m-1)!!}$. Furthermore, since m is even, we have $\mathrm{Vol}(S^{m-1}) = \dfrac{(2\pi)^{\frac{m}{2}}}{(m-2)!!}$, which leads to the following relationship:

$$a_{m,0}(m-1)! \, \mathrm{Vol}(S^{m-1}) = (2\pi)^{\frac{m}{2}}.$$

Applying the Poincaré–Hopf Theorem, we finally conclude that

$$\int_M \mathrm{Pf}(\Omega) - \int_{\partial M} \Phi = (2\pi)^{\frac{m}{2}} \sum_{i=1}^{q} \mathrm{Ind}(\xi) = (2\pi)^{\frac{m}{2}} \chi(M).$$

$\square$

Let us pause and take a short moment to listen to Chern's own account of his discovery, as he recalls it in an interview with Allyn Jackson in 1998 (see [23]).

> The Gauss–Bonnet formula is one of the important, fundamental formulas, not only in differential geometry, but in the whole of mathematics. Before I came to Princeton [in 1943] I had thought about it, so the development in Princeton was in a sense very natural. I came to Princeton and I met André Weil. He had just published his paper with Allendoerfer. Weil and I became good friends, so we naturally discussed the Gauss–Bonnet formula. And then I got my proof. I think this is one of my best works, because it solved an important, a fundamental, classical problem, and the ideas were very new. And to carry out the ideas you need some technical ingenuity. It's not trivial. It's not something where once you have the ideas you can carry it out. It is subtle. So I think this a very good piece of work.

4.4.4 Demystifying the Gauss–Bonnet Boundary Term Using Double Forms

In this section, we use the formalism of double forms to introduce two alternative formulations of the Gauss–Bonnet Formula for a compact, oriented Riemannian manifold (M, g) with non-empty boundary and even dimension m. The first reformulation involves the curvature tensor R of the oriented Riemannian manifold (M, g), coupled with the second fundamental form h of its boundary. The formula can be expressed as follows:

$$(2\pi)^{\frac{m}{2}} \chi(M) = \mathcal{K}_m(M) + \sum_{k=0}^{\frac{m}{2}-1} b_{m,k} \mathcal{Q}_k(R, h \mid \partial M, \bar{g}). \qquad (4.4.12)$$

Here $\bar{g}$ is the first fundamental form of ∂M as a hypersurface in (M, g). We will prove this formula and determine the coefficients $b_{m,k}$ below.

We work with the orientation of the boundary ∂M induced by the unit outer normal field $v \in T(\partial M)^{\perp}$. This means that a basis $v_1, \ldots, v_{m-1}$ of $T_p(\partial M)$ is positively oriented if and only if $v, v_1, \ldots, v_{m-1}$ is a positively oriented basis in $T_p M$. Referring to (A.3), we see that, under this orientation, the volume form on the boundary is given by

$$\mathrm{dvol}_{\partial M} = \varpi_v.$$

We next choose a vector field ζ of length 1, defined in some neighborhood of $\partial M \subset M$ and such that ζ is outward pointing and orthogonal to ∂M. Observe that $\zeta = v$ at every boundary point $p \in \partial M$. We then define Φ_k and h as previously, that is,

$$h(X, Y) = \langle X, \nabla_Y \zeta \rangle \quad \text{and} \quad \Phi_k = -\frac{2^k}{(m-1)!} \cdot C^{m-1}\left(R^k \oslash h^{m-1-2k}\right) \varpi_{\zeta},$$

and observe that the restriction of h to the boundary coincides with its second fundamental form.[5]

From Definition 4.3.6, we then have

$$\mathcal{Q}_k(R, h \mid \partial M) = \frac{1}{(m-1)!\,k!\,(m-2k-1)!} \int_{\partial M} C^{m-1}(R^k \oslash h^{m-2k-1})\,\mathrm{dvol}_{\partial M}$$

$$= \frac{-1}{2^k\,k!\,(m-2k-1)!} \int_{\partial M} \Phi_k.$$

So if we now define the coefficients $b_{m,k}$ by the following relation:

$$b_{m,k}\, \mathcal{Q}_k(R, h \mid \partial M) = -a_{m,k} \int_{\partial M} \Phi_k,$$

---

[5] In the literature, the second fundamental form is frequently defined using the opposite sign convention. The present convention yields positive principal curvatures for the boundary of the unit ball in Euclidean space.

we conclude that

$$\mathcal{K}_m(M) + \sum_{k=0}^{\frac{m}{2}-1} b_{m,k}\, \mathcal{Q}_k(R, h \mid \partial M, \bar{g}) = \int_M \mathrm{Pf}(\Omega)\, \mathrm{dvol}_M$$

$$- \int_{\partial M} \Phi = (2\pi)^{\frac{m}{2}} \chi(M).$$

Formula (4.4.12) is proved. Furthermore, the coefficients are calculated as follows:

$$b_{m,k} = a_{m,k} \cdot 2^k\, k!\, (m - 2k - 1)! = \frac{(m - 2k - 1)!}{(m - 2k - 1)!!} = (m - 2k - 2)!! \quad (4.4.13)$$

The second reformulation of the Gauss–Bonnet Formula expresses the boundary term using the intrinsic curvature tensor $\bar{R}$ of the boundary hypersurface

$$\mathcal{K}_m(M) + \sum_{k=0}^{\frac{m}{2}-1} c_{m,k}\, \mathcal{Q}_k(\bar{R}, h \mid \partial M, \bar{g}) = (2\pi)^{\frac{m}{2}} \chi(M), \quad (4.4.14)$$

where, again, the coefficients $c_{m,k}$ are to be determined. Using the Gauss Equation

$$R = \bar{R} - \frac{1}{2} h \otimes h,$$

together with the linearity of the contraction, the fact that $\bar{R}$ and $h \otimes h$ commute and the binomial formula, we see that the following relation holds on the boundary:

$$R^p \otimes h^{m-1-2p} = \sum_{k=0}^{p} \binom{p}{k} \left(-\frac{1}{2}\right)^{p-k} \bar{R}^k \otimes h^{m-1-2k},$$

therefore

$$C^{m-1}\left(R^p \otimes h^{m-1-2p}\right) = \sum_{k=0}^{p} \binom{p}{k} \left(-\frac{1}{2}\right)^{p-k} C^{m-1}\left(\bar{R}^k \otimes h^{m-1-2k}\right).$$

From (4.3.3) and the above relation we obtain:

$$\mathcal{Q}_p(R, h \mid \partial M, \bar{g}) = \sum_{k=0}^{p} w_{m,p,k} \mathcal{Q}_k(\bar{R}, h \mid \partial M, \bar{g}),$$

where

$$w_{m,p,k} = \left(-\frac{1}{2}\right)^{p-k} \frac{(m - 1 - 2k)!}{(m - 1 - 2p)!\, (p - k)!}$$

if $0 \le k \le p$. It will be convenient to also define $w_{m,p,k} = 0$ if $k > p$. We then have

$$\sum_{p=0}^{\frac{m}{2}-1} b_{m,p} \mathcal{Q}_p \left(R, h \mid \partial M, \bar{g} \right) = \sum_{k=0}^{\frac{m}{2}-1} c_{m,k} \mathcal{Q}_p \left(\bar{R}, h \mid \partial M, \bar{g} \right),$$

where $c_{m,k}$ is defined as

$$c_{m,k} = \sum_{p=k}^{\frac{m}{2}-1} w_{m,p,k} b_{m,p}.$$

We have thus proved that (4.4.12) holds with

$$c_{m,k} = (m - 2k - 1)! \sum_{p=k}^{\frac{m}{2}-1} (-1)^{p-k} \frac{(m - 2p - 2)!!}{2^{p-k}(m - 2p - 1)!\,(p - k)!}$$

$$= (-1)^{\frac{m}{2}-k-1}(m - 2k - 3)!! \tag{4.4.15}$$

Synthesizing our findings, we arrive at the following formulations of the Gauss–Bonnet–Chern Formula:

$$
\begin{aligned}
(2\pi)^{\frac{m}{2}} \chi(M) &= \int_M \mathrm{Pf}(\Omega) - \sum_{k=0}^{\frac{m}{2}-1} a_{m,k} \int_{\partial M} \Phi_k \\
&= \mathcal{K}_m(M) + \sum_{k=0}^{\frac{m}{2}-1} b_{m,k} \, \mathcal{Q}_k(R, h \mid \partial M, \bar{g}) \\
&= \mathcal{K}_m(M) + \sum_{k=0}^{\frac{m}{2}-1} c_{m,k} \, \mathcal{Q}_k(\bar{R}, h \mid \partial M, \bar{g}),
\end{aligned}
\tag{4.4.16}
$$

where the coefficients are

$$a_{m,k} = \frac{1}{(2k)!!\,(m - 2k - 1)!!}$$

$$b_{m,k} = (m - 2k - 2)!!$$

$$c_{m,k} = (-1)^{\frac{m}{2}-k-1}(m - 2k - 3)!!$$

We also remind the reader that

$$\int_M \mathrm{Pf}(\Omega) = \int_M \mathcal{R} \, \mathrm{dvol}_M = \mathcal{K}_m(M).$$

Remark 4.4.4 Perhaps one ought to assign a name to the tensors $R^k \otimes h^{m-1-2k}$ appearing in the preceding formulas. Although the term *Generalized Gauss–Kronecker Curvature* comes to mind, this terminology has already been adopted in [25] for the tensor R^k. An alternative suggestion could be the *Gauss–Kronecker–Chern Curvatures*.

It is noteworthy that the boundary term in the Gauss–Bonnet Formula appears as the integral of a weighted homogeneous polynomial of degree $m-1$ in R (the curvature tensor) and h (the second fundamental form), with weights of 1 assigned to h and 2 assigned to R.

Another point that might be worth the effort would be to try and find a more intuitive explanation for the values of the coefficients $b_{m,k}$ and $c_{m,k}$ (in particular the presence of double factorials here is no coincidence).

4.5   Examples and Applications

4.5.1   Flat Manifolds

Recall that a Riemannian manifold (M, g) is *flat* if its curvature tensor satisfies $R = 0$. Equivalently, every point $x \notin \partial M$ admits a neighborhood isometric to an open subset of $\mathbb{R}^n$.

Theorem 4.5.1 *If (M, g) is a compact flat oriented Riemannnian manifold with boundary of dimension m, then*

$$\int_{\partial M} K \, \mathrm{dvol}_{\partial M} = \mathrm{Vol}(S^{m-1}) \, \chi(M),$$

where $K = \det(h_{ij})$ is the Gauss–Kronecker curvature of the boundary ∂M.

Proof The proof splits in two cases according to whether m is even or odd. Consider first the case where m is even,. We have $\mathcal{K}_m(M) = 0$ since $R = 0$. We also have $\mathcal{Q}_k(R, h \mid \partial M, \bar{g}) = 0$ for any $k \geq 1$. Because $b_{m,0} = (m-2)!!$, the second formula in (4.4.16) reduces thus to

$$\int_{\partial M} K \, \mathrm{dvol}_{\partial M} = \mathcal{Q}_0(h \mid \partial M) = \frac{(2\pi)^{\frac{m}{2}}}{(m-2)!!} \chi(M) = \mathrm{Vol}(S^{m-1})\chi(M).$$

If m is odd, then $N = \partial M$ is an even-dimensional manifold and we know from Eq. (4.3.12) that

$$\int_N K \, \mathrm{dvol}_N = \frac{1}{2} \, \mathrm{Vol}(S^{m-1}) \, \chi(N)$$

(see also Proposition 4.3.5). We conclude the proof with the relation $\chi(\partial M) = 2\chi(M)$, which holds when $m = \dim(M)$ is odd.   □

An interesting outcome of this theorem is the following generalization of Corollary 4.2.2. Suppose that an immersion $f : M \to \mathbb{R}^m$ is given, where M is a smooth compact manifold of (even or odd) dimension m. We can then prescribe an orientation on M by requiring f to be a positively oriented map and then define the Gauss map $\nu : \partial M \to S^{m-1}$.

Corollary 4.5.2 *The degree of the Gauss map* $\nu : \partial M \to S^{m-1}$ *is equal to the Euler characteristic of* M:

$$\deg(\nu) = \chi(M).$$

Proof Observe first that the pullback g of the Euclidean metric by the map $f : M \to \mathbb{R}^m$ is a flat metric on M. The proof is now obvious from Theorem 4.5.1 since the Gauss–Kronecker curvature is the Jacobian of the Gauss map.   □

This result was independently discovered in 1960 by A. Haefliger and H. Samelson, see [18, Theorem 3bis] and [35, Theorem 1]. In the special case where M is a domain in $\mathbb{R}^n$, H. Hopf had already established it in [22, Satz VI.]. Some historical context is discussed in Gotlieb's essay [13].

4.5.2 Manifolds with Flat Boundary

Consider now the case of a compact oriented Riemannian manifold (M, g), of even dimension m, whose boundary is flat, that is, $\bar{R} = 0$. The second formula in (4.4.16) tells us that

$$\int_M \mathcal{R} \, \mathrm{dvol}_M + \int_{\partial M} K \, \mathrm{dvol}_{\partial M} = (2\pi)^{\frac{m}{2}} \chi(M), \qquad (4.5.1)$$

where $\mathcal{R}$ is the Gauss–Bonnet density of (M, g) and $K = \det(h_{ij})$ is the Gauss–Kronecker curvature of the boundary ∂M.

When $m = 2$, this recovers the classical Gauss–Bonnet Formula for compact surfaces with smooth boundary (up to a change in notation). In this case, we have indeed $\bar{R} = 0$ because the boundary is a one-dimensional manifold.

4.5.3   Space Forms

The Gauss–Bonnet density of a manifold (M, g) of constant sectional curvature a and even dimension m is a constant, equal to

$$\mathcal{R} = \left(\tfrac{a}{2}\right)^{\frac{m}{2}} \frac{m!}{\left(\frac{m}{2}\right)!} = \frac{(2\pi a)^{\frac{m}{2}}}{\frac{1}{2}\,\mathrm{Vol}(S^m)}. \tag{4.5.2}$$

If M is compact, then the Gauss–Bonnet Formula immediately implies the following identity, which was already proved in 1925 by Hopf in [21]:

$$a^{\frac{m}{2}}\,\mathrm{Vol}(M) = \frac{1}{2}\,\mathrm{Vol}(S^m)\chi(M). \tag{4.5.3}$$

In particular, when $a > 0$, the Euler characteristic $\chi(M)$ is positive, it vanishes in the flat case ($a = 0$), and its sign is $(-1)^{\frac{m}{2}}$ when $a < 0$. Note that Eq. (4.5.3) holds trivially for flat manifolds, while it becomes almost self-evident for spherical manifolds due to the fact that the universal cover of such a manifold is a sphere. A geometric proof, based on a geodesic triangulation, is presented in [33, Theorem 11.3.2]. Let us add a few comments on this result:

1. The formula (4.5.2) also works for the case of *non compact, complete hyperbolic manifolds of finite volume* of even dimension. This was first proved by Kellerhals and Zehrt [24] using an ideal geodesic triangulation of the manifold. It can also be seen as a consequence of (4.5.1). Here is a brief explanation of why this holds: such a manifold is the union of a compact manifold with horospherical boundary and a finite collection of cusps. Each connected component of the boundary is a compact flat manifold with constant Gauss–Kronecker curvature $K = +1$ (in fact we have $h = \bar{g}$, equivalently all the principal curvature are equal to $+1$). Progressively chopping off the manifold at increasing distances, it becomes evident that $\int_{\partial M} K\, \mathrm{dvol}_{\partial M} \to 0$ and we conclude by (4.5.1).
2. Our considerations prove that the volume of a complete hyperbolic manifold of even dimension m must be an integer multiple of $\frac{1}{2}\,\mathrm{Vol}(S^m)$. Furthermore, if the manifold has dimension 4, we know from the work of Ratcliffe and Tschantz [34] that every positive multiple of $\frac{4}{3}\pi^2$ is realized as the volume of a complete 4-dimensional hyperbolic manifold.
3. The situation is very different for odd-dimensional hyperbolic manifolds. A fundamental result by W. Thurston and O. Jørgensen, based on a construction involving hyperbolic Dehn surgeries, shows that the set of volumes of complete hyperbolic 3-manifolds is a closed and non-discrete subset of $\mathbb{R}$, containing infinitely many accumulation points, see [38] and [16].

4.5.4   Further Results on Non Compact Manifolds

An almost immediate consequence of (4.4.14) is the following

Theorem 4.5.3 *Let (M, g) be a non-compact Riemannian manifold of even dimension m and finite topological type. Assume g has an integrable Gauss–Bonnet density, and that there exists an exhausting sequence*

$$M_1 \subset \cdots \subset M_j \subset M_{j+1} \subset M = \bigcup_{j=1}^{\infty} M_j,$$

such that each M_j is a compact m-dimensional manifold with smooth boundary. If

$$\int_{\partial M_j} \left(\|\bar{R}_x\|^k \|h_x\|^{m-1-2k} \right) \mathrm{dvol}_g \to 0 \quad \text{as } j \to \infty \tag{4.5.4}$$

for any integer k such that $0 \le k \le \frac{m}{2} - 1$, then the Gauss–Bonnet Formula holds; that is, we have

$$(2\pi)^{\frac{m}{2}} \chi(M) = \mathcal{K}_m(M, g). \tag{4.5.5}$$

By an observation of M. Gromov, this result applies to the case of complete Riemannian manifolds with pinched negative sectional curvature, that is, whose sectional curvature satisfies $-a \le K \le -b < 0$, see [15, appendix 3]. When the manifold has non-positive curvature, the control of the geometry at infinity is more delicate but some explicit asymptotic conditions have been given by S. Rosenberg in [32]; see also [19] for the case of arithmetic manifolds. The paper [8] by Cheeger and Gromov contains a strikingly stronger result.

4.5.5   The Direct Product of a Ball and a Sphere

Our next example will illustrate that, even in a very simple case, keeping track of the boundary terms in the Gauss–Bonnet–Chern Formula can be tedious. Nevertheless, this example holds significance as it validates the accurate values of the coefficients $b_{m,k}$ in Formula (4.4.16).

Let us denote the standard metric on $\mathbb{R}^{p+1}$ as g_1 and the standard metric on $\mathbb{R}^{q+1}$ as g_2. Their respective restrictions to the unit spheres S^p and S^q are denoted by $\bar{g}_1$ and $\bar{g}_2$. The Riemannian metric on the manifold M is given by $g = g_1 + g_2$, and its restriction to ∂M is $\bar{g} = \bar{g}_1 + \bar{g}_2$. It is important to note that, in the algebra of double forms, $\bar{g}_1^j = 0$ if $j > p$ and $\bar{g}_2^j = 0$ if $j > q$.

With this notation, one can write the curvature tensor R of M as $R = \frac{1}{2}\bar{g}_2 \oslash \bar{g}_2$ (because B^{p+1} is flat, all the curvature comes from S^q). It follows immediately that the Gauss–Bonnet density defined in (4.3.9) vanishes on M:

$$\mathcal{R} = \frac{1}{m!(\frac{m}{2})!}\, C^m(R^{\frac{m}{2}}) = \frac{1}{2^m m!(\frac{m}{2})!}\, C^m(\bar{g}_2^m) = 0,$$

because $m > q$. Furthermore, the second fundamental form of the boundary $\partial M = S^p \times S^q$ is given by $h = \bar{g}_2$. We therefore have for any $0 \le k \le \frac{1}{2}(p+q)$

$$R^k \owedge h^{m-1-2k} = \left(\frac{1}{2}\right)^k \bar{g}_1^{2k} \owedge \bar{g}_2^{m-1-2k},$$

which vanishes if either $2k > p$ or $(m-1-2k) > q$. This implies that

$$R^k \owedge h^{m-1-2k} = \begin{cases} \left(\frac{1}{2}\right)^{\frac{p}{2}} \bar{g}_1^p \owedge \bar{g}_2^q, & \text{if } p = 2k, \\ 0 & \text{else.} \end{cases}$$

Note also that

$$\bar{g}^{m-1} = \bar{g}^{p+q} = \sum_{j=0}^{p+q} \binom{p+q}{j} \bar{g}_1^j \owedge \bar{g}_2^{p+q-j} = \frac{(p+q)!}{p!\,q!} \bar{g}_1^p \owedge \bar{g}_2^q.$$

We thus have $\mathcal{Q}_k(R, h \mid \partial M, \bar{g}) = 0$ if $k \ne \frac{p}{2}$ and

$$
\begin{aligned}
\mathcal{Q}_{\frac{p}{2}}(R, h \mid \partial M, \bar{g}) &= \frac{1}{(p+q)!\,(\frac{p}{2})!\,q!} \int_{\partial M} C^{p+q}\left(\left(\frac{1}{2}\right)^{\frac{p}{2}} \bar{g}_1^p \owedge \bar{g}_2^q\right) \mathrm{dvol}_{\partial M} \\
&= \frac{1}{(p+q)!\,(\frac{p}{2})!\,q!} \int_{\partial M} C^{p+q}\left(\frac{p!\,q!}{2^{\frac{p}{2}}(p+q)!}\, g^{p+q}\right) \mathrm{dvol}_{\partial M} \\
&= \frac{p!}{2^{\frac{p}{2}}(\frac{p}{2})!}\, \mathrm{Vol}(S^p)\,\mathrm{Vol}(S^q).
\end{aligned}
$$

The latter quantity can be explicitly computed. If p is even and q odd, we have

$$\mathrm{Vol}(S^p)\,\mathrm{Vol}(S^q) = \frac{2^{\frac{p}{2}+1}\left(\frac{p}{2}\right)!}{p!\,(q-1)!!} \cdot (2\pi)^{\frac{p+q+1}{2}}. \tag{4.5.6}$$

We finally conclude that

$$\mathcal{K}_m(M) + \sum_{k=0}^{\frac{m}{2}-1} b_{m,k}\, \mathcal{Q}_k(R, h \mid \partial M, \bar{g})$$

$$= 0 + b_{p+q+1,\frac{p}{2}} \cdot \mathcal{Q}_{\frac{p}{2}}(R, h \mid \partial M, \bar{g})$$

$$= (q-1)!! \, \frac{p!}{2^{\frac{p}{2}} \left(\frac{p}{2}\right)!} \, \mathrm{Vol}(S^p) \, \mathrm{Vol}(S^q)$$

$$= 2 \, (2\pi)^{\frac{p+q+1}{2}}$$

$$= (2\pi)^{\frac{m}{2}} \chi(M).$$

Remark 4.5.4 We emphasize that this calculation is independent of the proof of Proposition 4.4.1. Consequently, it offers an alternative approach to determining the exact value of the coefficients $b_{m,k}$. This example is inspired by the closing sentence in Volume 5 of Spivak's treatise (page 391). Spivak suggests working out the above computation and adds that "after performing the calculation, it should be fun to compare with Chern's paper".

4.5.6 Rotationally Symmetric Metrics

In our last example, we consider the case of a rotationally symmetric smooth Riemannian metric g in a closed ball B_r of radius r. Denoting by $o \in B_m(r)$ its center, we may identify $B^m \setminus \{o\}$ with $(0, r] \times S^{m-1}$, and the metric tensor can be represented as a warped product

$$g = dt^2 + f(t)^2 \, g_0,$$

where g_0 is the standard metric on the unit sphere S^{m-1}. The second fundamental form and the intrinsic curvature tensor of the boundary sphere $S_r = \partial B_r$ are then given by

$$h = \frac{f'(r)}{f(r)} \bar{g} = f'(r) f(r) \, g_0 \quad \text{and} \quad \bar{R} = \frac{g \, \oslash \, g}{2 \, f(r)^2} = \frac{1}{2} \, g_0 \, \oslash \, g_0.$$

The boundary contribution to the Gauss–Bonnet Formula (4.4.12) appears as a sum of terms of the following type:

$$c_{m,k} \, \mathcal{Q}_k(\bar{R}, h \mid S_r, \bar{g}) = \gamma_{m,k} \, \mathrm{Vol}\left(S^{m-1}\right) (f'(r))^{m-2k-1},$$

for some coefficient $\gamma_{m,k}$. The courageous reader will compute and find the value

$$\gamma_{m,k} = \frac{(m-1)! \, c_{m,k}}{2^k k! \, (m-1-2k)!} = \frac{(-1)^{\frac{m}{2}-k-1} (m-1)!}{2^{\frac{m}{2}-k} k! \, (m-2k-1) \left(\frac{m}{2}-k-1\right)!}.$$

The Gauss–Bonnet Formula for (B_r, g) is then the following identity:

$$\int_{B_r} \mathcal{R} \, dvol_g + \text{Vol}(S^{m-1}) \sum_{k=0}^{\frac{m}{2}-1} \gamma_{m,k} \, (f'(r))^{m-2k-1} = (2\pi)^m. \tag{4.5.7}$$

Let us look more closely at the cases of the Euclidean, spherical and hyperbolic balls. In the Euclidean case, we have $\mathcal{R} = 0$ and $f(t) = t$. Then $f'(t) = 1$ and (4.5.7) reduced to the following summation of the coefficients $\gamma_{m,k}$, which holds when m is an even integer

$$\sum_{k=0}^{\frac{m}{2}-1} \gamma_{m,k} = \frac{(2\pi)^{\frac{m}{2}}}{\text{Vol}(S^{m-1})} = (m-2)!! \tag{4.5.8}$$

Consider next the ball of radius r in the hyperbolic space $B_r \subset \mathbb{H}^m$ (with m even). We have $f(r) = \sinh(r)$, and therefore (4.5.7) gives us

$$\int_{B_r} \mathcal{R} \, dvol_g = \frac{(-2\pi)^{\frac{m}{2}}}{\frac{1}{2} \text{Vol}(S^m)} \text{Vol}_{\mathbb{H}^m}(B_r) = \frac{(-2\pi)^{\frac{m}{2}} \text{Vol}(S^{m-1})}{\frac{1}{2} \text{Vol}(S^m)} \int_0^r (\sinh(t))^{m-1} \, dt.$$

Using now (4.5.2), we see that the Gauss–Bonnet Formula (4.5.7) for the hyperbolic ball is equivalent to the identity

$$\frac{(-1)^{\frac{m}{2}} m!}{2^{\frac{m}{2}} \left(\frac{m}{2}\right)!} \int_0^r (\sinh(t))^{m-1} \, dt + \sum_{k=0}^{\frac{m}{2}-1} \gamma_{m,k} (\cosh(r))^{m-2k-1} = (m-2)!!$$

A direct derivation of this identity (without invoking the Gauss–Bonnet Formula) can be achieved through an induction-based argument, starting from the relation:

$$(m-1) \int_0^r (\sinh(t))^{m-1}(t) dt = \cosh(r) \left(\cosh(r)^2 - 1\right)^{\frac{m}{2}-1}$$
$$- (m-2) \int_0^r (\sinh(t))^{m-3}(t) dt,$$

which is easily verified with an integration by parts.

The argument and the calculations are similar in the spherical case. The corresponding integral identity being

$$\frac{m!}{2^{\frac{m}{2}} \left(\frac{m}{2}\right)!} \int_0^r (\sin(t))^{m-1} \, dt + \sum_{k=0}^{\frac{m}{2}-1} \gamma_{m,k} (\cos(r))^{m-2k-1} = (m-2)!!$$

We leave the detailed verification to the interested reader.

4.6 Miscellaneous Remarks

4.6.1 Remarks on Signs and Other Conventions

In the literature, including in Spivak's books, the curvature tensor R is often defined with the opposite sign. Nonetheless, we chose to follow the convention as in Gray's book [14], as it aligns well with the formalism of double forms. However, the crucial point to note is that the signs in our Lemma A.2 match those in [36] and [40].

These references have also a different convention on indices. They write the connection and curvatures forms as

$$\omega^i_j = \omega_{ij} \quad \text{and} \quad \Omega^i_j = \Omega_{ij}. \tag{4.6.1}$$

Our convention is justified by our choice to work with the covariant curvature tensor R_{ijkl} rather than the mixed one $R^i{}_{ijkl}$. Observe however that both index conventions yield the same quantities, as the general transformation rule

$$\Omega^i_j = \sum_k g_{ik} \Omega^k_j$$

reduces to (4.6.1) due to the orthonormality of the moving frame.

4.6.2 Comparison with Chern's Papers

We strongly encourage the reader to look into Chern's two original articles. Although not easily accessible, they are concise, elegant, and remarkably insightful. Both papers start with a succinct review of the moving frame method, in Élie Cartan's style. In this approach, given a moving frame $Pe_1 \cdots e_m$, the following equations are posed:

$$dP = \sum_i \omega_i e_i, \qquad \Omega_{ij} = d\omega_{ij} - \sum_k \omega_{ik} \omega_{kj},$$

$$de_i = \sum_j \omega_{ij} e_j, \qquad d\Omega_{ij} = \omega_{ik} \Omega_{kj} - \omega_{jk} \Omega_{ki}.$$

If, following Cartan, we interpret the symbol P (the "point P itself") as representing the identity map of the manifold to itself, then dP is the identity map in the tangent space. The first equation tells us that $\omega_i = \theta^i$ is the dual coframe. The second and third equations are the structure equations, with de_i interpreted as ∇e_i. The last equation represents the Bianchi identity. Cartan, and Chern after him, use these equations as *definitions* of the connection and curvature forms. Note however that these definitions yield the opposite sign compared to what appears in the present chapter:

$$\omega_{ij}^{\text{Chern}} = -\omega_{ij}, \qquad \Omega_{ij}^{\text{Chern}} = -\Omega_{ij}.$$

This differences in sign are reflected in several equations. Another important difference is that Chern does not state our Proposition 4.4.1 concerning manifolds with a vector field. Instead, he works with the unit tangent bundle SM of the manifold M and proves that the pullback of the Pfaffian $\text{Pf}(\Omega)$ is an exact form on SM. This operation is called a *transgression* of the corresponding differential form. The calculation is essentially the same as the one presented in our proof of Proposition 4.4.1.

Our presentation of the proof shows that working on the sphere bundle and transgressing the Pfaffian are not required to work out Chern's argument and prove the intrinsic Gauss–Bonnet Formula. However, the concept of transgression has played a central role in the theory of characteristic classes.

4.6.3   Double Factorials and the Volume of Spheres

Here we briefly discuss some quantities that have been used throughout the chapter. The *double factorial* is inductively defined for any integer $n \geq -1$ by the conditions $(-1)!! = 0!! = 1$ and

$$n!! = n(n-2)!!,$$

for $n \geq 1$. We thus have for even integers:

$$n!! = n \cdot (n-2) \cdot (n-4) \cdots 2 = 2^{\frac{n}{2}} \cdot \left(\tfrac{n}{2}\right)!$$

and for odd integers

$$n!! = n \cdot (n-2) \cdot (n-4) \cdot 1 = \frac{(n+1)!}{2^{\frac{n+1}{2}} \cdot \left(\frac{n+1}{2}\right)!} = \frac{2^{\frac{n+1}{2}} \Gamma\left(\frac{n+1}{2}\right)}{\sqrt{\pi}},$$

where $\Gamma(z) = \int_0^\infty t^{z-1} e^{-t} \, dt$ is the Euler Gamma function.

The double factorial has the following simple combinatorial interpretation: For a set with n elements, the number of different ways to form $\frac{n}{2}$ unordered pairs if n is even, or $\frac{n-1}{2}$ unordered pairs with one element left unmatched if n is odd, is given by

$$\begin{cases} (n-1)!! & \text{if } n \text{ is even,} \\ n!! & \text{if } n \text{ is odd.} \end{cases}$$

The other ubiquitous quantity is the volume (or should we call it the area?) of the unit sphere $S^d \subset \mathbb{R}^{d+1}$. It can be written in several different formalisms, starting with

$$\text{Vol}(S^d) = 2\,\frac{\pi^{\frac{d+1}{2}}}{\Gamma\left(\frac{d+1}{2}\right)}.$$

We see that for d odd we have

$$\text{Vol}(S^d) = 2\,\frac{\pi^{\frac{d+1}{2}}}{\left(\frac{d-1}{2}\right)!} = \frac{(2\pi)^{\frac{d+1}{2}}}{(d-1)!!},$$

and for even d:

$$\text{Vol}(S^d) = \frac{2^{d+1}\pi^{\frac{d}{2}}\left(\frac{d}{2}\right)!}{d!} = \frac{2\,(2\pi)^{\frac{d}{2}}}{(d-1)!!}.$$

4.7 Exercises

In conclusion of this Chapter, we propose a few exercises.

1. Verify that, in the case of a surface, each of the three formulations (4.4.16) yields the classical 2-dimensional Gauss–Bonnet Formula (this amounts essentially to identifying the boundary term as the integral of the geodesic curvature, with the correct orientation).

2. Recall that the Pfaffian of an $m \times m$ matrix A, where m is even, is the following sum involving $m!$ terms:

$$\text{Pf}(A) = \frac{1}{m!!} \sum_{\sigma \in S_m} (-1)^\sigma a_{\sigma(1)\sigma(2)} \cdot a_{\sigma(3)\sigma(4)} \cdots a_{\sigma(m-1)\sigma(m)}.$$

In this excercice we will show that the Pfaffian can be rewritten as a sum of only $(m-1)!!$ terms, each involving a different partition of the index set $\{1, \ldots, m\}$ in disjoint pairs. To this aim, we consider the following subset

$$\mathcal{P}_m = \{\sigma \in S_m \mid \sigma_{2i+2} > \max\{\sigma_{2i}, \sigma_{2i+1}\},\ 1 \leq i < \tfrac{m}{2}\}$$

of S_m. Your task is to first verify the following identity:

$$\frac{1}{m!!}|S_m| = (m-1)!! = |\mathcal{P}_m|.$$

Then prove that

$$\text{Pf}(A) = \sum_{\sigma \in \mathcal{P}_m} (-1)^{\sigma} a_{\sigma(1)\sigma(2)} \cdot a_{\sigma(3)\sigma(4)} \cdots a_{\sigma(m-1)\sigma(m)}.$$

3. Use the Gauss–Bonnet formula to prove that if M_1 and M_2 are two closed manifolds, then $\chi(M_1 \times M_2) = \chi(M_1)\chi(M_2)$. Extend this to the case where one of the manifolds has a non-empty boundary.
4. Generalize the Gauss–Bonnet Formula to the case of a closed non-orientable manifold. Can this be done for manifolds with non-empty boundaries?
5. Verify Eq. (4.4.4) in the proof of Proposition 4.4.1.
6. Prove that if N is a hypersurface in a Riemannian manifold (M, g), then the Gauss Equation can be written as

$$\bar{R} = R + \frac{1}{2} h \oslash h,$$

where R is the curvature tensor of the ambient manifold (M, g), $\bar{R}$ is the intrinsic curvature tensor of the submanifold N and h is the second fundamental form of N.
Two proofs (at least) are possible. One proof is to reduce the above formula to the Gauss Equation expressed in a formalism familiar to the reader. The other is to try and write a direct proof using the moving frame formalism.
7. Let M be a 4-dimensional manifold. Use formula (4.4.5) to calculate the forms Φ_0, Φ_1, Ψ_0, and Ψ_1. Find then an explicit primitive of the Pfaffian in this case (follow the steps in the proof of Proposition 4.4.1).
8. Our proof of the Gauss–Bonnet–Chern Formula uses the Poincaré–Hopf theorem. Show that, in fact, these two results are equivalent.
The argument will proceed in two steps. First, reverse the proof of Theorem 4.4.3 to demonstrate that the sum of the indices of a vector field is an invariant of the manifold. Then, identify this invariant by selecting a well-chosen vector field (for example, the gradient of a Morse function).
9. Let $h_1, h_2 \in \mathcal{D}^{1,1}(M)$ be two double forms of type $(1, 1)$ and set $A = h_1 \oslash h_2$. Show that if h_1 and h_2 are symmetric, then A satisfies the *first Bianchi identity:*

$$A(X, Y; U, V) + A(Y, U; X, V) + A(U, X; Y, V) = 0.$$

10. With the help of a computer algebra system such as Maple, Mathematica, or SageMath, verify the validity of some, or all, of the following identities appearing in the present chapter: (4.4.10), (4.4.11), (4.4.13), (4.4.15), (4.5.6) and (4.5.8). After mastering this task, you should be able to navigate through the computational details of Example 4.5.6.
11. In Example 4.5.5 we have discussed the Gauss–Bonnet Formula for the manifold $M = B^{p+1} \times S^q$ assuming p is even and q is odd. What happens if, instead, p is odd and q is even?

Acknowledgments The author extends his gratitude to Niky Kamran for his encouragements and meticulous review of the manuscript, as well as to Ruth Kellerhals for her comments on the case of space forms and for pointing to the precise reference in Ratcliffe's treatise. We also acknowledge Adrien Marcone for the valuable discussions we had regarding the Gauss–Bonnet Chern formula; his thesis [29] has been a helpful resource in preparing the current chapter. Special thanks are also due to Jean-Pierre Bourguignon for his comments and for drawing our attention to the notable paper [27] by Lanczos.

Appendix A: Some Background in Differential Geometry

In this appendix, we review some of the geometric concepts employed in this chapter and establish our notation. Additionally, we provide a detailed introduction to the moving frame method.

A.1   Some Linear Algebra on the (co)tangent Space

By its very definition, a Riemannian metric g on a smooth manifold M endows each tangent space with a Euclidean structure. The scalar product of two tangent vectors X and Y in T_pM will be denoted interchangeably as $g_p(X, Y)$ or $\langle X, Y \rangle_p$, and the Riemannian norm will be represented by $\|X\|_p = \sqrt{\langle X, Y \rangle_p}$. The subscript p will be omitted when there is no ambiguity. The same notation will be used when X and Y are smooth vector fields over M; in this case, the scalar product becomes a smooth function on M.

To any tangent vector $X \in T_pM$, we associate a covector $X^\flat \in T_p^*M$ defined by the relation $X^\flat(Y) = \langle X, Y \rangle$ for any $Y \in T_pM$. This yields a linear isomorphism from the tangent space T_pM to the cotangent space T_p^*M, and the inverse isomorphism maps a covector $\varphi \in T_p^*M$ to the unique tangent vector $\varphi^\sharp \in T_pM$ such that $\langle \varphi^\sharp, Y \rangle = \varphi(Y)$, again for any $Y \in T_pM$. The operators $\sharp$ and $\flat$ are sometimes called the *musical isomorphisms*.[6] A scalar product is then defined on $\Lambda^k T_p^*M$, first for $k = 1$ by

$$\langle \alpha, \beta \rangle = \langle \alpha^\sharp, \beta^\sharp \rangle$$

for any $\alpha, \beta \in T_p^*M$, then by the formula:

$$\langle \alpha_1 \wedge \cdots \wedge \alpha_k, \beta_1 \wedge \cdots \wedge \beta_k, \rangle = \det\left(\langle \alpha_i, \beta_j \rangle \right), \qquad (A.1)$$

---

[6] The name and notation have apparently been introduced by Marcel Berger in the 1970s.

where $\alpha_i, \beta_j \in T_p^*M$. The "volume element" at the point p is the unique element $\Theta = \Theta_p \in \Lambda^m T_p^*M$ representing the natural orientation and such that $\|\Theta\| = 1$. Its existence and uniqueness follows from the fact that $\dim\left(\Lambda^m T_p^*M\right) = 1$. It is characterized by the condition

$$\Theta_p(e_1, \ldots, e_m) = 1,$$

for any positively oriented orthonormal basis $\{e_1, \ldots, e_m\}$ of T_pM. The volume element is independent of the chosen moving frame, making it a globally defined m-form on the manifold. We will commonly use the notation:

$$\mathrm{dvol}_g = \Theta.$$

This notation, however, does not assume that Θ is a closed form. Instead, we view it as a measure on the manifold. In local coordinates, it can be expressed as:

$$\mathrm{dvol}_g = \sqrt{\det(g_{ij})}\, dx_1 \cdots dx_m,$$

where the g_{ij} are the components of the metric tensor.

We recall that the *Hodge star operator* is the isomorphism: $* : \Lambda^k T_p^*M \to \Lambda^{m-k} T_p^*M$ which is defined by the condition

$$\alpha \wedge (*\alpha) = \|\alpha\|^2 \Theta.$$

The dual basis of a positively oriented orthonormal basis $\{e_1, \ldots, e_m\} \subset T_pM$ will be denoted by $\{\theta^1, \ldots, \theta^m\} \subset T_p^*M$ of T_pM. It is given by $\theta^i = e_i^{\flat}$ and we have

$$\Theta = \theta^1 \wedge \cdots \wedge \theta^m.$$

It is easy to check that

$$*\,\theta^{i_1} \wedge \cdots \wedge \theta^{i_k} = \varepsilon\, \theta^{j_1} \wedge \cdots \wedge \theta^{j_{n-k}},$$

where $\{i_1, \ldots, i_k, j_1, \ldots, j_{m-k}\}$ is any permutation of $\{1, \ldots, m\}$ and $\varepsilon = \pm 1$ is its signature. Furthermore, the collection

$$\{\theta^{i_1} \wedge \cdots \wedge \theta^{i_k}\}_{i_1 < i_2 < \cdots < i_k}$$

forms an orthonormal basis of $\Lambda^k T_p^* M$. Using these operators, we define the isomorphism $\varpi : T_p M \to \Lambda^{n-1} T_p^* M$ by[7]

$$\varpi_X = *(X^\flat). \tag{A.2}$$

This isomorphism will play an important role in the sequel; note that ϖ_X is also characterized by the condition

$$X^\flat \wedge \varpi_X = \|X\|^2 \, \Theta.$$

As an important example, observe that if $\{e_1, \ldots, e_m\}$ is a positively oriented orthonormal basis of $T_p M$, then

$$\varpi_{e_m} = (-1)^{m-1} \theta^1 \wedge \cdots \wedge \theta^{m-1}. \tag{A.3}$$

A.2   Connection and Curvature

Riemannian geometry is of course much more than linear algebra in tangent spaces. To compare nearby tangent spaces effectively, we need the concept of connection. In Kozsul's formalism, the Levi-Civita connection of (M, g) is seen as the unique operator that associates to any pair of smooth vector fields X, Y on M a new vector field, denoted as $\nabla_X Y$, and satisfying the following conditions:

(i) $\nabla_X(Y_1 + Y_2) = \nabla_X Y_1 + \nabla_X Y_2$ and $\nabla_{X_1 + X_2} Y = \nabla_{X_1} Y + \nabla_{X_2} Y$.
(ii) $\nabla_{fX} Y = f \nabla_X Y$ and $\nabla_X (fY) = f \nabla_X Y + X(f)Y$ for any $f \in C^\infty(M)$.
(iii) $\nabla_X Y - \nabla_Y X = [X, Y]$ (the Lie bracket of X and Y).
(iv) $Z\langle X, Y \rangle = \langle \nabla_Z X, Y \rangle + \langle X, \nabla_Z Y \rangle$.

We will denote by R the associated covariant curvature tensor. It is the $(0, 4)$-tensor field defined by

$$R(X, Y; Z, W) = \langle \nabla_{[X,Y]} Z + \nabla_Y \nabla_X Z - \nabla_X \nabla_Y Z, W \rangle.$$

Its basic properties are:

(a) R is $C^\infty(M)$-multilinear.
(b) $R(X, Y, Z, W) = R(Z, W, X, Y) = -R(Y, X, Z, W) = -R(X, Y, W, Z)$.
(c) $R(X, Y, Z, W) + R(Y, Z, X, W) + R(Z, X, Y, W) = 0$ (the first Bianchi Identity).

---

[7] Using the interior product, one can also define it as $\varpi_X = \iota_X(\Theta)$.

A consequence of property (a) is that the value of $R(X, Y, Z, W)$ at a point $p \in M$ only depends on the values of the vectors $X_p, Y_P, Z_p, W_p \in T_pM$ at this point and not on the global, or even local, behavior of the vector fields. A consequence of (b) is that if $v, w \in T_pM$ are linearly independent, then the scalar

$$K(v, w) = \frac{R(v, w; v, w)}{\|v\|^2 \|w\|^2 - \langle u, v \rangle^2}$$

only depends on the 2-plane $\sigma = \mathrm{Span}(u, v) \subset T_pM$ spanned by these vectors. This value is the *sectional curvature* of σ.

A.3   Moving the Frame with Élie Cartan

The calculus of differential forms was developed in the 1920s by Élie Cartan, who employed it as a convenient formalism in Riemannian geometry.[8] Helpful references include Chapter 7 in [36, Vol II], Chapter 4 in Willmore's book [40], and the books [6, 7] by Cartan himself.

Here we briefly recall the basic definitions and we state and prove the main equations that are needed in this chapter. A *moving frame* in some open set U of the Riemannian manifold (M, g) is simply a collection of m smooth vector fields $e_1, \ldots, e_m : U \to TU$ defined on U such that $\langle e_i, e_j \rangle = \delta_{ij}$ at every point of U. We will always assume the moving frame to be positively oriented. The dual coframe $\theta^1, \ldots \theta^m$ will be denoted as $\theta^i = e_i^\flat$. It is defined by

$$\theta^i(X) = g(e_i, X).$$

Definition A.1 The *connection forms* associated with the moving frame $\{e_i\}$ are the 1-forms defined on U by:

$$\omega_{ij}(X) = g(e_i, \nabla_X e_j) = \theta^i(\nabla_X e_j),$$

and the *curvature forms* are the 2-forms:

$$\Omega_{ij}(X, Y) = R(X, Y; e_i, e_j).$$

Observe that the connection forms are characterized by the identity

$$\nabla_X e_j = \sum_{i=1}^m \omega_{ij}(X) e_i,$$

---

[8] Chern became acquainted with Cartan's calculus first through his interactions with E. Kähler in Hamburg, and later when he met Cartan in Paris. See his comments on Cartan in [23].

note also the skew symmetry:

$$\omega_{ij} + \omega_{ji} = \Omega_{ij} + \Omega_{ji} = 0.$$

The next lemma gives us three useful formulas representing the curvature in the moving frame formalism:

Lemma A.2 *Under the previously established notation for the moving coframe, the curvature forms Ω_{ij} can ve written as*

$$\Omega_{ij} = \frac{1}{2} \sum_{k,l=1}^{m} R_{ijkl}\, \theta^k \wedge \theta^l.$$

And the curvature tensor R is expressed as

$$R = \frac{1}{4} \sum_{i,j,k,l=1}^{m} R_{ijkl}\, \theta^i \wedge \theta^j \otimes \theta^k \wedge \theta^l$$

$$= \frac{1}{2} \sum_{i,j=1}^{m} \Omega_{ij} \otimes \theta^i \wedge \theta^j,$$

where $R_{ijkl} = R(e_i, e_j, e_k, e_l)$.

Proof From the symmetries of the curvature tensor we have $R_{klij} = R_{ijkl} = -R_{ijlk}$, therefore

$$\Omega_{ij} = \sum_{k,l=1}^{m} R_{klij}\, \theta^k \otimes \theta^l = \sum_{k,l=1}^{m} R_{ijkl}\, \theta^k \otimes \theta^l = \frac{1}{2} \sum_{k,l=1}^{m} R_{ijkl}\, \theta^k \wedge \theta^l.$$

This proves the first equation. The second equation is proved as follows

$$R = \sum_{i,j,k,l=1}^{m} R_{ijkl} \left(\theta^i \otimes \theta^j \right) \otimes \left(\theta^k \otimes \theta^l \right) = \sum_{i,j,k,l=1}^{m} R_{ijkl} \left(\theta^k \otimes \theta^l \right) \otimes \left(\theta^i \otimes \theta^j \right)$$

$$= \frac{1}{4} \sum_{i,j,k,l=1}^{m} R_{ijkl} \left(\theta^k \otimes \theta^l - \theta^l \otimes \theta^k \right) \otimes \left(\theta^i \otimes \theta^j - \theta^j \otimes \theta^i \right)$$

$$= \frac{1}{4} \sum_{i,j,k,l=1}^{m} R_{ijkl} \left(\theta^k \wedge \theta^l \right) \otimes \left(\theta^i \wedge \theta^j \right)$$

$$= \frac{1}{2} \sum_{k,l=1}^{m} \Omega_{ij} \otimes \theta^i \wedge \theta^j.$$

$\square$

Proposition A.3 (É. Cartan's Structure Equations) *The exterior derivative of the moving coframe and the connection forms are given by*

$$d\theta^i = -\sum_{j=1}^m \omega_{ij} \wedge \theta^j \quad \text{and} \quad d\omega_{ij} = \Omega_{ij} - \sum_{k=1}^m \omega_{ik} \wedge \omega_{kj}. \tag{A.4}$$

Proof We prove the first structure equation as follows:

$$
\begin{aligned}
d\theta^i(X, Y) &= X(\theta^i(Y)) - Y(\theta^i(X)) - \theta^i([X, Y]) \\
&= Xg(e_i, Y) - Yg(e_i, X) - g(e_i, [X, Y]) \\
&= g(\nabla_X e_i, Y) + g(e_i, \nabla_X Y) - g(\nabla_Y e_i, X) - g(e_i, \nabla_Y X) - g(e_i, [X, Y]) \\
&= g(\nabla_X e_i, Y) - g(\nabla_Y e_i, X) \\
&= \sum_{j=1}^m \big(g(\nabla_X e_i, e_j)g(e_j, Y) - g(\nabla_Y e_i, e_j)g(e_j, X) \big) \\
&= \sum_{j=1}^m \big(\omega_{ji}(X)\theta^j(Y) - \omega_{ji}(Y)\theta^j(X) \big). \\
&= -\sum_{j=1}^m \big(\omega_{ij} \wedge \theta^j \big)(X, Y)
\end{aligned}
$$

(we have used the symmetry of ∇ on the fourth line). To prove the second structure equation, we compute

$$
\begin{aligned}
d\omega_{ij}(X, Y) &= X(\omega_{ij}(Y)) - Y(\omega_{ij}(X)) - \omega_{ij}([X, Y]) \\
&= X(g(e_i, \nabla_Y e_j)) - Y(g(e_i, \nabla_X e_j)) - g(e_i, \nabla_{[X,Y]} e_j) \\
&= (g(\nabla_X e_i, \nabla_Y e_j) + g(e_i, \nabla_X \nabla_Y e_j)) - (g(\nabla_Y e_i, \nabla_X e_j) \\
&\quad - g(e_i, \nabla_Y \nabla_X e_j)) - g(e_i, \nabla_{[X,Y]} e_j) \\
&= g(\nabla_X e_i, \nabla_Y e_j) - g(\nabla_Y e_i, \nabla_X e_j) + g(e_i, \nabla_X \nabla_Y e_j \\
&\quad - \nabla_Y \nabla_X e_j - \nabla_{[X,Y]} e_j) \\
&= g(\nabla_X e_i, \nabla_Y e_j) - g(\nabla_Y e_i, \nabla_X e_j) - R(X, Y; e_j, e_i).
\end{aligned}
$$

Observe now that on the one hand

$$g(\nabla_X e_i, \nabla_Y e_j) = \sum_{k=1}^{m} \theta^k(\nabla_X e_i)\theta^k(\nabla_Y e_j) = \sum_{k=1}^{m} \omega_{k,i}(X)\omega_{k,j}(Y)$$

$$= -\sum_{k=1}^{m} \omega_{i,k}(X)\omega_{k,j}(Y),$$

therefore

$$g(\nabla_X e_i, \nabla_Y e_j) - g(\nabla_Y e_i, \nabla_X e_j) = -\sum_{k=1}^{m} \omega_{ik} \wedge \omega_{kj}(X, Y).$$

On the other hand we have

$$d\omega_{ij}(X, Y) = \Omega_{ij}(X, Y) - \sum_{k=1}^{m} \omega_{ik} \wedge \omega_{kj}(X, Y),$$

Because $-R(X, Y; e_j, e_i) = R(X, Y; e_i, e_j) = \Omega_{ij}(X, Y)$. □

Lemma A.4 (The Bianchi Identities) *The curvature forms also satisfy the following identities:*

$$\sum_{j=1}^{m} \Omega_{ij} \wedge \theta^j = 0 \quad \text{and} \quad d\Omega_{ij} = \sum_{k=1}^{m} \left(\Omega_{ik} \wedge \omega_{kj} - \omega_{ik} \wedge \Omega_{kj} \right). \tag{A.5}$$

Proof Since $d^2\theta^i = 0$ and $d^2\omega_{ij} = 0$, we have for $1 \le i \le m$,

$$0 = d\left(\sum_{j=1}^{m} \omega_{ij} \wedge \theta^j \right) = \sum_{j=1}^{m} d\omega_{ij} \wedge \theta^j - \sum_{j=1}^{m} \omega_{ij} \wedge d\theta^j$$

$$= \sum_{j=1}^{m} \left(\Omega_{ij} - \sum_{k=1}^{m} \omega_{ik} \wedge \omega_{kj} \right) \wedge \theta^j + \sum_{j=1}^{m} \omega_{ij} \wedge \left(\sum_{k=1}^{m} \omega_{jk} \wedge \theta^k \right)$$

$$= \sum_{j=1}^{m} \Omega_{ij} \wedge \theta^j + \sum_{j,k=1}^{m} \left(-\omega_{ik} \wedge \omega_{kj} \wedge \theta^j + \omega_{ij} \wedge \omega_{jk} \wedge \theta^k \right)$$

$$= \sum_{j=1}^{m} \Omega_{ij} \wedge \theta^j,$$

which proves the first equation. To prove the second equation, we compute

$$0 = d^2\omega_{ij} = d\left(\Omega_{ij} - \sum_{k=1}^{m}\omega_{ik} \wedge \omega_{kj}\right) = d\Omega_{ij} - \sum_{k=1}^{m}d\omega_{ik} \wedge \omega_{kj} + \sum_{k=1}^{m}\omega_{ik} \wedge d\omega_{kj}$$

$$= d\Omega_{ij} - \sum_{k=1}^{m}\left(\Omega_{ik} - \sum_{s=1}^{m}\omega_{is} \wedge \omega_{sk}\right) \wedge \omega_{kj} + \sum_{k=1}^{m}\omega_{ik} \wedge \left(\Omega_{kj} - \sum_{s=1}^{m}\omega_{ks} \wedge \omega_{sj}\right)$$

$$= d\Omega_{ij} - \sum_{k=1}^{m}\left(\Omega_{ik} \wedge \omega_{kj} - \omega_{ik} \wedge \Omega_{kj}\right)$$

$$+ \sum_{k,s=1}^{m}\left(\omega_{is} \wedge \omega_{sk} \wedge \omega_{kj} - \omega_{ik} \wedge \omega_{ks} \wedge \omega_{sj}\right)$$

$$= d\Omega_{ij} - \sum_{k=1}^{m}\left(\Omega_{ik} \wedge \omega_{kj} - \omega_{ik} \wedge \Omega_{kj}\right).$$

$\square$

We conclude this section by discussing the effect of changing the moving frame. Suppose that $e'_1, \ldots, e'_m$ is another orthonormal moving frame defined on U; then we have

$$e'_j = \sum_{i=1}^{m}a_{ij}e_i,$$

where $a = (a_{ij})$ is a smooth function defined on U with values in the orthogonal group $O(m)$. The corresponding moving coframe is

$$\theta'^j = \sum_{i=1}^{m}a_{ij}\theta^i$$

(indeed we have $\theta'^j(X) = g\left(e'_j, X\right) = \sum_{i=1}^{m}a_{ij}g(e_i, X) = \sum_{i=1}^{m}a_{ij}\theta^i(X)$).
About the connection and curvature forms, we have the

Lemma A.5 *Under a change of moving frame, the connection forms transform as*

$$\omega'_{ij} = \sum_{r,s=1}^{m}a_{si}a_{rj}\omega_{sr} + \sum_{r=1}^{m}a_{ri}da_{rj},$$

and the curvature forms transform as

$$\Omega'_{ij} = \sum_{r,s=1}^{m} a_{si} a_{rj} \Omega_{sr}.$$

Remark In this lemma we don't need to assume that our moving frames are positively oriented. In matrix notation, the transformation formula for the connection can be written as

$$\omega' = a^{-1} \omega a + a^{-1} da$$

(compare with [36, vol. 2, page 280]). This is the standard *gauge transformation formula* for connections in a principal bundle. Note however that in our case the inverse matrix a^{-1} is simply its transpose.

Proof Recall first that

$$\nabla_X e'_j = \nabla_X e'_j = \nabla_X \left(\sum_{r=1}^{m} a_{rj} e_r \right) = \sum_{r=1}^{m} a_{rj} \nabla_X e_r + \sum_{r=1}^{m} da_{rj}(X) e_r.$$

Therefore

$$\omega'_{ij}(X) = \theta'^i \left(\nabla_X e'_j \right) = \sum_{r,s=1}^{m} a_{si} a_{rj} \left(\theta^s \left(\nabla_X e_r \right) + da_{rj}(X) \theta^s (e_r) \right)$$

$$= \sum_{r,s=1}^{m} a_{si} a_{rj} \omega_{sr}(X) + \sum_{r=1}^{m} a_{ri} da_{rj}(X).$$

This proves the first formula. The proof of the second formula is immediate from the definition:

$$\Omega'_{ij}(X, Y) = R(X, Y; e'_i, e'_j) = \sum_{r,s=1}^{m} a_{si} a_{rj} R(X, Y; e_r, e_s) = \sum_{r,s=1}^{m} a_{si} a_{rj} \Omega_{sr}(X, Y),$$

because of the tensorial nature of R.

□

References

1. C.B. Allendoerfer, The Euler number of a Riemann manifold. Am. J. Math. **62**, 243–248 (1940)
2. C.B. Allendoerfer, A. Weil, The Gauss–Bonnet theorem for Riemannian polyhedra. Trans. Am. Math. Soc. **53**, 101–129 (1943)

3. A.L. Besse, *Einstein Manifolds*, Reprint of the 1987 edition. Classics in Mathematics (Springer-Verlag, Berlin, 2008)
4. J.P. Bourguignon, "Entretien avec un optimiste, S. S. Chern" [Interview with an optimist, S. S. Chern]. Gaz. Math. **48**, 5–10 (1991). An English translation was published in S. S. Chern: A great geometer of the twentieth century (1992). The video of the interview is available on the Youtube channel of IHES: https://www.youtube.com/watch?v=vConuqi5vT0
5. G. Bredon, *Topology and Geometry*. Graduate Texts in Mathematics (Springer-Verlag, Berlin, 1993)
6. É. Cartan, *La Géométrie des espaces de Riemann* (Gauthier-Villars, Paris, 1925)
7. É. Cartan, *Riemannian Geometry in an Orthogonal Frame* (World Scientific Publishing, River Edge, 2001)
8. J. Cheeger, M. Gromov, On the characteristic numbers of complete manifolds of bounded curvature and finite volume, in *Differential Geometry and Complex Analysis*, vol. dedicated to H. E. Rauch (Springer, Berlin/Heidelberg, 1985), pp. 115–154
9. S.S. Chern, A simple intrinsic proof of the Gauss–Bonnet formula for closed Riemannian manifolds. Ann. Math. (2) **45**, 747–752 (1944)
10. S.S. Chern, On the curvatura integra in a Riemannian manifold. Ann. Math. (2) **46**, 674–684 (1945)
11. G. De Rham, *Differentiable Manifolds: Forms, Currents, Harmonic Forms*. Grundlehren der mathematischen Wissenschaften (Springer-Verlag, Berlin, 1984)
12. W. Fenchel, On the total curvature of Riemannian manifolds, I. J. London Math. Soc. **15**, 15–22, (1940)
13. D.H. Gottlieb, All the Way with Gauss–Bonnet and the sociology of mathematics. Am. Math. Mon. **103**(6), 457–469, (1996)
14. A. Gray, *Tubes*, 2nd edn. Progress in Mathematics, vol. 221 (Birkhäuser Verlag, Basel, 2004)
15. M. Gromov, Volume and bounded cohomology. Inst. Hautes Études Sci. Publ. Math. **56**, 5–99, (1982)
16. M. Gromov, Hyperbolic manifolds according to Thurston and Jørgensen, in *Séminaire Bourbaki*, vol. 1979/1980. Lecture Notes in Mathematics, vol. 842 (Springer-Verlag, Berlin, 1981), pp. 40–53
17. V. Guillemin, A. Pollack, *Differential Topology* (Prentice-Hall, Englewood Cliffs, 1974)
18. A. Haefliger, Quelques remarques sur les applications différentiables d'une surface dans le plan. Ann. Inst. Fourier (Grenoble) **10**, 47–60 (1960)
19. G. Harder, A Gauss–Bonnet formula for discrete arithmetically defined groups. Ann. Sci. École Norm. Sup. (4), 409–455 (1971)
20. H. Hopf, Über die Curvatura integra geschlossener Hyperflächen. Math. Ann. **95**, 340–367 (1925)
21. H. Hopf, Die Curvatura Integra Clifford-Kleinscher Raumformen. Nachr. Ges.Wiss. Gottingen, Math.-Phys. Kl. 131–141 (1925)
22. H. Hopf, Vektorfelder in *n*-dimensionalen Mannigfaltigkeiten. Math. Ann. **96**, 225–250 (1926)
23. A. Jackson, Interview with Shiing-Shen Chern. Not. Am. Math. Soc. **45**(7), 860–865 (1998)
24. R. Kellerhals, T. Zehrt, The Gauss–Bonnet formula for hyperbolic manifolds of finite volume. Geom. Dedicata. **84**, 49–62 (2001)
25. R. Kulkarni, On the Bianchi Identities. Math. Ann. **199**, 175–204 (1972)
26. M. Labbi, Double forms, curvature structures and the (p,q)-curvatures. Trans. Am. Math. Soc. **357**, 3971–3992 (2005)
27. C. Lanczos, A remarkable property of the Riemann–Christoffel tensor in four dimensions. Ann. Math. (2) **39**, 842–850 (1938)
28. I. Madsen, J. Tornehave, *From Calculus to Cohomology: De Rham Cohomology and Characteristic Classes* (Cambridge University Press, Cambridge, 1997)
29. A. Marcone, *A Gauss–Bonnet Theorem for Asymptotically Conical Manifolds and Manifolds with Conical Singularities*. Ph.D. Thesis No. 9275, EPFL, 2019. Available at https://infoscience.epfl.ch/record/262901

30. J. Milnor, *Topology from the Differentiable Viewpoint* (The University Press of Virginia, Charlottesville, 1965)
31. J. Nash, The imbedding problem for Riemannian manifolds. Ann. Math. (2) **63**, 20–63 (1956)
32. S. Rosenberg, On the Gauss–Bonnet theorem for complete manifolds. Trans. Am. Math. Soc. **287**, 745–753 (1985)
33. J.G. Ratcliffe, *Foundations of Hyperbolic Manifolds*, 3rd expanded edn. Graduate Texts in Mathematics, vol. 149 (Springer, Berlin, 2019)
34. J. Ratcliffe, S. Tschantz, The volume spectrum of hyperbolic 4-manifolds. Exp. Math. **9**, 101–125, (2000)
35. H. Samelson, On immersion of manifolds. Can. J. Math.**12**, 529–534, (1960)
36. M. Spivak, *A Comprehensive Introduction to Differential Geometry*, vol. I–V, 3rd edn. (Publish or Perish, Inc., Lombard, 1999)
37. J.A. Thorpe, *Elementary Topics in Differential Geometry*. Undergraduate Texts in Mathematics (UTM) (Springer-Verlag, Berlin, 1979)
38. W.P. Thurston, *The Geometry and Topology of 3-Manifolds*. Lecture Notes (Princeton University, Princeton, 1979). Available at http://msri.org/publications/books/gt3m
39. H. Weyl, On the volume of tubes. Am. J. Math. **61**, 461–472 (1939)
40. T. Willmore, *Riemannian Geometry* (The Clarendon Press/Oxford University Press, New York, 1993)
41. H. Wu, Historical development of the Gauss–Bonnet Theorem. Sci. China Ser. A. **51**, 777–784 (2008)

Chapter 5
Quaternions, Monge–Ampère Structures and k-Surfaces

Graham Smith

Abstract In Labourie (Geom Funct Anal 7: 496–534, 1997) Labourie developed a theory of immersed surfaces of prescribed extrinsic curvature which has since found widespread applications in hyperbolic geometry, general relativity, Teichmüller theory, and so on. In this chapter, we present a quaternionic reformulation of these ideas. This yields simpler proofs of the main results whilst pointing towards the higher-dimensional generalisation studied by the author in Smith (Math Ann 335(1): 57–95, 2013).

Keywords Extrinsic curvature · Monge–Ampère equation · J-holomorphic curves

Classification AMS 53A05, 12E15, 35J96

5.1 Introduction

5.1.1 Introduction

Immersed surfaces of prescribed extrinsic curvature in 3-dimensional manifolds have fascinated mathematicians for almost two centuries. Following the pioneering work [16] of Labourie, remarkable developments have been made in our understanding of these objects, leading to striking applications across a broad range of mathematical theories.[1] In this chapter, we propose a quaternionic reformulation of Labourie's ideas. Not only will this yield simpler proofs of the main results, but

---

[1] The reader may consult, for example, [2–4, 15, 23, 26], for a selection of applications of these techniques.

G. Smith (✉)
Departamento de Matemática, Pontifícia Universidade Católica do Rio de Janeiro (PUC-Rio), Rio de Janeiro, Brazil

© The Author(s), under exclusive license to Springer Nature Switzerland AG 2024
A. Papadopoulos (ed.), *Surveys in Geometry II*,
https://doi.org/10.1007/978-3-031-43510-2_5

it will also point towards their higher-dimensional generalisations studied by the author in [25].

We first recall the main elements of Labourie's work as it pertains to immersed surfaces. Let $X := (X, h)$ be a complete, oriented, 3-dimensional Riemannian manifold, let TX denote its tangent bundle, and let $SX \subseteq TX$ denote its unit sphere bundle. We define an (oriented) *immersed surface* in X to be a pair (S, e), where S is an oriented surface, and $e : S \to X$ is a smooth immersion. Given such a pair, we denote by $\nu_e : S \to SX$ its unit normal vector field compatible with the orientation, by I_e, II_e and III_e its first, second and third fundamental forms respectively, by A_e its shape operator, and by $K_e := \mathrm{Det}(A_e)$ its *extrinsic curvature function*. We say that the immersed surface is *infinitesimally strictly convex (ISC)* whenever its second fundamental form is positive definite, we say that it is *quasicomplete* whenever it is complete with respect to the Riemannian metric $\mathrm{I}_e + \mathrm{III}_e$, and, given a smooth function $\kappa : SX \to \mathbb{R}$, we say that its extrinsic curvature is *prescribed* by κ whenever

$$K_e := \kappa \circ \nu_e.$$

We denote $\hat{e} := \nu_e$, and we call the immersed surface $(S, \hat{e})$ the *Gauss lift* of (S, e). Note that quasicompleteness of (S, e) is equivalent to completeness of its Gauss lift. Labourie's key insight is that the Gauss lift of any ISC immersed surface of prescribed extrinsic curvature is a pseudo-holomorphic curve for some suitable almost complex structure. This allows the powerful theory developed by Gromov in [8] to be applied. The first consequence is the following compactness result.

Theorem 5.1 (Labourie's Compactness Theorem) *Let $\kappa : SX \to \mathbb{R}$ be a smooth, positive function, and let (S_m, e_m, p_m) be a sequence of quasicomplete, pointed, ISC immersed surfaces in X of extrinsic curvature prescribed by κ. If the sequence $(e_m(p_m))$ is precompact in X, then the sequence $(S_m, \hat{e}_m, p_m)$ of Gauss lifts is precompact in the smooth Cheeger–Gromov topology.*

Remark 5.1 This is proven in Theorem 5.12. The smooth Cheeger–Gromov topology is described in Sect. 5.3.2.

Remark 5.2 We have stated this result in its simplest possible form. It may however be generalised in a number of ways. For example, the ambient space X as well as the function κ can be allowed to vary with m, the conditions of completeness of the ambient space and quasicompleteness of the immersed surfaces can also be relaxed, and so on. We refer the reader to Sect. 5.3.2, where potential generalisations are explained in greater detail. The reader may likewise consult [28] for recent developments concerning the case with non-trivial boundary.

Significantly, Theorem 5.1 is only of limited use without an understanding of degenerate limits, that is, those limits that are not Gauss lifts of immersed surfaces. This is addressed by Labourie in his second key result. For any complete geodesic Γ in X, we denote by $N\Gamma$ its unit normal bundle. We define a *tube* in SX to be an immersed surface $(S, \hat{e})$ which is a cover of $N\Gamma$ for some complete geodesic Γ.

Theorem 5.2 (Labourie's Dichotomy) *With the notation of Theorem 5.1, every accumulation point of the sequence (S_m, e_m, p_m) is either a tube or the Gauss lift of some quasicomplete, pointed, ISC immersed surface in X of extrinsic curvature prescribed by κ.*

Remark 5.3 This is proven in Theorem 5.12.

These two remarkable results and their variants have since become the basis of a rich theory of surfaces of prescribed extrinsic curvature in Riemannian and semi-Riemannian manifolds. It will be the object of this chapter to present a quaternionic framework within which they can be proved, allowing us, on the one hand, to emphasize their hyperkähler nature, and, on the other, to indicate their higher-dimensional generalisations.

5.2 Quaternions and Bernstein-Type Theorems

5.2.1 Quaternions

We begin our work with a detailed review of Hamilton's theory of quaternions. Introduced in 1843, this theory has enjoyed a striking revival over the past forty years on account of the remarkably simple approaches it provides to various deep mathematical and physical phenomena (see, for example, [7]). Its application to the study of prescribed curvature surfaces, which will be addressed in detail in the sequel, presents yet another instance of the ubiquity of this theory that makes it so intriguing.

Let $\mathbb{H}$ denote the algebra of quaternions. This is the associative, unital algebra over $\mathbb{R}$ generated by the 3 elements $\mathbf{i}$, $\mathbf{j}$ and $\mathbf{k}$, with the relations

$$\mathbf{i}^2 = \mathbf{j}^2 = \mathbf{k}^2 = \mathbf{i} \cdot \mathbf{j} \cdot \mathbf{k} = -1.$$

Given an element $x \in \mathbb{H}$ of the form

$$x := a + b\mathbf{i} + c\mathbf{j} + d\mathbf{k}, \tag{5.2.1}$$

its *conjugate* is defined by

$$\bar{x} := a - b\mathbf{i} - c\mathbf{j} - d\mathbf{k}.$$

Conjugation is an anti-involution of $\mathbb{H}$ in the sense that, for all x, y,

$$\overline{x \cdot y} = \bar{y} \cdot \bar{x}. \tag{5.2.2}$$

Its $(+1)$-eigenspace, the space of *real quaternions*, is a 1-dimensional subspace which we denote by $\mathcal{R}$. Its (-1)-eigenspace, the space of *imaginary quaternions*, is a 3-dimensional subspace which we denote by $\mathcal{I}$. We likewise denote by $\mathcal{R}$ and $\mathcal{I}$ the respective projections onto these subspaces.

By (5.2.2), for all $x \in \mathbb{H}$,

$$\overline{x \cdot \overline{x}} = \overline{\overline{x}} \cdot \overline{x} = x \cdot \overline{x},$$

so that $x \cdot \overline{x}$ is always real. We thus define an inner product by

$$\langle x, y \rangle := \mathcal{R}(x \cdot \overline{y}). \tag{5.2.3}$$

With x as in (5.2.1),

$$\|x\|^2 = a^2 + b^2 + c^2 + d^2,$$

so that (5.2.3) coincides with the standard inner product of 4-dimensional Euclidian space. Furthermore, for all $x, y \in \mathbb{H}$,

$$\|x \cdot y\|^2 = x \cdot y \cdot \overline{x \cdot y} = x \cdot y \cdot \overline{y} \cdot \overline{x} = \|x\|^2 \|y\|^2, \tag{5.2.4}$$

so that length is multiplicative.

For all $x, y, z \in \mathcal{I}$,

$$\mathcal{R}(x \cdot y \cdot z) = \mathcal{R}(\overline{x} \cdot \overline{y} \cdot z) = \mathcal{R}(\overline{y \cdot x} \cdot z) = \mathcal{R}(y \cdot x \cdot \overline{z}) = -\mathcal{R}(y \cdot x \cdot z).$$

We thus define an alternating 3-form over $\mathcal{I}$ by

$$\omega(x, y, z) := -\mathcal{R}(x \cdot y \cdot z). \tag{5.2.5}$$

This is the volume form of $\langle \cdot, \cdot \rangle$ with orientation chosen such that $(\mathbf{i}, \mathbf{j}, \mathbf{k})$ is a positive triple.

By (5.2.4), the sphere $\mathbb{S}^3$ of unit quaternions is a subgroup of $\mathbb{H}$ with inverse given by conjugation.

Lemma & Definition 5.1 *The homomorphism* $h : \mathbb{S}^3 \times \mathbb{S}^3 \to SO(\mathbb{H})$ *given by*

$$h(x, y)z := x \cdot z \cdot \overline{y}. \tag{5.2.6}$$

is a double cover of $SO(\mathbb{H})$. *In particular, it identifies* $\mathbb{S}^3 \times \mathbb{S}^3$ *with* $Spin(\mathbb{H})$.

Proof Since h is a continuous homomorphism between connected Lie groups of the same dimension, it suffices to show that its kernel is

$$\mathrm{Ker}(h) = \{(1, 1), (-1, -1)\}. \tag{5.2.7}$$

However, suppose that $h(x, y) = \text{Id}$. Substituting $z = 1$ into (5.2.6) yields $x \cdot \bar{y} = 1$, so that $x = y$. Thus, for all z,

$$z \cdot x = (h(x, x) \cdot z) \cdot x = x \cdot z \cdot \bar{x} \cdot x = x \cdot z,$$

so that

$$[x, z] = 0.$$

Since z is arbitrary, x is real, and since x has unit norm, it is equal to ± 1. This proves (5.2.7), and the result follows. □

5.2.2 Compatible Complex Structures

Recall that a *complex structure* over $\mathbb{H}$ is an $\mathbb{R}$-linear map $J : \mathbb{H} \to \mathbb{H}$ such that

$$J^2 = -\text{Id}.$$

We say, in addition, that J is *compatible* whenever it preserves the metric, that is, whenever, for all $x \in \mathbb{H}$,

$$\|Jx\| = \|x\|.$$

We now proceed to identify all compatible complex structures over $\mathbb{H}$.

Lemma 5.2 *The set of square roots of -1 in $\mathbb{H}$ is the sphere $\mathbb{S}^3 \cap \mathcal{I}$ of unit, imaginary quaternions.*

Proof Indeed, $x^2 = -1$ if and only if $x^{-1} = -x$. This holds if and only if $\|x\| = 1$ and $x = -\bar{x}$, as desired. □

For every unit, imaginary quaternion x, $h(x, 1)$ and $h(1, x)$ are trivially compatible complex structures over $\mathbb{H}$. We now verify that there are no others.

Lemma & Definition 5.3 *If $J : \mathbb{H} \to \mathbb{H}$ is a compatible complex structure, then J is given by multiplication either on the left or on the right by a unit, imaginary quaternion. We call the former* left complex structures *and the latter* right complex structures.

Remark 5.4 Every left complex structure trivially commutes with every right complex structure. Note also that quaternionic conjugation sends left complex structures into right complex structures and vice-versa.

Proof Since $J \in$ SO($\mathbb{H}$), by Lemma 5.1, there exists $(x, y) \in \mathbb{S}^3 \times \mathbb{S}^3$ such that $J = h(x, y)$. Since

$$- J^2 = h(x^2, -y^2) = \text{Id},$$

it follows that

$$x^2 = -y^2 = \pm 1.$$

Without loss of generality, we may suppose that $(x^2, y^2) = (1, -1)$. It then follows that $x = \pm 1$, $y \in \mathbb{S}^3 \cap \mathcal{I}$, and

$$J = h(\pm 1, y) = h(1, \pm y),$$

as desired. $\square$

In particular, we obtain the following algebraic characterisation of left and right complex structures.

Lemma 5.4 *Let $J : \mathbb{H} \to \mathbb{H}$ be a right complex structure. If $J' : \mathbb{H} \to \mathbb{H}$ is another compatible complex structure which commutes with J, then either J' is a left complex structure, or $J' = \pm J$.*

Remark 5.5 An analogous result trivially holds with the roles of left and right complex structures inverted.

Proof It suffices to show that if J' is a right complex structure which commutes with J, then $J' = \pm J$. However, let x and y be unit, imaginary quaternions such that $J = h(1, x)$ and $J' = h(1, y)$. Since J and J' commute, so too do x and y. Thus, since x and y are imaginary,

$$\overline{x \cdot y} = y \cdot \overline{x} = -y \cdot x = -x \cdot y = x \cdot \overline{y},$$

so that $x \cdot \overline{y}$ is real. Since x and y both have unit length, this holds if and only if $x = \pm y$, and the result follows. $\square$

5.2.3 Compatible Quaternionic Structures

Let $E := (E, g)$ be an inner-product space. We define a *compatible quaternionic structure* over E to be an algebra homomorphism $\rho : \mathbb{H} \to \text{End}(E)$ such that, for all $x \in \mathbb{H}$ and for all $u \in E$,

$$\|\rho(x)u\| = \|x\| \cdot \|u\|.$$

We now proceed to identify all compatible quaternionic structures over $\mathbb{H}$.

Lemma 5.5 *If* $\alpha : \mathbb{H} \to \mathbb{H}$ *is a non-trivial algebra homomorphism, then there exists a unit quaternion z such that, for all $x \in \mathbb{H}$,*

$$\alpha(x) = z \cdot x \cdot \bar{z}.$$

Proof Since $\mathbb{H}$ is a skew field, α is injective and $\alpha(\pm 1) = \pm 1$. In particular, α preserves $\mathcal{R}$. It likewise preserves the set $\mathbb{S}^3 \cap \mathcal{I}$ of square roots of -1, and therefore also $\mathcal{I}$. It follows that α preserves the decomposition $\mathbb{H} = \mathcal{R} \oplus \mathcal{I}$. It therefore also preserves conjugation, and thus also the metric (5.2.3). Finally, by (5.2.5), α preserves the orientation of $\mathcal{I}$ and thus also of $\mathbb{H}$, and is consequently an element of $\mathrm{SO}(\mathbb{H})$. It follows by Lemma 5.1 that there exist unit quaternions $z, w \in \mathbb{S}^3$ such that $\alpha = h(w, z)$. Finally, since $\alpha(1) = 1$, $w = z$, and the result follows. $\qquad\square$

Lemma & Definition 5.6 *If* $\alpha : \mathbb{H} \to End(\mathbb{H})$ *is a compatible quaternionic structure, then either*

(1) there exists $z \in \mathbb{S}^3$ such that, for all $x, y \in \mathbb{H}$,

$$\alpha(x) \cdot y = z \cdot x \cdot \bar{z} \cdot y, \text{ or}$$

(2) there exists $z \in \mathbb{S}^3$ such that, for all $x, y \in \mathbb{H}$,

$$\alpha(x) \cdot y = y \cdot z \cdot \bar{x} \cdot \bar{z}.$$

We call the former left quaternionic structures *and the latter* right quaternionic structures.

Remark 5.6 As before, every left quaternionic structure commutes with every right quaternionic structure. Furthermore, quaternionic conjugation sends left quaternionic structures to right quaternionic structures and vice-versa.

Proof Since $\mathbb{H}$ is a skew field, α is injective and $\alpha(\pm 1) = \pm \mathrm{Id}$. It follows that α maps the set of unit, imaginary quaternions to the set of compatible complex structures of $\mathbb{H}$. By connectedness, we may suppose without loss of generality that, for every unit, imaginary quaternion x, $\alpha(x)$ is a left complex structure. It follows that, for all $x, y \in \mathbb{H}$,

$$\alpha(x) \cdot y = \rho(x) \cdot y,$$

for some non-trivial algebra homomorphism $\rho : \mathbb{H} \to \mathbb{H}$. The result now follows by Lemma 5.5. $\qquad\square$

5.2.4 Calibrations

The utility of quaternions to the theory of partial differential equations arises from the theory of calibrations developed by Harvey–Lawson in [10]. We now explain how this applies in our setting. For every unit, imaginary quaternion x, we define

$$J_x := h(x, 1),$$

and we define the symplectic form ω_x over $\mathbb{H}$ by

$$\omega_x := \langle \cdot, J_x \cdot \rangle.$$

Lemma 5.7 *Let (x, y, z) be an orthonormal triplet of unit, imaginary quaternions. Let $P \subseteq \mathbb{H}$ be a real plane. The area form dA of P satisfies, for all $\xi, v \in P$,*

$$dA(\xi, v)^2 = \omega_x(\xi, v)^2 + \omega_y(\xi, v)^2 + \omega_z(\xi, v)^2. \tag{5.2.8}$$

Remark 5.7 In other words, the triplet $(\omega_x, \omega_y, \omega_z)$ forms a *calibration* of $\mathbb{H}$ in the sense of [10].

Proof We may suppose that ξ and v both have unit length. Since $(v, J_x v, J_y v, J_z v)$ is an orthonormal real basis of $\mathbb{H}$,

$$\langle \xi, v \rangle^2 + \langle \xi, J_x v \rangle^2 + \langle \xi, J_y v \rangle^2 + \langle \xi, J_z v \rangle^2 = \| \xi \|^2 = 1.$$

Thus, if θ denotes the angle between ξ and v, then

$$dA(\xi, v)^2 = \sin^2(\theta) = 1 - \langle \xi, v \rangle^2 = \omega_x(\xi, v)^2 + \omega_y(\xi, v)^2 + \omega_z(\xi, v)^2,$$

as desired.   □

In particular, we obtain the following result for real planes in $\mathbb{H}$.

Lemma 5.8 *Let x be a unit, imaginary quaternion. Let P be a real plane in $\mathbb{H}$. P is J_x-complex if and only if it is ω_y-Lagrangian for every unit, imaginary quaternion y orthogonal to x.*

Proof Let y and z be unit, imaginary quaternions such that (x, y, z) is an orthonormal triplet, and let dA denote the area form of P. P is J_x-complex if and only if, for all $\xi, v \in P$,

$$dA(\xi, v)^2 = \langle \xi, J_x v \rangle^2 = \omega_x(\xi, v)^2.$$

By (5.2.8), this holds if and only if ω_y and ω_z both vanish over P, as desired.   □

5.2.5 The Monge–Ampère Equation

We now apply our abstract framework to the study of solutions of the real Monge–Ampère equation

$$\text{Det}(\text{Hess}(u)) = 1. \tag{5.2.9}$$

To this end, we identify $\mathbb{C}$ with $\mathbb{R}^2$ in the natural manner, and we introduce an explicit compatible quaternionic structure over $\mathbb{C} \oplus \mathbb{C} = \mathbb{R}^2 \oplus \mathbb{R}^2$ as follows. First, let J_0 denote the operator of multiplication by i, so that

$$J_0 := \begin{pmatrix} 0 & -1 \\ 1 & 0 \end{pmatrix},$$

and let g denote the standard metric over $\mathbb{C} \oplus \mathbb{C}$, that is

$$g((z, w)^t, (z, w)^t) = |z|^2 + |w|^2.$$

We define

$$I := \begin{pmatrix} J_0 & 0 \\ 0 & -J_0 \end{pmatrix},$$

$$J := \begin{pmatrix} 0 & J_0 \\ J_0 & 0 \end{pmatrix}, \text{ and}$$

$$K := \begin{pmatrix} 0 & -\text{Id} \\ \text{Id} & 0 \end{pmatrix}.$$

Note that, in the spirit of Lemma 5.4, I, J and K are generators of the set of compatible complex structures over $\mathbb{C} \oplus \mathbb{C}$ which commute with

$$\hat{J}_0 := \begin{pmatrix} J_0 & 0 \\ 0 & J_0 \end{pmatrix}.$$

Let $(\omega_i, \omega_j, \omega_k)$ denote the triplet of symplectic forms corresponding to (I, J, K), that is

$$\omega_i := g(\cdot, I\cdot),$$
$$\omega_j := g(\cdot, J\cdot), \text{ and}$$
$$\omega_k := g(\cdot, K\cdot).$$

We now examine the algebraic properties of different types of Lagrangian subspaces of $\mathbb{C} \oplus \mathbb{C}$. To this end, let $P \subseteq \mathbb{C} \oplus \mathbb{C}$ be a real plane which is a graph over the first component, that is

$$P := \left\{ (x, Ax) \mid x \in \mathbb{R}^2 \right\},$$

for some matrix $A \in \text{End}(2)$. The following two useful relations for matrices $A \in \text{End}(2)$ may be verified by inspection.

$$A^t J_0 A J_0 = -\text{Det}(A)\text{Id}, \text{ and}$$

$$A - A^t = -\text{Tr}(A J_0) J_0. \tag{5.2.10}$$

Lemma 5.9

(1) P is ω_i-Lagrangian if and only if $\text{Det}(A) = 1$;
(2) P is ω_j-Lagrangian if and only if $\text{Tr}(A) = 0$; and
(3) P is ω_k-Lagrangian if and only if $A = A^t$.

Proof Indeed, P is ω_i-Lagrangian if and only if,

$$\langle x, J_0 y \rangle - \langle Ax, J_0 Ay \rangle = 0 \ \forall x, y \in \mathbb{R}^2$$

$$\Leftrightarrow \quad \langle x, (\text{Id} + A^t J_0 A J_0)y \rangle = 0 \ \forall x, y \in \mathbb{R}^2.$$

This holds if and only if

$$\Leftrightarrow \quad \text{Id} + A^t J_0 A J_0 = 0$$

$$\Leftrightarrow \quad \text{Det}(A) \qquad\quad = 1,$$

where the last equivalence follows by (5.2.10). In a similar manner, using (5.2.10) again, we show that P is ω_j-Lagrangian if and only if $\text{Tr}(A) = 0$. Finally, it is a standard result that P is ω_k-Lagrangian if and only if $A = A^t$, and this completes the proof. $\qquad\square$

We thus obtain the following key result which relates solutions of the 2-dimensional Monge–Ampère equation to pseudo-holomorphic curves.

Lemma 5.10 (Monge–Ampère = Pseudoholomorphic) *Let Ω be a simply-connected open subset of $\mathbb{R}^2$. Let $\alpha : \Omega \to \mathbb{R}^2$ be a smooth function and let $S \subseteq \Omega \times \mathbb{R}^2$ denote its graph. Then α is the derivative of a smooth solution $u : \Omega \to \mathbb{R}$ of the real Monge–Ampère equation (5.2.9) if and only if S is a J-holomorphic curve.*

Proof Indeed, by Lemma 5.8, S is J-holomorphic if and only if it is ω_i- and ω_k-Lagrangian. By Lemma 5.9, this holds if and only if $\text{Det}(D\alpha) = 1$ and α is closed. Since Ω is simply-connected, α is closed if and only if it is the derivative of some smooth function $u : \Omega \to \mathbb{R}$. The result now follows, since $\text{Det}(\text{Hess}(u)) = \text{Det}(D\alpha) = 1$. $\qquad\square$

5.2.6 Positivity I

We now express convexity in the quaternionic framework. To this end, let π_1, π_2 : $\mathbb{C} \oplus \mathbb{C} \to \mathbb{C}$ denote respectively the projections onto the first and second components, and denote

$$V := \mathrm{Ker}(\pi_1) = \{0\} \oplus \mathbb{C}.$$

We call V the *vertical subspace* of $\mathbb{C} \oplus \mathbb{C}$. Let m be the symmetric bilinear form defined over $\mathbb{C} \oplus \mathbb{C}$ by

$$m((z_1, w_1)^t, (z_2, w_2)^t) := \langle z_1, w_2 \rangle + \langle z_2, w_1 \rangle.$$

Note that m vanishes over V and that

$$m(\cdot, \cdot) = -m(I\cdot, I\cdot) = m(J\cdot, J\cdot) = -m(K\cdot, K\cdot). \tag{5.2.11}$$

It is worth noting that m is, up to sign, uniquely determined by these properties.

Lemma 5.11 *Up to sign, m is the unique symmetric bilinear form of unit norm satisfying (5.2.11) which vanishes over V.*

Proof Indeed, let m' be another symmetric bilinear form which vanishes over V and which satisfies (5.2.11). Since m' vanishes over V and is J-invariant, it is uniquely determined by the restriction of $\tilde{m}' := m(\cdot, J\cdot)$ to V. However, since m' is I-anti-invariant, and since I anticommutes with J, $\tilde{m}'$ is also I-invariant, and its restriction to V is thus unique up to a scalar factor. Finally, since m' has unit norm, it follows that $m' = \pm m$, as desired. $\qquad\square$

Define the subgroup $\Sigma \subseteq SO(4)$ by

$$\Sigma := \left\{ \begin{pmatrix} A & 0 \\ 0 & A \end{pmatrix} \,\middle|\, A \in SO(2) \right\}.$$

Lemma 5.12 *Σ is the stabiliser subgroup in $SO(4)$ both of (I, J, K, m) and of (V, J, ω), where ω as any orientation form of V.*

Remark 5.8 In particular, the prescription of a triplet of the form (V, J, ω) over any 4-dimensional inner-product space (E, g) is equivalent to the prescription of a compatible quaternionic structure (I, J, K) together with a symmetric bilinear form m satisfying (5.2.11).

Proof Suppose that $M \in SO(4)$ preserves (I, J, K, m). Define the involution α of $\mathbb{C} \oplus \mathbb{C}$ by $\alpha(z, w) = (z, \overline{w})$, and denote $\tilde{M} := \alpha M \alpha$. Since $(\alpha I \alpha)(z, w) = (iz, iw)$, $\tilde{M} \in U(2)$. Since, in addition, $\tilde{M}$ preserves $(\alpha K \alpha)(z, w) = (-\overline{w}, \overline{z})$,

$$M = \begin{pmatrix} a & b \\ -\overline{b} & \overline{a} \end{pmatrix},$$

for some $a, b \in \mathbb{C}$ such that $|a|^2 + |b|^2 = 1$. Finally, since $\tilde{M}$ preserves m, $b = 0$, and it follows that M is an element of Σ, as desired.

Suppose now that $M \in SO(4)$ preserves (V, J, ω). In particular, M takes the form

$$M = \begin{pmatrix} A & 0 \\ 0 & B \end{pmatrix},$$

for some $A, B \in SO(2)$. Since M preserves J_0, $A = B$, and it follows that M is an element of Σ, as desired. $\qquad\square$

The significance of the bilinear form m for the study of convexity is given by the following results.

Lemma & Definition 5.13 *The restriction of m to any J-complex line L in $\mathbb{C} \oplus \mathbb{C}$ is either positive-definite, negative-definite or null. Furthermore, this restriction is null if and only if L intersects V non-trivially, that is, if and only if π_1 has non-trivial kernel over L.*

We say that a J-complex line L in $\mathbb{C} \oplus \mathbb{C}$ is positive, negative or null according to whether the restriction of m to L is positive-definite, negative-definite or null.

Proof Let $L \subseteq \mathbb{R}^2 \oplus \mathbb{R}^2$ be a J-complex line. Since $m(J\cdot, J\cdot) = m$, m has well-defined sign over L. Since L is ω_k-Lagrangian, for all $\xi, v \in L$,

$$\langle \pi_1(\xi), \pi_2(v) \rangle = m(\xi, v) + \omega_k(\xi, v) = m(\xi, v).$$

It follows that m vanishes over L if and only if one of $\mathrm{Ker}(\pi_1) \cap L$ or $\mathrm{Ker}(\pi_2) \cap L$ is non-trivial. However, since $J_0 \pi_1 = \pi_2 J$, these spaces are either both trivial or both non-trivial, and the result follows. $\qquad\square$

Lemma 5.14 (Convexity = Positivity) *Let Ω be an open subset of $\mathbb{R}^2$. Let $u : \Omega \to \mathbb{R}$ be a smooth solution of the Monge–Ampère equation (5.2.9) and let $S \subseteq \Omega \times \mathbb{R}^2$ denote the graph of its derivative. The function u is strictly convex if and only if every tangent plane of S is positive.*

Proof Indeed, the tangent planes of S are positive if and only if, for all $x \in \Omega$ and for all $\xi \in \mathbb{R}^2$, $\langle \xi, \mathrm{Hess}(u)(x)\xi \rangle > 0$. $\qquad\square$

5.2.7 Positivity II: A Holomorphic Approach

Define $\phi, \psi : \mathbb{C} \oplus \mathbb{C} \to \mathbb{C}$ by

$$\phi(z, w) := z + w, \text{ and}$$
$$\psi(z, w) := -\bar{z} + \bar{w},$$

and define

$$\Phi := (\phi, \psi).$$

Note that Φ is $\mathbb{C}$-linear with respect to the complex structure J of the domain and the complex structure $\hat{J}_0$ of the codomain. In addition, for all $(z, w) \in \mathbb{C}^2$,

$$(\Phi_* m)((z, w)^t, (z, w)^t) = \frac{1}{2}(|z|^2 - |w|^2). \tag{5.2.12}$$

We thus consider ϕ and ψ as the respective projections of $\mathbb{C} \oplus \mathbb{C}$ onto the space and time components of m. This has two useful consequences. The first is the following Lipschitz property.

Lemma 5.15 *Let P be a real plane in $\mathbb{C} \oplus \mathbb{C}$. If*

$$m|_P \geq 0,$$

then the restriction of ϕ to P is $\sqrt{2}$-bilipschitz.

Proof Indeed, for all $(z, w) \in P$,

$$|\phi(z, w)| = |z + w| \leq \sqrt{2}\|z, w\|.$$

Conversely,

$$|\phi(z, w)|^2 \geq \|z, w\|^2 + m((z, w)^t, (z, w)^t) \geq \|z, w\|^2,$$

as desired. □

The second useful consequence is the following conformal description of positivity. First, let $\hat{\mathbb{C}}$ denote the extended complex plane and define the conformal diffeomorphism $\alpha : \mathbb{CP}^1 \to \hat{\mathbb{C}}$ by

$$\alpha([z : w]) := \frac{w}{z}.$$

Lemma 5.16 *Let L be a J-complex line in $\mathbb{C} \oplus \mathbb{C}$. L is positive if and only if $|(\alpha \circ \Phi)(L)| < 1$ and L is null if and only if $|(\alpha \circ \Phi)(L)| = 1$.*

Proof Choose $(z, w) \in \Phi(L)$. By (5.2.12), L is positive if and only if

$$|w|^2 < |z|^2,$$

which holds if and only if $|\alpha(L)| = |\alpha([z : w])| < 1$. Likewise, L is null if and only if

$$|w|^2 = |z|^2,$$

which holds if and only if $|\alpha(L)| = 1$. $\square$

5.2.8 Bernstein Type Theorems I

The quaternionic formalism that we have presented here now yields simple proofs of key Bernstein-type theorems. These results are not only interesting in their own right, but also provide the basis for the general compactness theorem that will be developed later.

Theorem 5.3 *If S is a complete, non-negative J-holomorphic curve in $\mathbb{C} \oplus \mathbb{C}$, then S is an affine plane.*

Proof Indeed, upon passing to the universal cover, we may assume that S is simply-connected. By Lemma 5.15, ϕ restricts to a bilipschitz, holomorphic map from S into $\mathbb{C}$. It is thus a conformal diffeomorphism, and S therefore has parabolic conformal type. Now define the meromorphic function $\tau : S \to \hat{\mathbb{C}}$ by

$$\tau(x) := (\alpha \circ \Phi)(T_x S).$$

By Lemma 5.16, τ takes values in the closed unit disk. By Liouville's theorem, τ is constant, and S is therefore an affine plane, as desired. $\square$

The Bernstein-type theorem [12] of Jörgens is an immediate corollary.

Theorem 5.4 (Jörgens) *If $u : \mathbb{R}^2 \to \mathbb{R}$ is a smooth solution of the real Monge–Ampère equation (5.2.9), then u is a quadratic function.*

Remark 5.9 Jörgens' theorem was a key result in the study of the real Monge–Ampère equations. It was generalised to Dimensions 3 and 4 by Calabi in [5], and then to arbitrary dimension by Pogorelov in [21]. The ideas originated in these papers were later applied to the study of Monge-Ampère equations in diverse settings. It is worth noting, however, that in the higher-dimensional case, there is no useful analogue of the quaternionic approach studied here. For the higher-dimensional case, we must instead look to the special Lagrangian potential equation, where a rich theory indeed exists (see [11] and [25]).

Proof Indeed, by Lemmas 5.10 and 5.13, the graph S of du is a positive J-holomorphic curve. In addition, being a graph over $\mathbb{R}^2$, it is complete, and the result now follows by Theorem 5.3.   □

Another nice application of Theorem 5.3 is the following mild improvement of the classification result of complete flat surfaces in $\mathbb{H}^3$ by Volkov–Vladimirova and Sasaki (see Chapter 7.F of [27]). We say that an immersed surface (S, e) in $\mathbb{H}^3$ is *flat* whenever it has constant extrinsic curvature equal to 1.[2]

Theorem 5.5 *The only quasicomplete, ISC, flat surfaces in $\mathbb{H}^3$ are the level sets of horofunctions and the level sets of distance functions to complete geodesics.*

Remark 5.10 When the surface is complete, we recover the result of Volkov–Vladimirova and Sasaki.

Proof By Gauss' Theorem, every flat surface is intrinsically flat and is thus locally isometric to $\mathbb{R}^2$. Let Ω be an open subset of $\mathbb{R}^2$, let $e : \Omega \to \mathbb{H}^3$ be an isometric immersion, and let II_e denote its second fundamental form. By the Codazzi–Mainardi equations, the derivative $(\mathrm{D}\mathrm{II}_e)_{ijk}$ of II_e is symmetric, so that

$$\mathrm{II}_e = \mathrm{Hess}(f),$$

for some smooth function f, unique up to addition of a linear function. We define the immersion $\alpha : \Omega \to \mathbb{R}^2 \oplus \mathbb{R}^2$ by

$$\alpha(x) := (x, df(x)).$$

Since $\mathrm{Det}(\mathrm{Hess}(f)) = \mathrm{Det}(\mathrm{II}_e) = 1$, it follows by Lemma 5.10 that α is J-holomorphic. Furthermore, $(\pi_1 \circ \alpha)$ is a local isometry and $\alpha^* m = 2\mathrm{II}_e$, so that α is also positive.

Now let (S, e) be a quasicomplete, flat immersed surface in $\mathbb{H}^3$. Upon taking the universal cover, we may suppose that S is simply connected. It then follows by the preceding discussion that there exists a J-holomorphic immersion $\alpha : S \to \mathbb{R}^2 \oplus \mathbb{R}^2$ such that $\beta := (\pi_1 \circ \alpha)$ is a local isometry and $\alpha^* m = 2\mathrm{II}_e$. Furthermore, since e is quasicomplete, α is complete, and it follows by Theorem 5.3 that (S, α) is an affine plane. In particular, β defines a global isometry from S into $\mathbb{R}^2$. Upon composing with a rotation if necessary, $e' := e \circ \beta^{-1}$ is an isometric immersion with second fundamental form given by

$$\mathrm{II}_{e'} := \begin{pmatrix} \lambda & 0 \\ 0 & 1/\lambda \end{pmatrix},$$

---

[2] Flat surfaces in $\mathbb{H}^3$ are studied in terms of their Gauss lifts in [13, 14] and [19]. The terminology used in these papers differs from our own: where they write "weakly complete", we write "quasicomplete"; and the objects which they refer to as "flat fronts" are for us the Gauss lifts of, possibly singular, flat surfaces.

for some constant $\lambda \in]0, 1]$. The result now follows by the fundamental theorem of surface theory: when $\lambda = 1$, S is the level set of a horofunction, and when $\lambda < 1$, S is the set of points lying at distance $\text{arctanh}(\lambda)$ from some complete geodesic in $\mathbb{H}^3$.

□

5.2.9 Bernstein Type Theorems II: Non-trivial Boundary

The case of curves with non-trivial boundary is similar, though more technical.

Theorem 5.6 *Let $P \subseteq \mathbb{C} \oplus \mathbb{C}$ be a plane over which m has signature $(1, 0)$. Let $(S, \partial S) \subseteq \mathbb{C} \oplus \mathbb{C}$ be a complete, non-negative J-holomorphic curve with boundary. If*

(1) $\partial S \subseteq P$; and
(2) there exists $\epsilon > 0$ such that if, for all $x \in \partial S$,

$$m|_{T_x S} \geq \epsilon g|_{T_x S}, \tag{5.2.13}$$

then S is an affine plane.

Proof Upon taking the universal cover, we may suppose that S is simply connected. Let $\mathcal{F}$ denote the foliation of P by null lines and let $L \subseteq P$ be any line transverse to this foliation. By Lemma 5.15, ϕ restricts to a linear isomorphism from P into $\mathbb{C}$. By (5.2.13), there exists $\theta > 0$ such that $\phi(\partial S)$ makes an angle of at least θ with $\phi_* \mathcal{F}$ at every point so that, by completeness, $\phi(\partial S)$ is a union of complete Lipschitz graphs over L. By Lemma 5.15 again, the restriction of ϕ to S is a conformal diffeomorphism onto its image. In particular $\phi(S)$ is a subset of $\mathbb{C}$ bounded by at most 2 complete Lipschitz graphs. $(S, \partial S)$ is therefore conformally equivalent to one of $\overline{\mathbb{D}} \setminus \{-1\}$ or $\overline{\mathbb{D}} \setminus \{\pm 1\}$. The Riemann surface $\tilde{S}$, obtained by doubling S along its boundary, is thus conformally equivalent to one of $\mathbb{C}$ or $\mathbb{C} \setminus \{0\}$. In both cases, $\tilde{S}$ is of parabolic type.

Define the meromorphic function $\tau : S \to \hat{\mathbb{C}}$ by

$$\tau(x) := (\alpha \circ \Phi)(T_x S).$$

By Lemma 5.16, τ takes values in the closed unit disk. Furthermore, by (5.2.13), there exists $\delta > 0$ such that, for all $x \in \partial S$,

$$|\tau(x)| \leq 1 - \delta.$$

Since the boundary ∂S of S has at most two isolated singularities, and since τ is bounded, it follows by the Cauchy integral formula that, for all $x \in S$,

$$|\tau(x)| \leq 1 - \delta.$$

For all $r > 0$, let D_r and C_r denote respectively the disk and circle in $\mathbb{C}$ of radius r about the origin. By definition of P, there exists a circle C, which is a strict interior tangent to C_1 at some point, such that, for all $x \in \partial S$,

$$\tau(x) \in C.$$

Let $R : \hat{\mathbb{C}} \to \hat{\mathbb{C}}$ denote the conformal reflection through C. By the Schwarz reflection principle, τ extends to a holomorphic function $\tilde{\tau} : \tilde{S} \to \hat{\mathbb{C}}$ such that, for all $x \in S$,

$$\tilde{\tau}(\bar{x}) := R\tau(x).$$

In particular,

$$\tilde{\tau}(\tilde{S}) \subseteq \overline{D}_{1-\delta} \cup R\overline{D}_{1-\delta} =: \Omega.$$

Since the complement of Ω in $\hat{\mathbb{C}}$ has non-trivial interior, it follows by Liouville's theorem that $\tilde{\tau}$, and therefore also τ, is constant. The surface S is thus an affine half-plane, and this completes the proof. $\qquad\square$

In terms of the real Monge–Ampère equation, this yields the following Bernstein-type theorem for functions defined over domains with boundary.

Theorem 5.7 *Let $\phi : \mathbb{R} \times [0, \infty[\to \mathbb{R}$ be a smooth solution of the real Monge–Ampère equation (5.2.9). If*

(1) ϕ restricts to a non-trivial quadratic function over $\mathbb{R} \times \{0\}$; and
(2) $\Delta\phi$ is bounded over $\mathbb{R} \times \{0\}$,

then ϕ is a quadratic function.

Proof Indeed, by Lemmas 5.10 and 5.13, the graph S of du is a positive J-holomorphic curve which satisfies the hypotheses of Theorem 5.6. The result follows. $\qquad\square$

5.2.10 Non-quadratic Solutions

We conclude this section with the following interesting open problem. It is not clear that Theorems 5.6 and 5.7 are optimal as stated. Although their first conditions are natural from the point of view of Liouville's theorem and the Schwarz reflection principle, their second conditions are not. Nevertheless, the first conditions alone are not sufficient, as the function

$$\phi(x, y) := \frac{x^2}{y} + \frac{y^3}{12} \tag{5.2.14}$$

shows. Indeed, this function is a convex solution of the real Monge–Ampère equation (5.2.9) over the open half-plane $\mathbb{R} \times]0, \infty[$ whose restriction to every horizontal line is quadratic, but which is itself trivially not quadratic.

The above function, together with many similar examples, is obtained by directly solving the Monge–Ampère equation (5.2.9) with the ansatz

$$f(x, y) = \alpha(y)x^2 + \beta(y)x + \gamma(y).$$

However, it is also worth reviewing its properties from the quaternionic point of view. Note first that

$$\Phi(V) = \{(z, \overline{z}) \mid z \in \mathbb{C}\}. \tag{5.2.15}$$

Let $\mathbb{H}^+$ denote the upper half-space in $\mathbb{C}$ and let $\overline{\mathbb{H}}^+$ denote its closure. Consider a holomorphic function $F := (f, g)^t : \overline{\mathbb{H}}^+ \to \mathbb{C} \oplus \mathbb{C}$, and suppose that

$$F(\mathbb{R}) \subseteq \Phi(V).$$

By (5.2.15), for all $x \in \mathbb{R}$,

$$f(x) = \overline{g(x)}.$$

It follows by the Schwarz reflection principle that f and g extend to entire functions such that, for all $z \in \mathbb{H}^+$,

$$g(z) = \overline{f(\overline{z})}.$$

In particular, it is sufficient to prescribe only the first component f. In order to obtain a complete immersed surface with boundary, it is also necessary that f define a conformal diffeomorphism of $\mathbb{H}^+$ onto its image. This holds, for example, when f is quadratic. We thus choose

$$f(z) := -i - i(z + i)^2,$$

so that

$$F(z) = (-i - i(z + i)^2, i + i(z - i)^2)^t.$$

Composing with Φ^{-1} yields

$$(\Phi^{-1} \circ F)(z) = (2xy + 2iy, 2x + i(y^2 - x^2))^t,$$

where $x + iy := z$, so that $(\Phi^{-1} \circ F)$ is the graph over $\mathbb{R} \times]0, \infty[$ of the 1-form

$$\alpha(x, y) := \frac{2x}{y} dx + \left(\frac{y^2}{4} - \frac{x^2}{y^2} \right) dy,$$

which integrates to (5.2.14).

With a view towards the optimal formulations of Theorems 5.6 and 5.7, it is tempting to conjecture that the only non-quadratic solutions of the Monge–Ampère equation over the half-plane which restrict to quadratic functions over the boundary are those described above. However, although we have not been able to find other examples, we also have no proof of this assertion.[3]

5.3 Monge–Ampère Structures and k-Surfaces

5.3.1 Monge–Ampère Structures

In [16] Labourie uses a non-integrable generalisation of the structures studied in Sect. 5.2 to derive compactness results for families of immersed surfaces of prescribed extrinsic curvature in 3-dimensional Riemannian manifolds. Labourie's work now underlies a large part of the modern theory of such surfaces, with applications in hyperbolic geometry, Teichmüller theory, general relativity, and so on. We now provide an elementary presentation of Labourie's ideas.

Let X be a manifold. Labourie defines a **Monge–Ampère structure** over X to be a quadruplet (W, g, V, J) where

(1) W is a smooth 4-dimensional subbundle of the tangent bundle of X furnished with a Riemannian metric g;
(2) V is a smooth, oriented 2-dimensional subbundle of W;
(3) J is a smooth section of $\text{End}(W)$ such that, for all x, J_x restricts to a compatible complex structure of the fibre W_x with respect to which the fibre V_x is real.

Remark 5.11 By Lemmas 5.11 and 5.12 and the subsequent remark, at each point x of X, the prescription of (V_x, J_x) is equivalent to the prescription of a compatible quaternionic structure (I_x, J_x, K_x), together with a unit-length symmetric bilinear form m_x satisfying (5.2.11). The above geometric structure can thus be expressed in purely quaternionic terms. This point of view is particularly useful for its extension

---

[3] After completion of this paper, it was brought to our attention by Nam Q. Le that this conjecture has already been resolved in the affirmative by A. Figalli in Section 5.1.4 of [6]. Figalli's work still leaves open the following question, more relevant to the present setting: with P as in Theorem 5.6, are the only complete, non-negative J-holomorphic curves with boundary in P the images of linear or quadratic polynomial functions?

to the higher-dimensional case (see [25]). We also highlight the parallels with the analogous theory of Monge–Ampère structures developed in [17, 18] and [22].

The geometric significance of V is easier to understand when described in conformal terms. Indeed, at every point x, the set of complex lines in W_x which intersect V_x non-trivially defines a circle C_x in the complex projective space of (W_x, J_x). Furthermore, the orientation of V_x in turn yields an orientation of this circle which therefore has a well-defined interior D_x. The prescription of V is thus equivalent to the prescription of a conformal disk bundle D in the complex projective bundle of (W, J). In the spirit of Lemma 5.16 this yields a notion of positivity of J_x-complex lines in V_x. We henceforth also denote by $\overline{D}$ the bundle whose fibre at x is the topological closure of D_x.

A J-holomorphic curve in X is a pair (S, ϕ) where S is a Riemann surface and $\phi : S \to X$ is a smooth function such that, for all x, $\mathrm{Im}(D\phi(x)) \subseteq W_{\phi(x)}$ and

$$J_{\phi(x)} D\phi(x) = D\phi(x) j_x,$$

where j here denotes the complex structure of S. We define a *Monge–Ampère surface* to be an immersed J-holomorphic curve (S, ϕ) in X such that

(1) (S, ϕ) is tangent to W;
(2) (S, ϕ) is complete with respect to g; and
(3) the tangent space to (S, ϕ) is at every point an element of $\overline{D}$.

We say that a Monge–Ampère surface is *positive* or *null* at some point according to whether its tangent space at that point is an element of D or $\partial D = \overline{D} \setminus D$. Following Labourie, we define a *curtain surface* to be a Monge–Ampère surface which is everywhere null.

Labourie's framework consists of two parts, namely a compactness result, which we will study in the next section, and a dichotomy result, given by the following theorem.

Theorem 5.8 *Let (S, ϕ) be a Monge–Ampère surface and let (S', ϕ') be a curtain surface. If there exists $x \in S$ and $x' \in S'$ such that $\phi(x) = \phi'(x')$ and $\mathrm{Im}(D\phi(x)) = \mathrm{Im}(D\phi'(x'))$, then there exists a neighbourhood U of x in S, a neighbourhood V of x' in S', and a conformal diffeomorphism $\alpha : U \to V$ such that*

$$\phi = \phi' \circ \alpha.$$

Theorem 5.8 motivates the following definition. We say that the Monge–Ampère structure is *integrable* whenever every point of $\partial D := \overline{D} \setminus D$ is tangent to some curtain surface.

Theorem 5.9 *If (W, V, J) is an integrable Monge–Ampère structure then every point of $\overline{D} \setminus D$ is tangent to a unique inextensible, simply-connected curtain surface.*

Proof This follows immediately from Theorem 5.8 and unique continuation. □

Theorem 5.10 (Labourie's Dichotomy) *If (W, V, J) is an integrable Monge–Ampère structure, then every Monge–Ampère surface whose tangent bundle meets $\partial D = \overline{D} \setminus D$ at some point is a curtain surface. That is, every Monge–Ampère surface is either everywhere positive, or everywhere null.*

Proof This likewise follows immediately from Theorem 5.8 and unique continuation. □

Proof of Theorem 5.8 Suppose the contrary. For $r > 0$, let $\mathbb{D}_r$ and $\mathbb{R}_r^m$ denote the open disk and the open ball, both of radius r, in $\mathbb{C}$ and $\mathbb{R}^m$ respectively. We parametrize a neighbourhood of $y := \phi(x) = \phi'(x')$ in X by $\mathbb{D}_r \times \mathbb{D}_r \times \mathbb{B}_r^m$ in such a manner that $\mathbb{D}_r \times \{(0, 0)\}$ identifies with a portion of $\phi'(S')$ about x' and, for all $z \in \mathbb{D}_r$,

$$W_{(z,0,0)} = \mathbb{C} \times \mathbb{C} \times \{0\}, \text{ and}$$

$$J_{(z,0,0)} = \begin{pmatrix} i & 0 & 0 \\ 0 & i & 0 \\ 0 & 0 & 0 \end{pmatrix}.$$

Since (S, ϕ) is tangent to (S', ϕ') at y, a portion of this surface about x identifies with the graph of a smooth function $F : \mathbb{D}_r \to \mathbb{D}_r \times \mathbb{B}_r^m$ which, by hypothesis, is non-zero. Since both $\mathbb{D}_r \times \{(0, 0)\}$ and the graph of F are J-holomorphic curves, it follows by Aronszajn's unique continuation theorem (see [1] and Theorem 2.3.4 of [20]) that

$$F(z) = (P(z), Q(z)) + O(z^{k+2}), \tag{5.3.1}$$

for real homogeneous polynomials P and Q, both of order $(k + 1)$, where $k \geq 1$. Furthermore, since the graph of F is tangent to W,

$$Q = 0, \tag{5.3.2}$$

and, by J-holomorphicity again,

$$P(z) = az^{k+1}, \tag{5.3.3}$$

for some $a \in \mathbb{C} \setminus \{0\}$.

From this point, the result follows by an elementary topological argument, although a certain amount of care is required for technical reasons. Let Ψ be a smooth function which, for all (z, w, ξ), sends the complex projective space of $(W_{(z,w,\xi)}, J_{(z,w,\xi)})$ to that of $(W_{(z,0,0)}, J_{(z,0,0)})$. We think of Ψ as sending the complex projective bundle of (W, J) to the trivial bundle with fibre $\mathbb{CP}^1$, and since W and J are constant over $\mathbb{D}_t \times \{(0, 0)\}$, we assume that Ψ is also constant over this set. For all z, let $\phi(z)$ denote the closest point of the boundary of $\Psi(D_{(z,F(z))})$ to the origin. We claim that

$$\phi(z) = \mathrm{O}(|z|^{k+1}). \tag{5.3.4}$$

Indeed, since $\mathbb{D}_r \times \{(0,0)\}$ is a curtain surface, for all $z \in \mathbb{D}_r$,

$$0 \in \Psi(D_{(z,0,0)}),$$

and, since $F = \mathrm{O}(|z|^{k+1})$, the assertion follows by smoothness of Ψ.

Consider now the curves $\gamma_\epsilon(\theta) := \epsilon e^{i\theta}$ and $\delta_\epsilon(\theta) := \phi(\epsilon e^{i\theta})$. For all θ, let $\gamma'_\epsilon(\theta)$ denote the image under Ψ of the tangent plane of the graph of F at the point $(\gamma_\epsilon(\theta), F(\gamma_\epsilon(\theta)))$. By (5.3.1), (5.3.2) and (5.3.3),

$$\gamma'_\epsilon(\theta) = a(k+1)\epsilon^k e^{ik\theta} + \mathrm{O}(\epsilon^{k+1}).$$

It follows by (5.3.4) that, for sufficiently small ϵ, $(\gamma'_\epsilon - \delta_\epsilon)$ winds k times around the origin. However, this is absurd, since the tangent plane of (S, ϕ) is at every point an element of D whilst δ_ϵ lies along its boundary. The result follows. $\square$

5.3.2 Compactness

We now prove a compactness property for families of Monge–Ampère surfaces. In order to correctly express the result, we require the concept of smooth Cheeger–Gromov convergence. First, we define a *pointed Riemannian manifold* to be a triplet (X, g, p), where X is a smooth manifold, g is a Riemannian metric and p is a point of X. We say that a sequence (X_m, g_m, p_m) of complete, pointed Riemannian manifolds converges to the complete, pointed Riemannian manifold $(X_\infty, g_\infty, p_\infty)$ in the *smooth Cheeger–Gromov sense* whenever there exists a sequence (Φ_m) of functions such that

(1) for all m, $\Phi_m : X_\infty \to X_m$ and $\Phi_\infty(p_\infty) = p_m$; and

for every relatively compact open subset Ω of X_∞, there exists M such that

(2) for all $m \geq M$, the restriction of Φ_m to Ω defines a smooth diffeomorphism onto its image; and

(3) the sequence $((\Phi_m|_\Omega)^* g_m)_{m \geq M}$ converges to $g_\infty|_\Omega$ in the C^∞ sense.

We call (Φ_m) a sequence of *convergence maps* of (X_m, g_m, p_m) with respect to $(X_\infty, g_\infty, p_\infty)$.

It is reassuring to observe that this mode of convergence defines a Hausdorff topology over the set of isometry classes of complete, pointed Riemannian manifolds.[4] Furthermore, although the convergence maps are trivially non-unique, any two sequences (Φ_m) and (Φ'_m) of convergence maps are equivalent in the sense that there exists an isometry $\Psi : X_\infty \to X_\infty$ preserving p_∞ such that, for any two relatively compact open subsets $U \subseteq \overline{U} \subseteq V$ of X_∞, there exists M such that

(1) for all $m \geq M$, the respective restrictions of Φ_m and $\Phi'_m \circ \Psi$ to U and V define smooth diffeomorphisms onto their images;
(2) for all $m \geq M$, $(\Phi'_m \circ \Psi)(U) \subseteq \Phi_m(V)$; and
(3) the sequence $((\Phi_m|_V)^{-1} \circ \Phi'_m \circ \Psi)$ converges in the C^∞ sense to the identity map over U.

This condition trivially defines an equivalence relation over the space of sequences of convergence maps.

We now return to the case of immersed submanifolds. We say that a sequence (S_m, e_m, p_m) of complete pointed immersed submanifolds in a complete Riemannian manifold (X, g) converges to the complete pointed immersed submanifold $(S_\infty, e_\infty, p_\infty)$ in the smooth Cheeger–Gromov sense whenever the sequence $(S_m, e_m^* g, p_m)$ converges to $(S_\infty, e_\infty^* g, p_\infty)$ in the smooth Cheeger–Gromov sense and, for one, and therefore for any, sequence (Φ_m) of convergence maps, the sequence $(e_m \circ \Phi_m)$ converges to e_∞ in the C^∞_{loc} sense.

Smooth Cheeger–Gromov convergence can also be characterised in terms of graphs. Indeed, let NS_∞ denote the normal bundle of (S_∞, e_∞) in $\phi_\infty^* TX$. Recall that the exponential map of X defines a smooth function $\text{Exp} : NS_\infty \to X$. In particular, given a suffciently small smooth section $f : \Omega \to NS_\infty$ defined over an open subset Ω of S_∞, the composition $\text{Exp} \circ f$ defines a smooth immersion of Ω in X which we call the *graph* of f. A straightforward but technical argument shows that the sequence (S_m, e_m, p_m) converges to $(S_\infty, e_\infty, p_\infty)$ in the smooth Cheeger–Gromov sense if and only if there exists a sequence (p'_m) of points in S_∞ and sequences of functions (f_m) and (α_m) such that

(1) (p'_m) converges to p_∞;
(2) for all m, f_m maps S_∞ into NS_∞, α maps S_∞ into S_m and $\alpha(p'_m) = p_m$; and

for every relatively compact open subset Ω of S_∞, there exists M such that

(3) for all $m \geq M$, the restriction of f_m to Ω defines a smooth section of NS_∞ over this set, the restriction of α_m to Ω defines a smooth diffeomorphism onto its image, and

$$\text{Exp} \circ f_m|_\Omega = e_m \circ \alpha|_\Omega; \text{ and}$$

(4) the sequence $(f_m)_{m \geq M}$ tends to zero in the C^∞ sense.

---

[4] Strictly speaking, the class of all complete, pointed Riemannian manifolds is not a set. However, the class of all embedded n-dimensional submanifolds of $\mathbb{R}^{2n+2}$ furnished with complete Riemannian metrics is. Whitney's theorem identifies these two classes up to isometry equivalence, so that the former can indeed be legitimately treated as a set.

These definitions are readily extended in a number of ways. For example, in the case of a sequence (X_m, g_m, p_m) of pointed Riemannian manifolds, the hypothesis of completeness is unnecessary. Instead, it is sufficient to assume that for all $R > 0$, there exists M such that, for all $m \geq M$, the closed ball of radius R about p_m in (X_m, g_m) is compact. Likewise, in the case of immersed submanifolds, the target space can be replaced with a sequence (X_m, g_m, p_m) of pointed Riemannian manifolds converging in the smooth Cheeger–Gromov sense to some complete pointed Riemannian manifold. Furthermore, it is not necessary to suppose that the Riemannian manifolds in this sequence are complete, and so on.

Elliptic regularity, together with the Arzela-Ascoli Theorem of [24], yields the following compactness result.

Lemma 5.17 *Let (X, g) be a complete Riemannian manifold furnished with a Monge–Ampère structure (W, V, J). Let (S_m, e_m, p_m) be a sequence of complete pointed J-holomorphic curves in (X, g). If*

(1) there exists a compact subset K of X such that $e_m(p_m) \in X$ for all m; and
(2) for every compact subset $L \subseteq X$, there exists a constant $B > 0$ such that, for all m, the norm of the second fundamental form of e_m is at every point of $e_m^{-1}(L)$ bounded above by B,

then there exists a complete pointed Monge–Ampère surface $(S_\infty, e_\infty, p_\infty)$ in (X, g) towards which this sequence subconverges in the smooth Cheeger–Gromov sense.

Remark 5.12 This result readily extends in many ways. For example, it is not necessary to assume that the J-holomorphic curves are complete. Likewise, the target manifold with its Monge–Ampère structure may be replaced by sequences converging to some limit, and so on.

We now recall a simplified version of the *quasi-maximum lemma* (see [9]).

Lemma 5.18 *Let X be a complete metric space and let $f : X \to [1, \infty[$ be an upper semi-continuous function. For all $x \in X$, there exists $y \in X$ such that*

(1) $f(y) \geq f(x)$; and
(2) for all $z \in B_{1/\sqrt{f(y)}}(y)$, $f(z) \leq 2f(y)$.

Proof Indeed, otherwise, we recursively construct a sequence (x_m) of points of X such that, for all m, $f(x_{m+1}) \geq 2f(x_m)$ and $d(x_{m+1}, x_m) \leq 1/\sqrt{f(x_m)}$. We readily verify that (x_m) is a Cauchy sequence which therefore converges to some limit x_∞, say. However, by upper semi-continuity,

$$f(x_\infty) \geq \underset{m \to \infty}{\text{LimSup}}\, f(x_m) = \infty,$$

which is absurd, and the result follows. $\qquad\square$

Lemma 5.19 *Let (X, g) be a complete Riemannian manifold furnished with a Monge–Ampère structure (W, V, J). For every compact subset K of X, there exists*

$B > 0$ *such that every Monge–Ampère surface* (S, e) *in* X *has second fundamental form bounded above by* B *at every point of* $e^{-1}(K)$.

Proof We prove this result using a standard blow-up argument. Indeed, suppose the contrary. There exists a sequence (S_m, e_m, p_m) of complete, pointed Monge–Ampère surfaces and a sequence (B_m) of positive numbers converging to $+\infty$ such that, for all m, $e_m(p_m) \in K$ and

$$\|\mathrm{II}_m(p_m)\| = B_m,$$

where, for all m, II_m denotes the second fundamental form of e_m. By Lemma 5.18, we may suppose that, for all m and for all $y \in B_{1/\sqrt{B_m}}(p_m)$,

$$\|\mathrm{II}_m(y)\| \le 2B_m.$$

We extend g to a Riemannian metric defined over the whole of X and, upon rescaling, we consider the sequence $(X, B_m^2 g, e_m(p_m))$ of pointed Riemannian manifolds. This sequence converges in the Cheeger–Gromov sense to $(X_\infty, g_\infty, 0)$, where $X_\infty := \mathbb{C} \times \mathbb{C} \times \mathbb{R}^m$ and g_∞ is the standard Euclidean metric. In addition, we may suppose that the Monge–Ampère structures of these manifolds converge to the constant Monge–Ampère structure $(W_\infty, V_\infty, J_\infty)$ given by

$$W_\infty := \mathbb{C} \times \mathbb{C} \times \{0\},$$

$$V_\infty := \mathbb{R} \times \mathbb{R} \times \{0\}, \text{ and}$$

$$J_\infty := \begin{pmatrix} 0 & i \\ i & 0 \end{pmatrix}.$$

For all m, let II'_m denote the shape operator of e_m with respect to the rescaled metric $B_m^2 g$. For all m,

$$\|\mathrm{II}'_m(p_m)\| = 1,$$

and, for all $y \in B_{\sqrt{B_m}}(x_m)$, $\|\mathrm{II}'_m(y)\| \le 2$. It follows by Theorem 5.12 and the subsequent remark that there exists a Monge–Ampère surface $(S_\infty, e_\infty, p_\infty)$ in X_∞ towards which (S_m, e_m, p_m) subconverges in the smooth Cheeger–Gromov sense. In particular, the norm of the second fundamental form of this surface at p_∞ is equal to 1. However, since this surface is contained in $\mathbb{C} \times \mathbb{C} \times \{0\}$, by Theorem 5.3, it is an affine plane. This yields the desired contradiction and the result follows. $\square$

Combining Lemmas 5.17 and 5.19 yields Labourie's compactness theorem.

Theorem 5.11 (Labourie's Compactness Theorem) *Let* (X, g) *be a complete Riemannian manifold furnished with a Monge–Ampère structure* (W, V, J). *Let*

$(S_m, e_m, p_m)_{m \in \mathbb{N}}$ *be a sequence of pointed Monge–Ampère surfaces in* (X, g). *If* $(e_m(p_m))_{m \in \mathbb{N}}$ *is precompact in* X, *then* $(S_m, e_m, p_m)_{m \in \mathbb{N}}$ *is precompact in the smooth Cheeger–Gromov topology.*

5.3.3   Applications I: k-Surfaces

Let (X, h) be a complete, oriented 3-dimensional Riemannian manifold, let SX denote its bundle of unit tangent spheres, and furnish this manifold with the Sasaki metric. We construct an integrable Monge–Ampère structure over SX as follows. First, let VSX denote the vertical subbundle of TSX, and let HSX denote the horizontal bundle of the Levi–Civita connection. Recall that, at the point $\xi := \xi_p \in SX$, we have the natural identifications

$$H_\xi SX = T_p X, \text{ and}$$

$$V_\xi SX = \langle \xi \rangle^{\perp},$$

where $\langle \xi \rangle^{\perp}$ here denotes the orthogonal complement of ξ_p in $T_p X$, so that $T_\xi SX$ decomposes as

$$T_\xi SX = T_p X \oplus \langle \xi_p \rangle^{\perp}. \tag{5.3.5}$$

With respect to this decomposition, we define the subbundles W and V by

$$W_\xi := \langle \xi_p \rangle^{\perp} \oplus \langle \xi_p \rangle^{\perp}, \text{ and}$$

$$V_\xi := \{0\} \oplus \langle \xi_p \rangle^{\perp}.$$

Since X is oriented, its tangent bundle carries a well-defined wedge product, given by

$$h(v_p, v'_p \wedge v''_p) := d\mathrm{Vol}(v_p, v'_p, v''_p).$$

Thus, for all $\xi := \xi_p \in SX$, $\langle \xi \rangle^{\perp}$ carries a well-defined complex structure j_ξ given by

$$j_\xi \cdot v_p = \xi_p \wedge v_p.$$

Given a smooth function $\phi : SX \to \mathbb{R}$, we define the complex structure J_ϕ over W by

$$J_{\phi, \xi} \cdot (v_p, \mu_p) := (e^\phi j_\xi \mu_p, e^{-\phi} j_\xi v_p).$$

Since, in addition, j_ξ defines an orientation over V, (W, V, J_ϕ) defines a Monge–Ampère structure over X. Note that the symmetric bilinear form m corresponding to this structure is, up to sign, given by

$$m_\xi((v_p, \mu_p), (v_p, \mu_p)) = 2\langle v_p, \mu_p \rangle. \tag{5.3.6}$$

We now verify integrability of this Monge–Ampère structure. For every complete geodesic Γ in X, let $N\Gamma$ denote its unit normal bundle in SX.

Lemma 5.20 *The Monge–Ampère structure (W, V, J_ϕ) is integrable and its curtain surfaces are the covers of those surfaces of the form $N\Gamma$, for some complete geodesic Γ in X.*

Proof Let Γ be a complete geodesic in X. With respect to the decomposition (5.3.5), given any $\xi := \xi_p \in N_p\Gamma$, the tangent space to $N\Gamma$ at this point is given by

$$T_\xi N\Gamma = \langle (\tau, 0), (0, \tau \wedge \xi) \rangle, \tag{5.3.7}$$

where τ is here a unit tangent vector to Γ at p. It immediately follows that $N\Gamma$ is a curtain surface. Conversely, given a point $\xi := \xi_p$ in SX and a vertical vector $(0, v_p) \in W_\xi$, let Γ denote the unique geodesic tangent to $\xi_p \wedge v_p$ at p. By (5.3.7), $N\Gamma$ is a curtain surface tangent to $(0, v_p)$ at ξ_p. Since ξ is arbitrary, integrability follows, and this completes the proof. $\square$

It remains only to study the geometry of positive Monge–Ampère surfaces. First, let (S, e) be an ISC immersed surface in X and let $(S, \hat{e})$ denote its Gauss lift. Recall from the introduction that, for any smooth function $\kappa : SX \to \mathbb{R}$, the extrinsic curvature K_e of (S, e) is said to be *prescribed* by κ whenever

$$K_e = \kappa \circ \hat{e}.$$

Lemma 5.21 *An immersed surface in SX is a positive Monge–Ampère surface if and only if it is the Gauss lift of a quasicomplete, ISC, immersed surface in X of extrinsic curvature prescribed by $e^{2\phi}$.*

Proof We first show that every positive Monge–Ampère surface is the Gauss lift of some immersed surface. Indeed, since V is the kernel of the canonical projection $\pi : SX \to X$, for every positive Monge–Ampère surface $(S, \hat{e})$, the function $e := \pi \circ \hat{e}$ is an immersion. Observe, furthermore, that $\hat{e}$ is normal to e, so that the orientation of S can be chosen in such a manner that $(S, \hat{e})$ is the Gauss lift of (S, e), as desired.

Now let (S, e) be an oriented, immersed surface in X and let $(S, \hat{e})$ denote its Gauss lift. Trivially (S, e) is quasicomplete if and only if $(S, \hat{e})$ is complete. Next, with respect to the decomposition (5.3.5), the derivative of $\hat{e}$ at a point $p \in S$ is given by

$$D\hat{e}(p) \cdot \xi_p = (De(p) \cdot \xi_p, De(p) \cdot A_e(p) \cdot \xi_p).$$

Thus, by Lemma 5.9, $(S, \hat{e})$ is a J-holomorphic curve if and only if (S, e) has extrinsic curvature prescribed by $e^{2\phi}$. Likewise, by (5.3.6), $(S, \hat{e})$ is positive if and only if A_e is everywhere positive-definite, that is, if and only if (S, e) is ISC. The result now follows.   □

The theory developed in Sects. 5.3.1 and 5.3.2 now yields the following result.

Theorem 5.12 (Labourie) *Let* $\phi : SX \to \mathbb{R}$ *be a smooth function. Let* (S_m, e_m, p_m) *be a sequence of quasicomplete, pointed, immersed surfaces of extrinsic curvature prescribed by* $e^{2\phi}$. *If* $(e_m(p_m))$ *is precompact in* X, *then the sequence* $(S_m, \hat{e}_m, p_m)$ *of Gauss lifts is precompact in the smooth Cheeger–Gromov sense. Furthermore, any accumulation point* $(S_\infty, \hat{e}_\infty, p_\infty)$ *of this sequence which is not a curtain surface is the Gauss lift of some quasicomplete, pointed immersed surface of constant extrinsic curvature prescribed by* $e^{2\phi}$.

Remark 5.13 As in Sect. 5.3.2, we have stated this result in its simplest form. More generally, the ambient space X and the function ϕ may be allowed to vary with m, the surfaces need not be quasicomplete, and so on.

5.3.4 Applications II: Isometric Immersions

We conclude this paper by reviewing a more sophisticated version of the construction of the preceding section developed in [15]. By incorporating the domain explicitly into the construction, we are able to study families of isometric immersions, as opposed to merely reparametrisation equivalence classes, as is the case in the preceding section.

We first review the corresponding linear framework. Let $\mathrm{Isom}(\mathbb{R}^2, \mathbb{R}^3)$ denote the subset of $\mathrm{Lin}(\mathbb{R}^2, \mathbb{R}^3)$ consisting of linear isometries $\alpha : \mathbb{R}^2 \to \mathbb{R}^3$. Note that this is an orbit of $\mathrm{SO}(3)$ in $\mathrm{Lin}(\mathbb{R}^2, \mathbb{R}^3)$ upon which this group acts with trivial stabiliser. It is therefore a smooth 3-dimensional submanifold, and its tangent bundle admits the trivialisation

$$\tau : \mathrm{Lin}(\mathbb{R}^2, \mathbb{R}^3) \times \mathfrak{so}(3) \to T\mathrm{Isom}(\mathbb{R}^2, \mathbb{R}^3); (\alpha, A) \mapsto (\alpha, A \circ \alpha).$$

In addition, viewing the wedge product as a linear isomorphism from $\mathbb{R}^3$ to $\mathfrak{so}(3)$, a useful alternative form of this trivialisation is

$$\sigma : \mathrm{Lin}(\mathbb{R}^2, \mathbb{R}^3) \times \mathbb{R}^3 \to T\mathrm{Isom}(\mathbb{R}^2, \mathbb{R}^3); (\alpha, x) \mapsto (\alpha, x \wedge \alpha).$$

We now return to the non-linear case. Let (S, g) be an oriented Riemannian surface, and let (X, h) be an oriented 3-dimensional Riemannian manifold. Over the cartesian product $S \times X$, we consider the bundle $\mathrm{Lin}(\pi_1^* TS, \pi_2^* TX)$, where π_1 and π_2 respectively denote the projections onto the first and second factors. Let $\mathrm{Isom}(S \times X)$ denote its subbundle whose fibre at the point (p, q) is $\mathrm{Isom}(T_p S, T_q X)$.

We construct a Monge–Ampère structure over the total space of this subbundle as follows. Let $\mathrm{VIsom}(S \times X)$ denote the vertical subbundle of $\mathrm{TIsom}(S \times X)$, and let $\mathrm{HIsom}(S \times X)$ denote the horizontal bundle of the Levi-Civita connection. Bearing in mind the discussion of the preceding paragraph, at the point $\alpha := \alpha_{(p,q)} \in \mathrm{Isom}_{(p,q)}(S \times X)$, we have the natural identifications,

$$H_\alpha \mathrm{Isom}(S \times X) := T_p X \oplus T_q X, \text{ and}$$

$$V_\alpha \mathrm{Isom}(S \times X) := T_q X,$$

so that $T_\alpha \mathrm{Isom}(S \times X)$ decomposes as

$$\mathrm{T}_\alpha \mathrm{Isom}(S \times X) = T_p S \oplus T_q X \oplus T_q X. \tag{5.3.8}$$

With respect to this decomposition, we define the subbundles W and V by

$$W_\alpha := \left\{ (\xi_p, \alpha \cdot \xi_p, \alpha \cdot v_p) \mid \xi_p, v_p \in T_p S \right\}, \text{ and}$$

$$V_\alpha := \left\{ (0, 0, \alpha \cdot v_p) \mid v_p \in T_p S \right\}.$$

We furnish W with the Sasaki metric g, and, given a smooth function $\phi : \mathrm{Isom}(S \times X) \to \mathbb{R}$, we define the complex structure J_ϕ over W by

$$J_{\phi,\alpha} \cdot (\xi_p, \alpha \cdot \xi_p, \alpha \cdot v_p) := (e^\phi j_p \cdot v_p, e^\phi \alpha \cdot j_p \cdot v_p, e^{-\phi} \alpha \cdot j_p \cdot \xi_p),$$

where j here denotes the complex structure of (S, g). Since the orientation of S yields an orientation of V, the quadruplet (W, g, V, J_ϕ) again defines a Monge–Ampère structure over $\mathrm{Isom}(S \times X)$. As before, we verify that the associated symmetric, bilinear form m is given by

$$m_\alpha((\xi_p, \alpha \cdot \xi_p, \alpha \cdot v_p), (\xi_p, \alpha \cdot \xi_p, \alpha \cdot v_p)) := 2\langle \xi_p, v_p \rangle. \tag{5.3.9}$$

We now show that this Monge–Ampère structure is integrable. To this end, let $\gamma : \mathbb{R} \to S$ and $\delta : \mathbb{R} \to X$ be unit speed parametrised geodesics and let $\alpha : T_{\gamma(0)} S \to T_{\delta(0)} X$ be an isometry such that

$$\alpha \cdot \dot{\gamma}(0) = \dot{\delta}(0).$$

We define $e_{\gamma,\delta,\alpha} : \mathbb{R}^2 \to \mathrm{Isom}(S \times X)$ by

$$e_{\gamma,\delta,\alpha}(s, t) := (\gamma(s), \delta(s), \mathrm{Exp}(t\dot{\delta}(s)\wedge) \cdot \tau_s \alpha),$$

where, for all s, τ_s here denotes parallel transport along (γ, δ) from $(\gamma(0), \delta(0))$ to $(\gamma(s), \delta(s))$.

Lemma 5.22 (W, V, J_ϕ) *is integrable, and its curtain surfaces are quotients of immersed surfaces of the form* $(\mathbb{R}^2, e_{\gamma,\delta,\alpha})$, *with* γ, δ *and* α *as above.*

Proof Indeed, with respect to the decomposition (5.3.8),

$$De_{\gamma,\delta,\alpha}(s, t) \cdot \partial_s = (\dot{\gamma}(s), \alpha \cdot \dot{\gamma}(s), 0), \text{ and}$$

$$De_{\gamma,\delta,\alpha}(s, t) \cdot \partial_t = (0, 0, \alpha \cdot \dot{\gamma}(s)), \tag{5.3.10}$$

so that $(\mathbb{R}^2, e_{\gamma,\delta,\alpha})$ is a curtain surface. Conversely, given a point $\alpha := \alpha_{(p,q)} \in$ Isom$(S \times X)$ and a vertical vector $(0, 0, \alpha \cdot \xi_p)$ at this point, let $\gamma : \mathbb{R} \to S$ and $\delta : \mathbb{R} \to X$ denote the unique unit speed geodesics such that

$$\dot{\gamma}(0) := \xi_p, \text{ and}$$

$$\dot{\delta}(0) := \alpha \cdot \xi_p.$$

By (5.3.10), $(\mathbb{R}^2, e_{\gamma,\delta,\alpha})$ is a curtain surface tangent to $(0, 0, \alpha\xi_p)$ at $\alpha_{(p,q)}$. Integrability follows, and this completes the proof.   □

It remains only to describe the geometric properties of positive Monge–Ampère surfaces. Let (S', g') be another Riemannian surface and consider an immersion $(e, f) : S' \to S \times X$. We say that $(S', (e, f))$ is *admissable* whenever both e and f are isometries. When this holds, we define $E_{e,f} : S' \to$ Isom$(S \times X)$ by

$$E_{e,f} := (e, f, Df \circ De^{-1}). \tag{5.3.11}$$

We call $(S', E_{e,f})$ the *Gauss lift* of $(S', (e, f))$. We say that $(S', (e, f))$ is *infinitesimally strictly complex* and *quasicomplete* whenever (S', f) has these properties and, given a smooth function $\kappa : $ Isom$(S \times X) \to \mathbb{R}$, we say that $(S', (e, f))$ has extrinsic curvature *prescribed* by κ whenever the extrinsic curvature K_f of f satisfies

$$K_f = \kappa \circ E_{e,f}.$$

Lemma 5.23 *An immersed surface in* Isom$(S \times X)$ *is a positive Monge–Ampère surface if and only if it is the Gauss lift of a quasicomplete, ISC, admissable immersed surface* $(S', (e, f))$ *in* $S \times X$ *of extrinsic curvature prescribed by* $e^{2\phi}$.

Proof As in the proof of Lemma 5.21, every positive Monge–Ampère surface is the Gauss lift of some immersed surface. Now let $(S', (e, f))$ be an admissable immersed surface and let $(S', E_{e,f})$ denote its Gauss lift. Let A_f denote the shape operator of f. Bearing in mind (5.3.11), with respect to the decomposition (5.3.8), the derivative of $E_{e,f}$ at a point $p \in S'$ is given by

$$DE_{e,f}(p) \cdot \xi_p = (De(p) \cdot \xi_p, Df(p) \cdot \xi_p, Df(p) \cdot A_f(p) \cdot \xi_p)$$
$$= (De(p) \cdot \xi_p, \alpha \cdot De(p) \cdot \xi_p, \alpha \cdot De(p) \cdot A_f(p) \cdot \xi_p),$$

where $\alpha := E_{e,f}(p)$. It follows by Lemma 5.9 that $(S', E_{e,f})$ is a J-holomorphic curve if and only if $(S', (e, f))$ has extrinsic curvature prescribed by $e^{2\phi}$. Likewise, by (5.3.9), $(S, \hat{e})$ is positive if and only if A_e is everywhere positive-definite, that is, if and only if $(S', (e, f))$ is ISC. The result now follows. $\qquad\square$

This construction can be applied to the study of isometric immersions in a variety of ways. Since the general result is technical and unenlightening, we shall content ourselves here with an illustrative example. Let $M := (M, h)$ be a 3-dimensional Cartan-Hadamard manifold, and let $\mathbb{H}^2$ denote 2-dimensional hyperbolic space with its metric g. Let (ω_m) be a sequence of bounded, smooth functions over $\mathbb{H}^2$ converging in the C^∞_{loc} sense to the bounded, smooth function ω_∞. For all m, denote $g_m := e^{2\omega_m} g$, $S_m := (\mathbb{H}^2, g_m)$, $X_m := \text{Isom}(S_m, M)$, and let $\phi_m : X_m \to \mathbb{R}$ be such that, for all $\alpha := \alpha_{p,q}$,

$$e^{2\phi_m(\alpha)} = \kappa_m(p) - \sigma(\text{Im}(\alpha)),$$

where κ_m here denotes the curvature of g_m, and σ denotes the sectional curvature of h. For all $m \in \mathbb{N}$, let $e_m : S_m \to M$ be an isometric immersion and denote

$$\hat{e}_m(x) := (x, e_m(x)).$$

Note that, for all m, $\hat{e}_m$ is admissable, quasicomplete, ISC, and of extrinsic curvature prescribed by $e^{2\omega_m}$. Thus, for all m, the Gauss lift E_m of $\hat{E}_m$ is a J_{ϕ_m}-holomorphic curve. The theory developed in Sects. 5.3.1 and 5.3.2 now yields the following result.

Theorem 5.13 (Labourie) *Let $x_0 \in \mathbb{H}^2$ be a fixed point. If $(e_m(x_0))$ is precompact, then the sequence (S_m, E_m, x_0) of Gauss lifts is precompact. Furthermore, any accumulation point $(S_\infty, E_\infty, x_0)$ is either a tube, or the Gauss lift of a quasicomplete, ISC surface of the form $(\Omega, \hat{e}_\infty, x_\infty)$, for some open subset Ω of $\mathbb{H}^2$.*

Remark 5.14 Note, in particular, that the domain of the limit immersion need not be the whole of $\mathbb{H}^2$.

Acknowledgments The author is grateful to Nam Q. Le for helpful comments made to an earlier draft of this paper. This paper was in part written whilst the author was visiting the Institut des Hautes Études Scientifiques. The author is grateful for the excellent working conditions enjoyed during that stay.

References

1. N. Aronszajn, A unique continuation theorem for elliptic differential equations or inequalities of the second order. J. Math. Pures Appl. **36**, 235–239 (1957)
2. T. Barbot, F. Béguin, A. Zeghib, Prescribing Gauss curvature of surfaces in 3-dimensional spacetimes: application to the Minkowski problem in the Minkowski space. Ann. Inst. Fourier **61**(2), 511—591 (2011)
3. F. Bonsante, G. Mondello, J.M. Schlenker, A cyclic extension of the earthquake flow I. Geom. Topol. **17**(1), 157–234 (2013)
4. F. Bonsante, G. Mondello, J.M. Schlenker, A cyclic extension of the earthquake flow II. Ann. Sci. Ec. Norm. Supér. **48**(4), 811—859 (2015)
5. E. Calabi, Improper affine hyperspheres of convex type and a generalisation of a theorem by K. Jörgens. Mich. Math. J. **5**, 105–126 (1958)
6. A. Figalli, *The Monge-Ampère Equation and its Applications*. Zurich Lectures in Advanced Mathematics, European Mathematical Society (EMS), Zürich (2017)
7. P.R. Girard, The quaternion group and modern physics. Eur. J. Phys. **5**, 25–32 (1984)
8. P.H. Gromov, Pseudo-holomorphic curves in symplectic manifolds. Invent. Math. **82**, 307–347 (1985)
9. M. Gromov, Foliated plateau problem, part II: harmonic maps of foliations. Geom. Func. Anal. **1**(3), 253–320 (1991)
10. R. Harvey, H.B. Lawson, Jr., Calibrated geometries. Acta. Math. **148**, 47–157 (1982)
11. F.R. Harvey, H.B. Lawson, Pseudoconvexity for the special Lagrangian potential equation. Calc. Var. PDEs. **60**(1), 37 (2021)
12. K. Jörgens, Über die Lösungen der Differentialgleichung $rt - s^2 = 1$. Math. Ann. **127**, 130–134 (1954)
13. M. Kokubu, W. Rossman, K. Saji, M. Umehara, K. Yamada, Singularities of flat fronts in hyperbolic space. Pac. J. Math. **221**(2), 303–351 (2005)
14. M. Kokubu, W. Rossman, M. Umehara, K. Yamada, Flat fronts in hyperbolic space and their caustics. J. Soc. Math. Japan **59**(1), 265–299 (2007)
15. F. Labourie, Métriques prescrites sur le bord des variétés hyperboliques de dimension 3. J. Differ. Geom. **35**(3), 609—626 (1992)
16. F. Labourie, Problèmes de Monge-Ampère, courbes pseudo-holomorphes et laminations. Geom. Funct. Anal. **7**, 496–534 (1997)
17. V.V. Lychagin, V.N. Rubtsov, Local classification of Monge-Ampère differential equations (in Russian). Dokl. Akad. Nauk SSSR **272**(1), 34–38 (1983)
18. V.V. Lychagin, V.N. Rubtsov, I.V. Chekalov, A classification of Monge-Ampère equations. Ann. Sci. École Norm. Sup. **26**(3), 281–308 (1993)
19. A. Martínez, F. Milán, Flat fronts in hyperbolic 3-space with prescribed singularities. Ann. Glob. Anal. Geom. **46**, 227–239 (2014)
20. D. McDuff, D. Salamon, *J-Holomorphic Curves and Quantum Cohomology*. University Lecture Series, vol. 6 (AMS, Providence, 1994)
21. A.V. Pogorelov, On the improper convex affine hyperspheres. Geom. Dedi. **1**, 33–46 (1972)
22. V. Rubtsov, Geometry of Monge-Ampère structures, in *Nonlinear PDEs, Their Geometry, and Applications* (Birkhäuser/Springer, Berlin, 2019), pp. 95–156
23. J.M. Schlenker, Hyperbolic manifolds with convex boundary. Invent. Math. **163**, 109–169 (2006)
24. G. Smith, An Arzela-Ascoli theorem for immersed submanifolds. Ann. Fac. Sci. Toulouse Math. **16**(4), 817–866 (2007)
25. G. Smith, Special Lagrangian curvature. Math. Ann. **335**(1), 57–95 (2013)

26. G. Smith, On the asymptotic geometry of finite-type k-surfaces in three-dimensional hyperbolic space Smith, J. Eur. Math. Soc. (JEMS) **26**(2), 407–467 (2024)
27. M. Spivak, *A Comprehensive Introduction to Differential Geometry*, vol. IV (Publish or Perish, 1999)
28. J. Toulisse, F. Labourie, M. Wolf, Plateau problems for maximal surfaces in pseudo-hyperbolic spaces. Ann. Sci. Éc. Norm. Supér (to appear). arXiv:2006.12190

Chapter 6
Lagrangian Grassmannians of Polarizations

Peter Kristel and Eric Schippers

Abstract This chapter is an introduction to polarizations in the symplectic and orthogonal settings. They arise in association to a triple of compatible structures on a real vector space, consisting of an inner product, a symplectic form, and a complex structure. A polarization is a decomposition of the complexified vector space into the eigenspaces of the complex structure; this information is equivalent to the specification of a compatible triple. When either a symplectic form or inner product is fixed, one obtains a Grassmannian of polarizations. We give an exposition of this circle of ideas, emphasizing the symmetry of the symplectic and orthogonal settings, and allowing the possibility that the underlying vector spaces are infinite-dimensional. This introduction would be useful for those interested in applications of polarizations to representation theory, loop groups, complex geometry, moduli spaces, quantization, and conformal field theory.

Keywords Lagrangian grassmannian · Polarizations · Symplectic form · Inner product · Siegel disk · Symplectic group · Orthogonal group

MSC 2020 15A66, 15B57, 32G15, 53D99, 81S10

6.1 Introduction

This chapter is a self-contained introduction to polarizations of complex vector spaces and Lagrangian Grassmannians of polarizations. These appear in geometry and algebra in many contexts, such as algebraic and complex geometry [3, 5, 23], moduli spaces [12, 23], loop groups [16, 18], and the metaplectic and spin

P. Kristel
Hausdorff Center for Mathematical Sciences, Bonn, Germany

E. Schippers (✉)
University of Manitoba, Winnipeg, MB, Canada
e-mail: eric.schippers@umanitoba.ca

representations [6, 15]. They also appear in physics in association with the latter objects [14, 30], and in conformal field theory [7, 19]. The chapter covers the two types of polarization: the Riemannian one, in which the polarization is an orthogonal decomposition; and the symplectic one, in which the polarization is a symplectic decomposition.

Our presentation is suitable for graduate students. We also hope that seasoned researchers might find it useful as a quick introduction. It could serve as preparation or as a reference for those encountering polarizations and Grassmannians in any of the topics above. Although these topics are not treated here, we included many examples with references for orientation.

Throughout, we have emphasized the symmetry of the Riemannian and symplectic point of view. We therefore chose a "middle ground" notationally, sampling from common notation in both fields, as well as from complex geometry.

At the heart of the idea of polarization is a triple of compatible structures: a complex structure, an inner product, and a symplectic form. A familiar example is that of a Kähler manifold, which is a manifold equipped with three compatible structures: (1) a Riemannian metric; (2) an integrable complex structure; (3) a symplectic form. Here the three structures vary from point to point on the manifold.

The compatibility condition ensures that when we are provided with two out of the three structures, we can reconstruct the third one, if it exists (which is not always the case). It then becomes natural to fix *one* of these structures, and study the space of structures of a second type, which allow for the reconstruction of a compatible structure of the third type. For example, we can fix a symplectic manifold (M, ω), and study the space of integrable complex structures on M, such that M admits a compatible Riemannian metric.

In this chapter we restrict our attention to structures on a fixed real vector space $M = H$. (In the example of Kähler manifold, one could restrict to a particular tangent space, or, one could consider the case that the manifold is just a complex vector space). In particular we do not discuss integrability. Instead, we consider deformations of these structures on the fixed vector space, or equivalently, the Grassmannian of polarizations on the complexification of H. We explore different models of the Grassmannians, and their manifold structures. We also simultaneously generalize the setting, by allowing (and indeed focusing on) the case that H is infinite-dimensional. Doing this requires only a little extra effort, and immediately broadens the scope. Some functional analytic issues rear their head when passing to the infinite-dimensional case, but these are easily dealt with. There are many interesting examples which fit in this context, many of which appear in the text. In general, most of the examples are completely accessible, but in a few cases where details would be distracting we indicate the relevant literature.

Here is an outline of the chapter. In Sect. 6.2 we recall the definition of compatible triples, and explore their basic properties and characterizations. We also take care of a few functional analytic issues necessary when dealing with the case of (infinite-dimensional) Hilbert spaces.

In Sect. 6.3, we consider both types of polarization. Briefly, if J is a complex structure on H, then the complexification $H_{\mathbb{C}}$ splits as a direct sum of the $\pm i$-eigenspaces of J. This establishes a correspondence between certain types of decompositions of $H_{\mathbb{C}}$ and complex structures on H. We exploit this relation to phrase our problem in the language of Grassmannians, in anticipation of Sects. 6.4 and 6.5. If g is a Riemannian metric, and J is orthogonal, then the corresponding decomposition of $H_{\mathbb{C}}$ is orthogonal. If ω is a symplectic form, and J is a symplectomorphism, then the corresponding decomposition of $H_{\mathbb{C}}$ is Lagrangian.

A Lagrangian, resp. orthogonal decomposition of $H_{\mathbb{C}}$ is part of the data required to carry out bosonic, resp. fermionic geometric quantization. In the current context, this procedure produces a representation of the Heisenberg, resp. Clifford algebra of H on the Fock space of the chosen subspace of $H_{\mathbb{C}}$. We give a cursory overview of this construction in Sect. 6.6 for the sake of motivation. For us, the most important point is that considerations of unitary equivalence of representations lead to a natural functional analytic condition, which picks out a subset of each of the Grassmannians under consideration. These are the "restricted" Grassmannians of representation theory.

In Sects. 6.4 and 6.5 we describe the symplectic and orthogonal Grasmannians respectively, giving different models and providing a logically complete exposition of their basic properties. These Grassmannians have the desirable property that they can be equipped with the structure of complex manifold. We give a detailed construction of an atlas of these manifolds. In the restricted case, this is modelled on the Hilbert space of Hilbert–Schmidt operators on some infinite-dimensional Hilbert space.

Section 6.7 explores a geometric interpretation of a particular Lagrangian Grassmannian in terms of sewing. This example arises in loop groups, conformal field theory, and Teichmüller theory. Finally, Sect. 6.8 contains a complete set of solutions to the exercises.

6.2 Compatible Triples

In this section, we introduce the notion of compatible triples. This is the information of an inner product, symplectic form, and complex structure, which are related in a sense defined shortly. We will see various equivalent ways of expressing this compatibility. The ubiquity of compatible triples is illustrated with examples. As promised, we give an exposition in the generality of Hilbert spaces.

We begin with a few considerations about Hilbert spaces. Those concerned only in the finite-dimensional case could skip this first page, and ignore a few extra lines of justification in some of the proofs, without losing the thread. This could profitably be maintained until the restricted Grassmannians are considered starting in Sect. 6.4.

Let H be a separable (possibly even finite-dimensional) real Hilbert space. That is, H is a separable topological vector space over $\mathbb{R}$, which admits an inner product with respect to which it is a Hilbert space. We recall the following three notions:

- A *strong inner product* on H is a continuous symmetric bilinear form on H, such that the map $\varphi_g : H \to H^*$ defined by the relation $\varphi_g(v)(w) = g(v, w)$ is an isomorphism, and such that $g(v, v) \geqslant 0$ for all $v \in H$.
- A *complex structure* J on H is a continuous map $J : H \to H$ with the property that $J^2 = -\mathbb{1}$.
- A *strong symplectic form* on H is a continuous anti-symmetric bilinear form ω on H, such that the map $\varphi_\omega : H \to H^*$ defined by the relation $\varphi_\omega(v)(w) = \omega(v, w)$ is an isomorphism.

The maps $\varphi_\omega, \varphi_g : H \to H^*$ are sometimes called the musical isomorphisms. Note that here, by isomorphism, we mean a bounded linear bijection. Thus the inverse must also be bounded. Of course in finite dimensions a linear bijection is automatically bounded.

When a vector space is equipped with a symplectic form ω, one may consider the *canonical commutation relation* algebra of H. This is the unital, associative algebra generated by H subject to the condition $vw - wv = \omega(v, w)$. Similarly, when one is given an inner product, one may consider the *canonical anti-commutation relation* algebra of H. This is the unital, associative algebra generated by H subject to the condition $vw + wv = g(v, w)$. Consideration of (certain variations of) these algebras and their modules motivates many of the definitions and results outlined in this chapter. We will give a sketch of some of this motivation in Sect. 6.6 ahead.

If (H, g) is a Hilbert space, then g is a strong inner product. In general, a strong inner product gives a Hilbert space structure equivalent to g, as the following proposition shows.

Proposition 6.1 *Let (H, g) be a Hilbert space. Let $\psi : H \times H \to \mathbb{R}$ be a bilinear pairing. Then ψ is a strong inner product if and only if (H, ψ) is a Hilbert space and ψ and g are equivalent.*

Proof Denote by $\|w\|_\psi$ the norm $\sqrt{\psi(w, w)}$. By assumption there is a K such that $\|w\|_\psi \leq K\|w\|$ for all w, and in particular $\|w\|_\psi$ is finite for any fixed w.

Since φ_ψ is an isomorphism, it is bounded below. Using this together with the Cauchy–Schwarz inequality for ψ we get

$$C\|w\| \leq \|\varphi_\psi(w)\| = \sup_{\|v\|=1} |\psi(w, v)| \leq \|w\|_\psi \|v\|_\psi \leq K\|w\|_\psi.$$

This shows that (H, ψ) is complete and thus a Hilbert space, and that the norms are equivalent.

Conversely, if (H, ψ) is a Hilbert space and the norms induced by ψ and g are equivalent, then they are also equivalent on H^*. Since $\varphi_\psi : (H, \psi) \to (H^*, \psi)$ is an isomorphism by the Riesz representation theorem, it is also a bounded isomorphism with respect to g.   □

Emboldened by Proposition 6.1 we occasionally refer to a Hilbert space without designating a particular metric as special.

It is worth pausing here to compare the infinite- and finite-dimensional settings. If $\psi : H \times H \rightarrow \mathbb{R}$ is a bilinear map, then ψ is *weakly non-degenerate* if the induced map $\varphi_\psi : H \rightarrow H^*$ defined by $\varphi_\psi(v)(w) = \psi(v, w)$ is injective. This is equivalent to the statement that for every $0 \neq v \in H$ there exists a $w \in H$ such that $\psi(v, w) \neq 0$. The map ψ is *strongly non-degenerate* if the induced map $\varphi_\psi : H \rightarrow H^*$ is an isomorphism. In finite dimensions, these notions coincide, but if H is infinite-dimensional, this is no longer true as Example 6.2 will show. On the other hand, in finite dimensions an inner product is automatically strong on account of its positive-definiteness.

Example 6.2 Consider the Hilbert space $\ell^2(\mathbb{R})$ of square-summable sequences, with its standard inner product $g(\{a_n\}, \{b_n\}) = \sum_{n \geqslant 1} a_n b_n$. We equip $\ell^2(\mathbb{R})$ with the symmetric bilinear map $\psi(\{a_n\}, \{b_n\}) = \sum_{n \geqslant 1} a_n b_n / n$. The map ψ is certainly weakly non-degenerate. It is however, not strongly non-degenerate, because φ_ψ is not surjective. Indeed, let $\xi : H \rightarrow \mathbb{R}$ be the map $\xi(\{a_n\}) = \sum_{n \geqslant 1} a_n / n$. The pre-image $\{x_n\}$ of ξ under the map φ_ψ (if it existed) would have to satisfy

$$\sum_{n \geqslant 1} a_n / n = \psi(x_n, a_n) = \sum_{n \geqslant 1} x_n a_n / n,$$

for all $\{a_n\} \in \ell^2(\mathbb{R})$. It follows that $x_n = 1$ for all n, but this sequence is not square-summable. △

We will not be concerned with weakly non-degenerate maps, so from now on, we assume that any inner product or symplectic form is strong. The term will be dropped except when needed for proof or emphasis.

Definition 6.3 A *compatible triple* (g, J, ω) consists of a strong inner product g, a complex structure J, and a strong symplectic form ω, such that $g(v, w) = \omega(v, Jw)$. △

Compatible triples can also be characterized as follows.

Proposition 6.4 *Let* (g, J, ω) *be an inner product, complex structure* J, *and symplectic form* ω *respectively. The following are equivalent.*

1. $g(v, w) = \omega(v, Jw)$ *for all* $v, w \in H$;
2. $\omega(v, w) = g(Jv, w)$ *for all* $v, w \in H$;
3. $J(v) = \varphi_g^{-1} \varphi_\omega(v)$ *for all* $v \in H$.

Proof Observe that properties 1 and 2 are equivalent to $\varphi_g = -\varphi_\omega J$ and $\varphi_\omega = \varphi_g J$ respectively. Now by multiplying with J from the right, we see that $\varphi_g = -\varphi_\omega J$ if and only if $\varphi_\omega = \varphi_g J$, and by multiplying with φ_g^{-1} from the left, we see that $\varphi_\omega = \varphi_g J$ if and only if $\varphi_g^{-1} \varphi_\omega = J$. □

Thus, any of the above could be taken as equivalent definitions of compatible triple.

Exercise 6.5 Assume that (g, J, ω) is a compatible triple. Show that J preserves both g and ω, i.e. $g(Jv, Jw) = g(v, w)$ and $\omega(Jv, Jw) = \omega(v, w)$. $\triangle$

Exercise 6.6 Show that

(a) If g and J are an inner product and a complex structure, then there exists a symplectic form ω such that (g, J, ω) is compatible if and only if J is skew-adjoint with respect to to g.
(b) If g and ω are an inner product and a symplectic form, then there exists a complex structure J such that (g, J, ω) is compatible if and only if $\varphi_g^{-1}\varphi_\omega = -\varphi_\omega^{-1}\varphi_g$.
(c) If J and ω are a complex structure and a symplectic form, then if $g(v, w) = \omega(v, Jw)$ defines an inner product, then J is skew-symmetric with respect to ω.

$\triangle$

Exercise 6.7 Show that there exist a symplectic form ω and a complex structure J, which is skew-symmetric with respect to ω, such that $g(v, w) = \omega(v, Jw)$ does not define an inner product. $\triangle$

Example 6.8 Let $H = \mathbb{R}^{2n}$. Let $\{e_i\}_{i=1,\ldots,2n}$ be the standard basis, and write $x_i = e_{2i-1}$ and $y_i = e_{2i}$. Let g be the standard inner product, and define J by $Jx_i = y_i$ and $Jy_i = -x_i$. The symplectic form ω such that (g, J, ω) is a compatible triple is given by

$$\omega(x_k, y_k) = -\omega(y_k, x_k) = 1, \quad k = 1, \ldots, n,$$

and ω is zero on all other pairs of basis vectors. $\triangle$

Example 6.9 Let M be a complex manifold; that is, a $2n$-dimensional manifold with an atlas of charts $\{\phi : U \to \mathbb{C}^n\}$ such that for any pair of charts ϕ_1, ϕ_2, it holds that $\phi_2 \circ \phi_1^{-1}$ is a biholomorphism on its domain. If $(z_1, \ldots, z_n)$ are coordinates on $\mathbb{C}^{2n}$ and $z_k = x_k + iy_k$, then a chart ϕ induces vector fields $\partial/\partial x_k$ and $\partial/\partial y_k$ in U. At a point $p \in U$, the real tangent space T_pM has a complex structure defined to be the real linear extension of

$$J_p \frac{\partial}{\partial x_k} = \frac{\partial}{\partial y_k}, \quad J_p \frac{\partial}{\partial y_k} = -\frac{\partial}{\partial x_k}. \tag{6.1}$$

It can be shown that this complex structure is independent of the choice of coordinates, using the Cauchy–Riemann equations. $\triangle$

Remark 6.10 In general, a smoothly varying choice of complex structure on each tangent space of a real $2n$-dimensional manifold M is called an almost complex structure. Not every manifold with an almost complex structure can be given an atlas making it a complex manifold with almost complex structure arising from Equation (6.1). If this can be done, the almost complex structure is called integrable. $\triangle$

Example 6.11 Let $\mathcal{R}$ be a compact Riemann surface, (i.e. a compact complex manifold of dimension 2). In local coordinates $z = x + iy$, any real one-form can be expressed as $\beta = a(z)dx + b(z)dy$. The Hodge star operator on one-forms is defined to be the complex linear extension of

$$* \, dx = dy, \quad *dy = -dx.$$

By the Cauchy–Riemann equations, this is coordinate-independent. A real or complex one-form β is said to be harmonic if it is closed and co-closed, that is, if $d\beta = 0$ and $d * \beta = 0$ respectively. Equivalently, in local coordinates $\beta(z) = h_1(z)dz + \overline{h_2(z)}d\bar{z}$ where h_1, h_2 are harmonic.

By the Hodge theorem, every de Rham cohomology class on $\mathcal{R}$ has a harmonic representative. So the set of real harmonic one-forms $\mathcal{A}_{\text{harm}}^{\mathbb{R}}(\mathcal{R})$ on $\mathcal{R}$ is a real $2g$-dimensional vector space. Define the pairing

$$g(\beta, \gamma) = \iint_{\mathcal{R}} \beta \wedge *\gamma. \tag{6.2}$$

It is easily checked that this is an inner product on $\mathcal{A}_{\text{harm}}^{\mathbb{R}}(\mathcal{R})$. Define also

$$\omega(\beta, \gamma) = \iint_{\mathcal{R}} \beta \wedge \gamma.$$

With $H = \mathcal{A}_{\text{harm}}^{\mathbb{R}}(\mathcal{R})$, $(g, *, \omega)$ is a compatible triple. $\triangle$

Remark 6.12 In the previous example, the restriction of the Hodge star operator to a single tangent space is the dual of the almost complex structure given in Example 6.9. $\triangle$

Exercise 6.13 Let $\mathcal{R}$ and ω be as in Example 6.11. Let $H_{\text{dR}}^1(\mathcal{R})$ denote the de Rham cohomology space of smooth closed one-forms modulo smooth exact one-forms on $\mathcal{R}$. Given $[\beta], [\gamma] \in H_{\text{dR}}^1(\mathcal{R})$, let $\widehat{\beta}, \widehat{\gamma}$ be the harmonic representatives of their respective equivalence classes. Show that for arbitrary representatives β, γ of $[\beta]$, $[\gamma]$

$$\omega(\beta, \gamma) = \omega(\widehat{\beta}, \widehat{\gamma}).$$

In particular, ω is a well-defined symplectic form on $H_{\text{dR}}^1(\mathcal{R})$. $\triangle$

Next, we introduce two examples that are of fundamental importance in both complex function theory and conformal field theory.

Example 6.14 Let $H^\flat$ denote the space of sequences

$$\left\{ \{a_n\}_{n \in \mathbb{Z}, n \neq 0} : a_n \in \mathbb{C}, \, a_{-n} = \overline{a_n}, \, \sum_{n=1}^{\infty} n|a_n|^2 < \infty \right\}.$$

Note that H^b is a subset of ℓ^2 (indexed doubly infinitely), and in particular the Fourier series

$$\sum_{n\in\mathbb{Z}\setminus\{0\}} a_n e^{in\theta}$$

converges almost everywhere on $\mathbb{S}^1$. The space H^b can be identified with the real homogeneous Sobolev space $\dot{H}_{\mathbb{R}}^{1/2}(\mathbb{S}^1)$. For the purpose of this example, we define $\dot{H}_{\mathbb{R}}^{1/2}(\mathbb{S}^1)$ to be the subset of $L_{\mathbb{R}}^2(\mathbb{S}^1)$ whose Fourier coefficients are in H^b.

This is a Hilbert space with respect to the following inner product: given

$$\{a_n\}_{n\in\mathbb{Z}\setminus\{0\}}, \quad \{b_n\}_{n\in\mathbb{Z}\setminus\{0\}}$$

we have

$$g(\{a_n\}, \{b_n\}) = 2\,\mathrm{Re}\sum_{n=1}^{\infty} n a_n \overline{b_n} = \sum_{n\in\mathbb{Z}\setminus\{0\}} |n| a_n \overline{b_n}. \tag{6.3}$$

Note that this is *not* the inner product in $L^2(\mathbb{S}^1)$. We also have the symplectic form

$$\omega(\{a_n\}, \{b_n\}) = 2\,\mathrm{Im}\sum_{n=1}^{\infty} n a_n \overline{b_n} = -i\sum_{n\in\mathbb{Z}\setminus\{0\}} n a_n b_{-n}. \tag{6.4}$$

The Hilbert transform $J : H^b \to H^b$ is defined to be the linear extension of the map

$$J(e^{in\theta}) = -i\,\mathrm{sgn}(n) e^{in\theta}$$

where $\mathrm{sgn}(n) = n/|n|$ for $n \neq 0$. With these choices (g, J, ω) is a compatible triple on H^b.   △

Exercise 6.15 In the previous example, show that if f_1 and f_2 are functions on $\mathbb{S}^1$ corresponding to $\{a_n\}$, $\{b_n\}$ respectively which happen to be smooth, then we can write the symplectic form as

$$\omega(\{a_n\}, \{b_n\}) = \frac{1}{2\pi}\int_{\mathbb{S}^1} f_1\,df_2.$$

                                                                                                          △

Remark 6.16 Fourier series arising from elements of H^{b} are precisely the set of functions on $\mathbb{S}^1$ arising as boundary values of harmonic functions h on $\mathbb{D} = \{z : |z| < 1\}$ in the real homogeneous Dirichlet space

$$\dot{\mathcal{D}}_{\mathbb{R},\,\mathrm{harm}}(\mathbb{D}) = \left\{ h : \mathbb{D} \to \mathbb{R} \text{ harmonic} : h(0) = 0, \ \iint_{\mathbb{D}} |\nabla h|^2 < \infty. \right\}.$$

The function h is obtained from the Fourier series by replacing $e^{in\theta}$ with z^n for $n > 0$, and with $\bar{z}^n$ for $n < 0$.

Equivalently, the Dirichlet space is the set of anti-derivatives of L^2 harmonic one-forms on $\mathbb{D}$, which vanish at 0. In summary

$$H_{\mathbb{R}}^{\mathsf{b}} \simeq \dot{H}_{\mathbb{R}}^{1/2}(\mathbb{S}^1) \simeq \dot{\mathcal{D}}_{\mathbb{R},\,\mathrm{harm}}(\mathbb{D}).$$

$\triangle$

Example 6.17 Let H^{f} be the space of complex-valued sequences

$$H^{\mathsf{f}} := \left\{ \{a_n\}_{n \in \mathbb{Z}} : \sum |a_n|^2 < \infty, a_{-n} = \overline{a_n} \right\}.$$

We equip H^{f} with a compatible triple, given by (g, J, ω)

$$g(\{a_n\}, \{b_n\}) = 2 \operatorname{Re} \sum_{n=0}^{\infty} a_n \overline{b_n}, \quad J\{a_n\} = \{ia_n\}, \quad \omega(\{a_n\}, \{b_n\}) = 2 \operatorname{Im} \sum_{n=0}^{\infty} a_n \overline{b_n}.$$

We note that H^{f} is a (real) Hilbert space with respect to g. We wish to view elements $f = \{a_n\}$ of H^{f} as *real*-valued functions on the circle, by setting

$$f(z) = \sum_{n=0}^{\infty} a_n z^{n+1/2} + \overline{a_n z^{n+1/2}},$$

where $z^{1/2}$ is the following choice of square root on S^1: $e^{i\theta} \mapsto e^{i\theta/2}$ for $0 \leqslant \theta < 2\pi$. This identifies H^{f} with the space of real-valued square-integrable functions on the circle. H^{f} arises naturally as a description of the space of L^2-sections of the odd spinor bundle on the circle [8, Section 2]. Equivalently, it can be identified with a space of L^2-half-densities on the circle.

The factor $z^{1/2}$ is necessary to identify these spaces as a space of functions. However, this identification has the feature that even analytically well-behaved elements of these spaces (e.g. $a_i = \delta_{i0}$) are represented by discontinuous functions. In fact, because the odd spinor bundle on the circle is non-trivial, there is no $C^{\infty}(S^1)$-equivariant way to identify its smooth sections with the smooth functions on the circle (see e.g. the Serre–Swan theorem [25]). $\triangle$

We say that a pairing $\langle -, - \rangle$ on a complex vector space H is *sesquilinear* if it is complex linear in the first entry, conjugate linear in the second entry, and $\langle v, w \rangle = \overline{\langle w, v \rangle}$ for all $v, w \in H$.

Remark 6.18 Suppose that a compatible triple (g, J, ω) is given, and suppose that H is a Hilbert space with respect to g. We denote by H_J the complex vector space H where complex multiplication is given by $Jv = iv$. Show that the form

$$\langle v, w \rangle = g(v, w) - i\omega(v, w) = g(v, w) - ig(Jv, w).$$

is sesquilinear. This turns H_J into a complex Hilbert space. $\triangle$

In the next section, we consider instead the complexification of a space H with compatible triple.

6.3 Polarizations

In this section, we introduce the notion of polarization. This is a decomposition of the complexification of the Hilbert space into conjugate subspaces. We consider two kinds of polarizations: those in which the subspaces are symplectic Lagrangian subspaces, and those in which the subspaces are orthogonal. The symplectic decomposition is assumed to satisfy an additional positivity condition (the asymmetry between the two cases is an artefact of the fact that symplectic forms are only required to be non-degenerate, whereas an inner product must also be positive definite). We refer to these as positive symplectic polarizations and orthogonal polarizations respectively. The terminology is not standard, but was chosen to easily keep track of the two varieties of polarization. For example, in the symplectic literature, the term positive polarization is used.

We outline the basic theory, with examples illustrating a few of the ways they arise. The main aim of this section is to show that a positive symplectic polarization defines a unique complex structure and complex inner product resulting in a compatible triple on the original real Hilbert space. Similarly, an orthogonal polarization defines a unique complex structure and symplectic form which restrict to a compatible triple on the real space. Thus in some sense the two concepts coincide.

However, the two perspectives are quite different when it comes to deformations of structures. In symplectic geometry the situation often arises that the symplectic form is fixed, and the inner product and complex structure vary; while in spin geometry, the situation where the inner product is fixed but the complex structure and symplectic form vary arises. We will explore those in subsequent sections.

We proceed as follows. First, we fix a compatible triple, and show how it naturally specifies a decomposition which is both a positive symplectic and an orthogonal polarization. Then, we show how each of the two types of polarizations naturally defines a compatible triple.

Let us write $H_\mathbb{C}$ for the complex Hilbert space $H \otimes_\mathbb{R} \mathbb{C}$ (not to be confused with H_J). The fact that $H_\mathbb{C}$ arises as the complexification of the real Hilbert space H is witnessed by the map $\alpha : H_\mathbb{C} \to H_\mathbb{C}$ defined to be the real-linear extension of the map $v \otimes \lambda \mapsto v \otimes \overline{\lambda}$. The map α is a *real structure*: it is a conjugate-linear, isometric involution. In applications, it is often more natural to start from a complex Hilbert space, which is then equipped with a real structure. The real Hilbert space can then be recovered as the vectors that are fixed by the real structure.

If $T : H \to H$ is a bounded linear operator, then we write $T : H_\mathbb{C} \to H_\mathbb{C}$ for its complex-linear extension. A complex-linear operator $T : H_\mathbb{C} \to H_\mathbb{C}$ corresponds to a linear operator $H \to H$ if and only if it commutes with α. We extend g to $H_\mathbb{C}$ sesquilinearly, i.e.

$$g(v \otimes \lambda, w \otimes \mu) = \lambda\overline{\mu}g(v, w),$$

and extend ω to $H_\mathbb{C}$ bilinearly. We then have

$$g(v, w) = \omega(v, J\alpha w), \qquad\qquad \omega(v, w) = g(Jv, \alpha w).$$

Any complex structure J on H induces a (complex direct-sum) decomposition $H_\mathbb{C} = L^+ \oplus L^-$, by setting

$$L^+ = \{v \in H_\mathbb{C} : Jv = iv\}, \qquad L^- = \{v \in H_\mathbb{C} : Jv = -iv\}. \tag{6.5}$$

We make the (somewhat trivial) observation that $L^\pm$ are *closed* subspaces because they can be realized as $L^\pm = \ker(J \mp i)$; this observation will be important later.

Remark 6.19 The decomposition $H_\mathbb{C} = L^+ \oplus L^-$ has the property that $\alpha(L^+) = L^-$. Indeed, let $v = x \otimes 1 + y \otimes i \in L^+$, we compute

$$-y \otimes 1 + x \otimes i = iv = Jv = J(x \otimes 1 + y \otimes i) = Jx \otimes 1 + Jy \otimes i,$$

and thus $Jx = -y$ and $Jy = x$. It follows that

$$J\alpha v = J(x \otimes 1 - y \otimes i) = -y \otimes 1 - x \otimes i = -i(x \otimes 1 - y \otimes i) = -i\alpha v,$$

whence $\alpha v \in L^-$, and thus $\alpha(L^+) \subseteq L^-$. Similarly, one proves $\alpha(L^-) \subseteq L^+$, which implies $L^- \subseteq \alpha(L^+)$, and thus $\alpha(L^+) = L^-$.

Conversely, suppose that we are given a direct-sum decomposition $H_\mathbb{C} = W^+ \oplus W^-$. Write $P_{W^\pm}$ for the projection of $H_\mathbb{C}$ onto $W^\pm$ along $W^\mp$. From the above, we know that for $J_W = i(P_{W^+} - P_{W^-})$ to restrict to a complex structure, a necessary condition is that $\alpha(L^+) = L^-$. One easily verifies that this is also a sufficient condition, by computing $[\alpha, J_W]$.   △

Fix now a compatible triple (g, J, ω) on H. A subspace $L \subset H$ is called *isotropic* (with respect to ω) if ω is identically zero on $L \times L$. A subspace $L \subset H$ is *Lagrangian* if it is isotropic, and there exists another isotropic subspace $L' \subset H$

such that $H = L \oplus L'$. Because H is taken to be a Hilbert space, we have the following characterization of Lagrangian subspace, taken from [29, Proposition 5.1].

Lemma 6.20 *If $L \subset H$ is a maximal isotropic subspace, then it is Lagrangian.*

The compatibility of the triple (g, J, ω) is reflected in the decomposition $H = L^+ \oplus L^-$ as follows.

Lemma 6.21 *The decomposition $H_{\mathbb{C}} = L^+ \oplus L^-$ induced by J has the following properties:*

- *it is orthogonal with respect to the sesquilinear extension of g;*
- *it is Lagrangian with respect to the complex-bilinear extension of ω.*

Proof We prove the orthogonality claim. Let $v \in L^+$ and $w \in L^-$. We then have $g(v, w) = g(Jv, Jw) = g(iv, -iw) = -g(v, w)$, whence $g(v, w) = 0$. The claim then follows from the fact that L^+ and L^- are closed subspaces of $H_{\mathbb{C}}$ (and thus $(L^\pm)^{\perp\perp} = L^\pm$). The claim that the decomposition is Lagrangian follows similarly. □

If $T : H_{\mathbb{C}} \to H_{\mathbb{C}}$ is either a complex-linear or conjugate-linear operator, then we write

$$
T = \begin{pmatrix} a & b \\ c & d \end{pmatrix} \begin{matrix} L^+ \\ \oplus \\ L^- \end{matrix} \to \begin{matrix} L^+ \\ \oplus \\ L^- \end{matrix}
$$

for the corresponding decomposition; where a, b, c, d are complex-linear, resp. conjugate-linear operators. It follows from Lemma 6.21 that with respect to the decomposition $H_{\mathbb{C}} = L^+ \oplus L^-$ we have

$$
\alpha = \begin{pmatrix} 0 & \alpha|_{L^-} \\ \alpha|_{L^+} & 0 \end{pmatrix}.
$$

A straightforward computation shows that the operator T commutes with α if and only if $\alpha a \alpha = a$, and $\alpha d \alpha = d$, and $\alpha b = c \alpha$.

In the sequel, we will have to compare the operator T to the operator $\alpha T \alpha$. In some sources, the notation $\alpha T \alpha = \overline{T}$ is used. This notation is justified by the observation that if $T_{ij} \in \mathbb{C}$ is some matrix component of T, with respect to a basis $\{e_i\}$ of α-fixed vectors, then $(\alpha T \alpha)_{ij} = \overline{T_{ij}}$. We eschew this notation, because we will usually be working with bases consisting of vectors that are not α-fixed; in particular, no non-zero element of $L^\pm$ is α-fixed.

Exercise 6.22 Consider the vector space $H = \mathbb{R}^{2n}$. Let $\{e_i\}_{i=1,\dots,2n}$ be the standard basis, and write $x_i = e_{2i-1}$ and $y_i = e_{2i}$. Let g be the standard inner product, and define J by $Jx_i = y_i$ and $Jy_i = -x_i$. Determine the symplectic form ω such that

(g, J, ω) is a compatible triple. Then, determine the splitting $H \otimes_{\mathbb{R}} \mathbb{C} = \mathbb{C}^{2n} = L^+ \oplus L^-$.   △

Exercise 6.23 Let H be a real Hilbert space with complex structure J. Let H_J be as in Remark 6.18. Show that

$$\Psi : H_J \to L^+, \quad v \mapsto \frac{1}{\sqrt{2}} (v - iJv)$$

is a complex vector space isomorphism.   △

Example 6.24 Let M be a complex manifold. For a point $p \in M$, let J_p be as in Example 6.9. The induced decomposition of the complexified tangent space

$$\mathbb{C} \otimes T_p M = L^+ \oplus L^-$$

is often denoted

$$L^+ = T_p^{(1,0)} M, \quad L^- = T_p^{(0,1)} M.$$

$T_p^{(1,0)} M$ is called the holomorphic tangent space.

By Exercise 6.23, Ψ is an isomorphism between the real tangent space $T_p M$ and the holomorphic tangent space $T_p^{(1,0)} M$. If $(z_1, \ldots, z_n)$ are local coordinates on M near p, then

$$T_p^{(1,0)} M = \text{span} \left\{ \frac{\partial}{\partial z_1}, \ldots, \frac{\partial}{\partial z_n} \right\}.$$

△

Exercise 6.25 Now assume that (g, J, ω) is a compatible triple for H in Exercise 6.23. Let g_+ denote the restriction to L^+ of the sesquilinear extension of g. Show that $\Psi : (H_J, \langle, \rangle) \to (L^+, g_+)$ is an isometry; that is, $g_+(\Psi(v), \Psi(w)) = \langle v, w \rangle$ for all $v, w \in H$.   △

Remark 6.26 In the previous exercise, $\langle -, - \rangle$ and g^+ are both often referred to as a Hermitian metric. The data J and a metric g such that $g(Jv, Jw) = g(v, w)$ for all v, w are sufficient to recover the compatible triple (g, J, ω) by Exercise 6.6.   △

Example 6.27 Let M be a finite-dimensional complex manifold. A Hermitian metric on M is a Hermitian metric on $T_p M$ for each $p \in M$ (or equivalently $T_p^{(1,0)} M$), which varies smoothly with p. If the associated form ω is closed (so that M is also a symplectic manifold), then M is called a Kähler manifold.   △

Example 6.28 Let $\mathcal{R}$ be a compact Riemann surface, and $H = \mathcal{A}_{\text{harm}}^{\mathbb{R}}(\mathcal{R})$ be the vector space of real harmonic one-forms on $\mathcal{R}$, and let $J = *$, ω, and g be as in Example 6.11. Then $H_{\mathbb{C}}$ is the vector space of complex-valued harmonic one forms, which we denote by $\mathcal{A}_{\mathrm{harm}}(\mathcal{R})$. The dimension of $\mathcal{A}_{\mathrm{harm}}(\mathcal{R})$ over $\mathbb{C}$ is twice the genus of $\mathcal{R}$. The real structure α is just complex conjugation.

Denote the set of holomorphic one-forms on $\mathcal{R}$ by $\mathcal{A}(\mathcal{R})$, and the anti-holomorphic one-forms by $\overline{\mathcal{A}(\mathcal{R})}$. We have the direct sum decomposition

$$\mathcal{A}_{\mathrm{harm}}(\mathcal{R}) = \mathcal{A}(\mathcal{R}) \oplus \overline{\mathcal{A}(\mathcal{R})}.$$

Explicitly, every complex harmonic one-form β has a unique decomposition $\beta = \gamma + \bar{\delta}$ where $\gamma, \delta \in \mathcal{A}(\mathcal{R})$. Locally these are written $\gamma = a(z)dz, \bar{\delta} = \overline{b(z)dz}$ where a and b are holomorphic functions of the coordinate z. It is easily checked that $J\gamma = -i\gamma$ and $J\bar{\delta} = i\bar{\delta}$. So

$$L^- = \mathcal{A}(\mathcal{R}), \quad L^+ = \overline{\mathcal{A}(\mathcal{R})}.$$

The corresponding sesquilinear extension of the inner product g in Example 6.11 is

$$g(\beta_1, \beta_2) = \iint_{\mathcal{R}} \beta_1 \wedge *\overline{\beta_2}. \tag{6.6}$$

$\triangle$

Example 6.29 (Sobolev-1/2 Functions on the Circle)   Let $H^{\flat}$ be as in Example 6.14. Treating $H^{\flat}$ as a set of functions on the circle (i.e. as $\dot{H}_{\mathbb{R}}^{1/2}(\mathbb{S}^1)$), its complexification $H_{\mathbb{C}}^{\flat}$ is the set of complex-valued functions in $L^2(\mathbb{S}^1)$ whose Fourier series

$$\sum_{n \in \mathbb{Z} \setminus \{0\}} a_n e^{in\theta}$$

satisfy the stronger condition

$$\sum_{n \in \mathbb{Z} \setminus \{0\}} |n||a_n|^2 < \infty.$$

that is, the Sobolev-1/2 space $\dot{H}^{1/2}(\mathbb{S}^1)$. Equivalently, one removes the condition $a_{-n} = \overline{a_n}$ in the sequence model of $H^{\flat}$:

$$H_{\mathbb{C}}^{\flat} = \left\{ \{a_n\}_{n \in \mathbb{Z} \setminus \{0\}} : \sum_{n \in \mathbb{Z} \setminus \{0\}} |n||a_n|^2 < \infty \right\}.$$

In this case, the real structure is given by complex conjugation of the function on $\mathbb{S}^1$; equivalently,

$$\alpha(\{a_n\}) = \{\overline{a_{-n}}\}.$$

We then see that L^- is given by the sequences $\{a_n\}$ such that $a_n = 0$ for $n < 0$ and L^+ is given by the sequences such that $a_n = 0$ for $n > 0$. Equivalently, L^- consists of those functions that extend to holomorphic functions on the unit disk while elements of L^+ extend anti-holomorphically. $\triangle$

Exercise 6.30 In the previous example, show that

$$\omega(\{a_n\}, \{b_n\}) = -i \sum_{n=-\infty}^{\infty} n a_n b_{-n}$$

and

$$g(\{a_n\}, \{b_n\}) = \sum_{n=-\infty}^{\infty} |n| a_n \overline{b_n}$$

and verify $g(v, w) = \omega(v, J\alpha w)$. Verify that $L^+ \oplus L^-$ is both Lagrangian and orthogonal. $\triangle$

Remark 6.31 The polarization in the previous example corresponds to the decomposition of complex harmonic functions in the homogeneous Dirichlet space on $\mathbb{D}$ into their holomorphic and anti-holomorphic parts. Defining

$$\dot{\mathcal{D}}(\mathbb{D}) = \left\{ h : \mathbb{D} \to \mathbb{C} \text{ holomorphic} : \iint_{\mathbb{D}} |h'|^2 < \infty \right\}$$

and

$$\dot{\mathcal{D}}_{\text{harm}}(\mathbb{D}) = \left\{ h : \mathbb{D} \to \mathbb{R} \text{ harmonic} : h(0) = 0, \iint_{\mathbb{D}} \left| \frac{\partial h}{\partial z} \right|^2 + \left| \frac{\partial h}{\partial \bar{z}} \right|^2 < \infty. \right\}$$

we have the decomposition

$$\dot{\mathcal{D}}_{\text{harm}}(\mathbb{D}) = \dot{\mathcal{D}}(\mathbb{D}) \oplus \overline{\dot{\mathcal{D}}(\mathbb{D})}.$$

Extending elements of $H_{\mathbb{C}}^{\flat}$ harmonically to $\mathbb{D}$ we can identify $L^- \simeq \dot{\mathcal{D}}(\mathbb{D})$, $L^+ \simeq \overline{\dot{\mathcal{D}}(\mathbb{D})}$. $\triangle$

Example 6.32 (Square-Integrable Functions on the Circle) Let H^{f} be as in Example 6.17. The complexification of H^{f} can be identified with the complex-valued square-integrable functions on the circle, which we in turn identify (by taking Fourier coefficients) with the space of complex-valued sequences $\{a_n\}$ indexed by $\mathbb{Z}$ such that the sum

$$\sum_n |a_n|^2$$

converges. We then see that L^+ is given by the sequences $\{a_n\}$ such that $a_n = 0$ for $n < 0$. Equivalently, L^+ consists of those functions f, such that $z \mapsto f(z)z^{-1/2}$ extends to a holomorphic function on the unit disk. $\triangle$

Remark 6.33 The set of holomorphic extensions $h(z)$ to the unit disk of functions $f(z)z^{-1/2}$ in the previous example is precisely the Hardy space. This follows from a classical characterization of the Hardy space as the class of holomorphic functions on the disk, whose non-tangential boundary values are in $L^2(\mathbb{S}^1)$. On the other hand L^- consists of those $f(z)z^{1/2}$ which extend anti-holomorphically into the unit disk. These extensions are precisely the conjugate of the Hardy space. $\triangle$

It turns out that the space of deformations of either g or ω has a rich geometric structure. In the rest of this section, we define two notions of polarization—those arising from a fixed inner product (orthogonal polarizations), and those arising from a fixed symplectic form (positive symplectic polarizations). Both lead to a compatible triple.

Definition 6.34 Let (H, g) be a real Hilbert space and let $H_\mathbb{C}$ be its complexification with real structure α. Extend g sesquilinearly to $H_\mathbb{C}$. An *orthogonal polarization* is an orthogonal decomposition

$$H_\mathbb{C} = W^+ \oplus W^-$$

such that $\alpha(W^+) = W^-$. $\triangle$

One can associate a complex structure and bilinear form to an orthogonal polarization.

Definition 6.35 Let (H, g) be a real Hilbert space and let $H_\mathbb{C}$ be its complexification with real structure α. Let $H_\mathbb{C} = W^+ \oplus W^-$ be an orthogonal polarization. Let $P_W^\pm : H_\mathbb{C} \to W^\pm$ be the orthogonal projections. The complex structure associated with the polarization is $J_W = i(P_W^+ - P_W^-)$, and the bilinear form on $H_\mathbb{C}$ associated with the polarization is

$$\omega_W(v, w) = g(J_W v, w).$$

                                                                                    $\triangle$

Observe that by the reasoning in Remark 6.19, the condition that $\alpha(W^+) = W^-$ is both necessary and sufficient for J_W to restrict to H.

An orthogonal polarization defines a compatible triple.

Proposition 6.36 *Let $H_\mathbb{C} = W^+ \oplus W^-$ be an orthogonal polarization. The complex structure J_W and symplectic form ω_W associated with this polarization restrict to H, and (g, J_W, ω_W) is a compatible triple. Moreover, W^+ and W^- are Lagrangian with respect to ω_W.*

Proof First, it is clear that $J_W^2 = -\mathbb{1}$. Now, let $v = (x^+, x^-) \in W^+ \oplus W^-$ be arbitrary. We then have $\alpha J_W v = \alpha(ix^+, -ix^-) = (i\alpha x^-, -i\alpha x^+) = J_W \alpha v$. Observe that $H \subseteq H_\mathbb{C}$ can be identified as the $+1$ eigenspace for α; it follows that if $v \in H$, then $\alpha J_W v = J_W \alpha v = J_W v$ and thus $J_W v \in H$. It immediately follows that $\omega_W(v, w) = g(J_W v, w)$ is real on H.

It remains to be shown that (g, J_W, ω_W) is a compatible triple. Of course, if ω_W is symplectic, then it is compatible. That ω_W is skew-symmetric follows from the computation $\omega_W(v, w) = g(J_W v, w) = -g(v, J_W w) = -\omega_W(w, v)$. The map $\varphi_{\omega_W} : H \to H^*$ is given by $\varphi_g J_W$, and is thus an isomorphism.

It is straightforward to show that W^+ and W^- are isotropic, and the assumption that $H_\mathbb{C} = W^+ \oplus W^-$ then implies that they are both Lagrangian. $\qquad\square$

Let $\mathrm{Pol}^g(H)$ denote the set of those closed subspaces $W^+ \subset H_\mathbb{C}$ such that $H_\mathbb{C} = W^+ \oplus \alpha(W^+)$ as an orthogonal direct sum. Observe that by Proposition 6.36 and Lemma 6.21 we may view $\mathrm{Pol}^g(H)$ as parameterizing the set of deformations of ω (compatible with g).

Next, we define positive symplectic polarizations.

Definition 6.37 Let (H, ω) be a real vector space with symplectic form ω, and let $H_\mathbb{C}$ be its complexification with real structure α. Extend ω complex bilinearly to $H_\mathbb{C}$.

A *positive symplectic polarization* is a Lagrangian decomposition

$$H_\mathbb{C} = W^+ \oplus W^-$$

such that $\alpha(W^+) = W^-$ and $-i\omega(v, \alpha w)$ is a positive-definite sesquilinear form on W^+. $\qquad\triangle$

In fact, this implies that $g_+(v, w) := -i\omega(v, \alpha w)$ is a strong inner product, which by Proposition 6.1 makes W_+ a Hilbert space. To see this, observe that $\varphi_\omega : H_\mathbb{C} \to H_\mathbb{C}^*$ restricts to an isomorphism $\varphi_\omega : W^+ \to (W^-)^*$. A direct computation shows that $\varphi_{g_+} : W^+ \to (W^+)^*$ is given by $\varphi_{g_+} = -i\alpha^*\varphi_\omega$, whence φ_{g_+} is an isomorphism. It follows that g_+ is a strong inner product if and only if it is positive-definite.

Exercise 6.38 Show that if $W^+ \oplus W^- = H_\mathbb{C}$ is a positive symplectic polarization, then the pairing $i\omega(v, \alpha w)$ is a positive-definite sesquilinear form on W^-, such that W^- is a Hilbert space with respect to this pairing. $\qquad\triangle$

As with orthogonal polarizations, a positive symplectic polarization is enough to define the other two pieces of data.

Definition 6.39 Let (H, ω) be a real Hilbert space, equipped with symplectic form ω. Let $H_\mathbb{C} = W^+ \oplus W^-$ be a symplectic polarization; let $P_W^\pm : H_\mathbb{C} \to W^\pm$ be the projections with respect to this decomposition. The complex structure associated with the polarization is $J_W = i(P_W^+ - P_W^-)$, and the sesquilinear form on $H_{\mathbb{C}}$ associated with the polarization is

$$g_W(v, w) = \omega(v, J\alpha w).$$

$$\triangle$$

The condition that the symplectic polarization $H_{\mathbb{C}} = W^+ \oplus W^-$ is positive is exactly the condition that is required to guarantee that (g_W, J_W, ω) is compatible.

Proposition 6.40 *Let $H_{\mathbb{C}} = W^+ \oplus W^-$ be a positive symplectic polarization. The associated complex structure J_W and sesquilinear form g_W restrict to a complex structure and inner product on H. Moreover, g_W is a positive definite sesquilinear form on $H_{\mathbb{C}}$, with respect to which $H_{\mathbb{C}}$ is a Hilbert space, and (g_W, J_W, ω) is a compatible triple.*

Proof It is clear that $J_W^2 = -\mathbb{1}$. Sesquilinearity of g_W follows immediately from the definition. let $g_{W,\pm}$ denote the restrictions of g_W to $W^\pm$. We have

$$g_{W,+}(v, w) = -i\omega(v, \alpha w) \quad v, w \in W^+$$

$$g_{W,-}(v, w) = i\omega(v, \alpha w) \quad v, w \in W^-.$$

It is easily checked that W^+ and W^- are orthogonal with respect to g_W, so we have

$$(H_{\mathbb{C}}, g_W) = (W^+, g_{W,+}) \oplus (W^-, g_{W,-})$$

and both summands are Hilbert spaces by Exercise 6.38.

The fact that $\alpha(W^+) = W^-$ implies that J_W commutes with α, which in turn implies that J_W restricts to H. To see that g_W is real on H, observe that any $v, w \in H$ can be written $(v' + \alpha v')/2$, $(w' + \alpha w')/2$ where $v', w' \in W^+$. Using the definition $g_W(v, w) = \omega(v, J\alpha w)$ and the fact that J commutes with complex conjugate, we have

$$g_W(v' + \alpha v', w' + \alpha w') = g_W(v', w') + g_W(\alpha v', \alpha w')$$

which is real. Finally, restricting to $v, w \in H$, we have

$$g_W(v, w) = \omega(v, Jw)$$

so (g_W, J_W, ω) is a compatible triple on H. $\qquad\square$

Denote the set W^+ such that $H_{\mathbb{C}} = W^+ \oplus \alpha(W^+)$ is a positive symplectic polarization by $\mathrm{Pol}^\omega(H)$. By Lemma 6.21 and Proposition 6.40, the positive symplectic polarizations can be viewed as parametrizing the deformations of g (compatible with ω).

Summarizing the section so far, we have seen that a vector space H, equipped with a compatible triple (g, J, ω), can equivalently be seen as

- A Hilbert space (H, g), equipped with an orthogonal polarization of $H_{\mathbb{C}}$; or
- A symplectic vector space (H, ω), equipped with a positive symplectic polarization of $H_{\mathbb{C}}$.

We write $\mathrm{GL}(H)$ for the group of invertible linear transformations from H to H. We then define the *symplectic group* of H to be

$$\mathrm{Sp}(H) := \{u \in \mathrm{GL}(H) : \omega(uv, uw) = \omega(v, w) \text{ for all } v, w \in H\}.$$

Similarly, we define the *orthogonal group* of H to be

$$O(H) := \{u \in \mathrm{GL}(H) : g(uv, uw) = g(v, w) \text{ for all } v, w \in H\}.$$

Define $u^\star g, u^\star J$, and $u^\star \omega$ by

$$u^\star g(v, w) = g(u^{-1}v, u^{-1}w), \quad u^\star J v = u J u^{-1} v, \quad u^\star \omega(v, w) = \omega(u^{-1}v, u^{-1}w).$$

We then have the following.

Proposition 6.41 *If $u \in \mathrm{GL}(H)$, then the triple $(u^\star g, u^\star J, u^\star \omega)$ is again compatible.*

Proof It is immediate that $u^\star J$ is a complex structure. To prove the other two claims, observe that if $\psi : H \times H \to \mathbb{R}$ is a bilinear pairing, then we have $\varphi_{u^\star\psi}(v)(w) = u^\star\psi(v, w) = \psi(u^{-1}v, u^{-1}w) = \varphi_\psi(u^{-1}v)(u^{-1}w)$. This implies that $\varphi_{u^\star\psi} = (u^{-1})^*\varphi_\psi u^{-1}$, where u^* is the adjoint map $H^* \to H^*$. It follows that φ_ψ is invertible if and only if $\varphi_{u^\star\psi}$ is invertible. Because pullback clearly preserves anti-symmetry, $u^\star\omega$ is a strong symplectic form. The fact that $u^\star g$ is a strong inner product would follow from $u^\star g(v, v) \geq 0$ by the discussion following Definition 6.37. But this follows from the positivity of g, since $u^\star g(v, v) = g(u^{-1}v, u^{-1}v) \geq 0$. $\square$

The groups $\mathrm{GL}(H)$, $O(H)$, and $\mathrm{Sp}(H)$ act on $H_{\mathbb{C}}$ by complex-linear extension. The decomposition $H_{\mathbb{C}} = L_u^+ \oplus L_u^-$ corresponding to the triple $(u^\star g, u^\star J, u^\star \omega)$ is given by $L_u^\pm = uL^\pm$ (this follows directly from Eq. (6.5)). The above guarantees that $\mathrm{Sp}(H)$ acts on $\mathrm{Pol}^\omega(H)$ by $(u, W^+) \mapsto uW^+$. Indeed, if (g_W, J_W, ω) is the triple corresponding to W^+, and $u \in \mathrm{Sp}(H)$, then $(u^\star g, u^\star J, \omega)$ is the triple corresponding to uW^+, which tells us that $uW^+ \in \mathrm{Pol}^\omega(H)$. Similarly, one sees that $O(H) \times \mathrm{Pol}^g(H) \to \mathrm{Pol}^g(H)$, $(u, W^+) \mapsto uW^+$ defines an action of $O(H)$ on $\mathrm{Pol}^g(H)$. The *unitary group* of H is defined to be the intersection of the symplectic and orthogonal group, i.e. $U(H) := \mathrm{Sp}(H) \cap O(H)$.

Exercise 6.42 Show that the unitary group $U(H)$ is equal to the unitary group of the complex Hilbert space H_J, which is defined in the usual manner. $\triangle$

The following lemma tells us that, at least as sets, we may view $\mathrm{Pol}^g(H)$ and $\mathrm{Pol}^\omega(H)$ as homogeneous spaces.

Lemma 6.43 *The maps* $O(H) \rightarrow \text{Pol}^g(H), u \mapsto u(L^+)$ *and* $Sp(H) \rightarrow \text{Pol}^\omega(H), u \mapsto u(L^+)$ *descend to bijections*

$$O(H)/U(H) \rightarrow \text{Pol}^g(H), \qquad Sp(H)/U(H) \rightarrow \text{Pol}^\omega(H).$$

Proof An operator $u \in O(H)$ preserves L^+ if and only if it commutes with J, which it does if and only if $u \in Sp(H)$; whence the stabilizer of $L^+ \in \text{Pol}^g(H)$ is $U(H)$. It thus remains to be shown that $O(H)$ acts transitively on $\text{Pol}^g(H)$. To this end, let $W^+ \in \text{Pol}^g(H)$ be arbitrary. Pick orthonormal bases $\{l_i\}$ and $\{e_i\}$ for L^+ and W^+ respectively. We obtain orthonormal bases $\{\alpha l_i\}$ and $\{\alpha e_i\}$ for $\alpha(L^+)$ and $\alpha(W^+)$ respectively. We let $u : H_\mathbb{C} \rightarrow H_\mathbb{C}$ be the complex-linear extension of the map that sends l_i to e_i and αl_i to αe_i. It is then easy to see that $u \in O(H)$ and $u(L^+) = W^+$. This proves that the map $O(H)/U(H) \rightarrow \text{Pol}^g(H)$ is a bijection.

The fact that the map $Sp(H)/U(H) \rightarrow \text{Pol}^\omega(H)$ is a bijection is proved similarly.   □

Exercise 6.44 Prove that the stabilizer of $L^+ \in \text{Pol}^\omega(H)$ is $U(H)$ and that $Sp(H)$ acts transitively on $\text{Pol}^\omega(H)$. Conclude that the map $Sp(H)/U(H) \rightarrow \text{Pol}^\omega(H)$ is a bijection.   △

6.4 The Grassmannian of Polarizations Associated with a Symplectic Form

In this section, we fix a symplectic form, and describe the associated Lagrangian Grassmannian $\text{Pol}^\omega(H)$ of positive symplectic polarizations. We will show that it is described by a space of operators called the Siegel disk, and that it is a complex Banach manifold. We also consider the "restricted" Siegel disk and restricted symplectic groups. These restricted objects require the extra analytic condition of Hilbert–Schmidt-ness, which plays a central role in representation theory. We also establish that the restricted Grassmannian is not just a Banach manifold, but a Hilbert manifold.

Let H be a real Hilbert space, and let (g, J, ω) be a compatible triple. We are interested in deformations of g, while keeping ω fixed. As explained in Sect. 6.2, we can study these deformations by considering the space of symplectic polarizations in $H_\mathbb{C}$.

We denote by $P^\pm$ the orthogonal projection $H_\mathbb{C} \rightarrow L^\pm$.

Proposition 6.45 *Suppose that $W^+ \subset H_\mathbb{C}$ is a Lagrangian such that $H_\mathbb{C} = W^+ \oplus \alpha(W^+)$ (not necessarily orthogonal); and set $W^- = \alpha(W^+)$. Then, W^+ is a positive symplectic polarization if and only if $\|P^+x\| > \|P^-x\|$ for all $0 \neq x \in W^+$. Moreover, if this holds, then $P^+|_{W^+} : W^+ \rightarrow L^+$ is a bijection.*

Proof We recall that the decomposition $H_\mathbb{C} = W^+ \oplus W^-$ allows us to define a complex structure J_W and a sesquilinear form g_W, see Definition 6.39. Let $x \in W^+$.

We then compute

$$g_W(x, x) = \omega(x, \alpha J_W x) = \omega(x, \alpha i x) = g(Jx, ix)$$
$$= g((P^+ - P^-)x, x) = \|P^+ x\|^2 - \|P^- x\|^2.$$

We thus see that g_W is positive definite if and only if $\|P^+ x\|^2 > \|P^- x\|^2$ for all $0 \neq x \in W^+$. Repeating the proof of Proposition 6.40 we see that J_W and g_W restrict to the real space. Furthermore, we have that the restriction of g_W to the real space is strong, because $\varphi_{g_W} = \varphi_\omega J_W$.

Now, suppose that $\|P^+ x\|^2 > \|P^- x\|^2$ for all $0 \neq x \in W^+$. We then have, for all $x \in W^+$

$$\|x\|^2 = \|P^+ x\|^2 + \|P^- x\|^2 \leqslant 2\|P^+ x\|^2,$$

whence $P^+|_{W^+}$ is bounded below, and thus injective with closed image. We now claim that $P^+ W^+$ is dense in L^+. Indeed, suppose that $y \in (P^+ W^+)^\perp \subseteq L^+$. We then have, for all $x \in W^+$

$$0 = g(y, P^+ x) = g(y, x) = g(y, (P_W^+ - P_W^-)x) = ig(y, J_u x) = \omega(y, \alpha x)$$

which implies that y is in the symplectic complement of W^-, which is equal to W^-. This means that $\alpha y \in L^- \cap W^+$, but this intersection is zero, because it is contained in the kernel of $P^+|_{W^+}$, which is zero. We conclude that $P^+|_{W^+}$ is a bijection, and thus invertible. $\qquad\square$

Exercise 6.46 Show that if $W^+ \subset H_\mathbb{C}$ is a positive symplectic polarization, then $W^+ = \text{graph}(Z)$, where $Z = P^-(P^+|_{W^+})^{-1} : L^+ \to L^-$. $\qquad\triangle$

Corollary 6.47 *If $W^+ \subset H_\mathbb{C}$ is a positive symplectic polarization, then the operator $Z = P^-(P^+|_{W^+})^{-1}$ is the unique operator such that $W^+ = \text{graph}(Z)$.*

Proof Any operator is uniquely defined by its graph, so the only question is existence; since we have provided an expression for Z, all that remains to be shown is that this does trick, which is Exercise 6.46. $\qquad\square$

Now, we would like to determine for which operators $Z : L^+ \to L^-$, we have that $W_Z = \text{graph}(Z) \in \text{Pol}^\omega(H)$.

Lemma 6.48 *The subspace W_Z is Lagrangian if and only if $Z = \alpha Z^* \alpha$.*

Proof First, assume that W_Z is Lagrangian. This means that it is in particular isotropic. We thus compute, for arbitrary $x, y \in L^+$,

$$0 = \omega(x + Zx, y + Zy) = \omega(x, Zy) + \omega(Zx, y)$$
$$= g(Jx, \alpha Zy) + g(JZx, \alpha y)$$
$$= -g(x, (J\alpha Z + Z^* J\alpha)y), \qquad (6.7)$$

and thus $Z = \alpha J Z^* J \alpha$. We have $\alpha J Z^* J \alpha = \alpha Z^* \alpha$, because the domain of Z is L^+.

Now, assume that $Z = \alpha J Z^* J \alpha$. The computation above then implies that W_Z is isotropic. It remains to be shown that W_Z is Lagrangian. To that end, suppose that $z = z^+ + z^- \in L^+ \oplus L^-$ satisfies $\omega(z, w) = 0$ for all $w \in W_Z$. We then have, for all $x \in L^+$,

$$
\begin{aligned}
0 = \omega(x + Zx, z^+ + z^-) &= \omega(Zx, z^+) + \omega(x, z^-) \\
&= g(JZx, \alpha z^+) + g(Jx, \alpha z^-) \\
&= -g(x, Z^* J \alpha z^+) - g(x, J \alpha z^-),
\end{aligned}
$$

thus $Z^* J \alpha z^+ + J \alpha z^- = 0$, which implies $z^- = \alpha J Z^* J \alpha z^+ = J z^+$, whence $z^+ + z^- \in W_Z$. This implies that W_Z is maximal isotropic, whence it is Lagrangian, by Lemma 6.20. $\qquad\square$

Lemma 6.49 *We have* $\mathbb{1} - Z^* Z > 0$ *if and only if* $W_Z \cap \alpha(W_Z) = \{0\}$ *and* $W_Z \in \mathrm{Pol}^\omega(H)$.

Proof First, assume that $\mathbb{1} - Z^* Z > 0$. We have, for all $0 \neq x \in L^+$, that

$$
x \neq Z^* Z x = \alpha Z \alpha Z x. \tag{6.8}
$$

Now, we have $\alpha(W_Z) = \mathrm{graph}(\alpha Z \alpha)$, from which it follows that if $z \in W_Z \cap \alpha(W_Z)$, then there exist $x \in L^+$ and $y \in L^-$ such that $(x, Zx) = z = (\alpha Z \alpha y, y)$. This implies that $x = \alpha Z \alpha Z x$, whence $x = 0$ by Eq. (6.8); and thus $W_Z \cap \alpha(W_Z) = \{0\}$.

Now, we compute, for arbitrary $0 \neq w = x + Zx \in W_Z$,

$$
\begin{aligned}
\|P^+ w\|^2 - \|P^- w\|^2 = \|x\|^2 - \|Zx\|^2 &= g(x, x) - g(x, Z^* Z x) \\
&= g(x, (\mathbb{1} - Z^* Z)x). \tag{6.9}
\end{aligned}
$$

Now, we wish to show that $W_Z \in \mathrm{Pol}^\omega(H)$. According to Proposition 6.45, it suffices to show that $\|P^+ w\|^2 > \|P^- w\|^2$ for all $0 \neq w \in W_Z$, this follows from the positivity of $\mathbb{1} - Z^* Z$ through Eq. (6.9).

Now, suppose that $W_Z \in \mathrm{Pol}^\omega(H)$ and set $g_Z = g_{W_Z}$ and $J_Z = J_{W_Z}$. Eq. (6.9) also implies (through Proposition 6.45) that, then $\mathbb{1} - Z^* Z$ is positive. $\qquad\square$

Set

$$
\mathfrak{D}(H) := \{Z \in \mathcal{B}(L^+, L^-) \mid Z = \alpha Z^* \alpha, \text{ and } \mathbb{1} - Z^* Z > 0\}. \tag{6.10}
$$

This is called the Siegel disk, originating with C. L. Siegel [22]. The Siegel disk in the setting of Hilbert spaces was defined by G. Segal [18]. There, the extra condition that Z is Hilbert–Schmidt was added, resulting in what we call the restricted Siegel disk below.

The preceding discussion is now nicely summarized by the following result.

Proposition 6.50 *The maps*

$$\text{Pol}^{\omega}(H) \to \mathfrak{D}(H), \qquad\qquad \mathfrak{D}(H) \to \text{Pol}^{\omega}(H),$$

$$W \mapsto P^{-}(P^{+}|_{W})^{-1}, \qquad\qquad Z \mapsto \text{graph}(Z),$$

are well-defined, and each others' inverses.

Proof By Proposition 6.45 we know that if $W \in \text{Pol}^{\omega}(H)$, then $(P^{+}|_{W})^{-1}$ exists. By Corollary 6.47 we then have that $W = \text{graph}(P^{-}(P^{+}|_{W})^{-1})$. From Lemmas 6.48 and 6.49 it then follows that $P^{-}(P^{+}|_{W})^{-1} \in \mathfrak{D}(H)$. Conversely, it follows from Lemmas 6.48 and 6.49 that if $Z \in \mathfrak{D}(H)$, then $\text{graph}(Z) \in \text{Pol}^{\omega}(H)$. That these maps are each others inverses is a straightforward consequence of Corollary 6.47. $\square$

Remark 6.51 Another model of $\text{Pol}^{\omega}(H)$ is given by the Siegel upper half space. This is equivalent but based on different conventions. Rather than viewing positive symplectic polarizations as a variation of a fixed polarization $L^{+} \oplus L^{-}$, instead one fixes a real Lagrangian decomposition. More precisely, we say that a Lagrangian subspace L in $H_{\mathbb{C}}$ is real if it is the complexification of a Lagrangian subspace in H. Fix a decomposition $H_{\mathbb{C}} = L \oplus L'$ where L and L' are transverse real Lagrangian spaces. It can be shown that any $W \in \text{Pol}^{\omega}(H)$ is transverse to L and L', and can be uniquely expressed as the graph of some $Z : L \to L'$. This Z satisfies $Z^{T} = Z$ and $\text{Im}(Z)$ is positive definite. The set of such matrices is called the Siegel upper half space $\mathfrak{H}(H)$. These two conditions can be given meaning either through the use of a specific basis, or by $Z^{T} := \alpha Z^{*} \alpha$ and $\text{Im}(Z) := (1/2i)(Z - \alpha Z \alpha)$. For details, see [2, 22, 27]. $\triangle$

Example 6.52 Here we refer to Example 6.28. Let $\mathcal{R}$ be a compact Riemann surface, and $\mathcal{A}(\mathcal{R}) \oplus \overline{\mathcal{A}(\mathcal{R})}$ the decomposition of the space of complex harmonic one-forms on $\mathcal{R}$. This polarization, induced by the complex structure on the Riemann surface, is traditionally represented by an element of the Siegel upper half space as follows. Choose simple closed curves $a_{1}, \ldots, a_{g}, b_{1}, \ldots, b_{g}$ which are generators of the first homology group of $\mathcal{R}$. These can be chosen so that their intersection numbers $\gamma_{1} \cdot \gamma_{2}$ satisfy $a_{k} \cdot b_{j} = -b_{j} \cdot a_{k} = \delta_{kj}$, and are zero otherwise. Intuitively, the intersection numbers count the number of crossings with sign, where the sign of the crossing depends on the relative direction. For a precise definition of intersection number, and proof of existence of such a basis see [5, 23].

It can be shown that there are holomorphic one-forms $\beta_{1}, \ldots, \beta_{g}$ such that

$$\int_{a_{k}} \beta_{j} = \delta_{jk}.$$

Then,

$$Z_{kj} = \int_{b_k} \beta_j$$

is a matrix representing an element of the Siegel upper half space. The symmetry of this matrix is referred to as the Riemann bilinear relations, and Z_{kj} is called the period matrix.

To see the relation to the operator specified in Remark 6.51, let η_j be the unique basis of harmonic one-forms on $\mathcal{R}$ dual to the curves $a_1, \ldots, a_\mathfrak{g}, b_1, \ldots, b_\mathfrak{g}$. Explicitly

$$\int_{a_k} \eta_j = \delta_{jk} \quad \text{and} \quad \int_{b_k} \eta_{j+\mathfrak{g}} = \delta_{jk}.$$

Let L be the span of $\{\eta_1, \ldots, \eta_\mathfrak{g}\}$ and L' be the span of $\{\eta_{\mathfrak{g}+1}, \ldots, \eta_{2\mathfrak{g}}\}$. Then it can be shown that

$$\beta_j = \eta_j + \sum_{k=1}^n Z_{jk}\eta_{k+\mathfrak{g}}, \quad j = 1, \ldots, \mathfrak{g}. \tag{6.11}$$

and

$$\mathcal{A}(\mathcal{R}) = \text{span}\{\beta_1, \ldots, \beta_\mathfrak{g}\}.$$

Thus if Z_{jk} are the components of an operator $Z : L \to L'$ with respect to the bases above, then Z is in the Siegel upper half plane and $\mathcal{A}(\mathcal{R})$ is the graph of Z. See [5, 23] for details.

Observe that an important role is played by the choice of homology basis (or marking). △

Exercise 6.53 The Hodge theorem says that every cohomology class in $H^1_{dR}(\mathcal{R})$ is represented by a unique harmonic one-form. Use the Hodge theorem to show that the one-forms Eq. (6.11) are indeed holomorphic and span $\mathcal{A}(\mathcal{R})$. △

Next, we return to the symplectic action on $\text{Pol}^\omega(H)$ and express it in terms of $\mathfrak{D}(H)$. Fix an element $u \in \text{Sp}(H)$, we obtain a new subspace $W_u^+ := u(L^+) \subset H_\mathbb{C}$. It is easy to see from the definition of $\text{Sp}(H)$ that W_u^+ is again a Lagrangian, which moreover satisfies $W_u^+ \cap \alpha W_u^+ = \{0\}$.

The equation $\omega(ux, uy) = \omega(x, y)$ implies that

$$-Ju^*Ju = \mathbb{1}. \tag{6.12}$$

Writing

$$u = \begin{pmatrix} a & \alpha b \alpha \\ b & \alpha a \alpha \end{pmatrix} : \begin{matrix} L^+ \\ \oplus \\ L^- \end{matrix} \rightarrow \begin{matrix} L^+ \\ \oplus \\ L^- \end{matrix} \tag{6.13}$$

in block form we obtain

$$\mathbb{1} = -\begin{pmatrix} i & 0 \\ 0 & -i \end{pmatrix} \begin{pmatrix} a^* & b^* \\ \alpha b^* \alpha & \alpha a^* \alpha \end{pmatrix} \begin{pmatrix} i & 0 \\ 0 & -i \end{pmatrix} \begin{pmatrix} a & \alpha b \alpha \\ b & \alpha a \alpha \end{pmatrix}$$

$$= \begin{pmatrix} a^*a - b^*b & a^*\alpha b \alpha - b^*\alpha a \alpha \\ -\alpha b^* \alpha a + \alpha a^* \alpha b & -\alpha b^* b \alpha + \alpha a^* a \alpha \end{pmatrix}. \tag{6.14}$$

By a symplectomorphism, we mean a bounded bijection which preserves the symplectic form. Recall that the inverse is also bounded.

Lemma 6.54 *Let u be a symplectomorphism in block form (6.13). Then a is invertible and*

$$u^{-1} = \begin{pmatrix} a^* & -b^* \\ -\alpha b^* \alpha & \alpha a^* \alpha \end{pmatrix}. \tag{6.15}$$

Proof The expression for u^{-1} follows from Eqs. (6.12) and (6.13). If $av = 0$, since by Eq. (6.14) $a^*a - b^*b = \mathbb{1}$ we see that for any $v \in L^+$

$$\|v\|^2 = \|av\|^2 - \|bv\|^2 = -\|bv\|^2$$

so $v = 0$. Thus a is injective. Applying this to u^{-1} we see that a^* is injective, so (denoting closure by cl)

$$\mathrm{cl\,range}(a) = \ker(a^*)^\perp = L^+,$$

i.e. a has dense range. Since $\|v\|^2 \le \|av\|^2$, a is bounded below, so it is invertible.
□

Exercise 6.55 Verify that u^{-1} is given by Eq. (6.15), completing the proof of the lemma. △

Corollary 6.56 *If $u \in \mathrm{Sp}(H)$ is given in block form by (6.13) and $Z \in \mathcal{D}(H)$, then $a + \alpha b \alpha Z$ is invertible. Under the bijection of Proposition 6.50, $\mathrm{Sp}(H)$ acts transitively on $\mathcal{D}(H)$ via*

$$Z \mapsto (b + \alpha a \alpha Z)(a + \alpha b \alpha Z)^{-1}.$$

Proof Fix $Z \in \mathcal{D}(H)$ and let W_+ be the graph of Z, so that

$$\Psi : L^+ \to W_+$$
$$x \mapsto x + Zx$$

is an isomorphism. Let $W'_+ = uW_+$ and observe that $u|_{W_+} : W_+ \to W'_+$ is an isomorphism. Since $P'_+|_{W'_+}$ is also an isomorphism by Corollary 6.47, we see that

$$a + \alpha b \alpha Z = P'_+ u \Psi : L^+ \to L^+$$

is invertible.

We have already seen in Lemma 6.43 that $\mathrm{Sp}(H)$ acts transitively on $\mathrm{Pol}^\omega(H)$, so the action induced on $\mathcal{D}(H)$ is transitive. Since W'_+ is the image of the graph of Z under u, it is the image of

$$\begin{pmatrix} a + \alpha b \alpha Z \\ b + \alpha a \alpha Z \end{pmatrix} = u \begin{pmatrix} 1 \\ Z \end{pmatrix} : L^+ \to \begin{matrix} L^+ \\ \oplus \\ L^- \end{matrix}.$$

This agrees with the image of

$$\begin{pmatrix} 1 \\ (b + \alpha a \alpha Z)(a + \alpha b \alpha Z)^{-1} \end{pmatrix}$$

which proves the claim.   □

Example 6.57 Let $H_{\mathbb{C}}^{\mathrm{b}}$ be as in Example 6.29. Assume that $\phi \in \mathrm{Diff}(\mathbb{S}^1)$. Modelling $H_{\mathbb{C}}^{\mathrm{b}}$ by the space of functions $\dot{H}^{1/2}(\mathbb{S}^1)$ and using the expression

$$\omega(f, g) = \int_{\mathbb{S}^1} f \, dg,$$

for smooth f, g, we see by change of variables that

$$C_\phi : \dot{H}^{1/2}(\mathbb{S}^1) \to \dot{H}^{1/2}(\mathbb{S}^1)$$
$$f \mapsto f \circ \phi - \frac{1}{2\pi} \int_{\mathbb{S}^1} f \, d\theta$$

preserves ω for smooth f, g. It can be shown that C_ϕ is bounded for any diffeomorphism ϕ. So the invariance of ω under C_ϕ extends to all of H^{b} by continuity of ω. Furthermore, $\phi^{-1} \in \mathrm{Diff}(\mathbb{S}^1)$ is a bounded inverse of C_ϕ so we see that C_ϕ is a symplectomorphism.

Letting $L^{\pm}$ be as in Example 6.29, we then have that

$$W_{\phi}^{\pm} = C_{\phi} L^{+}$$

defines an element of $\text{Pol}^{\omega}(H^{\flat})$. The stabilizer of L^{+} is precisely $\text{Möb}(\mathbb{S}^1)$, the set of Möbius transformations preserving the circle. In summary, we obtain a well-defined injection

$$\text{Diff}(\mathbb{S}^1)/\text{Möb}(\mathbb{S}^1) \to \text{Pol}^{\omega}(H^{\flat})$$

$$[\phi] \mapsto W_{\phi}$$

where $[\phi]$ denotes the equivalence class of a representative $\phi \in \text{Diff}(\mathbb{S}^1)$.

It can be shown that the operator $Z \in \mathfrak{D}(H^{\flat})$ associated with W_{ϕ} (see Corollary 6.47) is the Grunsky operator, which was introduced to complex function theory eighty years ago by H. Grunsky.   △

Remark 6.58 Examples 6.57, 6.14, and 6.29 originate with G. Segal [18]. In the same paper, he introduced the infinite Siegel disk. It was shown by Nag–Sullivan [12] and Vodopy'anov [28] that a homeomorphism of $\mathbb{S}^1$ is a symplectomorphism if and only if it is a quasisymmetry. Furthermore, the unitary subgroup is the set of Möbius transformations preserving the circle. (Equivalently, by Exercise 6.44, the unitary subgroup is the stabilizer of the polarization in Example 6.29).

The definition of quasisymmetries is beyond the scope of this chapter. We mention only that quasisymmetries modulo Möbius transformations $\text{QS}(\mathbb{S}^1)/\text{Möb}(\mathbb{S}^1)$ is a model of the universal Teichmüller space. Takhtajan and Teo showed that this gives a holomorphic embedding of the universal Teichmüller space $\text{QS}(\mathbb{S}^1)/\text{Möb}(\mathbb{S}^1)$ into the infinite Siegel disk. The fact that the operator Z is the Grunsky matrix was shown by Kirillov and Yuri'ev [9] in the smooth setting, and by Takhtajan and Teo [26] in the quasisymmetric case.   △

Next, we define the restricted Grassmannian and symplectic group. We first define a relation $\sim$ on $\text{Pol}^{\omega}(H)$ by saying that $W_1^{+} \sim W_2^{+}$ if the restriction to $W_1^{+} \subset H_{\mathbb{C}}$ of the projection operator $H_{\mathbb{C}} = W_2^{+} \oplus \alpha(W_2^{+}) \to \alpha(W_2^{+})$ is Hilbert–Schmidt. One might object that this definition is too vague, since it does not specify with respect to which inner product this operator is supposed to be Hilbert–Schmidt (reasonable options being g, g_{W_1}, and g_{W_2}). The following result tells us that it does not matter.

Lemma 6.59 *If H is a Hilbert space, equipped with two strong inner products, g_1 and g_2, then an operator $T : H \to H$ is Hilbert–Schmidt with respect to g_1 if and only if it is Hilbert–Schmidt with respect to g_2.*

Proof Choose bases $\{e_i\}$ and $\{f_i\}$ which are orthonormal for g_1 and g_2 respectively. Let $A : H \to H$ be the complex-linear extension of $e_i \mapsto f_i$. The map A is bounded linear, with bounded linear inverse, by Proposition 6.1. Moreover, A satisfies $g_1(v, w) = g_2(Av, Aw)$ for all $v, w \in H$.

Denote by $\| - \|_i$ the Hilbert–Schmidt norm w.r.t. g_i. Now, if T is Hilbert–Schmidt w.r.t g_1, then so is $A^{-1}TA$, and we compute

$$\|ATA^{-1}\|_1^2 = \sum_i g_1(A^{-1}TAe_i, A^{-1}TAe_i)$$

$$= \sum_i g_2(Tf_i, Tf_i)$$

$$= \|T\|_2^2.$$

$\square$

Lemma 6.60 *Let $u \in \mathrm{Sp}(H)$. Then $uL^+ \sim L^+$ if and only if b is Hilbert–Schmidt, where b is given by the decomposition Eq. (6.13).*

Proof The projection $uL^+ \to \alpha(L^+)$ is given by $(ax, bx) \mapsto bx$. Because the operator a is invertible (by Lemma 6.54), we have that this operator is Hilbert–Schmidt if and only if b is Hilbert–Schmidt. $\square$

Proposition 6.61 *The relation $\sim$ on $\mathrm{Pol}^\omega(H)$ is an equivalence relation.*

Proof It is clear that $\sim$ is reflexive, because the projection operator $W^+ \to \alpha(W^+)$ is identically zero.

Now, suppose that $W_1^+ \sim W_2^+$. We may, without loss of generality, assume that $W_2^+ = L^+$. There then exists an element $u \in \mathrm{Sp}(H)$ such that $uL^+ = W_2^+$.

Let $q : H_{\mathbb{C}} \to u(L^-)$ be the projection with respect to the splitting $H_{\mathbb{C}} = u(L^+) \oplus u(L^-)$. We determine the decomposition of q with respect to the splitting $H_{\mathbb{C}} = L^+ \oplus L^-$. Let $x \in H_{\mathbb{C}}$. Then, there exist $v^\pm \in u(L^\pm)$ such that $x = v^+ + v^-$; and we have $q(x) = v^-$. We apply u^{-1} to obtain $u^{-1}x = u^{-1}v^+ + u^{-1}v^-$. We observe that $u^{-1}v^\pm \in L^\pm$, thus projecting onto L^-, we obtain $P^-u^{-1}x = u^{-1}v^-$. Finally, we apply u to obtain $uP^-u^{-1}(x) = v^- = q(x)$. We now compute

$$uP^-u^{-1} = \begin{pmatrix} a & \alpha b\alpha \\ b & \alpha a\alpha \end{pmatrix} \begin{pmatrix} 0 & 0 \\ 0 & 1 \end{pmatrix} \begin{pmatrix} a^* & -b^* \\ -\alpha b^*\alpha & \alpha a^*\alpha \end{pmatrix} = \begin{pmatrix} 0 & \alpha b\alpha \\ 0 & \alpha a\alpha \end{pmatrix} \begin{pmatrix} a^* & -b^* \\ -\alpha b^*\alpha & \alpha a^*\alpha \end{pmatrix}$$

$$= \begin{pmatrix} -\alpha bb^*\alpha & \alpha ba^*\alpha \\ -\alpha ab^*\alpha & \alpha aa^*\alpha \end{pmatrix}.$$

The restriction of q to L^+ is simply given by the first column of this matrix. We thus see that if b is Hilbert–Schmidt, then q is Hilbert–Schmidt. The assumption is that $uL^+ \sim L^+$, which tells us that b is Hilbert–Schmidt, through Lemma 6.60. It follows that q is Hilbert–Schmidt, thus $L^+ \sim uL^+$.

Finally, suppose now that we have $W_1^+ \sim W_2^+$ and $W_2^+ \sim W_3^+$. Let $\iota_1^+ : W_1^+ \to H_{\mathbb{C}}$ and $\iota_2^+ : W_2^+ \to H_{\mathbb{C}}$ be the inclusions, and let $P_i^\pm : H_{\mathbb{C}} \to W_i^\pm$ be the projection with respect to the decompositions $H_{\mathbb{C}} = W_i^+ \oplus W_i^-$ for $i = 1, 2, 3$. We then have that the operators $P_2^- \iota_1^+$ and $P_3^- \iota_2^+$ are Hilbert–Schmidt. We have

$$P_3^- \iota_1^+ = P_3^- (\iota_2^+ P_2^+ + \iota_2^- P_2^-) \iota_1^+ = P_3^- \iota_2^+ P_2^+ \iota_1^+ + P_3^- \iota_2^- P_2^- \iota_1^+.$$

It follows that $P_3^- \iota_1^+$ is Hilbert–Schmidt, and we are done. $\square$

We introduce the *restricted symplectic Lagrangian Grassmannian*

$$\mathrm{Pol}_2^\omega(H) := \{W^+ \in \mathrm{Pol}^\omega(H) \mid W^+ \sim L^+\}$$

$$= \{W^+ \in \mathrm{Pol}^\omega(H) \mid L^+ \to \alpha(W^+) \text{ is Hilbert–Schmidt}\},$$

where $L^+ \to \alpha(W^+)$ is the restriction of the projection $H_{\mathbb{C}} = W^+ \oplus \alpha(W^+) \to \alpha(W^+)$ to $L^+ \subset H_{\mathbb{C}}$. Similarly, we may consider the *restricted Siegel disk*

$$\mathfrak{D}_2(H) := \mathfrak{D}(H) \cap \mathcal{B}_2(L^+, L^-),$$

where $\mathcal{B}_2(L^+, L^-)$ is the space of Hilbert–Schmidt operators from L^+ to L^-.

Lemma 6.62 *The isomorphism* $\mathrm{Pol}^\omega(H) \to \mathfrak{D}(H)$ *restricts to an isomorphism* $\mathrm{Pol}_2^\omega(H) \to \mathfrak{D}_2(H)$.

Proof First, the image of $\mathrm{Pol}_2^\omega(H)$ in $\mathfrak{D}(H)$ under the map above lies in $\mathfrak{D}_2(H)$, because P^- is Hilbert–Schmidt.

To prove the converse, we observe that the orthogonal projection $p : \mathrm{graph}(Z) \to \alpha(L^+)$ is given by $(x, Zx) \mapsto Zx$. If $\widehat{Z} : L^+ \to \mathrm{graph}(Z)$ is the invertible operator $x \mapsto (x, Zx)$ we thus have $Z = p\widehat{Z}$. This implies that if Z is Hilbert–Schmidt, then p must be Hilbert–Schmidt. In other words, $\mathrm{graph}(Z) \sim L^+$. The symmetry of $\sim$ then implies that $\mathrm{graph}(Z) \in \mathrm{Pol}_2^\omega(H)$. $\square$

The *restricted symplectic group* is

$$\mathrm{Sp}_2(H) := \{u \in \mathrm{Sp}(H) : u(L^+) \sim L^+\}.$$

Proposition 6.63 *Let* $u \in \mathrm{Sp}(H)$. *The following are equivalent.*

1. $u \in \mathrm{Sp}_2(H)$;
2. *If u is given in block form by (6.13) then b is Hilbert–Schmidt;*
3. $u^\star J - J$ *is Hilbert–Schmidt.*

Proof By Lemma 6.54 a is invertible. We have uL^+ is the image of $(a, \bar{b})^T L^+$, and thus is the graph of $\bar{b}a^{-1}$. Again since a is invertible, b is Hilbert–Schmidt if and only if Z is Hilbert–Schmidt. By Lemma 6.62 conditions 1 and 2 are equivalent. The equivalence of 2 and 3 follows from the computation

$$u^{-1} J u - J = 2i \begin{pmatrix} b^* b & b^* \alpha a \alpha \\ -\alpha b^* \alpha a & -\alpha b^* b \alpha \end{pmatrix}. \tag{6.16}$$

$\square$

Exercise 6.64 Show that Eq. (6.16) holds.   △

The actions of $\mathrm{Sp}(H)$ on the Siegel disk and Lagrangian Grassmannian restrict appropriately.

Proposition 6.65 *The actions of* $\mathrm{Sp}_2(H)$ *preserve* $\mathrm{Pol}_2^\omega(H)$ *and* $\mathfrak{D}_2(H)$.

Proof That $\mathrm{Sp}_2(H)$ preserves $\mathrm{Pol}_2^\omega(H)$ is clear by definition.

That $\mathrm{Sp}_2(H)$ preserves $\mathfrak{D}_2(H)$ follows from Corollary 6.56 together with part 2 of Proposition 6.63.   □

Theorem 6.66 *The Siegel disk* $\mathfrak{D}(H)$ *is an open subset of the Banach space of bounded linear operators on* H *satisfying* $\alpha Z \alpha = Z^*$, *and in particular is a complex Banach manifold. The restricted Siegel disk* $\mathcal{D}_2(H)$ *is an open subset of the Hilbert space of Hilbert–Schmidt operators on* H *satisfying* $\alpha Z \alpha = Z^*$, *and in particular is a complex Hilbert manifold.*

Proof Let X denote the set of operators satisfying $\alpha Z \alpha = Z^*$. X is a closed linear subspace of $\mathcal{B}(L^+, L^-)$, so it is itself a Banach space. We have that $I - Z^* Z$ is positive definite if and only if $\| Z \| < 1$. Thus $\mathcal{D}(H)$ is an open subset of $\mathcal{B}(L^+, L^-)$, and hence also an open subset of $X \cap \mathcal{B}(L^+, L^-)$.

The Hilbert–Schmidt norm controls the operator norm; that is, inclusion from the Hilbert–Schmidt operators into the bounded linear operators is a bounded map. Thus $\mathcal{D}_2(H)$ is an open subset of the space of Hilbert–Schmidt operators, and the remaining claims follow similarly.   □

By Propositions 6.50 and 6.65 this gives the Grassmannian and restricted Grassmannian Banach and Hilbert manifold structures respectively.

Example 6.67 Referring to Example 6.57 and Remark 6.58, it was shown by Segal that for smooth ϕ the operator $\mathcal{C}_\phi$ is an element of $\mathrm{Sp}_2(H)$. It was shown by Takhtajan and Teo that for a quasisymmetry ϕ, $\mathcal{C}_\phi$ is in $\mathrm{Sp}_2(H)$ if and only if ϕ is what is known as a Weil–Petersson quasisymmetry. The definition is beyond the scope of this chapter; we mention only that the Weil–Petersson quasisymmetries modulo the Möbius transformations of the disk, $\mathrm{QS}_{\mathrm{WP}}(\mathbb{S}^1)/\mathrm{M\ddot{o}b}(\mathbb{S}^1)$, is a model of the Weil–Petersson universal Teichmüller space.   △

6.5 The Grassmannian of Polarizations Associated with an Inner Product

In this section, we fix an inner product, and describe the Lagrangian Grassmannian of orthogonal polarizations $\mathrm{Pol}^g(H)$ associated with that inner product. We show that it is a complex Banach manifold. We also consider the "restricted" Grassmannian and orthogonal groups, and established that the restricted Grassmannian is a Hilbert manifold.

We assume again that H is equipped with a compatible triple (g, J, ω), and we write $H_{\mathbb{C}} = L^+ \oplus L^-$ for the decomposition w.r.t. J, i.e. $L^{\pm} = \ker(J \mp i)$. Now, choose an orthonormal basis $\{e_i\}_{i \geqslant 1}$ for L^+. For $i \leqslant 1$ we set $e_i = \alpha e_{-i}$, which yields an orthonormal basis $\{e_i\}_{i \leqslant 1}$ for $L^- = \alpha(L^+)$. Nothing material will depend on these choices, but they will be convenient for the proofs to come.

Motivated by Theorem 6.80 we introduce the *restricted Lagrangian Grassmannian*

$$\mathrm{Pol}_2^g(H) := \{W^+ \in \mathrm{Pol}^g(H) \mid L^+ \to \alpha(W^+) \text{ is Hilbert–Schmidt}\},$$

where $L^+ \to \alpha(W^+)$ is the restriction of the orthogonal projection $H_{\mathbb{C}} \to \alpha(W^+)$ to $L^+ \subset H_{\mathbb{C}}$. We define a relation $\sim$ on $\mathrm{Pol}^g(H)$ by saying that $W_1^+ \sim W_2^+$ if the orthogonal projection $W_1^+ \to \alpha(W_2^+)$ is Hilbert–Schmidt.

Exercise 6.68 Prove that $\sim$ defines an equivalence relation on $\mathrm{Pol}^g(H)$. (This exercise is challenging.)   △

We also introduce the *restricted orthogonal group*

$$O_2(H) := \{u \in O(H) \mid u(L^+) \in \mathrm{Pol}_2^g(H)\}.$$

Given an operator $u \in O(H)$, let us write

$$u = \begin{pmatrix} a & b \\ c & d \end{pmatrix},$$

with respect to the decomposition $H_{\mathbb{C}} = L^+ \oplus L^-$.

Lemma 6.69 *We have that $u \in O_2(H)$ if and only if c is Hilbert–Schmidt, if and only if b is Hilbert–Schmidt.*

Proof First, we observe that $\alpha(u(L^+)) = u(L^-)$. The set $\{ue_i\}_{i \leqslant 1} = \{(be_i, de_i)\}_{i \leqslant 1}$ is then an orthonormal basis for $u(L^-)$. The orthogonal projection of e_j onto $u(L^-)$, for $j \geqslant 1$ is given by

$$\sum_{i \leqslant 1} \langle e_j, be_i \rangle (be_i, de_i).$$

It follows that the Hilbert–Schmidt norm of the orthogonal projection $L^+ \to u(L^-)$ is given by

$$\sum_{j \geqslant 1} \| \sum_{i \leqslant 1} \langle e_j, be_i \rangle (be_i, de_i) \|^2 = \sum_{-i, j \geqslant 1} |\langle e_j, be_i \rangle|^2,$$

which is simply the Hilbert–Schmidt norm of the operator $b : L^- \to L^+$. It follows that $u(L^+) \in O_2(H)$ if and only if b is Hilbert–Schmidt. That b is Hilbert–Schmidt if and only if c is Hilbert–Schmidt follows from the equation $\alpha c = b\alpha$.   □

Corollary 6.70 *If $u \in O_2(H)$, then a and d are Fredholm.*

Proof It follows from the fact that $uu^* = \mathbb{1}$ that $aa^* + bb^* = \mathbb{1}$, and from $u^*u = \mathbb{1}$ that $a^*a + c^*c = \mathbb{1}$. Because b and c are Hilbert–Schmidt, we have that bb^* and c^*c are trace-class, thus in particular compact. It follows that a is Fredholm (and a^* is a "parametrix" for a). $\qquad\square$

Now, suppose that $Z : L^+ \to L^-$ is a linear operator. The computation

$$\langle (v, Zv), \alpha(w, Zw) \rangle = \langle v, \alpha Zw \rangle + \langle Zv, \alpha w \rangle = \langle v, (\alpha Z + Z^*\alpha)w \rangle,$$

shows that $\mathrm{graph}(Z)$ is perpendicular to $\alpha(\mathrm{graph}(Z))$ if and only if $\alpha Z\alpha = -Z^*$.

Exercise 6.71 Suppose that $\alpha Z\alpha = -Z^*$, and show that if $(x, y) \in H_{\mathbb{C}}$ is perpendicular to both $\mathrm{graph}(Z)$ and $\alpha(\mathrm{graph}(Z))$, then $(x, y) = 0$. $\qquad\triangle$

It follows from Exercise 6.71 that $\mathrm{graph}(Z) \in \mathrm{Pol}^g(H)$ if and only if $\alpha Z\alpha = -Z^*$. Following [16, Section 7.1], we equip $\mathrm{Pol}_2^g(H)$ with the structure of complex manifold. To be precise, we shall equip $\mathrm{Pol}_2^g(H)$ with an atlas modeled on the complex Hilbert space

$$\mathcal{B}_2^\alpha(L^+, L^-) = \{Z \in \mathcal{B}_2(L^+, L^-) \mid \alpha Z\alpha = -Z^*\},$$

such that the transition functions are holomorphic. For S a finite subset of the positive integers let W_S be the closed linear span of

$$\{e_k, e_{-l} \mid k \notin S, l \in S\}.$$

We see that $W_S \in \mathrm{Pol}_2^g(H)$. We shall use the set of finite subsets of the positive integers as index set for our atlas.

Lemma 6.72 *If $W^+ \in \mathrm{Pol}_2^g(H)$, then the orthogonal projection $P : W^+ \to L^+$ is a Fredholm operator, and $\mathrm{coker}(P) = \alpha \ker(P)$. In particular, the index of P is zero.*

Proof Let $W^+ \in \mathrm{Pol}_2^g(H)$, and write $P : W^+ \to L^+$ for the orthogonal projection. There exists an operator $u \in O_2(H)$ such that $u(L^+) = W^+$. As before, we write

$$u = \begin{pmatrix} a & b \\ c & d \end{pmatrix},$$

where $a : L^+ \to L^+$ is Fredholm. Now, we claim that the map

$$L^+ \to W^+, x \mapsto u(x)$$

restricts to a bijection from $\ker(a)$ to $\ker(P)$. Indeed, $x \in \ker(a) \subset L^+$ if and only if $u(x) = (0, cx)$. Similarly $v \in \ker(P) \subset W^+ = u(L^+)$ if and only if $v = (0, cx)$.

It thus follows that $\ker(P)$ is finite-dimensional, because $\ker(a)$ is (because a is Fredholm). Now if $x \in L^+$ is arbitrary, then we have $u(x) = (ax, cx)$, and thus $Pu(x) = ax$. it follows that the range of P is given by the range of a, which is closed, and has finite-dimensional cokernel, because a is Fredholm.

Observe that the kernel of P is given by $W^+ \cap \alpha(L^+)$. We claim that the cokernel of P is $\alpha(W^+) \cap L^+$. In other words $\mathrm{Im}(P)^\perp = \alpha(W^+) \cap L^+$. Let $v \in \alpha(W^+) \cap L^+$ and $w \in W^+$, and denote by $P^+ : H_\mathbb{C} \to L^+$ the orthogonal projection, so that $P^+|_{W^+} = P$. We then compute

$$\langle v, Pw \rangle = \langle v, P^+w \rangle = \langle P^+v, w \rangle = \langle v, w \rangle = 0.$$

So $v \in \mathrm{Im}(P)^\perp$. Conversely, if $v \in \mathrm{Im}(P)^\perp$, then $0 = \langle v, Pw \rangle = \langle v, w \rangle$ for all $w \in W^+$ so $v \in \alpha(W^+)$.   □

Corollary 6.73 *If $W_1^+ \sim W_2^+$, then the orthogonal projection $W_1^+ \to W_2^+$ is a Fredholm operator of index zero.*

Proof This follows from Lemma 6.72 together with the transitivity of the relation $\sim$, see Exercise 6.68.   □

Remark 6.74 Lemma 6.72 should be compared to Proposition 6.45. In that case, the projection operator is not just Fredholm, but even invertible.

Moreover, Lemma 6.72 should be compared to e.g. [16, Prop. 6.2.4]; there, any index can occur.   △

We define

$$\mathcal{O}_S := \{\mathrm{graph}(Z) \in \mathrm{Pol}_2^g(H) \mid Z \in \mathcal{B}_2^\alpha(W_S, W_S^\perp)\}.$$

The following result tells us that the sets $\mathcal{O}_S$ cover $\mathrm{Pol}_2^g(H)$, (see also [4, Section 4], and c.f. [16, Proposition 7.1.6]).

Lemma 6.75 *Any element of $\mathrm{Pol}_2^g(H)$ can be written as the graph of some operator $Z \in \mathcal{B}_2^\alpha(W_S, W_S^\perp)$.*

Proof Let $W^+ \in \mathrm{Pol}_2^g(H)$ be arbitrary. We claim that there exists a finite subset S of the positive integers such that the projection operator $P_S : W^+ \to W_S$ is an isomorphism. Once we have found such an S, it follows that $W^+ = \mathrm{graph}(P_S^\perp P_S^{-1})$, where $P_S^\perp$ is the orthogonal projection of W^+ onto $W_S^\perp$, (c.f. Exercise 6.46). The operator $P_S^\perp P_S^{-1}$ is Hilbert–Schmidt, because $P_S^\perp$ is.

Now, let $E_\emptyset \subset W^+$ be the kernel of the projection operator $P : W^+ \to L^+ = W_\emptyset$. Let $d < \infty$ be its dimension. If $d = 0$, it follows that P is an isomorphism, because it is Fredholm of index 0 by Corollary 6.73. So, assume that $d > 0$, and let $0 \neq v \in E_\emptyset$ be a unit vector. Choose a positive integer l such that $\langle v, e_{-l} \rangle \neq 0$. We now consider the orthogonal projection $P_{\{l\}} : W^+ \to W_{\{l\}}$; and we let $E_{\{l\}}$ be its kernel. We claim that $E_{\{l\}}$ is a strict subset of $E_\emptyset$, and thus that the dimension of $E_{\{l\}}$ is strictly smaller than the dimension of $E_\emptyset$. First, observe that $v \notin E_{\{l\}}$, while

$v \in E_\emptyset$. It thus remains to be shown that $E_{\{l\}}$ is a subset of $E_\emptyset$ in the first place. Suppose that $x \in E_{\{l\}}$, and write $x = \sum_k \lambda_k e_k$. The condition that $x \in E_{\{l\}}$ then implies that

$$\lambda_k = 0, \quad k \geqslant 1, k \neq l.$$

Now, we have $v = \sum_k \mu_k e_k$, where $\mu_k = 0$ for $k \geqslant 1$, and $\mu_{-l} = \langle v, e_{-l} \rangle \neq 0$. We have

$$\alpha(x) = \sum_k \overline{\lambda_{-k}} e_k.$$

We then use that $\alpha(x) \in W^- = (W^+)^\perp$ and compute

$$0 = \langle \alpha(x), v \rangle = \sum_k \overline{\lambda_{-k}} \mu_k = \overline{\lambda_l} \mu_{-l}.$$

The condition that $\mu_{-l} \neq 0$ then implies that $\lambda_l = 0$, whence $x \in E_\emptyset$.

By induction (using Corollary 6.73) on this dimension, we construct a finite subset of the positive integers S, such that $P_S : W^+ \to W_S$ is injective. Because this operator is Fredholm of index zero (again by Corollary 6.73), it follows that it is an isomorphism, which completes the proof.   □

Fix now finite subsets of the positive integers S_1 and S_2. We determine the "transition functions" corresponding to the charts $\mathcal{B}_2(W_{S_1}, W_{S_1}^\perp) \leftarrow \mathcal{O}_{S_1} \cap \mathcal{O}_{S_2} \to \mathcal{B}_2(W_{S_2}, W_{S_2}^\perp)$, following [16, Proposition 7.1.2]. Let a, b, c, d be operators such that

$$\begin{pmatrix} a & b \\ c & d \end{pmatrix} : \begin{pmatrix} W_{S_1} \\ W_{S_1}^\perp \end{pmatrix} \to \begin{pmatrix} W_{S_2} \\ W_{S_2}^\perp \end{pmatrix}$$

is the identity operator. A straightforward verification shows that a and d are Fredholm, and that b and c are Hilbert–Schmidt (indeed, even finite-rank).

Proposition 6.76 *The image of $\mathcal{O}_{S_1} \cap \mathcal{O}_{S_2}$ in $\mathcal{B}_2^\alpha(W_{S_1}, W_{S_1}^\perp)$ consists of those operators Z_1 such that $a + bZ_1$ has a bounded inverse. Moreover, the transition function is given by $Z_1 \mapsto (c + dZ_1)(a + bZ_1)^{-1}$.*

Proof Let $W^+ \in \mathcal{O}_{S_1} \cap \mathcal{O}_{S_2}$, we then have graph$(Z_1) = W^+ = $ graph(Z_2), for some $Z_i \in \mathcal{B}_2^\alpha(W_{S_i}, W_{S_i}^\perp)$. The orthogonal projection $P_i : W^+ \to W_{S_i}$ is invertible, with inverse $x \mapsto (x, Z_i x)$. This implies that the composition $P_2 P_1^{-1} : W_{S_1} \to W_{S_2}$ is invertible. The calculation

$$P_2 P_1^{-1} x = P_2(x, Z_1 x) = ax + bZ_1 x$$

shows that $P_2 P_1^{-1} = a + bZ_1$.

On the other hand, if $Z_1 \in \mathcal{B}_2^\alpha(W_{S_1}, W_{S_1}^\perp)$ is such that $a + bZ_1$ has a bounded inverse, then we consider the operator $(c + dZ_1)(a + bZ_1)^{-1} : W_{S_2} \to W_{S_2}^\perp$. This operator is Hilbert–Schmidt, because both c and Z_1 are Hilbert–Schmidt. Now, let $y \in W_{S_2}$ be arbitrary. Set $x = (a + bZ_1)^{-1}y \in W_{S_1}$. We then compute

$$
\begin{pmatrix} y \\ (c + dZ_1)(a + bZ_1)^{-1}y \end{pmatrix} = \begin{pmatrix} (a + bZ_1)x \\ (c + dZ_1)x \end{pmatrix} = \begin{pmatrix} a & b \\ c & d \end{pmatrix} \begin{pmatrix} x \\ Z_1 x \end{pmatrix}.
$$

This proves that the graph of Z_1 is equal to the graph of $(c + dZ_1)(a + bZ_1)^{-1}$. ☐

To prove that the transition functions form a complex atlas, we need an elementary lemma.

Lemma 6.77 *Let A and A' be bounded operators on a Banach space, and assume that A is invertible. If $\|A - A'\| < 1/(2\|A^{-1}\|)$ then A' is invertible and*

$$
\|(A')^{-1}\| \le 2\|A^{-1}\|.
$$

Proof Since $\|(A' - A)A^{-1}\| < 1/2$, the Neumann series provides an inverse for $I + (A' - A)A^{-1}$ and

$$
\left\| \left[I + (A' - A)A^{-1} \right]^{-1} \right\| \le \frac{1}{1 - \|(A' - A)A^{-1}\|} < 2.
$$

Thus

$$
(A')^{-1} = A^{-1} \left(I + (A' - A)A^{-1} \right)^{-1}
$$

and the claim follows directly. ☐

Theorem 6.78 *The transition functions in Proposition 6.76 are holomorphic.*

Proof Let $\Psi : Z \mapsto (c + dZ)(a + bZ)^{-1}$ be the transition map in the space of Hilbert–Schmidt operators. It is enough to show that Ψ is Gâteaux differentiable and locally bounded. Denote the Hilbert–Schmidt norm by $\|\cdot\|_{HS}$ and the usual operator norm by $\|\cdot\|$. Two facts we will use are that $\|AB\|_{HS} \le \|A\|\|B\|_{HS}$ and $\|A\| \le \|A\|_{HS}$.

It is easily seen that the map is Gâteaux holomorphic with Gâteaux derivative

$$
D\Psi(Z; W) = \left[dW + (c + dZ)(a + bZ)^{-1}bW \right](a + bZ)^{-1}.
$$

To see that Ψ is locally bounded, fix Z and assume that

$$
\|Z' - Z\| < \frac{1}{\|b\|\|(a + bZ)^{-1}\|}.
$$

By Lemma 6.77 with $A = a + bZ$, $A' = a + bZ'$ we then have

$$\left\| \left(a + bZ' \right)^{-1} \right\| \leq \frac{1}{2 \| (a + bZ)^{-1} \|}.$$

Thus

$$\left\| \left(c + dZ' \right) \left(a + bZ' \right)^{-1} \right\|_{HS} \leq \left(\| c \|_{HS} + \| d \| \| Z' \|_{HS} \right) \frac{1}{2 \| (a + bZ)^{-1} \|}$$

which proves the claim.   □

Remark 6.79 The Grassmannian $\mathrm{Pol}_2^g(H)$ is a rich geometric object. For example, it carries a "Pfaffian line bundle" [4].   △

6.6 Motivation for the Restricted Grassmannian

In this section, we give a sketch of the representation-theoretic and physical motivation for the restricted spaces.

Let H be a real Hilbert space, with inner product g. From this data, one can construct a C*-algebra, called the *Clifford C*-algebra*. We now give an overview of this construction, together with some aspects of the corresponding representation theory. A full account of the theory is given [15] and [1].

The complex Clifford algebra of H is the complex *-algebra generated by elements of $H_{\mathbb{C}}$, subject to the condition that $vw + wv = g(v, \alpha w)$ and $v^* = \alpha(v)$, for all $v, w \in H_{\mathbb{C}}$. This algebra can be completed to a C*-algebra, which we denote by $\mathrm{Cl}(H)$. For $W^+ \in \mathrm{Pol}^g(H)$ we define F_{W^+} to be the Hilbert completion of the exterior algebra of W^+:

$$F_{W^+} := \overline{\bigoplus_{n \geq 0} \wedge^n W^+}.$$

We define a map $\pi_{W^+} : H_{\mathbb{C}} \to \mathcal{B}(F_{W^+})$ by setting

$$\pi_{W^+}(x) w_1 \wedge \ldots \wedge w_n = x \wedge w_1 \wedge \ldots \wedge w_n$$

for $x, w_1, \ldots, w_n \in W^+$, and $\pi_{W^+}(y) = \pi_{W^+}(\alpha(y))^*$ for $y \in W^-$. The map π_{W^+} defined in this way admits a unique extension to a *-homomorphism $\pi_{W^+} : \mathrm{Cl}(H) \to \mathcal{B}(F_{W^+})$; this is the Fock representation of $\mathrm{Cl}(H)$ corresponding to W^+, ([15, Chapter 2]).

It is then natural to attempt to compare the Fock representations corresponding to two elements $W_1^+, W_2^+ \in \mathrm{Pol}^g(H)$. The following result tells us when two such representations should be viewed as equivalent.

Theorem 6.80 *There exists a unitary operator* $u : F_{W_1^+} \to F_{W_2^+}$ *with the property that*

$$\pi_{W_1^+}(a) = u^* \pi_{W_2^+}(a) u$$

for all $a \in \mathrm{Cl}(H)$ *if and only if the orthogonal projection* $H \to \alpha(W_2^+)$ *restricts to a Hilbert–Schmidt operator* $W_1^+ \to \alpha(W_2^+)$

A proof of this classical result, together with references to the expansive literature on the subject, can be found in [15, Chapter 3].

If H is a real Hilbert space with symplectic form ω, then one defines the Heisenberg (Lie) algebra to be $H \times \mathbb{R}$, equipped with the bracket

$$[(v, t), (w, s)] = (0, \omega(v, w)).$$

Given a positive symplectic polarization $H_{\mathbb{C}} = W^+ \oplus W^-$, one obtains a representation of the Heisenberg algebra on the Hilbert completion of the symmetric algebra of W^+, similar to the construction above. However, this representation is by *unbounded* operators. In spite of this, the situation is entirely analogous: If H is finite-dimensional, then all such representations are unitarily equivalent; this is the Stone–von Neumann theorem. The equivalence problem in the infinite-dimensional case was settled by D. Shale [21]. Indeed, the Stone–von Neumann theorem is superseded by the result that two positive symplectic polarizations $W_1^\pm$ and $W_2^\pm$ give unitarily equivalent representations, if and only if they are in the same Hilbert–Schmidt class. This statement is not quite precise, because we have not explained what a representation by unbounded operators is. There are several ways to deal with this issue, but this would take us too far afield. The reader can find the subject treated in for example [6, 11, 14].

6.7 Sewing and Diff($\mathbb{S}^1$)

In this section, we return to Example 6.57, and give the Grassmannian of polarizations a geometric interpretation in terms of sewing. This sewing interpretation arises in conformal field theory, see [7, 19]. This section can be treated as an extended example. This example appears in many contexts.

We first give a summary. Let $\overline{\mathbb{C}}$ denote the Riemann sphere,

$$\mathbb{D}_+ = \{z \in \mathbb{C} : |z| < 1\}, \quad \text{and} \quad \mathbb{D}_- = \{z \in \overline{\mathbb{C}} : |z| > 1\} \cup \{\infty\}.$$

Every diffeomorphism $\phi \in \mathrm{Diff}(\mathbb{S}^1)$ gives rise to a conformal map f from the unit disk into the sphere, obtained by sewing $\mathbb{D}_-$ to $\mathbb{D}_+$ using ϕ to identify points on their boundaries. The image of f is bounded by a smooth Jordan curve representing the seam generated by the choice of ϕ. If ϕ is the identity map, the seam is just the unit circle in the Riemann sphere.

Now let Σ be the complement of the closure of the image of f. Recall that ϕ induces a composition operator $\mathcal{C}_\phi$ on H^b, which is a bounded symplectomorphism. and that $W_\phi = \mathcal{C}_\phi W_+ \in \mathrm{Pol}^\omega(H^b)$. We then have that W_ϕ can be interpreted as pull-back by f of the set of boundary values of holomorphic functions on Σ.

To fill in this summary, we need a result known as conformal welding. It originated in quasiconformal Teichmüller theory in the 1960s (see [10] and references therein), and independently in other contexts including conformal field theory. We give the description in the smooth case because it allows a simpler presentation in terms of more well-known theorems.

The conformal welding theorem (in the smooth case) says the following.

Theorem 6.81 (Smooth Conformal Welding) *Let $\phi \in \mathrm{Diff}(\mathbb{S}^1)$. There are holomorphic one-to-one functions $f : \mathbb{D}_+ \to \overline{\mathbb{C}}$, $g : \mathbb{D}_- \to \overline{\mathbb{C}}$, which extend smoothly and bijectively to $\mathbb{S}^1$, such that $f(\mathbb{S}^1) = g(\mathbb{S}^1)$ and $\phi = g^{-1} \circ f|_{\mathbb{S}^1}$.*

These are unique up to post-composition by a Möbius transformation; that is, any other such pair of maps is given by $T \circ f$, $T \circ g$.

Proof We give a sketch of a proof. Given $\phi \in \mathrm{Diff}(\mathbb{S}^1)$, we treat it as a parametrization of the boundary of the Riemann surface $\mathbb{D}_-$. Sew on $\mathbb{D}_+$ by identifying points on $\partial\mathbb{D}_+$ and $\partial\mathbb{D}_-$ under ϕ. One then obtains a topological sphere $S = \mathrm{cl}\,\mathbb{D}_+ \sqcup \mathrm{cl}\,\mathbb{D}_-/\sim$, where $p \in \partial\mathbb{D}_+$ is equivalent to $q \in \partial\mathbb{D}_-$ if and only if $q = \phi(p)$. S can be given a unique complex manifold structure compatible with that on $\mathbb{D}_+$ and $\mathbb{D}_-$. By the uniformization theorem, there is a biholomorphism $\Psi : S \to \overline{\mathbb{C}}$. Set

$$f = \Psi|_{\mathbb{D}_+}, \quad g = \Psi|_{\mathbb{D}_-}.$$

It follows from continuity of Ψ together with the definition of the equivalence relation $\sim$ that $\phi = g^{-1} \circ f$.

Uniqueness can be obtained from the fact that any biholomorphism of the Riemann sphere is a Möbius transformation. $\square$

Remark 6.82 In fact, the original version in quasiconformal Teichmüller theory was valid more generally for quasisymmetries of $\mathbb{S}^1$ (see Remark 6.58). This is the theorem usually referred to as the conformal welding theorem. $\triangle$

The proof of the smooth conformal welding theorem shows that in general, given a disk $\mathbb{D}_-$ whose boundary is equipped with a parameterization $\phi \in \mathrm{Diff}(\mathbb{S}^1)$ one can sew on a disk $\mathbb{D}_+$. This resulting Riemann surface is the Riemann sphere, but the seam is now a smooth Jordan curve Γ. Let $\Sigma = \Psi(\mathbb{D}_-)$ denote the copy of $\mathbb{D}_-$ in $\overline{\mathbb{C}}$; the parameterization ϕ now is represented equivalently by the boundary values of f, which is a smooth function taking $\mathbb{S}^1$ to $\Gamma = f(\mathbb{S}^1) = g(\mathbb{S}^1)$.

Without loss of generality, assume that $\infty \in \Sigma$. Let

$$
\mathcal{D}_\infty(\Sigma) = \left\{ h : \Sigma \to \mathbb{C} \text{ holomorphic } : \iint_\Sigma |h'|^2 < \infty, \ h(\infty) = 0 \right\}
$$

be the Dirichlet space. Given $h \in \mathcal{D}_\infty(\Sigma)$, if we assume that h has a smooth extension to the boundary $\partial \Sigma$, then $h \circ f$ is a smooth function on $\mathbb{S}^1$, and in particular is in $H_{\mathbb{C}}^b$. In fact, it can be shown that h extends to the boundary and $h \circ f$ makes sense for any $h \in \mathcal{D}_\infty(\Sigma)$, and furthermore $h \circ f - h(f(0))$ is in $H_{\mathbb{C}}^b$ [24] (In fact this holds for any quasisymmetry ϕ, c.f. Remark 6.58.) We then have

Theorem 6.83 *For $\phi \in \mathrm{Diff}(\mathbb{S}^1)$, if $W_\phi = C_\phi L^+$, then*

$$
W_\phi = f^\star \mathcal{D}_\infty(\Sigma) := \{ h \circ f - h(f(0)) : h \in \mathcal{D}_\infty(\Sigma) \}.
$$

More generally, in his sketch of a definition of conformal field theory, Segal [19] considered the category whose objects are Riemann surfaces of genus g with n closed, boundary curves circles endowed with boundary parametrizations $(\phi_1, \ldots, \phi_n)$. After sewing on copies of the disk, one obtains a compact surface $\mathcal{R}$, holomorphic maps $f = (f_1, \ldots, f_n)$ representing the parametrizations, and Σ can be identified with the complement of the closures of the images of $f_1(\mathbb{D}_+), \ldots, f_n(\mathbb{D}_+)$. The sets $f^* \mathcal{D}(\Sigma)$ play a role in the construction; the boundary parameterizations are a means to obtain Fourier series from the boundary values of elements of $\mathcal{D}(\Sigma)$, and the induced polarizations represent the positive and negative Fourier modes.

The moduli space $\widetilde{\mathcal{M}}(g, n)$ space of surfaces with boundary is as follows. Two elements $(\Sigma, \phi_1, \ldots, \phi_n)$, $(\Sigma', \phi_1', \ldots, \phi_n')$ are equivalent if there is a conformal map $F : \Sigma \to \Sigma'$ such that $\phi_k' = F \circ \phi_k$ for $k = 1, \ldots, n$.

If the parameterizations are taken to be smooth in the definition of $\widetilde{\mathcal{M}}(g, n)$, then since all type $(0, 1)$ surfaces are equivalent to $\mathbb{D}_-$ we can identify

$$
\widetilde{M}(0, 1) \cong \mathrm{Diff}(\mathbb{S}^1)/\mathrm{M\ddot{o}b}(\mathbb{S}^1).
$$

The moduli space $\widetilde{\mathcal{M}}(0, 1)$, as well as the universal Teichmüller space can be embedded in the Grassmannian of polarizations $\mathrm{Pol}^\omega(H^b)$. To see this, recall from Remark 6.58 that the stabilizer of L^+ in $\mathrm{Diff}(\mathbb{S}^1)$ is the set $\mathrm{M\ddot{o}b}(\mathbb{S}^1)$ of Möbius transformations preserving $\mathbb{S}^1$. This gives the following.

Corollary 6.84 *We have a well-defined injective map*

$$
\widetilde{M}(0, 1) \to \mathrm{Pol}^\omega(H^b)
$$

$$
[(\mathbb{D}_-, \phi)] \mapsto W_\phi.
$$

The universal Teichmüller space $T(0, 1)$ is a moduli space containing the Teichmüller spaces of all Riemann surfaces covered by the disk. It is classically known to be modelled by $QS(\mathbb{S}^1)/M\ddot{o}b(\mathbb{S}^1)$. The embedding in Remark 6.58 is as follows.

Theorem 6.85 *There is a well-defined injective map [12] from* $T(0, 1) = QS(\mathbb{S}^1)/M\ddot{o}b(\mathbb{S}^1)$ *to* $\mathrm{Pol}^\omega(H^{\mathrm{b}})$.

$$T(0, 1) \to \mathrm{Pol}^\omega(H^{\mathrm{b}})$$

$$[\phi] \mapsto W_\phi.$$

This map is holomorphic with respect to the classical Banach manifold structures on $T(0, 1)$ *and* $\mathcal{D}(H^{\mathrm{b}})$ *[26, Theorem B1].*

Remark 6.86 If one uses quasisymmetric parametrizations in the definition of Segal's moduli space $\widetilde{M}(0, 1)$, one then sees that it can be identified naturally with the universal Teichmüller space $T(0, 1)$. The desirability of this extension to quasisymmetries is strongly motivated by Nag–Sullivan/Vodop'yanov's result that the quasisymmetries produce precisely the composition operators which are bounded symplectomorphisms (see Remark 6.58). It is remarkable that the link between these two spaces—a geometric fact—is established by an analytic condition. This analytic condition in turn is motivated by an algebraic requirement; namely, the requirement that the representation of quasisymmetries be bounded. Interestingly, this analytic condition was in place for independent reasons in Teichmüller theory decades before the result of Nag–Sullivan/Vodopy'anov.   △

Remark 6.87 The association between the Teichmüller space and the moduli space of Segal holds for arbitrary surfaces of genus g with n boundary curves. In general, the Segal moduli space $\widetilde{M}(g, n)$, if extended to allow quasisymmetric parametrizations, is a quotient of the Teichmüller space $T(g, n)$ by a finite-dimensional modular group [17]. (This modular group is trivial in the case $(g, n) = (0, 1)$).   △

Remark 6.88 The extension of the symplectic action of $\mathrm{Diff}(\mathbb{S}^1)$ to quasisymmetries (resulting in the embedding of universal Teichmüller space), was recognized by Nag and Sullivan [12]. The problem of determining which ϕ preserve the restricted Grassmannian has roots in work of Nag and Verjovsky [13], who considered the convergence of a natural generalization of the classical Weil–Petersson pairing on the tangent space to the universal Teichmüller space. This eventually gave birth to the Weil–Petersson universal Teichmüller space, now widely studied. Takhtajan and Teo [26] showed that the embeddings of the universal Teichmüller space and Weil–Petersson Teichmüller space into $\mathrm{Pol}^\omega(H^{\mathrm{b}})$ and $\mathrm{Pol}_2^\omega(H^{\mathrm{b}})$ are holomorphic. The interpretation in terms of pulling back boundary values of functions in the Dirichlet space requires some justification [24]. For the induced representation on symmetric Fock space, see [18] in the smooth case. This was extended to the Weil–Petersson class universal Teichmüller space by A. Serge'ev [20]. A quantization procedure on the classical universal Teichmüller space is also described there and in references therein.   △

6.8 Solutions

Exercise 6.5 By definition, we have

$$g(Jv, Jw) = \omega(Jv, J^2w) = \omega(w, Jv) = g(v, w)$$

and similarly using Proposition 6.4 part 2,

$$\omega(Jv, Jw) = g(-v, Jw) = -\omega(v, J^2w) = \omega(v, w).$$

Exercise 6.6

(a) A symplectic form that makes the triple (g, J, ω) compatible, if it exists, must be given by the equation $\omega(v, w) = g(Jv, w)$. It is clear that the induced map φ_ω is invertible, because $\varphi_\omega = \varphi_g J$. One sees that ω is anti-symmetric if and only if J is skew-adjoint with respect to g by the computation $\omega(v, w) = g(Jv, w) = -g(Jw, v) = \omega(w, v)$.
(b) In this case, the equation $J = \varphi_g^{-1}\varphi_\omega$ defines a complex structure.
(c) We compute $\omega(Jv, w) = -\omega(w, Jv) = -g(w, v) = -g(v, w) = -\omega(v, Jw)$.

Exercise 6.7 If (g, J, ω) is a compatible triple, then replacing J by $-J$ will give an example as required.

Exercise 6.13 Let δ be closed and $d\phi$ be exact. By Stokes' theorem, since $d\delta = 0$

$$\omega(\delta, d\phi) = \frac{1}{\pi} \iint_{\mathcal{R}} \delta \wedge d\phi = -\iint_{\mathcal{R}} d(\phi\delta) = 0$$

where the final equality is because $\mathcal{R}$ has no boundary. Thus given any closed one-forms β, γ, with harmonic one-form representatives $\widehat{\beta}, \widehat{\gamma}$ such that

$$\beta = \widehat{\beta} + d\phi, \quad \gamma = \widehat{\gamma} + d\psi,$$

we have $\omega(\widehat{\beta}, d\psi) = \omega(d\phi, \widehat{\gamma}) = \omega(d\phi, d\psi) = 0$ which proves the claim.

Exercise 6.15 One can work in the complex $L^2(\mathbb{S}^1)$ space. The Fourier series of df_2 is

$$df_2 = \sum_{n \in \mathbb{Z} \setminus \{0\}} inb_n e^{in\theta}.$$

This can be justified (for example) by using the fact that since df_2 is smooth, it converges uniformly; therefore, the Fourier series of f_2 can be derived from that of df_2 by integrating term by term. Since f_2 is real,

$$\omega(\{a_n\}, \{b_n\}) = \frac{1}{2\pi} \int f_1 d\overline{f_2}$$

$$= \frac{1}{2\pi} \int_0^{2\pi} \left(\sum_{n \in \mathbb{Z} \setminus \{0\}} a_n e^{in\theta} \right) \left(\sum_{m \in \mathbb{Z} \setminus \{0\}} -im\overline{b_m} e^{-im\theta} \right).$$

Now using $\overline{b_m} = b_{-m}$ and the fact that $\{e^{im\theta}\}_{n \in \mathbb{Z} \setminus \{0\}}$ is an orthonormal basis proves the claim.

Exercise 6.22 By Proposition 6.4 we have $\omega(v, w) = g(v, w)$. Thus

$$\omega(x_k, x_l) = g(x_k, Jx_l) = g(x_k, y_l) = 0$$

$$\omega(y_k, y_l) = g(y_k, Jy_l) = -g(y_k, x_l) = 0$$

$$\omega(x_k, y_l) = g(x_k, Jy_l) = -g(x_k, x_l) = -\delta_{kl}$$

where δ_{kl} is the Kronecker delta function.

Exercise 6.23 We first show that Ψ is complex linear:

$$\Psi(Jv) = \frac{1}{\sqrt{2}} \left(Jv - iJ^2v \right) = \frac{1}{\sqrt{2}} i (v - iJv) = i\Psi(v).$$

In other words Ψ is complex linear with respect to the complex structure H_J.

Clearly $\Psi(v) = 0$ implies $v = 0$. Dimension counting implies that it is surjective.

Exercise 6.25 By Exercise 6.5 $g(Jv, Jw) = g(v, w)$ and $\omega(Jv, Jw) = \omega(v, w)$ for all v, w. We compute

$$g_+(\Psi(v), \Psi(w)) = \frac{1}{2} g_+(v - iJv, w - iJw)$$

$$= \frac{1}{2} (g(v, w) + g(iJv, iJw)) - \frac{1}{2} (g(v, iJw) + g(iJv, w))$$

$$= \frac{1}{2} (g(v, w) + g(Jv, Jw)) + \frac{1}{2} (ig(v, Jw) - ig(Jv, w))$$

$$= g(v, w) - ig(Jv, w) = g(v, w) - i\omega(v, w)$$

$$= \langle v, w \rangle.$$

Exercise 6.30 The expression for ω is just the (implicit) complex linear extension of (6.4). Similarly, the expression for g is just the sesquilinear extension of (6.3). We have

$$J\{b_m\} = \{\widehat{b_m}\}$$

where

$$\widehat{b}_m = \begin{cases} -i\,\overline{b_{-m}} & m > 0 \\ i\,\overline{b_{-m}} & m < 0. \end{cases}$$

So we compute

$$\omega(\{a_n\}, J\alpha\{b_n\}) = -i \sum_{n=-\infty}^{\infty} n a_n \widehat{b}_{-n}$$

$$= -i \sum_{n=1}^{\infty} n a_n (i\overline{b_n}) - i \sum_{n=-\infty}^{-1} -i n a_n \overline{b_n}$$

$$= g(\{a_n\}, \{b_n\}).$$

The fact that $L^+ \oplus L^-$ is a direct sum decomposition follows from the definition of $H^\flat$. Assume that $a_n = 0$ for all $n < 0$ and $b_n = 0$ for all $n > 0$. Then

$$g(\{a_n\}, \{b_n\}) = \sum_{n=-\infty}^{\infty} n a_n \overline{b_n} = 0.$$

On the other hand, if $a_n = 0$, $b_n = 0$ for all $n < 0$ and $b_n = 0$ for all $n < 0$,

$$\omega(\{a_n\}, \{b_n\}) = \sum_{n=-\infty}^{\infty} n a_n b_{-n} = 0$$

and similarly if $a_n = 0$, $b_n = 0$ for all $n > 0$.

Exercise 6.38 Let $v, w \in W^-$, and observe that

$$i\omega(v, \alpha w) = \overline{-i\omega(\alpha v, w)}.$$

Since $\alpha v, \alpha w \in W^+$, the claim follows from the fact that g is positive-definite on W^+.

Exercise 6.42 Assume that $u \in U(H)$. Then for all $v, w \in H$

$$g_J(uv, uw) = g(uv, uw) - i\omega(uv, uw) = g(v, w) - i\omega(v, w) = g_J(u, v).$$

Conversely, if $g_J(uv, uw) = g_J(v, w)$ for all $v, w \in H$, then u must preserve the real and imaginary parts of g_J. So $u \in O(H) \cap \mathrm{Sp}(H)$.

Exercise 6.44 An operator $u \in \mathrm{Sp}(H)$ preserves L^+ if and only if it commutes with J, which it does if and only if $u \in O(H)$; whence the stabilizer of $L^+ \in \mathrm{Pol}^\omega(H)$ is $U(H)$. Let $W^+ \in \mathrm{Pol}^\omega(H)$ be arbitrary. Pick an orthonormal basis $\{e_i\}$ for W^+ and observe that $\{\alpha e_i\}$ is an orthonormal basis for $\alpha(W^+)$. Let u be the complex-linear extension of the map that sends l_i to e_i and αl_i to αe_i. It is straightforward to see that $u \in \mathrm{Sp}(H)$, and that $u(L^+) = W^+$.

Exercise 6.46 Let $x \in W^+$ be arbitrary. Set $y = P^+x$. We then have $x = (P^+x, P^-x) = (y, P^-(P^+|_{W^+})^{-1}y)$, thus $x \in \text{graph}(Z)$. On the other hand, let $y \in L^+$ be arbitrary, and set $x = (P^+|_{W^+})^{-1}$. We then have $(y, Zy) = (P^+x, ZP^+x) = (P^+x, P^-x) = x$.

Exercise 6.53 The one-forms η_k were assumed to be harmonic, so β_k are all harmonic. By definition of Z, for each $k = 1, \dots, \mathfrak{g}$ there is a holomorphic one-form whose periods agree with those of β_k. By uniqueness of the harmonic representative β_k is holomorphic.

It remains to be shown that β_k are linearly independent. But if

$$\sum_{k=1}^{\mathfrak{g}} \lambda_k \beta_k = 0,$$

then integrating over each of the curves a_k, using the definition of η_k we obtain that $\lambda_k = 0$ for $k = 1, \dots, \mathfrak{g}$.

Exercise 6.55 Since u is invertible, we need only show that the expression in Eq. (6.15) is a left inverse for u. This follows immediately from Eq. (6.14).

Exercise 6.64 We compute using Proposition 6.54 that

$$u^{-1}Ju - J = i \begin{pmatrix} a^*a + b^*b - \mathbb{1} & a^*\alpha b\alpha + b^*\alpha a\alpha \\ -\alpha b^*\alpha a - \alpha a^*\alpha b & -\alpha(b^*b + a^*a - \mathbb{1})\alpha \end{pmatrix}.$$

By Eq. (6.14) we have

$$a^*\alpha b\alpha = b^*\alpha a\alpha \quad \text{and} \quad a^*a - \mathbb{1} = b^*b.$$

Inserting these in the above proves the claim.

Exercise 6.68 We have that $W_1^+ \to \alpha(W_1^+)$ is the zero map, thus $\sim$ is reflexive. Suppose that $W_1^+ \sim W_2^+$. Denote by $P_i^\pm$ the orthogonal projection $H \to W_i^\pm$. We have that $\iota_i^\pm = (P_i^\pm)^* : W_i^\pm \to H$ is the inclusion. This means that the orthogonal projection $W_1^+ \to \alpha(W_2^+)$ factors as $P_2^- \iota_1^+$; by assumption, this operator is Hilbert–Schmidt, thus so is its adjoint $(P_2^- \iota_1^+)^* = P_1^+ \iota_2^-$. Conjugating this operator by α, we obtain another Hilbert–Schmidt operator

$$\alpha P_1^+ \iota_2^- \alpha = \alpha P_1^+ \alpha^2 \iota_2^- \alpha = P_1^- \iota_2^+,$$

but the expression on the right-hand side is nothing but the orthogonal projection $W_2^+ \to \alpha(W_1^+)$, so $W_2^+ \sim W_1^+$, i.e. the relation is symmetric. Now, suppose that $W_1^+ \sim W_2^+$ and $W_2^+ \sim W_3^+$. This implies that the operators $P_2^- \iota_1^+$ and $P_3^- \iota_2^+$ are Hilbert–Schmidt. We have

$$P_3^- \iota_1^+ = P_3^- (\iota_2^+ P_2^+ + \iota_2^- P_2^-)\iota_1^+ = P_3^- \iota_2^+ P_2^+ \iota_1^+ + P_3^- \iota_2^- P_2^- \iota_1^+.$$

It follows that $P_3^- \iota_1^+$ is Hilbert–Schmidt, and we are done.

Exercise 6.71 Indeed, if (x, y) is perpendicular to both graph(Z) and $\alpha(\text{graph}(Z))$ we have

$$
\begin{aligned}
0 &= \langle (x, y), (v, Zv) \rangle + \langle (x, y), (\alpha Zw, \alpha w) \rangle \\
&= \langle (x, y), (v, Zv) \rangle + \langle (x, y), (-Z^*\alpha w, \alpha w) \rangle \\
&= \langle x, v - Z^*\alpha w \rangle + \langle y, Zv + \alpha w \rangle
\end{aligned}
$$

for all $v, w \in L^+$. By setting $v = x$ and $\alpha w = -Zx$ we obtain

$$
0 = \langle x, x + Z^*Zx \rangle = \|x\|^2 + \|Zx\|^2.
$$

which implies that $x = 0$. Similarly, one shows that $y = 0$.

Acknowledgments PK gratefully acknowledges support from the Pacific Institute for the Mathematical Sciences, and from the Hausdorff Center for Mathematics. ES acknowledges the support of the Natural Sciences and Engineering Research Council of Canada (NSERC).

References

1. H. Araki, Bogoliubov automorphisms and Fock representations of canonical anticommutation relations, in *Operator Algebras and Mathematical Physics*. Contemporary Mathematics, vol. 62 (American Mathematical Society, Providence, 1987), pp. 23–141. http://www.ams.org/conm/062/
2. R. Berndt, R. Berndt, *An Introduction to Symplectic Geometry*. Graduate Studies in Mathematics, vol. 26 (American Mathematical Society, Providence, 2001)
3. C. Birkenhake, H. Lange, *Complex Abelian Varieties*. Grundlehren der mathematischen Wissenschaften, vol. 302 (Springer, Berlin, 2004). https://doi.org/10.1007/978-3-662-06307-1
4. D. Borthwick, The Pfaffian line bundle. Commun. Math. Phys. **149**(3), 463–493 (1992). https://projecteuclid.org/journals/communications-in-mathematical-physics/volume-149/issue-3/The-Pfaffian-line-bundle/cmp/1104251304.full
5. H.M. Farkas, I. Kra, *Riemann Surfaces*. Graduate Texts in Mathematics, vol. 71 (Springer, New York, 1992). https://doi.org/10.1007/978-1-4612-2034-3
6. K. Habermann, L. Habermann, *Introduction to Symplectic Dirac Operators*. Lecture Notes in Mathematics, vol. 1887 (Springer, Berlin, 2006). https://doi.org/10.1007/b138212
7. Y.-Z. Huang, *Two-Dimensional Conformal Geometry and Vertex Operator Algebras* (Birkhäuser, Boston, 1995). https://doi.org/10.1007/978-1-4612-4276-5
8. P. Kristel, K. Waldorf, Fusion of implementers for spinors on the circle. Adv. Math. **402**, 108325 (2022). https://doi.org/10.1016/j.aim.2022.108325
9. A.A. Kirillov, D.V. Yuriev, Representations of the Virasoro algebra by the orbit method. J. Geom. Phys. **5**(3), 351–363 (1988). https://doi.org/10.1016/0393-0440(88)90029-0
10. O. Lehto, *Univalent Functions and Teichmüller Spaces*. Graduate Texts in Mathematics, (Springer, New York, 1987). https://doi.org/10.1007/978-1-4613-8652-0_3
11. G. Lion, M. Vergne, *The Weil Representation, Maslov Index and Theta Series*. Progress in Mathematics, vol. 6 (Birkhäuser, Boston, 1980). https://doi.org/10.1007/978-1-4684-9154-8

12. S. Nag, D. Sullivan, Teichmüller theory and the universal period mapping via quantum calculus and the $H^{1/2}$ space on the circle. Osaka J. Math. **32**(1), 1–34 (1995). Publisher: Osaka University and Osaka Metropolitan University, Departments of Mathematics. https://projecteuclid.org/journals/osaka-journal-of-mathematics/volume-32/issue-1/Teichm%c3%bcller-theory-and-the-universal-period-mapping-via-quantum-calculus/ojm/1200785862.full

13. S. Nag, A. Verjovsky, Diff(S^1) and the Teichmüller spaces. Commun. Math. Phys. **130**(1), 123–138 (1990). Publisher: Springer. https://projecteuclid.org/journals/communications-in-mathematical-physics/volume-130/issue-1/rm-DiffS1-and-the-Teichm%c3%bcller-spaces/cmp/1104187932.full

14. J.T. Ottesen, *Infinite Dimensional Groups and Algebras in Quantum Physics*. Lecture Notes in Physics, vol. 27 (Springer, Berlin, 1995)

15. R.J. Plymen, P.L. Robinson, *Spinors in Hilbert Space*. Cambridge Tracts in Mathematics, vol. 114 (Springer, Berlin, 1994)

16. A. Pressley, G. Segal, *Loop Groups*. Oxford Mathematical Monographs (Clarendon Press, Oxford, 2003), repr. (with corr.) edition

17. D. Radnell, E. Schippers, Quasisymmetric sewing in rigged teichmüller space. Commun. Contemp. Math. **8**(4), 481–534 (2006). Publisher: World Scientific Publishing Co. https://doi.org/10.1142/S0219199706002210

18. G. Segal, Unitary representations of some infinite-dimensional groups. Commun. Math. Phys. **80**(3), 301–342 (1981). https://projecteuclid.org/journals/communications-in-mathematical-physics/volume-80/issue-3/Unitary-representations-of-some-infinite-dimensional-groups/cmp/1103919978.full

19. G.B. Segal, The definition of conformal field theory, in *Differential Geometrical Methods in Theoretical Physics* (Springer, Dordrecht, 1988), pp. 165–171. https://doi.org/10.1007/978-94-015-7809-7_9

20. A. Sergeev, *Lectures on Universal Teichmüller Space* (2014). ISBN: 9783037191415 9783037196410 ISSN: 2523-5176, 2523-5184. https://doi.org/10.4171/141

21. D. Shale, Linear symmetries of free boson fields. Trans. Am. Math. Soc. **103**(1), 149–167 (1962). https://doi.org/10.1090/S0002-9947-1962-0137504-6

22. C.L. Siegel, Symplectic geometry. Am. J. Math. **65**(1), 1–86 (1943). Publisher: Johns Hopkins University Press. https://doi.org/10.2307/2371774

23. C.L. Siegel, *Topics in Complex Function Theory. 2: Automorphic Functions and Abelian Integrals*, vol. 5, print edition (Wiley, New York, 1988)

24. E. Schippers, W. Staubach, Analysis on quasidisks: a unified approach through transmission and jump problems. EMS Surv. Math. Sci. **9**(1), 31–97 (2022). https://doi.org/10.4171/emss/53

25. R.G. Swan, Vector bundles and projective modules. Trans. Am. Math. Soc. **105**(2), 264–277 (1962). https://doi.org/10.1090/S0002-9947-1962-0143225-6

26. L.A. Takhtajan, L.-P. Teo, *Weil-Petersson Metric on the Universal Teichmüller Space*. Memoirs of the American Mathematical Society, vol. 183 (American Mathematical Society, Providence, 2006). ISSN: 0065-9266, 1947-6221 Issue: 861. https://doi.org/10.1090/memo/0861

27. I. Vaisman, *Symplectic Geometry and Secondary Characteristic Classes*. Progress in Mathematics (Birkhäuser, Boston, 1987). https://doi.org/10.1007/978-1-4757-1960-4

28. S.K. Vodop'yanov, Mappings of homogeneous groups and imbeddings of functional spaces. Siberian Math. J. **30**(5), 685–698 (1989). https://doi.org/10.1007/BF00971258

29. A. Weinstein, Symplectic manifolds and their Lagrangian submanifolds. Adv. Math. **6**(3), 329–346 (1971). https://doi.org/10.1016/0001-8708(71)90020-X

30. P. Woit, *Quantum Theory, Groups and Representations* (Springer, Cham, 2017). https://doi.org/10.1007/978-3-319-64612-1

Chapter 7
Metric Characterizations
of Projective-Metric Spaces

Árpád Kurusa

Abstract This chapter is concerned with the study of projective-metric spaces, that is, metrics on open subsets of projective space whose geodesics are the intersection of this open set with the lines of the ambient space. The stress is on the effect of additional conditions on these so-called "projective-metric spaces", which lead to some characterization of special geometries. We rely heavily on the work of Herbert Busemann in this domain. We formulate many open problems on this subject.

Keywords Metric characterizations · Projective-metric spaces · Mikowski geometry · Hilbert geometry · Busemann spaces · Constant curvature spaces · Ellipses · Hyperbolas · Conics

1991 Mathematics Subject Classification 52A41, 53C60, 51F99, 53A40

7.1 Introduction

At the International Congress of Mathematicians in Paris in 1900, D. Hilbert raised numerous far reaching problems of mathematics. The fourth problem, in the language of today's mathematics, was to construct all the projective-metric spaces and to give a systematic geometric treatment of these spaces [39]. Due to the works in [8, 13, 14, 37, 79, 84], today we can construct all the projective-metric spaces

The research leading to these results has received funding from the national projects TKP2021-NVA-09 and NKFIH-1279-2/2020. Project no. TKP2021-NVA-09 has been implemented with the support provided by the Ministry of Innovation and Technology of Hungary from the National Research, Development and Innovation Fund, financed under the TKP2021-NVA funding scheme. Project no. NKFIH-1279-2/2020 has been implemented with the support provided by the Ministry for Innovation and Technology of Hungary (MITH) under grant NKFIH-1279-2/2020.

Á. Kurusa (✉)
Bolyai Institute, University of Szeged, Szeged, Hungary
e-mail: kurusa@math.u-szeged.hu

225

through the integral geometric way called Blaschke–Busemann-construction [84]. However this does not help much in giving a systematic geometric description of these spaces. Instead, it shows how immense is the number of such spaces. As Busemann noted, "the second part of the problem is not a well posed question and has inevitably been replaced by the investigation of special, or special classes of, interesting geometries." [18]

Instead of investigating "special, or special classes of, interesting" projective-metric spaces, in this chapter, we investigate the effect of additional conditions on the projective-metric spaces, which usually lead to some characterization of special geometries. There are numerous such characterizations among special classes of projective-metric spaces (see for instance [2, 83] and [36]), also characterizations of special classes among projective-metric spaces (see for instance [12, 19]), but our goal now is to survey those metric characterizations of the projective-metric spaces which are based on metric properties of geometric or algebraic objects, like pairs of points or lines, triangles or quadrics.

From a different point of view, our aim can also be described as to assess how far the known metric descriptions of objects in Euclidean geometry extend among the projective-metric spaces. Since Euclidean geometry has an endless set of such descriptions, this survey leads inevitably to a set of open problems which are formulated in almost every topic.

Earlier researches in this direction subsided for various reasons, but returning to this circle of problems is nowadays justified by the new tools that had come to light in recent decades, and by the appearance of a new geometric point of view. The most important new methods come from geometric tomography [33], whose results and tools shed new light on many issues [57]. The novelty of the geometric approach (see, for instance, [51, 52, 52, 55] etc.) is to regard each classical geometry as the reference for the appropriate type of projective-metric spaces. For the projective-metric spaces of the hyperbolic, parabolic, and elliptic type, the reference classical geometries are the hyperbolic, the Euclidean, and the elliptic geometry, respectively.

The plan of this chapter is the following.

In the Sect. 7.2, a very concise description of the most important geometries is presented. This is for ensuring a uniform interpretation of the terms used later. Then, we consider the metric properties of the spaces in Sect. 7.3, and of the objects that are tied to pairs of points and lines in Sect. 7.4. Section 7.5 contains results about triangles in projective-metric spaces, and Sect. 7.6 deals with such metric descriptions of the quadrics known from Euclidean geometry.

7.2 Preliminaries

Points of $\mathbb{R}^n$ ($n \in \mathbb{N}$) are denoted as $A, B, \dots$, vectors are $\overrightarrow{AB}$ or $\boldsymbol{a}, \boldsymbol{b}, \dots$; however we use these latter notation also for points if the origin is fixed. The open segment with endpoints A and B is denoted by $\overline{AB}$. The open ray starting from A passing through B is $\overline{A}B$, and the line through A and B is denoted by AB. The Euclidean scalar product is $\langle \cdot, \cdot \rangle$, and the norm coming from it is $| \cdot |$.

The *affine ratio* $(A, B; C)$ of the collinear points A, $B \neq A$ and $C \neq B$ satisfies $(A, B; C)\overrightarrow{BC} = \overrightarrow{AC}$. The *cross ratio* of the collinear points A, $B \neq A$ and $C \neq B$, $D \neq B$, A, is $(A, B; C, D) = (A, B; C)/(A, B; D)$ [19, page 243].

We call a closed set *regular* if it is the closure of its nonempty open interior, and we call a compact, regular set a *body*. For a convex body $\mathcal{D} \in \mathbb{R}^2$ containing the origin $O \in \mathcal{D}$ we usually *polar parameterize* the boundary $\partial \mathcal{D}$ by $\boldsymbol{r} : [-\pi, \pi) \to \mathbb{R}^2$ of the form $\boldsymbol{r}(\varphi) = r(\varphi)\boldsymbol{u}_\varphi$, where $r = r_{O,\mathcal{D}}$ is the *radial function* of $\mathcal{D}$ at the point O, and $\boldsymbol{u}_\varphi = (\cos \varphi, \sin \varphi)$.

We call a curve *analytic* if its coordinates depend on its arc length analytically. The bounded open interior of a simple closed plane curve $\mathcal{C}$ is denoted by $\text{Int}(\mathcal{C})$.

Let $(\mathcal{M}, d)$ be a metric space structure on a set $\mathcal{M}$ with the metric d. If $\mathcal{M}$ is the real projective space $\mathbb{P}^n$ or an affine space $\mathbb{R}^n \subset \mathbb{P}^n$ or a proper open convex subset of $\mathbb{R}^n$ for some $n \in \mathbb{N}$, and if the metric d is complete and continuous with respect to the standard topology of $\mathbb{P}^n$, and if the non-empty intersection of $\mathcal{M}$ with the projective lines are *exactly*[1] the geodesic lines of d, then the metric d is called *projective*, and the pair $(\mathcal{M}, d)$ is called a *projective-metric space* of dimension n (see [19, p. 115] and [84, p. 188]). Projective-metric spaces are called of *elliptic*, *parabolic* or *hyperbolic type* according to whether $\mathcal{M}$ is $\mathbb{P}^n$, $\mathbb{R}^n$, or a proper convex subset of $\mathbb{R}^n$. The projective-metric spaces of the latter two types are called *straight* [12, p. 1].

The geodesic lines of a straight projective-metric space are isometric to a Euclidean straight line.

The geodesic lines of a projective-metric space of elliptic type are isometric to Euclidean circles, and these circles have equal lengths by Busemann [12, (31.1) Theorem], so we can, after multiplying the metric with an appropriate $\kappa > 0$, say that

the length of every geodesic in a projective-metric space of elliptic type is π.
An isometry between two projective-metric spaces maps the geodesic lines onto geodesic lines, therefore it is the restriction of a collineation [23, Theorem 3.1]. A collineation is, by von Staudt's theorem [40, (ii) Fundamental theorem of projective geometry, p. 30], a projective map of $\mathbb{P}^n$, so we obtain that

$$\textit{isometries of projective} - \textit{metric spaces}$$

$$\textit{are restrictions of projective maps.} \tag{7.2.1}$$

We call the projective-metric spaces of dimension 2 *projective-metric planes* and, if not explicitly said otherwise,

we shall confine ourselves to projective-metric planes.

---

[1] Some authors use the weaker condition that the segments are geodesics that makes every Hilbert metric a projective-metric space.

If ℓ is an $\mathcal{M}$-*line*, i.e. a non-empty intersection of a projective line with $\mathcal{M}$, and if the point $O \in \mathcal{M}$ is outside of ℓ, then we say that the point $F \in \ell$ is the *d-foot of* O on ℓ if $d(O, F) \leq d(O, P)$ for every $P \in \ell$. Although this definition allows more d-foots in general, by Busemann and Kelly [19, (21.9) pp. 122] we know that any given point has a unique foot on any given line if and only if the circles of $(\mathcal{M}, d)$ are *strictly convex*, meaning that the discs of the circles are strictly convex. From now on

we consider only projective-metric planes that have only strictly convex circles.

Perpendicularity of $\mathcal{M}$-lines is defined in [19, pp. 119–121]: an $\mathcal{M}$-line ℓ' intersecting the $\mathcal{M}$-line ℓ in a point F is said to be *d-perpendicular to* ℓ if F is a d-foot of O on ℓ for every $O \in \ell' \setminus \{F\}$. We denote this relation, which is not necessarily a symmetric one, by $\ell' \perp_d \ell$. Since $(\mathcal{M}, d)$ has only strictly convex circles, i.e., if every point has a unique foot on any $\mathcal{M}$-line, then for any given point $P \in \mathcal{M}$ and for an $\mathcal{M}$-line ℓ there exists a unique $\mathcal{M}$-line ℓ' such that it goes through P and $\ell' \perp_d \ell$ [19, (21.7) pp. 121].

7.2.1 Some Classes of Projective-Metric Spaces

Before presenting the most important projective-metric spaces, we construct now a particular projective-metric space to demonstrate how general this concept is.

Let $\mathcal{M} = \{(x, y) \in \mathbb{R}^2 : -\pi/2 < x < \pi/2\}$. Let $d \colon \mathcal{M} \times \mathcal{M} \to \mathbb{R}_+$ be defined by $d((x, y), (z, t)) = e((x, y), (z, t)) + |\tan x - \tan z|$, where e is the usual Euclidean distance. Then $(\mathcal{M}, d)$ is a projective-metric plane [19, p. 116], and from the proof it is clear that the metric e could be any Minkowski metric.

The most important examples of projective-metric spaces are clearly the spaces of constant curvature. Next to them stand their straight generalizations, the Minkowski and Hilbert geometries. These are the only noncompact projective-metric spaces with compact spheres in which every isometry of one geodesic on another or itself is a projectivity [15, (3) Theorem on p. 38].

7.2.1.1 Classical Geometries

By Beltrami's theorem [6] Riemannian projective-metric spaces have constant curvature, so, after normalizing the curvatures, they are models of the spaces of constant curvature $\kappa = +1, 0, -1$. It is easy to see [53] that the gnomonic projection $\tilde{\tau}$ [35] of the well-known quadratic models $\mathbb{K}_\kappa^n$ sends a point of $\mathbb{K}_\kappa^n$ given in polar coordinates $(\boldsymbol{u}, r)$ at a point $O \in \mathbb{K}_\kappa^n$ to the point $(\boldsymbol{u}, \tau_\kappa(r))$, where τ_κ is called the *projector function*, given in polar coordinates of the tangent space $T_O \mathbb{K}_\kappa^n \simeq \mathbb{R}^n$. Further, $\mathbb{K}_\kappa^n$ is a *rotational manifold* [41], so it is determined by the *size function* σ_κ giving the radius $\sigma_\kappa(r)$ of the Euclidean sphere that is isometric to the geodesic sphere of radius r in $\mathbb{K}_\kappa^n$. This determines the projector function

$\tau_\kappa : \mathbb{R}_+ \ni r \mapsto \tau_\kappa(r) = \frac{\sigma_\kappa(r)}{\eta_\kappa(r)}$, where $\eta_\kappa(r) = \sqrt{1 - \kappa \sigma_\kappa^2(r)}$. The projector function of every constant curvature space is given in the following table, where also the *injectivity radius* $\imath$, the upper limit of r until which the polar coordinatization keeps injectivity, is indicated.

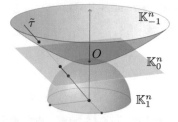

$\mathbb{K}_\kappa^n$	$\imath$	$\sigma_\kappa(r)$	$\tau_\kappa(r)$	$\eta_\kappa(r)$
$\kappa = -1$	∞	$\sinh r$	$\tanh r$	$\cosh r$
$\kappa = 0$	∞	r	r	1
$\kappa = 1$	$\pi/2$	$\sin r$	$\tan r$	$\cos(r)$

Now it is clear that the gnomonic projection of the quadratic models of constant curvature spaces transforms the constant curvature spaces into projective-metric spaces. For instance, this transforms hyperbolic geometry into the Beltrami–Cayley–Klein model [22, § 7]. We call these simply a model of the classical geometries or more specifically a model of Euclidean geometry, model of hyperbolic geometry, or model of elliptic geometry.

It is worth noticing that the metric d in these quadratic models, in the projective-metric spaces, satisfies

$$\eta_\kappa(d(\boldsymbol{x}, \boldsymbol{y})) = \frac{(1 + \kappa \langle \boldsymbol{x}, \boldsymbol{y} \rangle)^2}{(1 + \kappa |\boldsymbol{x}|^2)(1 + \kappa |\boldsymbol{y}|^2)} \quad \text{if } \kappa \neq 0,$$

$$d(\boldsymbol{x}, \boldsymbol{y}) = |\boldsymbol{x} - \boldsymbol{y}| \quad \text{if } \kappa = 0. \tag{7.2.2}$$

∎

7.2.1.2   Minkowski Geometries

Such geometries were first constructed by Minkowski in [72], and in general these are the real finite-dimensional Banach spaces. Let $\mathcal{I}$ be the open interior of a strictly convex body in $\mathbb{R}^n$, symmetric with respect to the origin $\boldsymbol{0}$. The function $d \colon \mathbb{R}^n \times \mathbb{R}^n \to \mathbb{R}$ defined by

$$d(\boldsymbol{x}, \boldsymbol{y}) = \begin{cases} 0, & \text{if } \boldsymbol{x} = \boldsymbol{y}, \\ |(\boldsymbol{x} - \boldsymbol{y}, \boldsymbol{a}; \boldsymbol{0})|, \text{ where } \overline{\boldsymbol{ab}} = \mathcal{I} \cap \boldsymbol{0}(\boldsymbol{x} - \boldsymbol{y}), & \text{if } \boldsymbol{x} \neq \boldsymbol{y}, \end{cases}$$

is a metric on $\mathbb{R}^n$ [19, IV.24], and is called *Minkowski metric*. Such a pair $(\mathbb{R}^n, d)$ is called *Minkowski geometry (plane* if $n = 2$), and $\mathcal{I}$ is called the *indicatrix*.

A Minkowski geometry is a model of the Euclidean geometry if and only if $\partial \mathcal{I}$ is an ellipse [19, 25.4].

By the Mazur–Ulam theorem [71] Minkowski geometries are isomorphic if and only if their indicatrices are affinely equivalent.

Let $\mathcal{I}_x = \mathcal{I} + x$ for every $x \in \mathbb{R}^n$. Then, by Busemann and Kelly [19, p. 137], ℓ' is the unique line through x, which is d-perpendicular to the line ℓ, if the lines tangent to $\mathcal{I}_x$ at points $\partial \mathcal{I}_x \cap \ell'$ are parallel to ℓ.

We say that a Minkowski plane is *analytic* if $\partial \mathcal{I}$ is an analytic curve.

For further reading we suggest [88] and [69].

■

7.2.1.3   Hilbert Geometries

Such geometries were first constructed by Hilbert in [38]. Let $\mathcal{M}$ be an open, strictly convex set in $\mathbb{R}^n$. The function $d \colon \mathcal{M} \times \mathcal{M} \to \mathbb{R}$ defined by $d(X, X) = 0$, and by

$$d(X, Y) = \frac{1}{2}\bigl|\ln(X, Y; A, B)\bigr|, \text{ where } \overline{AB} = \mathcal{M} \cap XY, \text{ if } X \neq Y,$$

is a metric on $\mathcal{M}$ [19, page 297], and is called *Hilbert metric*. Such a pair $(\mathcal{M}, d_{\mathcal{M}})$ is called *Hilbert geometry* (*plane* if $n = 2$), and $\mathcal{M}$ is called the *domain of the geometry*.

It is known that a Hilbert geometry is a model of Bolyai's hyperbolic geometry if and only if the boundary of its domain is an ellipsoid [19, (29.3)].

It is known, by Busemann and Kelly [19, (28.11)], that in a Hilbert plane a geodesic line ℓ' is d-perpendicular to the geodesic line ℓ if the Euclidean line containing ℓ passes through the intersection of those tangents of $\mathcal{M}$, which touch $\mathcal{M}$ at the points $\ell' \cap \partial \mathcal{M}$.

We say that a Hilbert plane is *twice differentiable* or *analytic* if the boundary $\partial \mathcal{M}$ is a twice differentiable or analytic curve.

For further reading we suggest [76].

■

7.2.1.4   Finslerian Projective-Metric Spaces

We call a projective-metric space Finslerian, if the metric d is generated by a Finsler function [5].

The Minkowski geometries are clearly Finslerian manifolds, but there are many more Finslerian projective-metric spaces—compare this to the fact that only the constant curvature spaces are Riemannian projective-metric spaces, by Beltrami [6]. For instance, each Hilbert geometry is a Finslerian manifold too by Busemann and Kelly [19, (29.6)]. In more details, identifying the tangent spaces $T_P\mathcal{M}$ with $\mathbb{R}^n$ through the mapping $\iota_P \colon v \mapsto P + v$, the Finsler function $F_{\mathcal{M}} \colon \mathcal{M} \times \mathbb{R}^n \to \mathbb{R}$

associated with the Hilbert metric d at a point $P \in \mathcal{M}$ is given by

$$F_{\mathcal{M}}(P, \boldsymbol{v}) = \frac{1}{2}\left(\frac{1}{\lambda_{\boldsymbol{v}}^-} + \frac{1}{\lambda_{\boldsymbol{v}}^+}\right), \tag{7.2.3}$$

where $\boldsymbol{v} \in T_P\mathcal{M}$, and $\lambda_{\boldsymbol{v}}^{\pm} \in (0, \infty]$ is such that $P_{\boldsymbol{v}}^{\pm} := P \pm \lambda_{\boldsymbol{v}}^{\pm}\boldsymbol{v} \in \partial\mathcal{M}$ [19, (50.4)].[2]   ■

7.2.2   Reference Functions and Ratios

According to our geometric approach, we define the *reference functions*, σ, τ and η, for each projective-metric space $(\mathcal{M}, d)$ as follows:

Type of $(\mathcal{M}, d)$	ι	$\sigma(r)$	$\tau(r)$	$\eta(r)$
Hyperbolic	∞	$\sinh r$	$\tanh r$	$\cosh r$
Parabolic	∞	r	r	1
Elliptic	$\pi/2$	$\sin r$	$\tan r$	$\cos r$

Clearly, $\sigma(r)$, $\tau(r)$ and $\eta(r)$ are $\sigma_\kappa(r)$, $\tau_\kappa(r)$ and $\eta_\kappa(r)$, respectively, according to the curvature κ of the classical geometry being in the respective type of the projective-metric space. We keep the names and call $r \mapsto \sigma(r)$ the *size function*, and $r \mapsto \tau(r)$ the *projector function*.

Let A, B be different points in a projective-metric space $(\mathcal{M}, d)$ and let $C \in (AB \cap \mathcal{M}) \setminus \{B\}$. We define different ratios of the triple (A, B, C) as follows:

$$\langle A, B; C\rangle_d = \begin{cases} \frac{d(A,C)}{d(C,B)} & \text{if } C \in \overline{AB}, \\ -\frac{d(A,C)}{d(C,B)} & \text{otherwise} \end{cases} \tag{7.2.4}$$

is the *metric ratio*;

$$\langle A, B; C\rangle_d^\circ = \begin{cases} \frac{\sigma(d(A,C))}{\sigma(d(C,B))} & \text{if } C \in \overline{AB}, \\ -\frac{\sigma(d(A,C))}{\sigma(d(C,B))} & \text{otherwise} \end{cases} \tag{7.2.5}$$

is the *size-ratio* [55];

$$\langle A, B; C\rangle_d' = \begin{cases} \frac{\tau(d(A,C))}{\tau(d(C,B))} & \text{if } C \in \overline{AB}, \\ -\frac{\tau(d(A,C))}{\tau(d(C,B))} & \text{otherwise} \end{cases} \tag{7.2.6}$$

is the *tangential ratio* [61].

---

[2] If $\lambda_{\boldsymbol{v}}^{\pm} = \infty$, then $P_{\boldsymbol{v}}^{\pm}$ is an ideal point.

Observe that in Minkowski geometries all these ratios coincide with the affine ratio. In the classical geometries the tangential ratio $\langle A, B; O \rangle'_d$ is nothing else but the affine ratio of the gnomonically projected points $\tilde{\tau}(A)$, $\tilde{\tau}(B)$, and $\tilde{\tau}(O) = O$, i.e. $\langle A, B; O \rangle'_d = (\tilde{\tau}(A), \tilde{\tau}(B); O)$.

7.3 Metric Properties of the Space

7.3.1 Curvature

The *curvature in the sense of Busemann at a point O in a projective-metric space* $(\mathcal{M}, d)$ is said to be *positive, non-negative, non-positive,* and *negative* respectively if there exists a neighborhood $\mathcal{U}$ of O such that for every pair of points $P, Q \in \mathcal{U}$ we be

$$2d(\hat{P}, \hat{Q}) > d(P, Q), \qquad 2d(\hat{P}, \hat{Q}) \geq d(P, Q),$$

$$2d(\hat{P}, \hat{Q}) \leq d(P, Q), \qquad 2d(\hat{P}, \hat{Q}) < d(P, Q),$$

respectively, where $\hat{P}, \hat{Q}$ are the respective midpoints of the segments $\overline{OP}$ and $\overline{OQ}$. We say that $(\mathcal{M}, d)$ has *zero curvature in the sense of Busemann* at the point O if it has both *non-negative* and *non-positive* curvature in the sense of Busemann at O. If neither of the cases is satisfied in any neighborhood of O, then $(\mathcal{M}, d)$ has *indeterminate curvature* at O. For the relations of the Busemann curvature to other terms of curvature see [1]. From now on, we deal only with the curvature in the sense of Busemann, so we simply call it curvature.

It is clear that every Minkowski geometry has zero curvature, but the situation changes considerably for Hilbert geometries, because, although the curvature cannot be positive or non-negative at any point [56, Theorem 4.1] (see also [49]), there may be some points of indeterminate curvature. Still, we have the following improvement of [47, the second statement of Theorem].

Theorem 7.3.1 *If the border of a Hilbert plane is twice differentiable and has non-positive curvature at two points, then it is a Cayley–Klein model of the hyperbolic geometry.*

Proof By Kurusa [56, Theorem 4.1] (an improvement of [47, the first statement of Theorem]), a Hilbert geometry $(\mathcal{M}, d)$ has non-positive curvature at a point O if and only if O is a *projective center* of $\mathcal{M}$, i.e. there is a projectivity ϖ such that $\varpi(O)$ is the center of $\varpi(\mathcal{M})$.

It is clear that a point is a metric center of a Hilbert plane $(\mathcal{M}, d)$ if and only if it is a projective center of $\mathcal{M}$.

By Kelly and Straus [48, Theorem 3], a Hilbert plane $(\mathcal{M}, d)$ with twice differentiable boundary has at least two metric centers if and only if it is a Cayley–Klein model of the hyperbolic geometry.   □

To further emphasize the importance of the curvature, we note that there are Hilbert planes $(\mathcal{M}, d)$ with points of indeterminate curvatures in which there are two geodesics with parameterizations $g\colon (a, b) \to \mathcal{M}$ and $f\colon (c, d) \to \mathcal{M}$ proportional to arc length such that the mapping $(a, b) \times (c, d) \ni (x, y) \mapsto (g(x), f(y)) \in \mathbb{R}$ is not convex.[3] This also seems to happen at the boundary [24].   ∎

7.3.2   Ptolemaic Projective-Metric Spaces

Ptolemy's inequality [26, § 2.6] states in the Euclidean plane that

$$d(P, Q)d(R, S) + d(Q, R)d(S, P) \geq d(P, R)d(Q, S) \tag{7.3.1}$$

holds for any four points P, Q, R, and S, and equality occurs if and only if the points P, Q, R, and S lie in cyclic order on a circle.

A metric space $(\mathcal{M}, d)$, and so a projective-metric space too, is called *Ptolemaic* if (7.3.1) is satisfied for all points $P, Q, R, S \in \mathcal{M}$.

It is known (see [11, Proposition 3.1]) that the hyperbolic plane is Ptolemaic, but, a bit surprisingly in this context, the sphere is not even locally Ptolemaic [9, p. 80, exercise 5].

It is well-known that a Ptolemaic Minkowski geometry is a model of Euclidean geometry [82], and a Ptolemaic Hilbert geometry is a model of the hyperbolic geometry [44]. We can generalize these to the Finslerian projective-metric spaces using the idea of Kay [43].

Theorem 7.3.2 *An everywhere locally Ptolemaic Finslerian projective-metric space is a model of Euclidean or hyperbolic geometry.*

Proof First, we show that the tangent spaces inherit the ptolemaic property from the underlying Finslerian projective-metric space $(\mathcal{M}, d)$.

Let O be any point in $\mathcal{M}$ such that it has a convex neighborhood $\mathcal{N}$ about O in which d is ptolemaic. Let δ be the Minkowski metric in the tangent space $T_O \mathcal{M}$.

Let $P = \exp(\boldsymbol{p})$, $Q = \exp(\boldsymbol{q})$, $R = \exp(\boldsymbol{r})$, and $S = \exp(\boldsymbol{s})$ be four points in $\mathcal{N}$, and define for every $\lambda \in (0, 1)$ the points $P_\lambda = \exp(\lambda \boldsymbol{p})$, $Q_\lambda = \exp(\lambda \boldsymbol{q})$, $R_\lambda = \exp(\lambda \boldsymbol{r})$, and $S_\lambda = \exp(\lambda \boldsymbol{s})$. Then we have

$$d(P_\lambda, Q_\lambda)d(R_\lambda, S_\lambda) + d(Q_\lambda, R_\lambda)d(S_\lambda, P_\lambda) \geq d(P_\lambda, R_\lambda)d(Q_\lambda, S_\lambda)$$

---

[3] This means that *Busemann convexity* [75, Definition 8.1.1] is not satisfied in this space.

which implies

$$\frac{d(P_\lambda, Q_\lambda)}{\delta(\lambda p, \lambda q)} \frac{d(R_\lambda, S_\lambda)}{\delta(\lambda r, \lambda s)} \frac{\delta(\lambda p, \lambda q)\delta(\lambda r, \lambda s)}{\delta(\lambda p, \lambda r)\delta(\lambda q, \lambda s)}$$

$$+ \frac{d(Q_\lambda, R_\lambda)}{\delta(\lambda q, \lambda r)} \frac{d(S_\lambda, P_\lambda)}{\delta(\lambda s, \lambda p)} \frac{\delta(\lambda q, \lambda r)\delta(\lambda s, \lambda p)}{\delta(\lambda p, \lambda r)\delta(\lambda q, \lambda s)}$$

$$\geq \frac{d(P_\lambda, R_\lambda)}{\delta(\lambda p, \lambda r)} \frac{d(Q_\lambda, S_\lambda)}{\delta(\lambda q, \lambda s)}$$

According to the Busemann–Mayer theorem [20, Theorem 4.3], taking the limit as $\lambda \to 0$ yields

$$\frac{\delta(\lambda p, \lambda q)\delta(\lambda r, \lambda s)}{\delta(\lambda p, \lambda r)\delta(\lambda q, \lambda s)} + \frac{\delta(\lambda q, \lambda r)\delta(\lambda s, \lambda p)}{\delta(\lambda p, \lambda r)\delta(\lambda q, \lambda s)} \geq 1,$$

the desired inequality.

Since, by Schoenberg [82, Theorem 1], a Ptolemaic Minkowski geometry is a model of the Euclidean geometry, we conclude that $(T_O\mathcal{M}, \delta)$ is a model of the Euclidean geometry, hence $(\mathcal{M}, d)$ is a Riemannian projective-metric space. By Beltrami's theorem [6] this means that $(\mathcal{M}, d)$ is a model of a classical geometry. Since the sphere is not Ptolemaic, the statement of the theorem follows. □

Open Question 7.3.3 *Is every Ptolemaic projective-metric space a model of Euclidean or hyperbolic geometry?*

According to [32, Theorem 1.3] if a projective-metric space is Ptolemaic and Busemann convex, then it should be a CAT(0) metric space [10]. ∎

7.3.3 Symmetry of the Perpendicularity of Lines

It is very well known that perpendicularity of lines in the classical geometries is a symmetric relation. However, this is not the case for more general projective-metric spaces.

Radon curves were introduced and constructed by Radon in [80], where he proved that these are the boundaries of the indicatrices in Minkowski planes with symmetric perpendicularity.

Theorem 7.3.4 ([68]) *In an n-dimensional Minkowski geometry* $(n \geq 2)$, *the perpendicularity is a symmetric relation if and only if either* $n \geq 3$ *or* $n = 2$ *and the boundary of the indicatrix is a Radon curve.*

For a survey on *Radon planes*, Minkowski planes with an indicatrix bounded by a Radon curve, see [69, Section 6] and [4].

In 1952 a more rigid result arrived.

Theorem 7.3.5 ([46, Theorem 2]) *In a Hilbert plane perpendicularity is a symmetric relation if and only if the Hilbert plane is a model of hyperbolic plane.*

Later it was proved in [44, Theorem 2] that perpendicularity in a Hilbert geometry is symmetric for two lines if and only if the perpendicularity of these two lines is also symmetric with respect to the Minkowski geometry in the tangent space at the intersection of the lines. This can be generalized to the Finslerian projective-metric planes.

Theorem 7.3.6 *In a Finslerian projective-metric plane $(\mathcal{M}, d)$ perpendicularity at any point coincides with perpendicularity in the tangent Minkowski plane.*

Proof Let O be any point in $\mathcal{M}$ such that it has a convex neighborhood $\mathcal{N}$ about O. Let δ be the Minkowski metric in the tangent space $T_O\mathcal{M}$.

Let $x, y \in T_O\mathcal{M}$, and let $\ell_x = \{\lambda x : \lambda \in \mathbb{R}\}$ and $\ell_y = \{\lambda y : \lambda \in \mathbb{R}\}$. Define the geodesic lines $\mathcal{L}_x = \exp \ell_x$ and $\mathcal{L}_y = \exp \ell_y$ in $\mathcal{M}$, on which the points $X_\lambda = \exp(\lambda x)$ and $Y_\lambda = \exp(\lambda y)$ lie for every $\lambda \in (0, 1)$.

Observe that the triangles $\mathbf{0}(\lambda x)(\lambda y)\triangle$ are all similar to $\mathbf{0}x y\triangle$ in the Minkowski plane $(T_O\mathcal{M}, \delta)$, hence $\frac{\delta(x, \mathbf{0})}{\delta(\lambda x, \mathbf{0})} = \frac{\delta(x, y)}{\delta(\lambda x, \lambda y)}$.

Assume that $\mathcal{L}_x$ is perpendicular to $\mathcal{L}_y$. This means $d(X_\lambda, O) \leq d(X_\lambda, Y_\lambda)$ for every $\lambda \in (0, 1)$. Then also

$$d(X_\lambda, O)\frac{\delta(x, \mathbf{0})}{\delta(\lambda x, \mathbf{0})} \leq \frac{\delta(x, y)}{\delta(\lambda x, \lambda y)}d(X_\lambda, Y_\lambda).$$

By the Busemann–Mayer theorem [20, Theorem 4.3], taking the limit of this as $\lambda \to 0$ yields $\delta(x, \mathbf{0}) \leq \delta(x, y)$. Thus, ℓ_x is perpendicular to ℓ_y.

Assume now that ℓ_x is perpendicular to ℓ_y. Let $\mathcal{L}'_x$ be the geodesic line in $(\mathcal{M}, d)$ that is perpendicular to $\mathcal{L}_y$. If $\mathcal{L}'_x = \exp(\ell'_x)$, then we proved above that ℓ'_x is perpendicular to ℓ_y. But perpendicularity is unequivocal in the Minkowski plane $(T_O\mathcal{M}, \delta)$, so $\ell'_x = \ell_x$. Thus $\mathcal{L}'_x = \mathcal{L}_x$ is perpendicular to $\mathcal{L}_y$.

The theorem is proved. □

As a corollary, we deduce that a Finslerian projective-metric space of dimension $n \geq 3$ with symmetric perpendicularity is Riemannian and hence of constant curvature, that is, it is a model of a classical geometry.

Open Question 7.3.7 *Is a Finslerian projective-metric plane with symmetric perpendicularity either a Radon plane, or a Riemannian projective-metric plane?*

Compare this question with Theorem 7.6.5 about the Riemannian points. ∎

7.3.4 Erdős Ratio

In the Euclidean plane for a point O in a triangle $ABC\triangle$ the *Erdős-ratio* [28] is

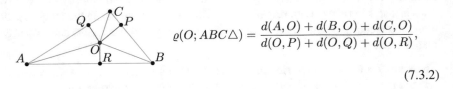

$$\varrho(O; ABC\triangle) = \frac{d(A,O) + d(B,O) + d(C,O)}{d(O,P) + d(O,Q) + d(O,R)},$$

$$(7.3.2)$$

where P, Q, R are the feet of the perpendiculars from O upon the sides $\overline{BC}$, $\overline{CA}$, $\overline{AB}$ respectively, where d is the Euclidean metric. The *Erdős–Mordell inequality* [73] states that $\varrho(O; ABC\triangle) \geq 2$ holds for any point O in any triangle $ABC\triangle$. For more details on this inequality see [34].

Assume $d(A, O), d(B, O), d(C, O) < \iota$ for a configuration $O \in ABC\triangle$ in a classical geometry. The gnomonic projection $\tilde{\tau}$ into the tangent space at O maps this configuration into a configuration $O \in A'B'C'\triangle$. Observe that the gnomonic projection $\tilde{\tau}$ keeps perpendicularity with respect to the lines through O, so it is natural to define the *Erdős-ratio* for classical geometries as

$$\varrho(O; ABC\triangle) = \frac{\tau(d(A, O)) + \tau(d(B, O)) + \tau(d(C, O))}{\tau(d(O, P)) + \tau(d(O, Q)) + \tau(d(O, R))}.$$

$$(7.3.3)$$

The following result is new for hyperbolic geometry only (for the elliptic case see [74]). Its proof is an obvious application of the Erdős–Mordell inequality to the configuration $O \in A'B'C'\triangle$.

Theorem 7.3.8 *If $d(A, O), d(B, O), d(C, O) < \iota$ holds for a configuration $O \in ABC\triangle$ in a classical geometry, then $\varrho(O; ABC\triangle) \geq 2$, and equality occurs if and only if the geodesic triangle $ABC\triangle$ is regular, and O is its center.*

Extending (7.3.3) to the projective-metric spaces raises the question how big or small the Erdős-ratio can be in general projective-metric spaces.

Since every τ is a strictly increasing function, and since P, Q, R, by definition, are the points closest to O in the respective sides of $ABC\triangle$, we have the natural lower bound 1 for the Erdős ratio, because

$$\varrho(O; ABC\triangle) = \frac{\tau(d(A, O)) + \tau(d(B, O)) + \tau(d(C, O))}{\tau(d(O, P)) + \tau(d(O, Q)) + \tau(d(O, R))}$$

$$\geq \frac{\tau(d(A, O)) + \tau(d(B, O)) + \tau(d(C, O))}{\tau(d(O, C)) + \tau(d(O, A)) + \tau(d(O, B))} = 1.$$

$$(7.3.4)$$

In a Minkowski plane, if $r_{\min}$ and $r_{\max}$ denote the Euclidean minimum and maximum radii of the indicatrix, then, by the observation of Ghandehari and Martini in [34], for every configuration the Erdős ratio is at least $2r_{\min}/r_{\max}$, which is better than the natural lower estimate (7.3.4) if $2r_{\min} > r_{\max}$.

We call the infimum of the Erdős ratios in a projective-metric plane $(\mathcal{M}, d)$ the *Erdős ratio of the projective-metric plane* $(\mathcal{M}, d)$, which we denote by $\rho(\mathcal{M}, d)$. By the above the Erdős ratio of a Minkowski plane is at least $\max(1, 2r_{\min}/r_{\max})$, and, by Theorem 7.3.8, it is 2 if the metric is Euclidean.

Conjecture 7.3.9 *The Erdős ratio of a Minkowski plane is* $2r_{\min}/r_{\max}$.

This is supported by the example in [34, Fig. 3] which shows that the Minkowski plane of the l_1 norm has a configuration such that $\varrho(O; ABC\triangle) = 3/2$ while $2r_{\min}/r_{\max} = \sqrt{2}$.

Notice that if Conjecture 7.3.9 is true, then only Euclidean geometry has the maximal Erdős ratio of 2 among the Minkowski planes. The analogous intriguing question for the Hilbert plane is the following.

Open Question 7.3.10 *Does the hyperbolic plane have the maximal Erdős ratio among the Hilbert planes?*

Since a tangent plane of a Hilbert plane is a Minkowski plane, an affirmative answer to the previous question would certainly support the conjecture that the answer is affirmative for the former question.

Similar questions seem quite hard for the general projective-metric planes. ∎

7.3.5 Regular Polygons

An n-gon $P_1 P_2 \ldots P_n$ in Euclidean plane $(n \in \mathbb{N})$ is called k-equilateral if $d(P_i, P_j) = d(P_{1+h}, P_{j+h})$ for all $h = 1, 2, \ldots, n - 1$ and for $1 < j \leq k + 1$ where the indices are taken mod n. An $(n-1)$-equilateral n-gon is said to be *totally equilateral*, or *regular* [50].

It is obvious that for any pair of points A and B, there are two regular n-gons with A and B as adjacent vertices in the Euclidean plane for every $n \in \mathbb{N}$, because there are two points C and C' forming isosceles triangles with $\overline{AB}$ as base such that the angles at C and C' are $2\pi/n$. This construction also works in the hyperbolic plane, but in the elliptic plane there are obvious constraints.

If two points A and B are given in a straight projective-metric plane $(\mathcal{M}, d)$, let $\mathcal{C}_A(B)$ be the circle with center A and radius $d(A, B)$. By convexity, the circle $\mathcal{C}_A(B)$ intersects the line AB in a point, say B', different from B. Considering a point C moving from B to B' along one of the arcs of $\mathcal{C}_A(B)$, we see that $d(C, A) = d(A, B)$ and $d(C, B)$ changes continuously from $0 = d(B, B)$ to $d(B, B') = 2d(A, B)$, hence there is a point C such that the triangle $ABC\triangle$ is equilateral. So there are at least two regular triangles with any two given vertices in every straight projective-metric plane.

Similar reasoning seems to work also for constructing regular quadrangles in every straight projective-metric plane $(\mathcal{M}, d)$. Let two points A and B be given, let E be one of the points in $\mathcal{C}_A(B) \cap \mathcal{C}_B(A)$, and let $A' \neq A$ and $B' \neq B$ be the points where $\mathcal{C}_B(A)$ and $\mathcal{C}_A(B)$ intersect the line AB, respectively. Considering a point C moving from E to A' along the arc of $\mathcal{C}_B(A)$ outside of $\mathcal{C}_A(B)$, we see that $d(A, C)$ changes continuously from $d(A, E) = d(B, E) = d(A, B)$ to $d(A, B') = 2d(A, B)$. Further, a point D moving from E to B' along the arc of $\mathcal{C}_A(B)$ outside of $\mathcal{C}_B(A)$, we see that $d(B, D)$ changes continuously from $d(B, E) = d(A, E) = d(A, B)$ to $d(B, A') = 2d(A, B)$. At the same time, $d(D, C)$ changes continuously from 0 to $d(A', B') = 3d(A, B)$. Thus, moving D so that $d(B, D) = d(A, C)$ holds is possible, and $d(C, D) = d(A, B)$ at a position of C, hence $ABCD$ is a regular quadrangle.

Open Question 7.3.11 *Is a projective-metric space Euclidean or hyperbolic if for every $n \in \mathbb{N}$ and any pair of points A and B, there is a regular n-gon with A and B as adjacent vertices?*[4]   ∎

7.4 Metrically Defined Objects

Most classical objects, like bisectors, medians etc., that are metrically determined by point-sets in classical geometries, have numerous different, but equivalent definitions. However, these definitions lead to different objects with different properties in most projective-metric spaces. In this section we review some of these.

7.4.1 Bisector of Point Pairs

Given two different points A and B in metric spaces we call the set of those points which are equidistant with respect to A and B the *bisector* of A and B.

It is so clear that the bisectors are flat in Euclidean geometry, that in 1849 Leibniz suggested in [64, p. 166] to define a plane as the locus of points equidistant from two given points in the space. However, the effect that such a definition would cause turned out only in 1955.

In a long proof of [12, Theorem (47.4)], Busemann showed that if each bisector contains the segments of its any two points, then the projective-metric space is a model of the Euclidean, hyperbolic, or elliptic geometry.   ∎

---

[4] To investigate regular polygons in projective-metric spaces was an idea of Gábor Korchmáros raised amid an oral conversation in 2018.

7.4.2 Median Rays of Point Pairs

Take non-collinear triple of points A, B, and C, and let M be the midpoint of A and B. We call the ray $\overline{M}C$ the *median ray* of the triple (A, B, C).

In any projective-metric plane, obviously, for any point $P \in \overline{M}C$ there are unique points A' and B' on $\overline{P}M$ such that $d(A', P) = d(P, A)$ and $d(B', P) = d(P, B)$, respectively. It is clear, that A' is the intersection of the ray $\overline{P}M$ and the circle $\mathcal{C}_P(A)$ centered at P and going through A, and $B' = \overline{P}M \cap \mathcal{C}_P(B)$ too.

Consider a Minkowski plane with differentiable indicatrix. Then $\mathcal{C}_P(A)$ has a tangent line t_A through A which intersects the line $\overline{P}M$, say in the point A^*. Let B^* be the point where the tangent t_B of $\mathcal{C}_P(B)$ through B intersects $\overline{P}M$. It is clear, that as P tends to infinity AP and BP tend to lines through A and B that are parallel to MC. This means that B^* and B' tend to the same point, say $B^\#$, furthermore A^* and A' tend to the same point, say $A^\#$. Further, the lines BB^* and AA^* tend also to parallel lines, because AP and BP tend to lines that are parallel to MC. Since M is the midpoint of the segment $\overline{AB}$, it is also the midpoint of $\overline{A^\#B^\#}$, so $d(M, P) - d(P, A)$ and $d(M, P) - d(P, B)$ are of different sign if $d(M, P)$ is big enough. Thus, if P is far enough from M and tends to infinity on $\overline{M}C$, then

$$|2d(M, P) - d(P, A) - d(P, B)| = |d(M, A') - d(M, B')|$$
$$\to |d(M, A^\#) - d(M, B^\#)| = 0. \tag{7.4.1}$$

Using trigonometry [67] in curved classical geometries, it is easy to show that (7.4.1) fails in the curved classical geometries, so the following is not a surprise.[5]

Theorem 7.4.1 ([21, Theorem 3]) *If (7.4.1) holds for every median ray in a straight projective-metric plane $(\mathcal{M}, d)$, then $(\mathcal{M}, d)$ is a Minkowski plane with differentiable circles.*

To consider every projective-metric plane in the spirit of this theorem, we need an equation similar to (7.4.1) in every classical geometry.

Lemma 7.4.2 *In the classical geometries*

$$2\sigma^2\left(\frac{d(M, P)}{2}\right) - \sigma^2\left(\frac{d(A, P)}{2}\right) - \sigma^2\left(\frac{d(B, P)}{2}\right) = -2\eta(d(M, P))\sigma^2\left(\frac{d(M, A)}{2}\right)$$
$$\tag{7.4.2}$$

---

[5] Notice however that the proof does not need, instead implies the parallel axiom.

Proof To shorten the formulas, let $r = d(M, A) = d(M, B)$, $p = d(M, P)$, $x = d(A, P)$, and $y = d(B, P)$ Using the law of cosine in the triangles $AMP\triangle$ and $BMP\triangle$ to get x and y, then summing them up gives

$$\begin{cases} \eta_\kappa(x) + \eta_\kappa(y) = 2\eta_\kappa(p)\eta_\kappa(r) & \text{if } \kappa \neq 0, \\ x^2 + y^2 = 2(r^2 + p^2) & \text{if } \kappa = 0. \end{cases}$$

Substituting $\eta_\kappa(t) = 1 - 2\kappa\sigma_\kappa^2(t/2)$ in the first equation on the left, and rearranging the equations we obtain

$$\begin{cases} 2\kappa\sigma_\kappa^2\left(\frac{p}{2}\right) - \kappa\sigma_\kappa^2\left(\frac{x}{2}\right) - \kappa\sigma_\kappa^2\left(\frac{y}{2}\right) = \eta_\kappa(p)\left(\eta_\kappa(r) - 1\right) & \text{if } \kappa \neq 0, \\ 2\left(\frac{p}{2}\right)^2 - \left(\frac{x}{2}\right)^2 - \left(\frac{y}{2}\right)^2 = -2\left(\frac{r}{2}\right)^2 & \text{if } \kappa = 0. \end{cases} \tag{7.4.3}$$

This proves the lemma. $\qquad\square$

So the following question arises.

Open Question 7.4.3 *Is a projective-metric plane* $(\mathcal{M}, d)$ *the model of a classical geometry if (7.4.2) holds for every median ray?*

Observe that for $p = r$ Eq. (7.4.2) becomes

$$\sigma^2\left(\frac{x}{2}\right) + \sigma^2\left(\frac{y}{2}\right) = 2\sigma^2\left(\frac{r}{2}\right)\left(1 + \eta(r)\right) = \sigma^2(r), \tag{7.4.4}$$

which is an alternative metric description of the circle of radius r resembling to Thales's theorem.

If $(\mathcal{M}, d)$ is of parabolic type, then (7.4.4) gives $(2r)^2 = x^2 + y^2$. This is the parallelogram law [2, (1.1)] in a Minkowski plane, hence such a Minkowski metric comes from an inner product, so the Minkowski plane is a model of the Euclidean geometry. This supports the idea behind the following question.

Open Question 7.4.4 *Is a projective-metric plane* $(\mathcal{M}, d)$ *the model of a classical geometry if (7.4.4) holds for every circle?* $\qquad\blacksquare$

7.4.3 Equidistants of Lines

Let ℓ be a geodesic line in a projective-metric space $(\mathcal{M}, d)$. For a point P, the set of points X equidistant with P with respect to ℓ, i.e. $\{X \in \mathcal{M} : d(X, \ell) = d(P, \ell)\}$, is called an *equidistant*, and denoted by $e(\ell, P)$. Obvious, but important observations are that $e(\ell, Q) = e(\ell, R)$ follows from $R \in e(\ell, Q)$, and that $e(\ell, Q)$ intersects ℓ if and only if $Q \in \ell$, because at the intersection the distance vanishes.

Equidistants are not geodesics in the classical geometries except in the Euclidean one, so the following result, which was proved in a more general setup by Phadke in [77], is not a surprise.

Theorem 7.4.5 *A projective-metric space is a Minkowski plane if every equidistant is a geodesic.*

To show how things work out for such a result, for the interested reader we present a linearly reasoning step-by-step proof in [60, Theorem 4.5] which is an adaptation of Phadke's original proof.

In closing this topic, it is worth noticing that the reference functions would not shed light on more details, because the equidistants are defined by the equality of distances. This is the reason behind the fact that much of the proof of Theorem 7.4.5 had to be spent on investigating $\mathcal{M}$ in order to exclude the non-parabolic projective-metric planes. ∎

7.4.4 Circumcenter and Orthocenter

Existence of the *circumcenter* of a triangle is a well-known property in Euclidean plane. It follows from the fact that in the Euclidean plane the perpendicular median line of a pair of points is in fact the equidistant (see 7.4.3) of that pair of points.

The situation is very similar in the curved classical geometries. For an easy formulation we need the term of *pencil*, which means any set of geodesic lines such that their corresponding projective lines in the projective model are concurrent. In the 2-dimensional classical geometries the perpendicular median line of a pair of points is the equidistant of that pair of points (this can be easily seen using the Pythagoras theorem of these geometries), hence the perpendicular median lines of the sides of any triangle form a pencil.

Remember that in a projective-metric space the perpendicular median line of a pair of points is the equidistant of that pair of points, i.e., their bisector if and only if it is a model of a classical geometry (see 7.4.1).

Since perpendicularity is defined through the minima of distances, and the size functions are strictly monotone, it would not make any difference if we used the reference functions to generalize the term of circumcenter for projective-metric spaces. However, there is an important natural difference: perpendicularity is not symmetric in general (see 7.3.3).

In a projective-metric plane we call a geodesic line ℓ' reverse perpendicular to geodesic line ℓ, if ℓ is perpendicular to ℓ'. With this terminology we have the following results about the circumcenter in projective-metric spaces.

Theorem 7.4.6 ([51, Theorem 3.1 and 4.1]) *A Minkowski plane is a model of Euclidean plane if either the perpendicular median lines or the reverse perpendicular median lines of every triangle are concurrent.*

Theorem 7.4.7 ([52, Theorem 5.1]) *A Hilbert plane is a model of hyperbolic plane if the perpendicular median lines of every triangle form a pencil.*

There is no known result for the reverse perpendicular median lines of triangles in Hilbert planes.

Open Question 7.4.8 *Is a Hilbert geometry a model of hyperbolic plane if the reverse perpendicular median lines of every triangle form a pencil?*

In Euclidean geometry the point where the three altitudes of a triangle intersect is called *orthocenter*. It is known in the classical geometries that the altitudes of a triangle form a pencil [42, Theorem 3 & §8].

We obtain the following result from Theorem 7.4.6 by affine reasoning.

Theorem 7.4.9 ([51, Theorem 3.2]) *A Minkowski plane is a model of the Euclidean geometry if the altitudes of every triangle are concurrent.*

Proof The medians of a triangle $ABC\triangle$ are concurrent (see Theorem 7.5.2), and their intersection is the centroid P. The centroid divides the medians in the ratio $1:2$.

Observe that the altitudes are parallel to the appropriate perpendicular median lines, and so the homothety χ_P of ratio $-1/2$ maps every altitude to the appropriate perpendicular median line. So χ_P does the same with the intersection of the altitudes, hence the perpendicular median lines are concurrent. Then Theorem 7.4.6 implies the theorem.   □

There is no such easy way to get a similar result for the Hilbert plane through Theorem 7.4.7, but it is proved in [52, Theorem 5.2].

Theorem 7.4.10 ([52, Theorem 5.2]) *A Hilbert plane is a model of hyperbolic plane if the altitudes of every triangle form a pencil.*

Again, by the lack of symmetry of perpendicularity, we introduce the reverse altitudes and prove the following theorem in the same way as Theorem 7.4.9 was done.

Theorem 7.4.11 ([51, Theorem 4.2]) *A Minkowski plane is a model of Euclidean plane if the reverse altitudes of every triangle are concurrent.*

There is no known result for the reverse perpendicular altitudes of triangles in Hilbert planes.

Open Question 7.4.12 *Is a Hilbert geometry a model of hyperbolic geometry if the reverse altitudes of every triangle form a pencil?*

It is worth noting that the above results about Hilbert planes are based on an especially interesting ellipse characterization [52, Theorem 4.2]: The differentiable border $\partial \mathcal{K}$ of a strictly convex body $\mathcal{K}$ in the plane is an ellipse if and only if for any three points $K_1, K_2, K_3 \in \partial \mathcal{K}$ the lines $f_i \ni K_i$ ($i \in \{1, 2, 3\}$) are concurrent exactly when $(\ell_j, \ell_k; f_i, t_i) = -1$, where $\{i, j, k\} = \{1, 2, 3\}$, $\ell_i = K_j K_k$, and t_i is the tangent of $\mathcal{K}$ at K_i. Notice that the dual of this statement is, via the theorems of Menelaus and Ceva, equivalent to Segre's result in [81, §3], which in fact does not use the finiteness of the geometry but only the commutativity of the underlying field, stating that the boundary $\partial \mathcal{K}$ of a strictly convex body $\mathcal{K}$ is an ellipse if and only if every circimscribed and inscribed triangle, such that the touching points of the former triangle are the vertices of the latter triangle, are perspective.   ∎

7.4.5   Angular Bisector

Given two different rays ℓ_+ and ℓ_-, with common starting point O in a projective-metric plane, we call the set of points which are equidistant with respect to ℓ_+ and ℓ_- the *angular bisector* of the angle $\ell_+\ell_-\angle$.

It is clear that angular bisectors in the classical geometries are straight lines, because the quadratic models are rotational manifolds with base point O, and the projector mapping at point O is angle preserving.

Open Question 7.4.13 *Does it follow that a projective-metric plane is the model of a classical geometry if every angular bisector is a geodesic?*

A kind of answer to this question can be achieved using the *angular bisector theorem* [26, Theorem 1.33] which, in the Euclidean plane, states that if X is the point in a triangle $ABC\triangle$ where the angular bisector of the angle at A intersects the side $\overline{BC}$, then

$$\frac{d(B, X)}{d(C, X)} = \frac{d(A, B)}{d(A, C)}. \tag{7.4.5}$$

Starting from this, Busemann and Phadke proved in a series of articles [16, 17, 78] that a straight projective-metric plane is a Minkowski plane if and only if for every two non-collinear rays $\overline{A}Z$ and $\overline{A}Y$ there exists a ray $\overline{A}D$ such that (7.4.5) holds for any points $B \in \overline{A}Z$ and $C \in \overline{A}Y$, where $X = \overline{A}D \cap \overline{B}C$.

Notice however, that (7.4.5) is a too restrictive condition to consider the angular bisectors in projective-metric planes, because (7.4.5) does not hold in the curved classical geometries. However, the law of sines in the classical geometries gives

$$\frac{\sigma_\kappa(d(B, X))}{\sigma_\kappa(d(C, X))} = \frac{\sigma_\kappa(d(A, B))}{\sigma_\kappa(d(A, C))}, \tag{7.4.6}$$

which raises the following:

Open Question 7.4.14 *Does it follow that a projective-metric plane is the model of a classical geometry if for every two non-collinear rays $\overline{A}Z$ and $\overline{A}Y$ there exists a ray $\overline{A}D$ such that (7.4.6) holds for any points $B \in \overline{A}Z$ and $C \in \overline{A}Y$, where $X = \overline{A}D \cap \overline{B}C$?*   ∎

We close this section by a question arising when considering the angular bisectors and the bisectors of point pairs:

Open Question 7.4.15 *Does it follow that a projective-metric plane is the model of a classical geometry if the set of angular bisectors coincide the set of bisectors?*

7.5 Metric Properties of Triangles

The most classical metric property of point sets in general position is the *triangle inequality* that holds, by definition, in every projective-metric space for any three points. Next to this is the statement that if there is an isometry for any two isometric triangles which takes the first triangle into the second one, then the projective-metric space is a model of the classical geometries [12, p. 307].

7.5.1 Ceva and Menelaus Property

In a projective-metric plane $(\mathcal{M}, d)$, we call a set of three points $Z \in AB$, $X \in BC$, and $Y \in CA$ a *triplet* (Z, X, Y) *of a non-degenerate triangle* $ABC\triangle$. A triplet (Z, X, Y) of $ABC\triangle$ is called

(p1) a *Menelaus triplet* if the points Z, X, and Y are collinear, and
(p2) a *Ceva triplet* if the *Cevian* geodesics AX, BY, and CZ are concurrent.

A 3-tuple (α, β, γ) of real numbers is called

(n1) *Menelaus tuple* if $\alpha \cdot \beta \cdot \gamma = -1$, and
(n2) *Ceva tuple* if $\alpha \cdot \beta \cdot \gamma = +1$.

We say that a projective-metric space has the *Menelaus property* or the *Ceva property* in metric spaces if every triplet (Z, X, Y) of any triangle $ABC\triangle$ is a Menelaus or Ceva triplet if and only if $(\langle A, B; Z \rangle_d^\circ, \langle B, C; X \rangle_d^\circ, \langle C, A; Y \rangle_d^\circ)$ is a Menelaus or Ceva tuple, respectively.

Theorem 7.5.1 ([70]) *Classical geometries have Menelaus and Ceva properties.*

Notice that in the curved classical geometries this theorem is very different from Menelaus' and Ceva's theorems in the Euclidean plane, because a reference function, more precisely the size function, is used in the size-ratios.

Theorem 7.5.2 ([55, Theorems 4.1]) *A projective-metric plane has the Ceva property if and only if it is a Minkowski plane, or a model of hyperbolic or elliptic geometry.*

To show how things work out for such a result, for the interested reader we present a linearly reasoning proof in the supplement to this chapter [60], which is an adaptation of the original proof.

By a very much similar proof of [55, Theorem 4.2], we also know that a projective-metric plane has the Menelaus property if and only if it is a Minkowski plane, or a model of the hyperbolic geometry, or a model of the elliptic geometry. ∎

7.5.2 Euler's Ratio-Sum

L. Euler's classical ratio-sum formula [29] for triangles $ABC\triangle$ in the Euclidean plane states that

$$\frac{d(A, O)}{d(O, X)} + \frac{d(B, O)}{d(O, Y)} + \frac{d(C, O)}{d(O, Z)} + 2 = \frac{d(A, O)}{d(O, X)} \cdot \frac{d(B, O)}{d(O, Y)} \cdot \frac{d(C, O)}{d(O, Z)}, \quad (7.5.1)$$

where O is any point in the interior of $ABC\triangle$, and $X = AO \cap BC$, $Y = BO \cap CA$, $Z = CO \cap AB$.

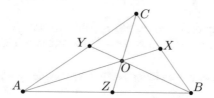

If $ABC\triangle$ is in a 2-dimensional classical geometry, then the gnomonic projection $\tilde{\tau}$ into the tangent space at point O, maps the triangle $ABC\triangle$ into the triangle $A'B'C'\triangle$ and the Cevians of $ABC\triangle$ into the Cevians of $A'B'C'\triangle$, where (7.5.1) gives

$$\langle A, X; O \rangle_d' + \langle B, Y; O \rangle_d' + \langle C, Z; O \rangle_d' + 2$$
$$= \langle A, X; O \rangle_d' \cdot \langle B, Y; O \rangle_d' \cdot \langle C, Z; O \rangle_d'. \quad (7.5.2)$$

This is the appropriate generalization of (7.5.1) to the classical geometries [61].

Now the method presented in the proof of Theorem 7.5.2 gives the following.

Theorem 7.5.3 ([61, Theorem 4.5]) *A projective-metric plane fulfills (7.5.2) for every triangle if and only if it is a Minkowski plane, or a model of the hyperbolic or the elliptic geometry.*

We notice, as a fun fact, that using the polarity on the sphere and (7.5.2), a Menelaus type version of Euler's ratio sum can be drawn too. It states that if a straight line ℓ intersects the sidelines of a triangle $ABC\triangle$, then

$$\frac{\tan(a\ell\angle)}{\tan(\ell x\angle)} + \frac{\tan(b\ell\angle)}{\tan(\ell y\angle)} + \frac{\tan(c\ell\angle)}{\tan(\ell z\angle)} - 2 = \frac{\tan(a\ell\angle)}{\tan(\ell x\angle)}\frac{\tan(b\ell\angle)}{\tan(\ell y\angle)}\frac{\tan(c\ell\angle)}{\tan(\ell z\angle)},$$

where $a = BC$, $b = CA$, and $c = AB$, and $X = a \cap \ell$, $Y = b \cap \ell$, $Z = c \cap \ell$, and $x = AX$, $y = BY$, and $z = CZ$.

■

7.6 Metric Constructions Describing Quadratic Curves

We call a set of point *quadratic* if it is part of a quadric. A curve in $\mathbb{R}^2$ is called a *quadric* if it is the zero set of an irreducible polynomial of degree two in 2 variables. A quadric in the plane has the equation of the form

$$\mathcal{Q}_\mathfrak{s}^\sigma := \left\{ (x, y) : \begin{array}{ll} 1 = x^2 + \sigma y^2 & \text{if } \sigma \in \{-1, 1\}, \\ x = y^2 & \text{if } \sigma = 0, \end{array} \right\} \tag{D_q}$$

in a suitable affine coordinate system $\mathfrak{s}$, and we call it *elliptic*, *parabolic*, or *hyperbolic*, if $\sigma = 1$, $\sigma = 0$, or $\sigma = -1$, respectively (for more details see [3, §VI]).

By (7.2.1), quadraticity is an inner geometrical property, so the following general question arises.

Open Question 7.6.1 *Which of the metric constructions in a projective-metric space describe quadratic sets of points?*

For example in Euclidean plane, the set $\mathcal{E}$ of the points C satisfying $d(A, C) + d(C, B) - d(A, B) = b$ for a constant $b > 0$ and for two fixed different points A and B is an ellipse, hence a quadric, as it is well known.

This shows that to find answers to the above question one can first try the metric constructions known in Euclidean plane corresponding to quadratic curves.

7.6.1 Ellipses, Circles and Hyperbolas

We work from now on in projective-metric planes $(\mathcal{M}, d)$. If the different points $F_1, F_2 \in \mathcal{M}$, called *focuses*, and a number $a > 0$, called *radius*, are given, then

(D_1) $\mathcal{E}_{d; F_1, F_2}^a := \{X : 2a = d(F_1, X) + d(X, F_2)\}$ is called an *ellipse* if $a > d(F_1, F_2)/2$, and

(D_2) $\mathcal{H}_{d; F_1, F_2}^a := \{X : 2a = |d(F_1, X) - d(F_2, X)|\}$ is called a *hyperbola* if $a < d(F_1, F_2)/2$.

The distance $f := d(F_1, F_2)/2$ is called the *(linear) eccentricity*. The metric midpoint M of the focuses F_1, F_2 is called the *metric center* of the ellipse and the hyperbola, and the segment of length $2a$ symmetric to M on line $F_1 F_2$ is called the *major axis* of the ellipse. The *numerical eccentricity* is $\varepsilon = \sigma(f)/\sigma(a)$.

Notice that allowing eccentricity vanishing in 7.6.1 defines the *circles* as a special class $\mathcal{E}^a_{d;F,F}$ of ellipses.

The following lemma (which is probably a folklore result) implies that the metric constructions 7.6.1 and 7.6.1 describe quadrics in the classical geometries.

Lemma 7.6.2 *If O is the metric center of the ellipse $\mathcal{E}^a_{d;F_1,F_2}$ or the hyperbola $\mathcal{H}^a_{d;F_1,F_2}$ in a 2-dimensional classical geometry, then the polar equation of $\mathcal{E}^a_{d;F_1,F_2}$ or $\mathcal{H}^a_{d;F_1,F_2}$, respectively, in the geodesic polar coordinatization with base point O is*

$$\frac{\eta^2(a)}{\sigma^2(r(\omega))} = \frac{\cos^2 \omega}{\tau^2(a)} + \frac{\sin^2 \omega}{\tau^2(a) - \tau^2(f)}. \tag{7.6.1}$$

On the other hand, the circles in Minkowski planes are quadrics if and only if the plane is Euclidean [19, IV.25.4], and, by the following theorem, the circles in Hilbert planes are quadrics if and only if the plane is hyperbolic.

Theorem 7.6.3 *A Hilbert plane is a model of hyperbolic plane if and only if at least one circle is a quadric.*

Proof Consider the Hilbert plane $(\mathcal{M}, d)$ in which a circle $\mathcal{E}^a_{d;F,F}$ is a quadric.

Let O be the center of $\mathcal{E}^a_{d;F,F}$, let ℓ be any straight line through O, and let A, B and I, J be the intersections of ℓ with $\mathcal{E}^a_{d;F,F}$ and $\partial \mathcal{M}$, respectively, so that $A \in \overline{IB}$. Let t_A and t_B be the tangents of $\mathcal{E}^a_{d;F,F}$ at A and B, respectively, and let t_I and t_J are the tangents of $\mathcal{M}$ at I and J, respectively.

It follows that ℓ is d-perpendicular to t_A and t_B, hence, by Busemann and Kelly [19, (28.11)], t_A and t_B pass through the (maybe ideal) intersection of t_I and t_J.

Let d_e be the Euclidean metric for which the quadric $\mathcal{E}^a_{d;F,F}$ is a unit circle. Then O is the Euclidean center of the Euclidean circle $\mathcal{E}^a_{d;F,F}$, hence $t_A \perp \ell \perp t_B$. This implies, by the previous paragraph, that $t_I \parallel t_J \perp \ell$ holds too.

Thus, every ℓ intersects $\partial \mathcal{M}$ orthogonally, that implies that $\partial \mathcal{M}$ is a circle with respect to d_e. The theorem is proved. $\qquad\square$

It seems that the situation is the same for both the metric constructions of the ellipses 7.6.1 and the hyperbolas 7.6.1, but it is proved only for the Minkowski plane yet:

Theorem 7.6.4 *A Minkowski plane is Euclidean if and only if*

(1) at least one of its ellipses is a quadric [58, Theorem 4.3], or
(2) at least one of its hyperbolas is a quadric [63, Theorem 4.3].

It is natural to conjecture[6] that a Hilbert plane is a model of hyperbolic plane if and only if at least one of the ellipses or at least one of the hyperbolas is a quadric.

Notice that all these statements are related to the assertion that the underlying projective-metric plane is analytic if and only if there is an analytic ellipse or analytic hyperbola in the projective-metric plane.

If the underlying projective-metric plane is analytic, then the methods of geometric tomography appear naturally in the proofs of the above statements (See [58, 63]).

Finally, we mention the following result, where infinitesimal ellipses play the main role.

Theorem 7.6.5 ([57, Theorem 4.4]) *If a twice differentiable Hilbert plane has two Riemannian points,[7] then it is a model of hyperbolic plane.*

The original proof of this theorem uses [57, Theorem 2.1], a version of the stable manifold theorem, which was introduced to geometric tomography by Falconer in [30, 31]. Instead of the above theorem our proof here is less sharp, but still shows well how things work out for this type of results.

Let the *infinitesimal circle*, determined by $F_\mathcal{M}(P, \cdot) \equiv 1$ (for the definition of $F_\mathcal{M}$ see (7.2.3)), at point P in the Hilbert plane $(\mathcal{M}, d)$ be denoted by $\mathcal{C}_P^\mathcal{M}$.

Theorem 7.6.6 ([57, Theorem 5.3]) *If two Hilbert planes $(\mathcal{L}, d_\mathcal{L})$ and $(\mathcal{M}, d_\mathcal{M})$ in the plane $\mathbb{R}^2$ with boundaries of class $C^{2+\delta}(\mathcal{S}^1)$, where $\delta > 0$, have two common infinitesimal circles $\mathcal{C}_P^\mathcal{L} \equiv \mathcal{C}_P^\mathcal{M}$ and $\mathcal{C}_Q^\mathcal{L} \equiv \mathcal{C}_Q^\mathcal{M}$, for $P.Q \in \mathcal{L} \cap \mathcal{M}$. and have equal curvatures at the points where the line PQ intersects the boundaries, then $\mathcal{L} \equiv \mathcal{M}$.*

Proof For a convex set $\mathcal{K}$ in $\mathbb{R}^2$ the (-1)-*chord function* $\rho_{P,\mathcal{K}} \colon \mathcal{S}^1 \to \mathbb{R}$ at a point P in the interior of $\mathcal{K}$ is defined [33, Definition 6.1.1] by $\rho_{P,\mathcal{K}}(\boldsymbol{u}) = r_{P,\mathcal{K}}^{-1}(\boldsymbol{u}) + r_{P,\mathcal{K}}^{-1}(-\boldsymbol{u})$, where $r_{P,\mathcal{K}}$ is the radial function of $\mathcal{K}$ at the point P.

Observe that, according to (7.2.3), the radial function of the infinitesimal circle $\mathcal{C}_P^\mathcal{K}$ is the (-1)-chord function $\rho_{P,\mathcal{K}}$. Further, one can specialize [33, Theorem 6.2.14, p. 247] to the following statement:

> Let $\mathcal{L}$ and $\mathcal{M}$ be bounded convex open domains in $\mathbb{R}^2$ with boundaries $\partial\mathcal{L}$ and $\partial\mathcal{M}$ belonging to $C^{2+\delta}$ for some $\delta > 0$. Let P and Q be in $\mathcal{L} \cap \mathcal{M}$, and suppose that $\mathcal{L}$ and $\mathcal{M}$ have equal (-1)-chord functions at these points. Then line PQ intersects $\partial\mathcal{L} \cap \partial\mathcal{M}$ in two points I and J. If $\partial\mathcal{L}$ and $\partial\mathcal{M}$ have equal curvatures at I and J, then $\mathcal{L} = \mathcal{M}$.

$$(7.6.2)$$

Combining our observations and (7.6.2) proves the theorem.   □

Based on these results, we formulate the following conjectures as questions.

---

[6] It seems well-founded in two manuscripts [59, 62] of the author.

[7] A Riemannian point of a Hilbert plane is where the tangent plane is Riemannian.

Open Question 7.6.7 *Are the ellipses in a projective-metric plane quadrics if and only if the projective-metric plane is a model of a classical geometry?*

Open Question 7.6.8 *Are the hyperbolas in a projective-metric plane the inter-sections of hyperbolic quadrics with the projective-metric plane if and only if the projective-metric plane is a model of a classical geometry?* ∎

7.6.2 Conics

We work again in projective-metric planes $(\mathcal{M}, d)$. If a geodesic line $\mathcal{L}$, called the *leading line*, a point $F \notin \mathcal{L}$, called the *focus*, and a number $\varrho > 0$, called the *radius*, are given, then

3. $\mathcal{C}^{\varrho}_{d;F,\mathcal{L}} := \{X \in \mathcal{M} : \varrho d(X, \mathcal{L}) = d(F, \mathcal{L})d(F, X)\}$ is called a *conic*.

A conic is said to be *elliptic*, *parabolic*, or *hyperbolic*, if $\varrho < d(F, \mathcal{H})$, $\varrho = d(F, \mathcal{H})$, or $\varrho > d(F, \mathcal{H})$, respectively.

It is shown in [86] that if the elliptic conics coincide with the ellipses in a Minkowski plane, then the Minkowski plane is a model of the Euclidean geometry. Furthermore, if a Finsler plane satisfies the same condition, then it is a Riemannian plane. Finally, [87, Theorem 1] states that if the elliptic conics coincide with the ellipses in a Riemannian plane, then the Riemannian plane is a model of the Euclidean geometry.

So, since ellipses are quadrics in the classical geometries, it follows that elliptic conics are quadrics in a classical geometry if and only if the classical geometry is the Euclidean geometry. It is proved in [65, 66] that hyperbolic conics are quadrics in a classical geometry if and only if the classical geometry is the Euclidean geometry. Interestingly, the parabolic conics are quadrics in the hyperbolic plane [27, (15)] and in the elliptic plane [66] too.

It is proved in [54, Theorem 4.2 and 4.3], that the only Minkowski plane in which a centrally symmetric conic exists is a model of the Euclidean plane. Based on this, the following result needs to be proved only for parabolic conics.

Theorem 7.6.9 ([54, Theorem 5.1]) *A Minkowski plane is a model of the Euclidean plane if and only if at least one conic is a quadric.*

Open Question 7.6.10 *Can a non-parabolic conic in a Hilbert plane be a quadric?*

Open Question 7.6.11 *Is a Hilbert plane a model of the hyperbolic plane if a parabolic conic is a quadric?*

Open Question 7.6.12 *Is a projective-metric plane a model of a classical geometry if every parabolic conic is the intersection of the projective-metric plane with a parabolic quadric?* ∎

There are numerous further metric descriptions of the quadrics in the Euclidean plane, so the question to what extent quadrics have metric counterparts in general projective-metric planes remains open.

7.7 Discussions and Further Open Questions

While most of the results in this paper use some characterizations of the ellipse, we must stress that they themselves are also characterizations of ellipses in several cases.

Despite the fact that most of the results in this paper are formulated in projective-metric planes, most of them can be generalized to higher dimensions using ellipsoid characterizations like the following two important ones.

Theorem 7.7.1 ([7, Theorem 1]) *Let $\mathcal{K}$ be a convex body in $\mathbb{R}^d$ ($d \geq 3$) and let δ be a real continuous function on $\mathcal{S}^{d-1}$ such that for any $\boldsymbol{u} \in \mathcal{S}^{d-1}$ the hyperplane $\mathcal{H} = \{\boldsymbol{x} : \langle \boldsymbol{x}, \boldsymbol{u} \rangle = \delta(\boldsymbol{u})\}$ meets the interior of $\mathcal{K}$. If the intersection $\mathcal{H} \cap \partial \mathcal{K}$ is an ellipsoid in $\mathcal{H}$, for any $\boldsymbol{u} \mathcal{S}^{d-1}$, then the boundary $\partial \mathcal{K}$ is an ellipsoid.*

Theorem 7.7.2 ([12, (16.12), p. 91]) *Let $\mathcal{K}$ be a convex body in $\mathbb{R}^d$ ($d \geq 3$), and let $k \in (1, d)$ be an integer. If every k-plane through an inner point of $\mathcal{K}$ intersects $\partial \mathcal{K}$ in a k-dimensional ellipsoid, then the boundary $\partial \mathcal{K}$ is an ellipsoid.*

Although we mentioned quite a number of topics, the set of open problems is not exhausted at all. We show three problems before closing this paper.

In the Euclidean plane the set of points for which the ratio of the distances from two fixed points is a constant different from 1, is a circle. This circle is the so-called *Appollonius circle* [25, §6.6].

In a projective-metric space $(\mathcal{M}, d)$ let $\mathcal{A}_{A,B}^r$ be the set of points C satisfying $d(A, C)/d(C, B) = r \neq 1$ for the constant $r > 0$.

Open Question 7.7.3 *In which of the projective-metric planes are the circles of the form $\mathcal{A}_{A,B}^r$?*

In the Euclidean plane let $\mathcal{K}$ be a convex body and let P be a point. If the product $d(P, K_1)d(P, K_2)$ does not depend on the distinct points $K_1, K_2 \in \partial \mathcal{K}$ collinear with P, then P is called *equiproduct point* of $\mathcal{K}$ [33, p. 255]. If P is an equiproduct point, then the *power with respect to $\mathcal{K}$* is $\Pi_{\mathcal{K}}(P) := d(P, K_1)d(P, K_2)$ if P is outside $\mathcal{K}$, and $\Pi_{\mathcal{K}}(P) := -d(P, K_1)d(P, K_2)$ if P is inside $\mathcal{K}$.

By the *power of a point theorem* [26, Theorem 2.11], if $\mathcal{K}$ is a circular disc, then every point is an equiproduct point of $\mathcal{K}$, and $\Pi_{\mathcal{K}}(P) = d^2(O, P) - \varrho^2$, where O is the center of $\mathcal{K}$ and ϱ is radius of $\mathcal{K}$. It is proved in [45] (see also [33, Theorem 6.3.13]) that a differentiable convex curve having two equiproduct points P and Q is a circle (see [89] if differentiability is not provided).

Open Question 7.7.4 *Is there a convex curve in a Minkowski plane with at least two equiproduct points? If so, is the Minkowski plane a model of the Euclidean geometry?*

In the Euclidean plane *Carnot's theorem* [85] states that in a triangle $ABC\triangle$ the points Z_A, Z_B on the side line AB, X_B, X_C on the side line BC, Y_C, Y_A on the side line CA lie on a common quadric if and only if

$$(A, B; Z_A)(A, B; Z_B)(B, C; X_B)(B, C; X_C)(C, A; Y_C)(C, A; Y_A) = 1$$
$$(7.7.1)$$

Every factor in the left-hand side of (7.7.1) is an affine ratio, so Carnot's theorem clearly remains valid in Minkowski plane after putting (7.7.1) in the form

$$\langle A, B; Z_A \rangle_d^\circ \langle A, B; Z_B \rangle_d^\circ \langle B, C; X_B \rangle_d^\circ \langle B, C; X_C \rangle_d^\circ \langle C, A; Y_C \rangle_d^\circ \langle C, A; Y_A \rangle_d^\circ = 1.$$
$$(7.7.2)$$

Theorem 7.7.5 *Carnot's theorem with (7.7.2) is valid for the circles in every 2-dimensional classical geometry.*

Proof In Euclidean geometry we have $\langle A, B; Z_A \rangle_d^\circ \langle A, B; Z_B \rangle_d^\circ = \Pi_C(A)/\Pi_C(B)$ which, using the analogous equations for the side lines BC and CA, implies (7.7.2).

We prove the theorem in the curved 2-dimensional classical geometries in the following way. Assume that the circle C of radius $\varrho > 0$ intersects every sides of the triangle $ABC\triangle$ in the points Z_A, Z_B on the side line AB, X_B, X_C on the side line BC, Y_C, Y_A on the side line CA in clockwise order. Let F_c, F_a, F_b be the feet of the center O of C on the side lines AB, BC, CA, respectively. Then $OF_cZ_A\triangle$, $OF_cZ_B\triangle$ and $OF_cA\triangle$ are right triangles, hence

$$\eta(d(F_c, Z_A)) = \eta(d(F_c, Z_B)) = \frac{\eta(\varrho)}{\eta(d(O, F_c))} \quad \text{and} \quad \eta(d(F_c, A)) = \frac{\eta(d(O, A))}{\eta(d(O, F_c))}.$$

Then

$$\sigma(d(A, Z_A))\sigma(d(A, Z_B))$$

$$= \sigma(d(A, F_c) - d(Z_A, F_c))\sigma(d(A, F_c) + d(F_c, Z_B))$$

$$= \frac{-\kappa}{2}(\eta(2d(A, F_c)) - \eta(2d(F_c, Z_A))) = -\kappa(\eta^2(d(A, F_c)) - \eta^2(d(F_c, Z_A)))$$

$$= -\kappa \frac{\eta^2(d(O, A)) - \eta^2(\varrho)}{\eta^2(d(O, F_c))}.$$

This immediately gives

$$\langle A, B; Z_A \rangle_d^\circ \langle A, B; Z_B \rangle_d^\circ = \frac{\eta^2(d(O, A)) - \eta^2(\varrho)}{\eta^2(d(O, B)) - \eta^2(\varrho)}.$$

The product of this and the same formulas for the side lines BC and CA is therefore clearly 1, proving (7.7.2) for the circles. □

Notice that

$$\langle A, B; Z_A \rangle_d^{\circ} \langle A, B; Z_B \rangle_d^{\circ} = \frac{\eta^2(d(F_c, A)) - 1}{\eta^2(d(F_c, B)) - 1} = \frac{\sigma^2(d(F_c, A))}{\sigma^2(d(F_c, B))} = (\langle A, B; F_c \rangle_d^{\circ})^2.$$

Open Question 7.7.6 *Is Carnot's theorem with (7.7.2) valid in a projective-metric plane other than the Minkowski plane?*

Acknowledgments It is a pleasure to thank Athanase Papadopoulos for inviting me to contribute a chapter to Surveys in Geometry Vol. II.

I thank József Kozma, and Athanase Papadopoulos for their help.

References

1. L.M. Alabdulsada, L. Kozma, On non-positive curvature properties of the Hilbert metric. J. Geom. Anal. **29**(1), 569–576 (2019). https://doi.org/10.1007/s12220-018-0011-9
2. D. Amir, Characterizations of inner product spaces, in *Operator Theory: Advances and Applications*, vol. 20 (Birkhäuser Verlag, Basel, 1986). https://doi.org/10.1007/978-3-0348-5487-0
3. M. Audin, *Geometry*. Universitext (Springer-Verlag, Berlin, 2003). https://doi.org/10.1007/978-3-642-56127-6. Translated from the 1998 French original
4. V. Balestro, H. Martini, *Minkowski Geometry-Some Concepts and Recent Developments*. Surveys in Geometry I, (Springer, Cham, 2022), pp. 49–95. https://doi.org/10.1007/978-3-030-86695-2_3
5. D. Bao, S.-S. Chern, Z. Shen, *An Introduction to Riemann-Finsler Geometry*. Graduate Texts in Mathematics, vol. 200 (Springer, New York, 2000). https://doi.org/10.1007/978-1-4612-1268-3
6. E. Beltrami, Risoluzione del problema: riportare i punti di una superficie sopra un piano in modo che le linee geodetiche vengano rappresentate da linee rette. Opere **I**, 262–280 (1865)
7. G. Bianchi, P.M. Gruber, Characterizations of ellipsoids. Arch. Math. **49**(4), 344–350 (1987). https://doi.org/10.1007/BF01210721
8. W. Blaschke, Integralgeometrie 11. Abh. Math. Sem. Univ. Hamburg **11**, 359–366 (1936)
9. L.M. Blumenthal, *Theory and Applications of Distance Geometry*, 2nd edn. (Chelsea Publishing, New York, 1970)
10. M.R. Bridson, A. Haefliger, *Metric Spaces of Non-positive Curvature*. Grundlehren der mathematischen Wissenschaften, vol. 319 (Springer, Berlin, 1999). https://doi.org/10.1007/978-3-662-12494-9
11. S.M. Buckley, K. Falk, D.J. Wraith, Ptolemaic spaces and CAT(0). Glasg. Math. J **51**(2), 301–314 (2009). https://doi.org/10.1017/S0017089509004984
12. H. Busemann, *The Geometry of Geodesics* (Academic Press, New York, 1955)
13. H. Busemann, Areas in affine spaces. III. The integral geometry of affine area. Rend. Circ. Mat. Palermo **9**, 226–242 (1960). https://doi.org/10.1007/BF02854583
14. H. Busemann, Geometries in which the planes minimize area. Ann. Mat. Pura Appl. **55**(4), 171–189 (1961). https://doi.org/10.1007/BF02412083
15. H. Busemann, *Recent Synthetic Differential Geometry*. Ergebnisse der Mathematik und ihrer Grenzgebiete, Band 54 (Springer, New York, 1970)

16. H. Busemann, Planes with analogues to Euclidean angular bisectors. Math. Scand. **36**, 5–11 (1975). https://doi.org/10.7146/math.scand.a-11556
17. H. Busemann, Remark on: "Planes with analogues to Euclidean angular bisectors" (Math. Scand. 36 (1975), 5–11). Math. Scand. **38**(1), 81–82 (1976). https://doi.org/10.7146/math.scand.a-11618
18. H. Busemann, *Problem IV: Desarguesian Spaces*. Mathematical Developments Arising from Hilbert Problems (Proceedings of Symposia in Pure Mathematics, Northern Illinois University, De Kalb, 1974) (American Mathematical Society, Providence, 1976), pp. 131–141. Proc. Sympos. Pure Math., Vol. XXVIII
19. H. Busemann, P.J. Kelly, *Projective Geometry and Projective Metrics* (Academic Press, New York, 1953)
20. H. Busemann, W. Mayer, On the foundations of calculus of variations. Trans. Am. Math. Soc. **49**, 173–198 (1941). https://doi.org/10.2307/1990020
21. H. Busemann, B.B. Phadke, Minkowskian geometry, convexity conditions and the parallel axiom. J. Geom. **12**(1), 17–33 (1979). https://doi.org/10.1007/BF01920230
22. J.W. Cannon, W.J. Floyd, R. Kenyon, W.R. Parry, Hyperbolic geometry, in *Flavors of Geometry* (Cambridge University Press, Cambridge, 1997), pp. 59–115
23. A. Cap, M.G. Cowling, F. de Mari, M. Eastwood, R. McCallum, The Heisenberg group, SL(3,$\mathbb{R}$), and rigidity, in *Harmonic Analysis, Group Representations, Automorphic Forms and Invariant Theory*. Lecture Notes Series, Institute for Mathematical Sciences, National University of Singapore, vol. 12 (World Scientific, Hackensack, 2007), pp. 41–52. https://doi.org/10.1142/9789812770790_0002
24. C. Charitos, I. Papadoperakis, G. Tsapogas, Convexity of asymptotic geodesics in Hilbert Geometry. Beitr. Algebra Geom. **63**, 809–828 (2021). Posted on 10 October 2021, https://doi.org/10.1007/s13366-021-00601-3
25. H.S.M. Coxeter, *Introduction to Geometry*. Wiley Classics Library, 2nd edn. (Wiley, New York, 1989). https://archive.org/details/introductiontogeometry2ndedcoxeter1969/page/n1/mode/2up. Reprint of the 1969 edition
26. H.S.M. Coxeter, S.L. Greitzer, *Geometry Revisited*. New Mathematical Library, vol. 19 (The Mathematical Association of America, New York, 1967). https://www.maa.org/press/maa-reviews/geometry-revisited
27. G. Csima, J. Szirmai, Isoptic curves of conic sections in constant curvature geometries. Math. Commun. **19**(2), 277–290 (2014). https://hrcak.srce.hr/129576
28. P. Erdos, Problem 3740. Am. Math. Month. **42**, 396 (1935). https://doi.org/10.2307/2301373
29. L. Euler, Geometrica et sphaerica quaedam. Memoir. l'Acad. Sci. Saint-Petersbourg **5**, 96–114 (1815). http://eulerarchive.maa.org/docs/originals/E749.pdf
30. K.J. Falconer, On the equireciprocal point problem. Geom. Ded. **14**(2), 113–126 (1983). https://doi.org/10.1007/BF00181619
31. K.J. Falconer, Differentiation of the limit mapping in a dynamical system. J. Lond. Math. Soc. **27**(2), 356–372 (1983). https://doi.org/10.1112/jlms/s2-27.2.356
32. T. Foertsch, A. Lytchak, V. Schroeder, Nonpositive curvature and the Ptolemy inequality. Int. Math. Res. Not. IMRN **22**, Art. ID rnm100, 15 (2007). https://doi.org/10.1093/imrn/rnm100. [Erratum in Int. Math. Res. Not.IMRN 24(2007) Art. ID rnm160, 1]
33. R.J. Gardner, *Geometric Tomography,* 2nd edn., Encyclopedia of Mathematics and its Applications, vol. 58 (Cambridge University Press, New York, 2006). https://doi.org/10.1017/CBO9781107341029
34. M. Ghandehari, H. Martini, On the Erdős-Mordell inequality for normed planes and spaces. Stud. Sci. Math. Hungar. **55**(2), 174–189 (2018). https://doi.org/10.1007/s00022-007-1961-4
35. P.P. Gilmartin, Showing the shortest routes - great circles, in *Matching the Map Projection to the Need*, ed. by A.H. Robinson, J.P. Snyder. Special Publication ... of the American Cartographic Association, vol. 3 (American Congress on Surveying and Mapping, Bethesda, 1991), pp. 18–19

36. R. Guo, Characterizations of hyperbolic geometry among hilbert geometries, in *Handbook of Hilbert Geometry*. IRMA Lectures in Mathematics and Theoretical Physics, vol. 22 (European Mathematical Society, Zürich, 2014), pp. 147–158. http://sites.science.oregonstate.edu/~guoren/docs/survey-Hilbert.pdf

37. G. Hamel, Über die Geometrieen, in denen die Geraden die Kürzesten sind (in German). Math. Ann. **57**(2), 231–264 (1903). https://doi.org/10.1007/BF01444348

38. D. Hilbert, Über die gerade Linie als kürzeste Verbindung zweier Punkte (in German). Math. Ann. **46**, 91–96 (1895). https://doi.org/10.1007/BF02096204.

39. D. Hilbert, Mathematische probleme. Nachr. Ges. Wiss. Göttingen Math.-Phys. Kl. 253–297 (1900). http://aleph0.clarku.edu/~djoyce/hilbert/problems.html. [*Archiv der Math. Physik (3)* **1** (1901), 44–63, 213–237]

40. J.W.P. Hirschfeld, *Projective Geometries Over Finite Fields*. Oxford Mathematical Monographs (The Clarendon Press/Oxford University Press, New York, 1979)

41. W.-Y. Hsiang, On the laws of trigonometries of two-point homogeneous spaces. Ann. Global Anal. Geom. **7**(1), 29–45 (1989). https://doi.org/10.1007/BF00137400

42. N.V. Ivanov, Arnol'd, the Jacobi identity, and orthocenters. Am. Math. Month. **118**(1), 41–65 (2011). https://doi.org/10.4169/amer.math.monthly.118.01.041

43. D.C. Kay, Ptolemaic metric spaces and the characterization of geodesics by vanishing metric curvature. Ph.D. Thesis, Michigan State University (1963)

44. D.C. Kay, The ptolemaic inequality in Hilbert geometries. Pac. J. Math. **21**, 293–301 (1967). https://doi.org/10.2140/pjm.1967.21.293

45. J.B. Kelly, Power points. Am. Math. Month. **53**(7), 395–396 (1946). https://doi.org/10.2307/2305862

46. P.J. Kelly, L.J. Paige, Symmetric perpendicularity in Hilbert geometries. Pac. J. Math. **2**, 319–322 (1952). https://projecteuclid.org/euclid.pjm/1103051777

47. P. Kelly, E.G. Straus, Curvature in Hilbert geometries. Pac. J. Math. **8**, 119–125 (1958). https://projecteuclid.org/journalArticle/Download?urlId=pjm%2F1103040248

48. P. Kelly, E.G. Straus, On the projective centres of convex curves. Can. J. Math. **12**, 568–581 (1960). https://doi.org/10.4153/CJM-1960-050-7

49. P. Kelly, E.G. Straus, Curvature in Hilbert geometries. II. Pac. J. Math. **25**, 549–552 (1968). https://projecteuclid.org/journalArticle/Download?urlId=pjm%2F1102986149

50. G. Korchmáros, J. Kozma, Regular polygons in higher dimensional Euclidean spaces. J. Geom. **105**(1), 43–55 (2014). https://doi.org/10.1007/s00022-013-0191-1

51. J. Kozma, Characterization of Euclidean geometry by existence of circumcenter or orthocenter. Acta Sci. Math. **81**(3–4), 685–698 (2015). https://doi.org/10.14232/actasm-015-518-0

52. J. Kozma, Á. Kurusa, Hyperbolic is the only Hilbert geometry having circumcenter or orthocenter generally. Beitr. Algebra Geom. **57**(1), 243–258 (2016). https://doi.org/10.1007/s13366-014-0233-3

53. Á. Kurusa, Support theorems for totally geodesic Radon transforms on constant curvature spaces. Proc. Am. Math. Soc. **122**(2), 429–435 (1994). https://doi.org/10.2307/2161033

54. Á. Kurusa, Conics in Minkowski geometries. Aequationes Math. **92**(5), 949–961 (2018). https://doi.org/10.1007/s00010-018-0592-1

55. Á. Kurusa, Ceva's and Menelaus' theorems in projective-metric spaces. J. Geom. **110**(2), 39, 12 (2019). https://doi.org/10.1007/s00022-019-0495-x

56. Á. Kurusa, Curvature in Hilbert geometries. Int. J. Geom. **9**(1), 85–94 (2020) https://ijgeometry.com/wp-content/uploads/2020/03/10..pdf

57. Á. Kurusa, Hilbert geometries with Riemannian points. Ann. Mat. Pura Appl. **199**(2), 809–820 (2020). https://doi.org/10.1007/s10231-019-00901-5

58. Á. Kurusa, Quadratic ellipses in Minkowski geometries. Aequationes Math. **96**, 567–578 (2022). https://doi.org/10.1007/s00010-021-00839-1

59. Á. Kurusa, Quadratic ellipses in Hilbert geometries. Aequationes Math. **96**, 567–578 (2022)

60. Á. Kurusa, Supplement to "Metric Characterization of Projective-metric Spaces", In: *Surveys in Geometry* II, Springer, Cham, 2024. posted on 2023, 1–11 (2024). https://doi.org/10.48550/arXiv.2312

61. Á. Kurusa, J. Kozma, Euler's ratio-sum formula in projective-metric spaces. Beitr. Algebra Geom. **60**(2), 379–390 (2019). https://doi.org/10.1007/s13366-018-0422-6
62. Á. Kurusa, J. Kozma, Quadratic hyperbolas in Hilbert geometries. Ann. Matematica Pura Appl. 199, 809–820 (2021)
63. Á. Kurusa, J. Kozma, Quadratic hyperboloids in Minkowski geometries. Mediterr. J. Math. **19**, 106 (2022). https://doi.org/10.1007/s00009-022-02002-9
64. G.W. Leibniz, *Mathematische Schriften* (zweite Abteilung I, Berlin, 1849)
65. A.M. Mahdi, Quadratic conics in hyperbolic geometry. Int. J. Geom. **8**(2), 60–69 (2019). https://ijgeometry.com/product/ahmed-mohsin-mahdi-quadraticconics-in-hyperbolic-geometry/
66. A.M. Mahdi, Conics on the sphere. Int. J. Geom. **9**(2), 5–14 (2020). https://ijgeometry.com/product/ahmed-mohsin-mahdi-conics-on-the-sphere/
67. G.E. Martin, *The Foundations of Geometry and the Non-Euclidean Plane*. Undergraduate Texts in Mathematics (Springer, New York, 1996). https://doi.org/10.1007/978-1-4612-5725-7. Corrected third printing of the 1975 original
68. H. Martini, K.J. Swanepoel, Antinorms and Radon curves. Aequationes Math. **72**(1–2), 110–138 (2006). https://doi.org/10.1007/s00010-006-2825-y
69. H. Martini, K. J. Swanepoel, G. Weiss, The geometry of Minkowski spaces—a survey. I. Expo. Math. **19**(2), 97–142 (2001). https://doi.org/10.1016/S0723-0869(01)80025-6. [Erratum in *Expo. Math.* **19**:4 (2001), 364; doi: 10.1016/S0723-0869(01)80021-9]
70. L.A. Masal'tsev, Incidence theorems in spaces of constant curvature. Ukrain. Geom. Sb. **35**, 67–74, 163 (1992). https://doi.org/10.1007/BF01249519 (Russian, with Russian summary); English transl., *J. Math. Sci.* **72** (1994), no. 4, 3201–3206
71. S. Mazur, S. Ulam, Sur le transformations isométriques d'espaces vectoriels, normés. C. R. Acad. Sci. Paris **194**, 946–948 (1932). https://gallica.bnf.fr/ark:/12148/bpt6k31473/f950.item
72. H. Minkowski, Sur les propriétés des nombres entiers qui sont dérivées de l'intuition de l'espace. Nouvelles Ann. Math. 3e série **15**, 393–403 (1896). http://eudml.org/doc/101072. Also in Gesammelte Abhandlungen, 1. Band, XII, pp. 271–277
73. L.J. Mordell, D.F. Barrow, Solution to 3740. Am. Math. Month. **44**, 252–254 (1937). https://doi.org/10.2307/2300713
74. A. Oppenheim, Some inequalities for a spherical triangle and an internal point. Publ. Fac. Electrotech. Univ. Belgrade, Ser. Math. Phys. **203**, 13–16 (1967)
75. A. Papadopoulos, *Metric Spaces, Convexity and Non-positive Curvature*. IRMA Lectures in Mathematics and Theoretical Physics, vol. 6, 2nd edn. (European Mathematical Society (EMS), Zürich, 2014). https://doi.org/10.4171/132
76. A. Papadopoulos, M. Troyanov (eds.), *Handbook of Hilbert Geometry*. IRMA Lectures in Mathematics and Theoretical Physics, vol. 22 (European Mathematical Society (EMS), Zurich, 2014). https://doi.org/10.4171/147
77. B.B. Phadke, Conditions for a plane projective metric to be a norm. Bull. Austral. Math. Soc. **9**, 49–54 (1973). https://doi.org/10.1017/S0004972700042854
78. B.B. Phadke, The theorem of Desargues in planes with analogues to Euclidean angular bisectors. Math. Scand. **39**(2), 191–194 (1976). https://doi.org/10.7146/math.scand.a-11656
79. A.V. Pogorelov, A complete solution of Hilbert's fourth problem. Soviet Math. Dokl. **14**, 46–49. [*Dokl. Akad. Nauk SSSR* **208** (1973), 48–51]
80. J. Radon, Über eine besondere Art ebener Kurven. Ber. Verh. Sächs. Ges. Wiss. Leipzig. Math.-Phys. Kl. **68**, 23–28 (1916)
81. B. Segre, Ovals in a finite projective plane. Can. J. Math. **7**, 414–416 (1955). https://doi.org/10.4153/CJM-1955-045-x
82. I.J. Schoenberg, A remark on M. M. Day's characterization of inner-product spaces and a conjecture of L. M. Blumenthal. Proc. Am. Math. Soc. **3**, 961–964 (1952). https://doi.org/10.2307/2031742
83. V. Soltan, Characteristic properties of ellipsoids and convex quadrics. Aequationes Math. **93**(2), 371–413 (2019). https://doi.org/10.1007/s00010-018-0620-1

84. Z.I. Szabó, Hilbert's fourth problem. I. Adv. Math. **59**(3), 185–301 (1986). https://doi.org/10.1016/0001-8708(86)90056-3
85. Z. Szilasi, Two applications of the theorem of Carnot. Ann. Math. Inf. **40**, 135–144 (2012). https://ami.uni-eszterhazy.hu/uploads/papers/finalpdf/AMI_40_from135to144.pdf
86. L. Tamássy, K. Bélteky, On the coincidence of two kinds of ellipses in Minkowskian spaces and in Finsler planes. Publ. Math. Debrecen **31**(3–4), 157–161 (1984)
87. L. Tamássy, K. Bélteky, On the coincidence of two kinds of ellipses in Riemannian and in Finsler spaces, in *Topics in Differential Geometry, Vols. I–II (Debrecen, 1984)*. Colloquia Mathematica Societatis János Bolyai, vol. 46 (North-Holland, Amsterdam, 1988), pp. 1193–1200
88. A.C. Thompson, *Minkowski Geometry*. Encyclopedia of Mathematics and its Applications, vol. 63 (Cambridge University Press, Cambridge, 1996). https://doi.org/10.1017/CBO9781107325845
89. L. Zuccheri, Characterization of the circle by equipower properties. Arch. Math. **58**(2), 199–208 (1992). https://doi.org/10.1007/BF01191886

Chapter 8
Supplement to "Metric Characterization of Projective-Metric Spaces"

Árpád Kurusa

Abstract This chapter is a supplement to the chapter "Metric Characterization of Projective-metric Spaces" and has no other goal than providing step-by-step and linearized proofs for two theorems.

Keywords Metric characterizations · Projective-metric spaces · Mikowski geometry · Hilbert geometry · Busemann spaces · Constant curvature spaces · Ellipses · Hyperbolas · Conics

2020 Mathematics Subject Classification 52A41, 53C60, 51F99, 53A40

8.1 Introduction

This chapter is a supplement to the chapter "Metric Characterization of Projective-metric Spaces" [6] that is included in the present volume, and it has no other goals just providing step-by-step and linearized proofs for Theorems 45 and 52 in that chapter.

The research leading to these results has received funding from the national projects TKP2021-NVA-09 and NKFIH-1279-2/2020. Project no. TKP2021-NVA-09 has been implemented with the support provided by the Ministry of Innovation and Technology of Hungary from the National Research, Development and Innovation.

Á. Kurusa (✉)
Bolyai Institute, University of Szeged, Szeged, Hungary
e-mail: kurusa@math.u-szeged.hu

8.2   Equidistants of Lines

Let ℓ be a geodesic line in a projective-metric space $(\mathcal{M}, d)$. For a point P, the set of points X equidistant with P with respect to ℓ, i.e. $\{X \in \mathcal{M} : d(X, \ell) = d(P, \ell)\}$, is called an *equidistant*, and denoted by $e(\ell, P)$. Obvious, but important observations are that $e(\ell, Q) = e(\ell, R)$ follows from $R \in e(\ell, Q)$, and that $e(\ell, Q)$ intersects ℓ if and only if $Q \in \ell$, because at the intersection the distance vanishes.

Equidistants are not geodesics in the classical geometries except in the Euclidean one, so the following result, which was proved in a more general setup by Phadke in [8], is not a surprise.

Theorem 8.2.1 *A projective-metric space is a Minkowski plane if every equidistant is a geodesic.*

We present a linearly reasoning proof, which is an adaptation of Phadke's original proof.

Proof Let $\mathcal{L}$ be a geodesic, $Q \notin \mathcal{L}$ be a point, $\mathcal{Q} = e(\mathcal{L}, Q)$, and let $Q_{\mathcal{L}}^{\perp}$ be the foot of Q on $\mathcal{L}$. Let $P \in \mathcal{L}$ be such that $d(P, \mathcal{Q}) = \min_{X \in \mathcal{L}} d(X, \mathcal{Q})$. If X and Q are on different sides of $\mathcal{L}$, then

$$d(X, \mathcal{Q}) = d(X, X_{\mathcal{Q}}^{\perp}) = d(X, L) + d(L, X_{\mathcal{Q}}^{\perp})$$

$$= d(X, L) + d(L, \mathcal{Q}) \geq d(X, L) + d(P, \mathcal{Q}),$$

where $X_{\mathcal{Q}}^{\perp}$ is the foot of X on $\mathcal{Q}$, and $L = \mathcal{L} \cap \overline{XX_{\mathcal{Q}}^{\perp}}$. Thus, $d(X, \mathcal{Q}) > d(P, \mathcal{Q})$, hence $X \in P$, so P does not have any point on the side of $\mathcal{L}$ which does not contain $\mathcal{Q}$, so $P = \mathcal{L}$.

Thus, if $\mathcal{Q} = e(\mathcal{L}, Q)$ and $P \in \mathcal{L}$, then $\mathcal{L} = e(\mathcal{Q}, P)$ and $d(\mathcal{L}, Q) = d(\mathcal{Q}, P)$, hence the foot of $P_{\mathcal{Q}}^{\perp}$ on $\mathcal{L}$ is P.

Let $\mathcal{L}$ be a geodesic, $P, Q \notin \mathcal{L}$ points, $\mathcal{P} = e(\mathcal{L}, P)$, and $\mathcal{Q} = e(\mathcal{L}, Q)$. Let X be outside $\mathcal{P}$ and $\mathcal{Q}$. Let $Y = X_{\mathcal{L}}^{\perp}$ be the foot of X on $\mathcal{L}$, and let $\mathcal{H} = XX_{\mathcal{L}}^{\perp}$. Let $X'_{\mathcal{P}} = \mathcal{H} \cap \mathcal{P}$. Then Y is also the foot of $X'_{\mathcal{P}}$, hence $Y_{\mathcal{P}}^{\perp} = X'_{\mathcal{P}}$, and also $Y_{\mathcal{Q}}^{\perp} = X'_{\mathcal{Q}}$, where $X'_{\mathcal{Q}} = \mathcal{H} \cap \mathcal{Q}$.

Thus, if X lies outside $\mathcal{P} = e(\mathcal{L}, P)$ and $\mathcal{Q} = e(\mathcal{L}, Q)$, then there is a geodesic $\mathcal{H}$ such that $\mathcal{H}$ is perpendicular to $\mathcal{P}$ and $\mathcal{Q}$.

Now we consider the projective-metric plane $(\mathcal{M}, d)$.

If $(\mathcal{M}, d)$ is of elliptic type, then any two geodesics intersect each other, and so do any two equidistants, because, by the condition of Theorem 8.2.1, every equidistant is geodesic. This is a contradiction, hence the equidistants in a projective-metric plane of elliptic type cannot be geodesics.

Thus, $(\mathcal{M}, d)$ is a straight projective-metric space, hence $\mathcal{M}$ is either the real plane or a proper convex subset of the real plane.

Assume that $\mathcal{M}$ is a proper convex subset of the real plane. Then $\mathcal{M}$ clearly does not contain two intersecting affine lines.

If $\mathcal{M}$ contains an affine line, then it is either a half plane or a strip bounded by two parallel affine lines [2, Exercise [17.8]]. Since a half plane and a strip are projectively equivalent, we can assume in this case that $\mathcal{M}$ is a strip.

If $\mathcal{M}$ does not contain any affine line, then every supporting line ℓ at any point M of $\partial\mathcal{M}$ can intersect $\partial\mathcal{M}$ only in a point, a segment or a ray. Let ℓ^+ be an affine line parallel to ℓ that is on the other side of ℓ than $\mathcal{M}$ is. Then the projectivity that takes ℓ^+ to the line at infinity maps $\mathcal{M}$ to a convex bounded subset of the real plane, hence in this case we can assume that $\mathcal{M}$ is bounded.

So we need to consider only three cases: $\mathcal{M}$ is a bounded convex subset of the real plane, $\mathcal{M}$ is a strip of the real plane, $\mathcal{M}$ is the real plane.

Assume first that $\mathcal{M}$ is a bounded convex subset of the real plane. Then $\mathcal{M}$ has at least three extreme points A, B, and C, say. If $ABC\triangle$ is a proper subset of $\mathcal{M}$, then one of the sides of the triangle, say AB, separates $\mathcal{M}$. Let point P and line $\mathcal{L}$ lie on different sides of AB. The equidistant $e(\mathcal{L}, P)$ cannot contain both of the points A and B, because $P \notin AB$. However, every point X between $\mathcal{L}$ and $e(\mathcal{L}, P)$ is collinear with its feet $X_{\mathcal{L}}^{\perp}$ and $X_{e(\mathcal{L},P)}^{\perp}$, so the points of $\mathcal{M}$ between $\mathcal{L}$ and $e(\mathcal{L}, P)$ constitutes the convex hull $\mathcal{M}^{\|}$ of $\mathcal{L} \cup e(\mathcal{L}, P)$. Since either A or B must belong to $\mathcal{M}^{\|}$ either A or B is not an extreme point, which contradicts the hypothesis. Thus $\mathcal{M}$ is the triangle $ABC\triangle$.

If a geodesic $\mathcal{L}$ passes through C, then its equidistants on one side of $\mathcal{L}$ must again pass through C, because C is an extreme point that cannot lie between two equidistants. Let $e(\mathcal{L}, Q)$ be an equidistant to $\mathcal{L}$ that passes through C, and choose two points K and L, such that $d(K, \mathcal{L}) \leq d(K, L) < d(\mathcal{L}, e(\mathcal{L}, Q))$, K is in the side of $\mathcal{L}$ that does not contain $e(\mathcal{L}, Q)$, and $L \in \mathcal{L}$. Let $\mathcal{K}$ be the geodesic through K passing C. Then the equidistant $e(\mathcal{K}, L)$ is $\mathcal{L}$, because otherwise it intersects $e(\mathcal{L}, Q)$ which contradicts $d(K, L) < d(\mathcal{L}, e(\mathcal{L}, Q))$.

So all geodesics through any of the vertices of $ABC\triangle$ are mutually equidistant.

Choose $K \in ABC\triangle$, and let $\mathcal{K}$ be a geodesic through K not passing through any vertex of $ABC\triangle$. Let $\mathcal{L}$ be perpendicular to $\mathcal{K}$ at K. If $\mathcal{L}$ does not pass through any vertex of $ABC\triangle$, there is a side, say AB, of $ABC\triangle$ intersected by $\mathcal{K}$ at the point X and by $\mathcal{L}$ at the point Y. The geodesics CX and CY are both equidistants. Let $\mathcal{L}_K$ be the ray starting from K and passing through Y. For any point $Z \in \mathcal{L}_K$, let the line of $Z_{CY}^{\perp}$ and $Z_{CX}^{\perp}$ intersect $\mathcal{K}$ in W. Then $d(Z, K) \leq d(Z, W) \leq d(Z_{CX}^{\perp}, Z_{CY}^{\perp}) = d(CX, CY)$ which is a constant. Since a ray cannot be of finite length, this is a contradiction. Thus, $\mathcal{L}$ goes through a vertex of $ABC\triangle$. This contradicts the fact that as $\mathcal{K}$ revolves freely about K, the perpendicular line $\mathcal{L}$ should change continuously with it.

Thus, $\mathcal{M}$ cannot be a triangle.

Assume that $\mathcal{M}$ is a strip bounded by the parallel affine lines $\partial\mathcal{M}_l$ and $\partial\mathcal{M}_r$. Let $O \in \mathcal{M}$ and let $\mathcal{L}$ and $\mathcal{K}$ be geodesics such that $\mathcal{K}$ is perpendicular to $\mathcal{L}$ at O.

Fig. 8.1 The strip bounded
by the parallel affine lines
$\partial \mathcal{M}_l$ and $\partial \mathcal{M}_r$

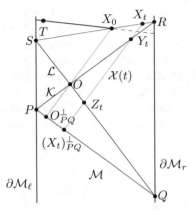

Let P and R be the points where $\mathcal{K}$ meets $\partial \mathcal{M}_l$ and $\partial \mathcal{M}_r$, respectively. Let Q and S be the points where $\mathcal{L}$ meets $\partial \mathcal{M}_r$ and $\partial \mathcal{M}_l$, respectively. Let $X_0 = OO_{PQ}^{\perp} \cap RS$ and let the corresponding arc length parameterization of $\overline{X_0 R}$ be $X : [0, \infty) \ni t \mapsto X_t \in \overline{X_0 R}$. Let $\mathcal{X}(t) = X_t (X_t)_{PQ}^{\perp}$, $Y_t = \mathcal{X}(t) \cap \mathcal{K}$, and $Z_t = \mathcal{X}(t) \cap \mathcal{L}$. See Fig. 8.1.

Then $d(X_t, PQ) > d(Y_t, Z_t) \ge d(Y_t, O) \to \infty$ as $t \to \infty$. This means that if a point on the geodesic RS tends to R or S, then its distance from the geodesic PQ tends to infinity. However, the equidistant $e(PQ, X_0)$, as it is a geodesic, has a ray starting from X_0 which is completely outside of the quadrangle $PQRS$. Thus, if a point (T on Fig. 8.1) on that ray $\overline{T X_0}$ approaches the boundary of $\mathcal{M}$, i.e. $\partial \mathcal{M} \cap T X_0$, then its distance from the geodesic PQ tends to infinity. This cannot happen to an equidistant, so $\mathcal{M}$ cannot be a strip.

Thus, $\mathcal{M}$ is the whole real plane, hence every equidistant lays on affine line.

Let $\mathcal{C}_{\varrho}(O)$ be the (d-metric) circle of radius $\varrho > 0$ centered at $O \in \mathcal{M}$. Let $\overline{AB}$ be a diameter of $\mathcal{C}_{\varrho}(O)$, and let the line $\mathcal{L}_A$ support $\mathcal{C}_{\varrho}(O)$ at A. Then AB is perpendicular (in the sense of d) to $\mathcal{L}_A$, and so the equidistant line $\mathcal{L}_B = e(\mathcal{L}_A, B)$ supports $\mathcal{C}_{\varrho}(O)$ at B. Since $\mathcal{L}_A$ and $\mathcal{L}_B$ do not intersect, they are (affine) parallel, hence $\mathcal{C}_{\varrho}(O)$ possesses parallel supporting lines at the endpoints of every chord through O. Then, by Busemann [1, (16.7)], O is the affine center of $\mathcal{C}_{\varrho}(O)$. Since every segment is a diameter of a (d-metric) circle, we deduce that the metric and affine centers of every segment coincide, hence, by the definition in [1, §17 p. 94], $\mathcal{C}_1(O)$ is an indicatrix of the Minkowski metric d. □

To close this topic, it is worth noticing that the reference functions would not shed light on more details, because the equidistants are defined by the equality of distances. This is the reason behind the fact that much of the proof of Theorem 8.2.1 had to be spent on investigating $\mathcal{M}$ in order to exclude the non-parabolic projective-metric planes. ∎

8.3 Ceva and Menelaus Property

In a projective-metric plane $(\mathcal{M}, d)$, we call a set of three points $Z \in AB$, $X \in BC$, and $Y \in CA$ a *triplet* (Z, X, Y) *of a non-degenerate triangle* $ABC\triangle$. A triplet (Z, X, Y) of $ABC\triangle$ is called

(p1) a *Menelaus triplet* if the points Z, X, and Y are collinear, and
(p2) a *Ceva triplet* if the *Cevian* geodesics AX, BY, and CZ are concurrent.

A 3-tuple (α, β, γ) of real numbers is called

(n1) *Menelaus tuple* if $\alpha \cdot \beta \cdot \gamma = -1$, and
(n2) *Ceva tuple* if $\alpha \cdot \beta \cdot \gamma = +1$.

We say that a projective-metric space has the *Menelaus property* or the *Ceva property* if every triplet (Z, X, Y) of any triangle $ABC\triangle$ is a Menelaus or Ceva triplet if and only if $(\langle A, B; Z\rangle_d^\circ, \langle B, C; X\rangle_d^\circ, \langle C, A; Y\rangle_d^\circ)$ is a Menelaus or Ceva tuple, respectively.

Theorem 8.3.1 ([7]) *The classical geometries have the Menelaus and Ceva properties.*

Notice that in the curved classical geometries this theorem is very different from Menelaus' and Ceva's theorems in the Euclidean plane, because a reference function, more precisely the size function, is used in the size-ratios.

Theorem 8.3.2 ([5, Theorems 4.1]) *A projective-metric plane has the Ceva property if and only if it is a Minkowski plane, or a model of hyperbolic or elliptic geometry.*

We present a linear reasoning proof, which is an adaptation and improved version of the original proof.

Proof Let $(\mathcal{M}, d)$ be a projective-metric plane. If it is of elliptic type, then let us cut out a projective line $\mathcal{L}$ and consider the remaining part with the inherited metric (this is a restriction of d, but we denote it with the same letter d). In this way we can consider the triangles in an affine plane independently from the type of $(\mathcal{M}, d)$.

Let us take a segment $\overline{AZ}$ and a point C within complement of the line AZ. Let the point $B \in AZ$ be such that $(A, B; Z) = (Z, A; R)$, let $X \in \overline{BC}$ be such that $(B, C; X) = (B, Z; Q)$, and let $Y \in \overline{CA}$ be such that $(C, A; Y) = (Z, A; R)$. Then, by the affine Ceva theorem, $\overline{AX}$, $\overline{BY}$ and $\overline{CZ}$ intersect each other in a common point, say M.

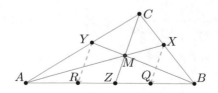

Since $(\mathcal{M}, d)$ has the Ceva property, we also have

$$\langle C, A; Y \rangle_d^\circ \langle B, C; X \rangle_d^\circ \langle A, B; Z \rangle_d^\circ = 1. \tag{8.3.1}$$

Mapping the triangle $ABC\triangle$ continuously into the degenerate triangle $AZB\triangle$ via the axial affinity with axis CZ so that C is moving along the segment $\overline{CZ}$, i.e. $C \to Z$, defines the points $Q, R \in \overline{AB}$ by $X \to Q$ and $Y \to R$. Then, by continuity of the metric, we obtain from (8.3.1) that $\langle Z, A; R \rangle_d^\circ \langle B, Z; Q \rangle_d^\circ \langle A, B; Z \rangle_d^\circ = 1$. By the additivity of d, this can be written in the equivalent form

$$\frac{\overrightarrow{ZB} - \overrightarrow{ZQ}}{\overrightarrow{ZA} - \overrightarrow{ZR}} \frac{\overrightarrow{RZ}}{\overrightarrow{ZQ}} \frac{\overrightarrow{AZ}}{\overrightarrow{ZB}} = 1 \Leftrightarrow \frac{\sigma(d(Z, B)) - d(Z, Q))}{\sigma(d(A, Z)) - d(R, Z))} \frac{\sigma(d(R, Z))}{\sigma(d(Z, Q))} \frac{\sigma(d(A, Z))}{\sigma(d(Z, B))} = 1 \tag{8.3.2}$$

for collinear points in the order $A \prec R \prec Z \prec Q \prec B$.

Let $P \colon \mathbb{R} \to e$ be the linear parameterization of the line $e = AZ$ such that $Z = P(0)$, $A = P(a)$, $R = P(r)$, $Q = P(q)$, $B = P(b)$, where $a < r < 0 < q < b$. Further, let $\varsigma \colon RQ \to \mathbb{R}$ be such that $\varsigma(s) = \sigma(d(P(s), Z))$.

In what follows, we investigate the projective-metric plane type by type.

(E): $(\mathcal{M}, d)$ is of *elliptic type, hence* $\sigma(\cdot) = \sin(\cdot)$.
Then, using the angle sum and difference identities for the sine function, (8.3.2) gives

$$\frac{b - q}{b} \frac{-a}{r - a} \frac{-r}{q} = 1 \Leftrightarrow \frac{\varsigma(b)\cos(d(Z, Q)) - \cos(d(Z, B))\varsigma(q)}{\varsigma(a)\cos(d(R, Z)) - \cos(d(A, Z))\varsigma(r)} \frac{\varsigma(r)}{\varsigma(q)} \frac{\varsigma(a)}{\varsigma(b)} = 1.$$

After some easy simplifications, this turns out to be

$$\frac{1}{q} - \frac{1}{b} = \frac{1}{a} - \frac{1}{r} \Leftrightarrow \cot(d(Z, Q)) - \cot(d(Z, B))$$

$$= \cot(d(R, Z)) - \cot(d(A, Z)). \tag{8.3.3}$$

Fixing the points R and Z, let $b \to \infty$ and $a \to -\infty$. From the left-hand equation of (8.3.3) we get $q \to -r$. From the right-hand equation of (8.3.3) we obtain $\cot(d(Z, Q)) = \cot(d(R, Z))$, hence $d(Z, Q) = d(R, Z)$. Thus, $q = -r$ is equivalent to $d(Z, Q) = d(R, Z)$, hence ς is an even function.

If $f \colon \mathbb{R} \to \mathbb{R}_+$ is defined by $f(x) := \cot(d(Z, P(x)))$, then (8.3.3) reads as

$$f\left(\frac{abr}{ar + br - ab}\right) = f(b) + f(r) - f(a).$$

Putting $r = -b$ (hence also $a < -b$ as can be read off from Figure above), this gives

$f\left(\frac{ab}{2a+b}\right) = 2f(b) - f(a)$, because, by the evenness of ς, f is an even function. Let $g : \mathbb{R} \backslash \{0\} \to \mathbb{R}$ be defined by

$$g(x) = \begin{cases} f(1/x) & \text{if } x > 0, \\ -f(1/x) & \text{if } x < 0, \end{cases}$$

which is an odd function. Then, as $2a + b < a < 0 < b$, we get

$$g\left(\frac{2}{b} + \frac{1}{a}\right) = 2g\left(\frac{1}{b}\right) + g\left(\frac{1}{a}\right). \tag{8.3.4}$$

For the moment let $b = -a/2$. Then (8.3.4) gives $g\left(\frac{-3}{a}\right) = 2g\left(\frac{-2}{a}\right) + g\left(\frac{1}{a}\right)$. This, by $a \to -\infty$, results in the continuous extension $g(0) = 0$ of $g : \mathbb{R} \to \mathbb{R}$.

Now, by the continuity of g, $a \to -\infty$ in (8.3.4) gives $g(2/b) = 2g(1/b)$. Substituting this into (8.3.4) we arrive at Cauchy's functional equation [9] for the continuous function g, $g(x) = \gamma_e x$ for every x and a $\gamma_e > 0$ determined by the line e. So, by the definition of g, $f(x) = \gamma_e/x$, which gives $\tan(d(P(x), P(0))) = |x/\gamma_e|$. The substitution $x = \gamma_e \tan s$ results in $d(P(\gamma_e \tan s), P(0)) = s$ meaning that the parameterization $s \mapsto P(\gamma_e \tan s)$ is by arc length in the projective metric, hence the point $P(\infty)$ should be on the projective line $\mathcal{L}$, which was cut out for the sake of ease the calculation. So we have $d(P(\infty), P(0)) = d(P(\gamma_e \tan(\pi/2)), P(0)) = \pi/2$ too.

Let B be the point such that $d(B, \mathcal{L}) = \pi/2$, and let C be a point on $\mathcal{L}$. Let A be a point such that $d(A, C) = \pi/2$, and let C_s be the point on the ray $\overline{AC}$ such that $d(A, C_s) = s$. Then the lengths of the sides of $AC_s B \triangle$ in the Euclidean metric are $a'_s = |AC_s| = \gamma_{AC} \tan(s)$, $b'_s = |BC_s| = \gamma_{BC_s} \tan(s)$, and a constant $c' = |AB| > 0$. If $s \to \pi/2$, then $\gamma_{BC_s} \to \gamma_{BC}$ by the continuity of the metrics, $b'_s, a'_s \to \infty$. Since $\left|\frac{a'_s}{b'_s} - 1\right| < \frac{c'}{b'_s}$ by the triangle inequality $|a'_s - b'_s| < c'$ in the Euclidean metric, $\frac{c'}{b'_s} \to 0$, hence $\frac{a'_s}{b'_s} \to 1$. So $1 = \lim_{s \to \pi/2} \frac{\gamma_{AC}}{\gamma_{BC_s}} = \frac{\gamma_{AC}}{\gamma_{BC}}$. Thus, the coefficients are equal for all geodesics which intersects $\mathcal{L}$ in a common point. This clearly implies that all the coefficients are the same, say, γ. So $(\mathcal{M}, d)$ is the elliptic plane.

(P): *Assume that* $(\mathcal{M}, d)$ *is of parabolic type, hence* $\sigma(\cdot) = \cdot$. Then (8.3.2) gives

$$\frac{b-q}{b} \frac{-a}{r-a} \frac{-r}{q} = 1 \Leftrightarrow \frac{\varsigma(b) - \varsigma(q)}{\varsigma(a) - \varsigma(r)} \frac{\varsigma(r)}{\varsigma(q)} \frac{\varsigma(a)}{\varsigma(b)} = 1.$$

After some easy simplifications, this shows

$$\frac{1}{q} - \frac{1}{b} = \frac{1}{a} - \frac{1}{r} \Leftrightarrow \frac{1}{\varsigma(q)} - \frac{1}{\varsigma(b)} = \frac{1}{\varsigma(r)} - \frac{1}{\varsigma(a)}. \tag{8.3.5}$$

Fixing R and Z, let $a \to -\infty$ and $b \to \infty$. Then (8.3.5) gives

$$\frac{1}{q} = -\frac{1}{r} \iff \frac{1}{\varsigma(q)} = \frac{1}{\varsigma(r)},$$

hence the affine and the d-metric midpoint of any segment coincide. So, d is a Minkowski metric by the definition in [1, §17 pp. 94].

(H): *Assume that* $(\mathcal{M}, d)$ *is of hyperbolic type, hence* $\sigma(\cdot) = \sinh(\cdot)$. Then, using the angle sum and difference identities for the hyperbolic sine function, (8.3.2) gives

$$\frac{b-q}{b} \frac{-a}{r-a} \frac{-r}{q} = 1 \iff \frac{\varsigma(b)\cosh(d(Z, Q)) + \cosh(d(Z, B))\varsigma(q)}{\varsigma(a)\cosh(d(R, Z)) + \cosh(d(A, Z))\varsigma(r)} \frac{\varsigma(r)}{\varsigma(q)} \frac{\varsigma(a)}{\varsigma(b)}$$

$$= 1.$$

After some easy simplifications this shows

$$\frac{1}{q} - \frac{1}{b} = \frac{1}{a} - \frac{1}{r} \iff \coth(d(Z, Q)) + \coth(d(Z, B))$$

$$= \coth(d(R, Z)) + \coth(d(A, Z)). \tag{8.3.6}$$

The intersection of a straight line and the domain $\mathcal{M}$ can be one of three types: a whole affine line AB, a ray $\overline{A_\infty B}$, or a segment $\overline{A_\infty B_\infty}$. Now we consider these cases one after the other.

Fixing points R and Z on the affine line AB, and letting $b \to \infty$ and $a \to -\infty$, imply that $q \to -r$ by the left-hand equation of (8.3.3). From the right-hand equation of (8.3.3) we get that $\coth(d(Z, Q)) = \coth(d(R, Z))$, hence $d(Z, Q) = d(R, Z)$. Thus, $q = -r$ is equivalent to $d(Z, Q) = d(R, Z)$, hence ς is an even function. Moreover, the map $\rho_{d;e;z}\colon P(z-x) \leftrightarrow P(z+x)$ is a d-isometric point reflection of e for every $P(z) \in e$, hence

$$\tau_{d;e;z,t} := \rho_{d;e;t} \circ \rho_{d;e;z}\colon P(y) \to P(2z-y) \to P(2(t-z)+y))$$

is a d-isometric translation. So $d(P(x), P(y)) = d(P(0), P(y-x))$, hence

$$d(P(0), P(y-x)) + d(P(0), P(z-y)) = d(P(x), P(y)) + d(P(y), P(z))$$

$$= d(P(x), P(z)) = d(P(0), P(z-x)).$$

Thus, the continuous function $f(x) = d(P(0), P(x))$ satisfies Cauchy's functional equation [9] hence a constant $\gamma_e > 0$ exists such that $d(P(x), P(y)) = \gamma_e |x - y|$ for every $x, y \in \mathbb{R}$.

Fixing points R and Z on the ray $e = \overline{A_\infty B}$, where $A_\infty = P(a_\infty)$, and letting $b \to \infty$ and $a \to a_\infty$, imply that

$$\frac{1}{q} = \frac{1}{a_\infty} - \frac{1}{r} \iff \coth(d(Z, Q)) = \coth(d(R, Z)) \tag{8.3.7}$$

by (8.3.6). Reparameterizing ray e by the linear map $\bar{P} \colon \mathbb{R} \to RQ$ such that $A_\infty = \bar{P}(0)$, $R = \bar{P}(r)$, $Z = \bar{P}(z)$, $Q = \bar{P}(q)$, we can reformulate the equivalence in (8.3.7) to

$$\frac{1}{q - z} = \frac{1}{-z} - \frac{1}{r - z} \iff d(Z, Q) = d(R, Z),$$

where $0 < r < z < q$. Thus, the map $\rho_{d;e;z} \colon \bar{P}(r) \leftrightarrow \bar{P}(z^2/r)$ is a d-isometric point reflection on ray e for every $\bar{P}(z) \in e$, hence

$$\tau_{d;e;z,t} := \rho_{d;e;t} \circ \rho_{d;e;z} \colon \bar{P}(r) \to \bar{P}(z^2/r) \to \bar{P}(rt^2/z^2)$$

is a d-isometric translation. So $d(\bar{P}(r), \tau_{d;e;z,t}(\bar{P}(r)))$ does not depend on r, hence it is a real function δ of t/z. As d is additive, this implies $\delta(x) + \delta(y) = \delta(xy)$, so, by the solution of Cauchy's functional equation [9] we have a constant $\bar{\gamma}_e > 0$ such that $\delta(x) = 2\gamma_e |\ln(x)|$, hence for every $x, y \in \mathbb{R}$ we have

$$d(\bar{P}(x), \bar{P}(y)) = d(\bar{P}(x), \tau_{d;e;1,\sqrt{y/x}}(P(x))) = \delta(\sqrt{y/x}) = \bar{\gamma}_e |\ln(y/x)|.$$

This is the Hilbert metric $d(\bar{P}(x), \bar{P}(y)) = \bar{\gamma}_e |\ln(A_\infty, \infty; \bar{P}(y), \bar{P}(x))|$ on ray e.

Fixing points R and Z on the segment $e = \overline{A_\infty B_\infty}$, where $A_\infty = P(a_\infty)$ and $B_\infty = P(b_\infty)$, and letting $b \to b_\infty$ and $a \to a_\infty$, implies that

$$\frac{1}{q} - \frac{1}{b_\infty} = \frac{1}{a_\infty} - \frac{1}{r} \iff \coth(d(Z, Q)) = \coth(d(R, Z)).$$

by (8.3.6). Reparameterizing segment e by the linear map $\bar{P} \colon \mathbb{R} \to RQ$ such that $A_\infty = \bar{P}(0)$, $R = \bar{P}(r)$, $Z = \bar{P}(z)$, $Q = \bar{P}(q)$, and $B_\infty = \bar{P}(1)$, we can reformulate the equivalence in (8.3.7) to

$$\frac{1}{q - z} - \frac{1}{1 - z} = \frac{1}{-z} - \frac{1}{r - z} \iff d(Z, Q) = d(R, Z),$$

where $0 < r < z < q < 1$. Thus, the map $\rho_{d;e;z} \colon \bar{P}(r) \leftrightarrow \bar{P}\left(\frac{z^2(1-r)}{z^2-r(2z-1)}\right)$ is a d-isometric point reflection on segment e for every $\bar{P}(z) \in e$, hence

$$\tau_{d;e;z,t} := \rho_{d;e;t} \circ \rho_{d;e;z} \colon \bar{P}(r) \to \bar{P}\left(\frac{z^2(1-r)}{z^2 - r(2z-1)}\right)$$

$$\to \bar{P}\left(\frac{1}{1 + \frac{1-r}{r}\frac{z^2}{(1-z)^2}\frac{(1-t)^2}{t^2}}\right)$$

is a d-isometric translation. So $d(\bar{P}(r), \tau_{d;e;z,t}(\bar{P}(r)))$ does not depend on r, hence it is a real function δ of $\frac{z^2}{(1-z)^2}\frac{(1-t)^2}{t^2}$. As d is additive, this implies $\delta(x) + \delta(y) = \delta(xy)$, so, by the solution of Cauchy's functional equation [9] we have a constant $\bar{\gamma}_e > 0$ such that $\delta(x) = 2\bar{\gamma}_e|\ln(x)|$, hence

$$d(\bar{P}(x), \bar{P}(y)) = d(\bar{P}(x), \tau_{d;e;1,\frac{x}{1-x}\frac{1-y}{y}}(\bar{P}(x)))$$

$$= \delta\left(\sqrt{\frac{x}{1-x}\frac{1-y}{y}}\right) = \bar{\gamma}_e\left|\ln\left(\frac{x}{1-x}\frac{1-y}{y}\right)\right|.$$

This is $d(\bar{P}(x), \bar{P}(y)) = \bar{\gamma}_e|\ln(A_\infty, B_\infty; \bar{P}(y), \bar{P}(x))|$, i.e. a Hilbert metric on segment e.

Having the metric for every possible domain of a projective-metric space of hyperbolic type, we are ready to step forward by considering the properties of the domain $\mathcal{M}$.

If $\mathcal{M}$ contains a whole affine line, then, by Busemann and Kelly [2, Exercise [17.8]], it is either a half plane or a strip bounded by two parallel lines, because it is not the whole plane. Thus, $\mathcal{M}$ is either $\mathcal{P}_{(0,\infty)} := \{(x, y) \in \mathbb{R}^2 : 0 < x\}$ or $\mathcal{P}_{(0,b)} := \{(x, y) \in \mathbb{R}^2 : 0 < x < b\}$ in suitable linear coordinates. As the perspective projectivity $\varpi \colon (x, y) \mapsto \left(\frac{x}{x+1}, \frac{y}{x+1}\right)$ maps $\mathcal{P}_{(0,\infty)}$ onto $\mathcal{P}_{(0,1)}$ bijectively, it is enough to consider the case $\mathcal{M} = \mathcal{P}_{(0,1)}$.

By the above, we know about the metric that $d((x, y), (x, z)) = c(x)|z - y|$ for a continuous function $c \colon (0, 1) \to \mathbb{R}_+$, and

$$d((x, \lambda+\sigma x), (\mu x, \lambda+\mu\sigma x)) = \bar{c}(\lambda, \sigma)\left|\ln\left(0, \frac{1}{x}; 1, \mu\right)\right| = \bar{c}(\lambda, \sigma)\left|\ln\frac{1-\mu x}{\mu(1-x)}\right|,$$

where $\bar{c} \colon \mathbb{R} \times \mathbb{R}_+ \to \mathbb{R}_+$ is also a continuous function. Putting these together gives

$$d((x, 0), (s, y)) = \begin{cases} \bar{c}\left(\frac{-yx}{s-x}, \frac{y}{s-x}\right)\left|\ln\frac{x(1-s)}{s(1-x)}\right| & \text{if } x \neq s, \\ c(x)|y| & \text{if } x = s, \end{cases}$$

for every $x, s \in (0, 1)$ and $y \in \mathbb{R}$. For $y = k(s - x) > 0$, where $k \geq 0$, this gives

$$kc(x) = \lim_{s \to x} \frac{d((x, 0), (x, s - x))}{s - x} = \bar{c}(-kx, k) \lim_{s \to x} \left| \frac{\ln \frac{x(1-s)}{s(1-x)}}{s - x} \right|$$

$$= \bar{c}(-kx, k) \lim_{s \to x} \left| \frac{\ln \left(1 - \frac{1}{s(1-x)/(s-x)} \right)^{s(1-x)/(s-x)}}{s(1 - x)} \right| = \frac{\bar{c}(-kx, k)}{x(1 - x)}.$$

Thus, $0 = \lim_{k \to 0} \bar{c}(-kx, k)$, hence continuity implies $\bar{c}(0, 0) = 0$, a contradiction.

Thus, $\mathcal{M}$ does not contain a whole affine line, hence it is either bounded or contains some rays. Then the metric on every chord $\ell \cap \mathcal{M}$ is of the form $\gamma_\ell \delta$, where δ is the Hilbert metric on $\mathcal{M}$. Coefficient γ_ℓ depends on ℓ continuously, because d and δ are continuous. Given non-collinear points $A, B, C \in \mathcal{M}$, the strict triangle inequality gives that $|\delta(A, C) - \delta(B, C)| < \delta(A, B)$ and

$$|\gamma_{AC}\delta(A, C) - \gamma_{BC}\delta(B, C)| = |d(A, C) - d(B, C)| < d(A, B) = \gamma_{AB}\delta(A, B).$$

These imply

$$\left| \frac{\delta(A, C)}{\delta(B, C)} - 1 \right| < \frac{\delta(A, B)}{\delta(B, C)}, \quad \text{and} \quad \left| \gamma_{AC} \frac{\delta(A, C)}{\delta(B, C)} - \gamma_{BC} \right| < \gamma_{AB} \frac{\delta(A, B)}{\delta(B, C)}.$$

If C tends to a point ∞ on the boundary $\partial \mathcal{M}$ of $\mathcal{M}$, then the first inequality implies $\frac{\delta(A,C)}{\delta(B,C)} \to 1$, so from the second inequality $\gamma_{A\infty} = \gamma_{B\infty}$ follows. Thus γ_ℓ is the same for every line with common point on $\partial \mathcal{M}$. This clearly implies that γ_ℓ does not depend on ℓ, i.e. constant, hence $(\mathcal{M}, d)$ is a Hilbert plane.

Now we are proving that a Hilbert plane is a Cayley–Klein model of hyperbolic plane if and only if it has the Ceva property [4]. If $\partial \mathcal{M}$ is an ellipse, then the Hilbert plane is a Cayley–Klein model of hyperbolic plane, hence it has the Ceva property.

In order to arrive at a contradiction, we assume that $\partial \mathcal{M}$ is not an ellipse, and the Hilbert plane $(\mathcal{M}, d)$ has the Ceva property.

By Gruber and Schuster [3, Theorem 2] of F. John, there is an ellipse $\mathcal{E}$ containing $\mathcal{M}$ and having at least 3 contact points, i.e., $|\partial \mathcal{M} \cap \partial \mathcal{E}| \geq 3$. Since $\partial \mathcal{M}$ is not an ellipse, $\text{Int}(\mathcal{E}) \setminus \mathcal{M}$ is a non-empty open set.

If $|\partial \mathcal{M} \cap \mathcal{E}| \leq 5$, then shrinking the ellipse $\mathcal{E}$ with a homothety χ at a center in $\mathcal{M}$ with coefficient $1 - \varepsilon$, where $\varepsilon > 0$ is small enough, moves the points of $\partial \mathcal{H} \cap \mathcal{E}$ into $\mathcal{M}$, and the resulting ellipse $\mathcal{E}_\chi := \chi(\mathcal{E})$ is such that $\mathcal{E}_\chi \cap \partial \mathcal{M}$ has at least six different points, and $\text{Int}(\mathcal{E}_\chi) \setminus \mathcal{M}$ is a non-empty open set.

Let $\mathcal{U}$ be a small open disc in $\text{Int}(\mathcal{E}_\chi) \setminus \mathcal{M}$, let P_0 be the center of $\mathcal{U}$, and let P_i ($i = 1, \ldots, 6$) be different points of $\mathcal{E}_\chi$. The lines $P_0 P_i$ ($i = 1, 2, 3, 4, 5$) are clearly pairwise different, therefore exactly one of them separates the four remaining points so that exactly two of those points are on both sides of it. Assume that the indexes were chosen so that this separating line is $P_0 P_3$, points P_1 and P_2 are on its left side, P_4 and P_5 are on its right side.

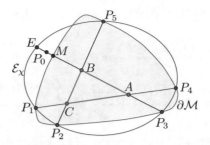

Let $A = \overline{P_1P_4} \cap \overline{P_2P_5}$, $B = \overline{P_0P_3} \cap \overline{P_2P_5}$, and $C = \overline{P_1P_4} \cap \overline{P_0P_3}$. Then $ABC\triangle$ is clearly a non-degenerate triangle. If any one of these points lies in any of the lines P_0P_i ($i = 1, 2, 3, 4, 5$), then move the point P_0 a little bit over so that it remains in $\mathcal{U}$, it separates $\overline{P_1P_2}$ and $\overline{P_4P_5}$, and neither of the points A, B, C is on any of P_0P_i ($i = 1, 2, 3, 4, 5$).

Observe that the open segments $\overline{P_1P_4}$ and $\overline{P_2P_5}$ are geodesic lines of the Hilbert geometry $(\mathcal{M}, d)$ and of the Cayley–Klein model $(\mathrm{Int}(\mathcal{E}_\chi), \delta)$ of hyperbolic geometry too. Furthermore, the open segment $\overline{P_3E} = \mathcal{H} \cap P_3P_0$ is a geodesic line of the hyperbolic geometry $(\mathrm{Int}(\mathcal{E}_\chi), \delta)$, and the open segment $\overline{P_3M} = \mathcal{M} \cap \overline{P_3P_0}$ is a geodesic line of the Hilbert geometry $(\mathcal{M}, d)$. Moreover, $\overline{P_3M} \subset \overline{P_3E}$.

If (Z, X, Y) is a Ceva triplet of $ABC\triangle$, then the Ceva property implies that

$$(\langle B, C; X\rangle_d^\circ, \langle C, A; Y\rangle_d^\circ, \langle A, B; Z\rangle_d^\circ) \text{ is a Ceva tuple,} \qquad (8.3.8)$$

where $\langle B, C; X\rangle_d^\circ = \langle B, C; X\rangle_\delta^\circ$ and $\langle C, A; Y\rangle_\delta^\circ = \langle C, A; Y\rangle_\delta^\circ$ by coincidence.

Let the linear parameterization $P: \mathbb{R} \to \ell$ of line $\ell = P_3E$ be such that $P_3 = P(0)$, $A = P(1)$, $B = P(b)$, $Z = P(z)$, $M = P(m)$, and $E = P(e)$, where we have $1 < b < m < e$ and $z \in (0, m)$. Then a long and tedious calculation gives

$$|\langle A, B; Z\rangle_d^\circ| = \frac{|z - b|}{|z - 1|\sqrt{b}}\sqrt{1 + \frac{b - 1}{m - b}} > \frac{|z - b|}{|z - v1|\sqrt{b}}\sqrt{1 + \frac{b - 1}{e - b}}$$

$$= |\langle A, B; Z\rangle_\delta^\circ|. \qquad (8.3.9)$$

Thus, (8.3.8) implies that

$$(\langle B, C; X\rangle_\delta^\circ, \langle C, A; Y\rangle_\delta^\circ, \langle A, B; Z\rangle_\delta^\circ) \text{ is } \textit{shape not} \text{ a Ceva tuple,}$$

which contradicts the Ceva property of the hyperbolic plane.

The proof of the Theorem 8.3.2 is now complete.   □

By a very much similar proof of [5, Theorem 4.2], we also know that a projective-metric plane has the Menelaus property if and only if it is a Minkowski plane, or a model of the hyperbolic geometry, or a model of the elliptic geometry.   ∎

References

1. H. Busemann, *The Geometry of Geodesics* (Academic Press Inc., New York, 1955)
2. H. Busemann, P. J. Kelly, *Projective Geometry and Projective Metrics* (Academic Press Inc., New York, 1953)
3. P.M. Gruber, F.E. Schuster, An arithmetic proof of John's ellipsoid theorem. Arch. Math. (Basel) **85**(1), 82–88 (2005). https://doi.org/10.1007/s00013-005-1326-x
4. J. Kozma, Á. Kurusa, Ceva's and Menelaus' theorems characterize the hyperbolic geometry among Hilbert geometries. J. Geom. **106**(3), 465–470 (2015). https://doi.org/10.1007/s00022-014-0258-7
5. Á. Kurusa, Ceva's and Menelaus' theorems in projective-metric spaces. J. Geom. **110**(2), Paper No. 39, 12 (2019). https://doi.org/10.1007/s00022-019-0495-x
6. Á. Kurusa, Metric characterization of projective-metric spaces, in *Surveys in Geometry II* (Springer, Cham, 2024)
7. L.A. Masal'tsev, Incidence theorems in spaces of constant curvature. Ukrain. Geom. Sb. **35**, 67–74, 163 (1992). https://doi.org/10.1007/BF01249519 (Russian, with Russian summary); English transl., J. Math. Sci. **72**(4), 3201–3206 (1994)
8. B.B. Phadke, Conditions for a plane projective metric to be a norm. Bull. Aust. Math. Soc. **9**, 49–54 (1973). https://doi.org/10.1017/S0004972700042854
9. D. Reem, Remarks on the Cauchy functional equation and variations of it. Aequationes Math. **91**(2), 237–264 (2017). https://doi.org/10.1007/s00010-016-0463-6

Chapter 9
Metric Problems in Projective and Grassmann Spaces

Boumediene Et-Taoui

Abstract In this chapter, several metric problems in projective and Grassmann spaces are presented, such as the determination of their congruence order and their superposability order. For that aim, among others, we investigate sets of equiangular lines and sets of equi-isoclinic subspaces in F^r, where $F = \mathbb{R}, \mathbb{C}$ or $\mathbb{H}$. It turns out that the construction of these sets is obtained from the construction of some classes of real, complex or quaternionic square matrices.

Keywords Projective space · Grassmann space · Congruence order · Superposability order · Equiangular lines · Equi-isoclinic subspaces · Conference matrices

AMS Codes 51F20, 51M15, 51M20, 51K99, 15B33, 15B57

9.1 Introduction

Let n be an integer such that $n \geq 2$. We study the properties of structures of lines and subspaces, as equiangular lines and equi-isoclinic n-subspaces in a general setting that includes real, complex and quaternionic spaces. Let $\mathbb{F} = \mathbb{R}, \mathbb{C}$ or $\mathbb{H}$ be the real, complex or quaternionic field. Let p and r be integers such that $p \geq 3$ and $r \geq 2$. A p-set of equiangular lines in $\mathbb{F}^r$ is a set of lines such that the angle between each pair of lines is the same. A p-set of equi-isoclinic n-subspaces with parameter λ in $\mathbb{F}^r$ is a set of p n-planes spanning $\mathbb{F}^r$ each pair of which has the same non-zero angle $\arccos \sqrt{\lambda}$. These structures of some kind of regularity have been studied, under various names, in several fields as discrete geometry, combinatorics, harmonic analysis, frame theory, coding theory and quantum information theory. Unfortunately due to the usage of different terminologies many results have been

B. Et-Taoui (✉)
Université de Haute Alsace: IRIMAS, Mulhouse, France
e-mail: boumediene.ettaoui@uha.fr

© The Author(s), under exclusive license to Springer Nature Switzerland AG 2024
A. Papadopoulos (ed.), *Surveys in Geometry II*,
https://doi.org/10.1007/978-3-031-43510-2_9

rediscovered independently. In this chapter we give a survey of the development of these structures in the last 70 years.

In the present chapter we wish to report on several problems which may be interpreted as metric problems in projective and Grassmann spaces. Any metric space M is isometrically embeddable in the Euclidean plane whenever each 5 element subset of M is isometrically embeddable in the Euclidean plane. Any M is isometrically embeddable in Euclidean r-space whenever each $(r + 3)$-subset of M is. This has been proved by Menger [42]. The number $r + 3$ is the smallest number having this property and is called the congruence order of Euclidean r-space. Analogously, Blumenthal [6] has determined the congruence order of hyperbolic r-space, and of spherical r-space. Both numbers turn out to be $r + 3$. For elliptic space, the situation is different. The congruence order of the elliptic plane turns out [49] to be 7 and not 5, and that of elliptic $(r - 1)$-space, for $r > 3$, is unknown but much larger than $r + 3$. In order to explain this different behavior recall that in the vector space $\mathbb{R}^r$ elliptic points are the lines through the origin. The elliptic distance between two elliptic points is defined to be the angle between the corresponding lines. Notice that all distances are at most $\frac{\pi}{2}$. Thus the elliptic $(r - 1)$-space is defined in terms of $\mathbb{R}^r$. The 6 diagonals of an icosahedron in $\mathbb{R}^3$ yield 6 elliptic points in the elliptic plane $\mathbb{R}P^2$ whose distances are all equal to $\arccos \frac{1}{\sqrt{5}}$. This explains why the congruence order of the elliptic plane cannot be 6. Indeed, let the metric space M consist of 100 points all of whose distances are $\arccos \frac{1}{\sqrt{5}}$. Then each 6-subset of M is isometrically embeddable in $\mathbb{R}P^2$, but M is not. For the congruence order in $\mathbb{R}P^{r-1}$ we should at least know the maximum number of elliptic points all of whose distances are equal. Seidel [40] denoted this number $N(r)$ and he conjectured that the congruence order of $\mathbb{R}P^{r-1}$ equals $N(r) + 1$ [49]. The determination of the congruence order of other metric spaces is out of reach, so in this chapter we restrict ourselves to equidistant points in projective and Grassmann spaces. Investigations on equiangular lines in Euclidean spaces were initiated by Blumenthal [6] and Haantjes [32], in the terminology of elliptic geometry. Van Lint and Seidel [40] covered the dimensions up to 7, and stressed the relations with discrete mathematics. In particular, in 1973, Lemmens and Seidel [40] determined the sequence $(N(r))_{r \in \mathbb{N}}$ consecutively for r up to 13. (See Sequence A002853 in The On-Line Encyclopedia of Integer Sequences). Equiangular line systems correspond to a variety of objects in different mathematical disciplines: regular two-graphs in group theory [51], equilateral point-sets in elliptic geometry [40], and optimal Grassmannian frames in frame theory [33]. Here are our main problems:

Problem 9.1.1 How many equiangular lines can be placed in $\mathbb{F}^r$?

Problem 9.1.2 How many equi-isoclinic n-subspaces can be placed in $\mathbb{F}^r$?

In this chapter we discuss several problems which are very close. In Sect. 9.2 we present general results on equiangular lines and equi-isoclinic subspaces. It is seen in Sect. 9.3 that Problem 9.1.2 amounts to finding the maximum number p for which there exists a real symmetric matrix of order np, partitioned into square blocks of order n with zero blocks on the diagonal and orthonormal blocks elsewhere, having a prescribed value and multiplicity for its smallest eigenvalue. It is also seen that these block matrices, called Seidel matrices, can sometimes be constructed from other square matrices both real and complex, called conference matrices. In Sect. 9.4 we report on almost all known results on real equiangular lines; among others the asymptotic behavior of $N(r)$ is discussed. Section 9.5 is devoted to complex equiangular lines and the determination of $v(2, 2r, \mathbb{R})$, the maximum number of equi-isoclinic planes in $\mathbb{R}^{2r}$. In Sect. 9.6 we present the known results concerning the superposability order of some projective and Grassmann spaces. The superposability order of a metric space is the smallest integer k such that each two subsets are congruent whenever corresponding k- tuples of the subsets are congruent. Section 9.7 presents all the known results concerning real and complex conference matrices, which are important for our geometric problems. In Sect. 9.8 we report on some particular complex equiangular lines called equiangular tight frames which may be considered as error-correcting codes. In Sects. 9.9 and 9.10 we construct an infinite family of square complex symmetric conference matrices of odd orders and an infinite family of equi-isoclinic planes in Euclidean odd dimensional spaces. In Sect. 9.11 we list all the quadruples, quintuples and sextuples of equi-isoclinic planes in $\mathbb{C}^4$ and state that no 7-tuple of equi-isoclinic planes can exist in $\mathbb{C}^4$. Finally in Sect. 9.12 we prove that there is a one-to-one correspondence between congruence classes of p-tuples of equi-isoclinic planes in $\mathbb{C}^{2r}$ with parameter λ whose associated Seidel matrices contain zero blocks on the diagonal, and blocks in $SU(2)$ elsewhere and congruence classes of p-tuples of equiangular lines in $\mathbb{H}^r$ with angle arccos $\sqrt{\lambda}$. Based on that we conclude by listing all the p-tuples of equiangular lines in $\mathbb{H}^2$.

9.2   Equiangular Lines and Equi-Isoclinic Subspaces

Investigating equiangular lines in spaces over the complex or quaternion numbers is also very difficult as the real case. However, the results for equiangular lines in $\mathbb{F}^r$, $\mathbb{F} = \mathbb{R}, \mathbb{C}$ or $\mathbb{H}$ have their consequences for equi-isoclinic subspaces in Euclidean spaces and conversely results for equi-isoclinic subspaces in Euclidean $2r$-dimensional spaces have their consequences for equiangular lines in $\mathbb{F}^r$, $\mathbb{F} = \mathbb{C}, \mathbb{H}$. First recall that the angles between subspaces of a linear space over the skewfield $\mathbb{F}$ are the stationary values of the angles of two lines, one in each subspace. Notice that the Riemannian distance between two subspaces is equal to the square root of the sum of the squares of their angles. Two subspaces of equal dimension are said to be *isoclinic* if all of their angles coincide and have value, say arccos $\sqrt{\lambda}$.

Let $v_\lambda(n, r, \mathbb{F})$ denote the maximum number of *equi-isoclinic* (i.e., pairwise *isoclinic*) n-subspaces in $\mathbb{F}^r$ with parameter λ, and $v(n, r, \mathbb{F})$ the maximum number of equi-isoclinic n-subspaces (or n-planes) in $\mathbb{F}^r$. Notice that a set of equi-isoclinic n-subspaces of $\mathbb{R}^r$ is a set of equidistant points in the Grassmann manifold $G(n, r)$.

Lemmens and Seidel [38] derived an upper bound for $v_\lambda(n,r,\mathbb{R})$:

$$v_\lambda(n, r, \mathbb{R}) \leq \frac{r - r\lambda}{n - r\lambda}, \tag{9.1}$$

for $n - r\lambda > 0$.

Moreover, they proved that

$$v(n, r, \mathbb{R}) \leq \frac{1}{2}r(r + 1) - \frac{1}{2}n(n + 1) + 1. \tag{9.2}$$

They also determined the value of $v(n, 2n, \mathbb{R})$ for all n, in particular they showed that $v(2, 4, \mathbb{R}) = 4$.

Hoggar [34] proved that the last formula applies to vector spaces over $\mathbb{F} = \mathbb{C}$ or $\mathbb{H}$. In fact he gave the following bound for $v(n, r, \mathbb{F})$, where $\mathbb{F} = \mathbb{R}, \mathbb{C}$ or $\mathbb{H}$,

$$v(n, r, \mathbb{F}) \leq f(r) - f(n) + 1, \tag{9.3}$$

where
$f(x) = x + \frac{1}{2}cx(x - 1)$ and $c = \dim_{\mathbb{R}} \mathbb{F}$.
For instance,

$$v(1, r, \mathbb{C}) \leq r^2, \tag{9.4}$$

and

$$v(1, r, \mathbb{H}) \leq 2r^2 - r. \tag{9.5}$$

In the following linear transformations and matrices are identified with each other. Let us outline the proof of (3). It was proved by Hoggar [35] that two n-planes in $\mathbb{F}^r$ with projections P, Q are isoclinic with parameter λ if and only if $PQP = \lambda P$. Consider p pairwise isoclinic n-planes with parameter λ and $P_1, \ldots, P_p$ their corresponding projections. Clearly we have for $i = 2, \ldots, p$, $P_1 P_i P_1 = \lambda P_1$ and $P_1^3 = P_1$. Hence the independent $P_1, \ldots, P_p$ belong to the linear subspace of Hermitian square matrices Q having the property $P_1 Q P_1$ is a multiple of P_1. We just need to compute the dimension d of this linear subspace. Since P_1 is of rank n we may assume $P = \begin{pmatrix} 0 & 0 \\ 0 & I_n \end{pmatrix}$, now putting $Q = \begin{pmatrix} Q_1 & T \\ T^* & Q_2 \end{pmatrix}$, where Q_1 and Q_2 are square matrices of orders $r - n$ and n respectively. We deduce from the equation $P_1 Q P_1 = \lambda P_1$ that $Q_2 = \lambda I_n$. Hence $d = r - n + \frac{c(r-n)(r-n-1)}{2} + c(r - n)n + 1$, which is equal to $f(r) - f(n) + 1$.

Zauner has conjectured that r^2 equiangular lines in $\mathbb{C}^r$ exists for each r, but his conjecture remains tantalizingly out of reach [1, 53]. These tuples are called tight 2-designs in [53], they are analogues of "symmetric, informationally complete, positive operator-valued measures" (SIC-POVMs). SIC-POVMs are r^2 equidistant points in $\mathbb{C}P^{r-1}$, which play an important role in quantum information theory [45]. Examples of SIC-POVMs have been algebraically constructed in low dimensions (up to d = 16 and a few larger cases) and numerically approximated for $d \leq 67$, but no infinite families are known [47]. The analogous question in quaternionic projective spaces has an entirely different character: tight 2-designs in $\mathbb{H}P^{r-1}$ do not seem to exist for $r > 3$ [11], however, they exist for $r = 2$ [22] and $r = 3$ [11].

We can view $\mathbb{H}$ as a right vector space over $\mathbb{C}$ of dimension two by writing quaternions in the form $z + jw$ $(z, w \in \mathbb{C})$, where $j^2 = -1$ and $ja = \bar{a}j$ for every a in $\mathbb{C}$. Applying this to linear transformations Hoggar showed that if Γ and Δ are isoclinic n-planes in $\mathbb{H}^r$ with parameter λ then $\Phi_{r,r}\Gamma$ and $\Phi_{r,r}\Delta$ are isoclinic $2n$-planes in $\mathbb{C}^{2r}$ with parameter λ, where $\Phi_{r,r}$ is the injective function from square quaternionic matrices of order r to square complex matrices of order $2r$ which replaces each entry $z + jw$ by the block

$$\begin{pmatrix} z & -\bar{w} \\ w & \bar{z} \end{pmatrix} \text{ [35].}$$

Thus, in an obvious notation $v_\lambda(n, r, \mathbb{H}) \leq v_\lambda(2n, 2r, \mathbb{C})$. Tensoring one has for any integer $k \geq 2$, $v_\lambda(n, r, \mathbb{F}) = v_\lambda(kn, kr, \mathbb{F})$. Hoggar also extended the determination of $v(n, 2n, \mathbb{F})$ to $\mathbb{F} = \mathbb{C}$ or $\mathbb{H}$. More precisely he proved that $v(n, 2n, \mathbb{F}) = v_{\frac{h(n)}{2(h(n)+1)}}(n, 2n, \mathbb{F}) = h(n) + 2$, where

$$h(n) = \begin{cases} 8a + 2^b \text{ if } \mathbb{F} = \mathbb{R} \\ 8a + 2b + 2 \text{ if } \mathbb{F} = \mathbb{C} \\ 8a + 2^b + \frac{1}{2}(b+2)(3-b) \text{ if } \mathbb{F} = \mathbb{H} \end{cases}$$

and $n = u2^{4a+b}$, u odd, $0 \leq b \leq 3$, are the Hurwitz numbers.

For instance, for any odd integer n, $v(n, 2n, \mathbb{F}) = 3, 4, 6$ respectively for $\mathbb{F} = \mathbb{R}, \mathbb{C}, \mathbb{H}$. This shows that $\mathbb{C}$ and $\mathbb{H}$ do add new possibilities.

Sets of equi-isoclinic subspaces with parameter λ that achieve equality in (1) are a type of optimal Grassmannian codes, being a way to arrange the subspaces so that the minimal chordal distance between any pair of them is as large as possible [12]. Notice that the chordal distance between two subspaces is equal to the square root of the sum of the squares of the sines of their angles. These optimal codes are called equi-isoclinic tight fusion frames in [27]. It became clear that all these geometric problems are related to a great variety of other subjects, both of pure and applied nature. As such we mention finite simple groups, finite fields, spherical harmonics, error correcting codes, electrical networks and statistical designs. Algebraic coding theory has developed from rudiments to a mature field.

9.3   Introduction to Seidel Matrices

The main tool for our query is the notion of Seidel matrices. If the equi-isoclinic
n-planes $\Gamma_1, \Gamma_2, \ldots, \Gamma_p$ with parameter λ, generate $\mathbb{C}^r$ and are each provided with a
fixed orthonormal basis, then the matrix A_{ij} built up by the Hermitian products of
vectors in the basis of Γ_i with vectors in the basis of Γ_j satisfies $A_{ij} A_{ij}^* = c^2 I$,
$i \neq j = 1, \ldots, p, c^2 = \lambda$ [38]. Hence the block matrix

$$
A = \begin{pmatrix}
I & A_{12} & \cdots & & A_{1p} \\
A_{21} & I & \ddots & & \vdots \\
\vdots & \ddots & \ddots & & A_{p-1\,p} \\
A_{p1} & \cdots & A_{p\,p-1} & & I
\end{pmatrix},
$$

of order np, is Hermitian positive semi-definite and of rank r. Thus, in order to
investigate p-tuples of equi-isoclinic n-planes with parameter c in $\mathbb{C}^r$ we ask for
Hermitian block matrices, that are positive semi-definite with rank r, i.e., that have
the smallest eigenvalue 0 with multiplicity $np - r$. In other words, we ask for
Hermitian matrices $M = \frac{1}{c}(A - I)$ whose smallest eigenvalue has multiplicity
$np - r$.

The matrix M is then partitioned into square blocks (M_{ij}) of order n with $M_{ii} =$
0 for all $i = 1, \ldots, p$, and $M_{ij} \in U(n)$ for all $i \neq j$.

Such matrices are called Seidel matrices. Conversely, a Seidel matrix with
smallest eigenvalue μ_0 whose multiplicity is $np - r$ leads to p equi-isoclinic n-
planes in $\mathbb{C}^r$ with parameter $c = -\frac{1}{\mu_0}$ [23].

However p-tuples of equi-isoclinic planes are not characterized by a single
matrix. We will say that two Seidel matrices M and M' are *equivalent* if and only if
there exist p matrices $P_1, \ldots, P_p \in U(n)$ such that

$$
P_i^* M_{ij} P_j = M_{ij}'.
$$

Two Seidel matrices are equivalent if and only if they are associated with the same
p-tuple.

By multiplication of each block column $j = 2, \ldots, p$ by M_{j1} and the
corresponding row by M_{1j} the Seidel matrix can be brought into one Seidel matrix
which has besides the zero block, I's in the first row. This matrix is called a normal
form and denoted by $\mathcal{M}(M_{ij})_{2 \leq i, j \leq p}$ with $M_{ij} \in U(n)$.

Extending a result of [23] we have

Proposition 9.3.1 *If (Γ_j) and $(\widetilde{\Gamma}_j)$ are two p-tuples of equi-isoclinic n-planes in
$\mathbb{C}^r$ with parameter $c \in \,]0, 1]$ and respective normal forms $\mathcal{M}(M_{ij})_{2 \leq i, j \leq p}$ and
$\mathcal{M}(\widetilde{M}_{ij})_{2 \leq i, j \leq p}$, then the two p-tuples are congruent if and if only if M_{ij} and $\widetilde{M}_{ij}$
are simultaneously conjugate, that is, there exists a matrix $U \in U(n)$ such that
$U^* M_{ij} U = \widetilde{M}_{ij}$, for all $i, j = 2, \ldots, p$.*

This proposition was often used in several articles in the complex setting [22] and in the real setting [18, 19, 23].

It is seen in this chapter that if $(\Gamma_1, \ldots, \Gamma_p)$ is a p-tuple of equi-isoclinic planes in $\mathbb{R}^{2r}$, respectively $\mathbb{C}^{2r}$ with parameter λ such that its associated Seidel matrix contains zero blocks on the diagonal and blocks of the form $\begin{pmatrix} z & -\overline{w} \\ w & \overline{z} \end{pmatrix}$ elsewhere, then the planes of the p-tuple can be viewed as a p-tuple of equiangular lines in $\mathbb{C}^r$, respectively $\mathbb{H}^r$ with angle $\arccos \sqrt{\lambda}$.

9.4 Real Equiangular Lines

Recall that in 1973, the value of $N(r)$ was known for all $r \leq 13$, $r = 15$, and $r = 21, 22, 23$. Recently, the problem was solved for dimensions 14 and 16 in [30], and for dimension 17 in [28].

In Table 9.1 below, we give the currently known values or lower and upper bounds for $N(r)$ for r at most 43. For instance, $N(2) = 3$ is easily obtained, $N(3)$, $N(4)$ were obtained by Haantjes in [32], $N(5)$, $N(6)$ were obtained by Van Lint and Seidel in [40], $N(7), \ldots, N(14)$ were obtained by Lemmens and Seidel in [39], for other values of $N(r)$ we refer to [39].

Note that this table appeared first in the seminal paper [39] of Lemmens and Seidel in 1973 with some errors which have been corrected very recently in [28, 30]. In these last articles the authors showed the nonexistence of equiangular line systems of cardinality p in $\mathbb{R}^r$ for certain pairs (p, r), by proving the nonexistence of their corresponding Seidel matrices. They take advantage of modular constraints on the coefficients of the characteristic polynomial of a Seidel matrix [28]. The eigenvalues of a Seidel matrix which corresponds to an equiangular line system of large cardinality relative to its ambient space are subject to strong geometric constraints. Their approach is to combine these modular and geometric constraints together which enable them to enumerate each possible characteristic polynomial for a putative Seidel matrix. Once an exhaustive list of the possible characteristic polynomials has been found, they apply spectral methods to show that no Seidel matrix can exist having the corresponding characteristic polynomials. To produce their exhaustive lists of possible characteristic polynomials, they use a polynomial enumeration algorithm, which they have implemented in SageMath.

The asymptotic behaviour of $N(r)$ is quadratic in r with a general upper bound of $r(r + 1)/2$ and a general lower bound of $(32r^2 + 328r + 296)/1089$ [29]. First,

Table 9.1 Bounds for the sequence $N(r)$ for $2 \leqslant r \leqslant 43$. A single number is given in the cases where the exact number is known

r	2	3	4	5	6	7	...	14	15	16	17	18	19	20
$N(r)$	3	6	6	10	16	28	...	28	36	40	48	57-60	72-74	90-94

r	21	22	23	..	41	42	43
$N(r)$	126	176	276	...	276	276-288	344

Lemmens and Seidel proved that for any $r \geq 6$, $N(r) \geq r\sqrt{r}$. This lower bound was improved by de Caen [14], who proved that $N(r) \geq \frac{2}{9}(r+1)^2$ for r of the form $r = 6 \times 4^i - 1$, where i is an integer. Motivated by a conjecture of Bukh [10], the asymptotic behavior of $v_\lambda(1, r, \mathbb{R})$ was recently shown to be linear in r [2, 37]. The main results of [37] and [2] use graph theory. Lemmens and Seidel proved in [38] that for any $r \geq 15$, $N_{\frac{1}{3}}(r) = 2r - 2$; however, very recently the following theorem was proved in [2] .

Theorem 9.4.1 *For a fixed α and r sufficiently large relative to α, the maximum number of equiangular lines in $\mathbb{R}^r$ with angle* arccos α *is exactly $2r - 2$ if $\alpha = \frac{1}{3}$ and 1.93r otherwise.*

The complete proof of this theorem can be found in [2]. The authors use Ramsey's theorem to find a large positive clique in its associated complete edge-labelled graph G_C. They then negate some vertices outside of this clique, in order to obtain a particularly advantageous graph, for which they show that almost all vertices attach to this positive clique entirely via positive edges. They then project this large set onto the orthogonal complement of the positive clique. Next they observe that the resulting graph contains few negative edges, which implies that the diagonal entries of the Gram matrix of the projected vectors are significantly larger in absolute value than all other entries. Combining this with an inequality which bounds the rank of such matrices already gives them a bound of $(2 + o(1))n$. To prove the exact result, they use more carefully the semi-definiteness of the Gram matrix together with some estimates on the largest eigenvalue of a graph.

On the other hand here is the main result of [37].

Theorem 9.4.2 *Fix α. Let $\lambda = \frac{1-\alpha}{\alpha}$ and $k = k(\lambda)$ be its spectral radius order. The maximum number $N_\alpha(r)$ of equiangular lines in $\mathbb{R}^r$ with common angle* arccos α *satisfies*

1. $N_\alpha(r) = \lfloor \frac{k(r-1)}{k-1} \rfloor$ *for all sufficiently large r if k is finite.*
2. $N_\alpha(r) = r + o(r)$ *as r tends to infinity if k is infinite.*

A key ingredient for the proof of this theorem is a new result in spectral graph theory: the adjacency matrix of a connected bounded degree graph has sublinear second eigenvalue multiplicity. For the complete proof we refer to [37].

9.5 Complex Equiangular Lines

Triangles (i.e. triples) of equi-isoclinic planes in $\mathbb{R}^6$ are obtained from 3 block row symmetric matrices. Now using our equivalence relation each of these matrices can be brought into one of the two following forms.

$$\begin{pmatrix} 0 & I & I \\ I & 0 & R_\omega \\ I & R_{-\omega} & 0 \end{pmatrix}, \text{ where } R_\omega = \begin{pmatrix} \cos \omega & -\sin \omega \\ \sin \omega & \cos \omega \end{pmatrix}$$

with $\omega \in [0, 2\pi[$,

$$\begin{pmatrix} 0 & I & I \\ I & 0 & S_0 \\ I & S_0 & 0 \end{pmatrix}, \text{ where } S_0 = \begin{pmatrix} 1 & 0 \\ 0 & -1 \end{pmatrix}.$$

We get the existence of two families of triangles of equi-isoclinic planes denoted by $T_{\lambda,\omega}^{(1)}$ and $T_\lambda^{(2)}$. Thus we arrive to the following proposition proved in [18].

Proposition 9.5.1 *There exist two families of congruence classes of triangles of equi-isoclinic planes in $\mathbb{R}^6$ denoted by $T_{\lambda,\omega}^{(1)}$ and $T_\lambda^{(2)}$, depending on two and one real parameters respectively.*

1. *For a triangle $T_{\lambda,\omega}^{(1)}$ the parameters satisfy $\cos\omega \geq \frac{1}{2\lambda\sqrt{\lambda}}(3\lambda - 1)$, equality holding if and only if the triangle spans a subspace $\mathbb{R}^4$.*
2. *For a triangle $T_\lambda^{(2)}$, we have $0 < \lambda \leq \frac{1}{4}$.*
   If $\lambda = \frac{1}{4}$ then $T_\lambda^{(2)}$ spans a subspace $\mathbb{R}^5$.

As a consequence, if $\lambda > \frac{1}{4}$ then the non-diagonal blocks of a Seidel matrix are all in $SO(2)$. Using our equivalence relation we can assume that a Seidel matrix of order $2v$ is partitioned into 2-order blocks (R_{ij}) with $R_{ii} = \begin{pmatrix} 0 & 0 \\ 0 & 0 \end{pmatrix} = 0$ for all $i = 1, \ldots, v$, $R_{1j} = I_2$ for all $j = 2, \ldots, v$ and $R_{ij} \in SO(2)$ for all $i < j$, $i = 2, \ldots, v - 1, j = 3, \ldots, v$.

In [19] the following theorems are proved.

Theorem 9.5.2 *There is a one-to-one correspondence between v-sets of equi-isoclinic planes with parameter λ spanning $\mathbb{R}^{2r}$ whose associated matrices are matrices with non-diagonal blocks all in $SO(2)$ and v-sets of equiangular lines with angle $\arccos\sqrt{\lambda}$ spanning $\mathbb{C}^r$.*

This theorem and the above consequence lead to the following result.

Theorem 9.5.3 *If $\lambda > \frac{1}{4}$ then $v_\lambda(2, 2r, \mathbb{R}) = v_\lambda(1, r, \mathbb{C})$.*

From this theorem and the relative bound of Lemmens and Seidel we get the following.

Corollary 9.5.4 *If $\lambda > \frac{1}{4}$ then $v_\lambda(2, 2r, \mathbb{R}) \leq r^2$.*

Now using our Corollary with $r = 3$, the list of 9 equiangular lines in $\mathbb{C}^3$ [19] and the fact that, if $0 < \lambda \leq \frac{1}{4}$ then $v_\lambda(2, 6, \mathbb{R}) \leq 9$, we arrive to this result.

Theorem 9.5.5 *$v(2, 6, \mathbb{R}) = 9$.*

9.6 Superposability Order of Projective and Grassmann Spaces

In [23] the aim was to find necessary and sufficient conditions for the congruence of two p-tuples of equi-isoclinic n-subspaces with the same parameter λ of $\mathbb{R}^r$.

Definition 9.6.1 A p-tuple of equi-isoclinic subspaces of the Grassmann manifold $G(n, r)$ is called regular if its symmetry group is isomorphic to the symmetric group $\mathbb{S}_p$.

Since the symmetry group can be rather difficult to compute, another attempt to check the regularity could be the use of the following notion of *pseudo-regularity*, defined in [23]. We first recall the notion of holonomy of a p-tuple $\left(\Gamma_1, \ldots, \Gamma_p\right)$ of equi-isoclinic n-planes with parameter c, $c = \sqrt{\lambda}$. The *holonomy* along a k-cycle $\gamma = (i_1 \ldots i_k) \in S_p$ is by definition the conjugation class (in $O\left(\Gamma_{i_1}\right)$) of $c^{-k}\pi_{i_1 i_2} \circ \ldots \circ \pi_{i_{k-1} i_k} \circ \pi_{i_k i_1}$ where $\pi_{ij} : \Gamma_j \to \Gamma_i$ is the orthogonal projection. Ones can check that this conjugation class actually does not depend on the choice of the first index in the parenthesis. If we identify $O\left(\Gamma_{i_1}\right)$ and $O(n)$, the holonomy is also the conjugation class of the matrix product

$$c^{-k} M_{i_1 i_2} \cdot \ldots \cdot M_{i_{k-1} i_k} \cdot M_{i_k i_1},$$

where M_{ij} are the blocks of one Seidel matrix associated to the p-tuple. It is also easy to see that different Seidel matrices give the same conjugation class.

For instance, the holonomies along a transposition are trivial. The holonomies along 3-cycles for a quadruple with associated Seidel matrix $M(A, C^T, B)$ ($A, B, C \in O(n)$) are (the conjugation classes of) A, B, C and ABC, and those along 4-cycles are AB, BC and CA. The notation $M(A, C^T, B)$ means that our Seidel matrix is of the form

$$\begin{pmatrix} 0 & I & I & I \\ I & 0 & A & C^T \\ I & A^T & 0 & B \\ I & C & B^T & 0 \end{pmatrix}.$$

It is clear that a regular p-tuple has the same holonomies as that obtained by any permutation of its n-planes, but in general the condition is not sufficient. This leads to the following definition. A p-tuple (Γ_i) is called pseudo-regular if for any permutation $\sigma \in S_p$ the p-tuples (Γ_i), $(\Gamma_{\sigma(i)})$ have the same holonomies. It is equivalent to say that the holonomies along k-cycles are equal, for each integer $k \leq p$ [23].

We obtain in [23] the following result :

Theorem 9.6.2

1. *Two quadruples of $G(2, 8)$ are congruent if and only if they have the same parameter λ and the same holonomies. In particular a pseudo-regular quadruple of $G(2, 8)$ is necessarily regular.*
2. *There exist in $G(3, 12)$ non-congruent quadruples with the same parameter and holonomies.*

Definition 9.6.3 A p-tuple of planes is called of type I if all its 3-holonomies are in O_2^+, of type II if all its 3-holonomies are in O_2^- and of mixed type if its 3-holonomies are in O_2^+ and in O_2^-.

The following result is proved in [23].

Theorem 9.6.4

1. *Two p-tuples of type I are congruent if and only if the corresponding subquadruples are congruent.*
2. *Two p-tuples of type II are congruent if and only if the corresponding subquadruples are congruent.*
3. *There exist non-congruent quadruples of mixed type such that the corresponding subquadruples are congruent.*

Definition 9.6.5 The superposability order of a metric space is the smallest integer k such that each two subsets are congruent whenever corresponding k-tuples of the subsets are congruent.

For every r, the Euclidean, spherical, and hyperbolic r-dimensional spaces have minimum superposability order 2.

Blumenthal showed that the elliptic space has minimum superposability order 3 for the class of non-degenerated tuples [7].

Brehm and Et-Taoui [8] proved that for the class of non-degenerate tuples the superposability order of $\mathbb{C}P^{r-1}$ is 4, and that one of $\mathbb{H}P^{r-1}$ is at least 6 [9].

The above theorem shows that the superposability order of $G(2, r)$ is at least 5 for the class of sets of equi-isoclinic planes.

Concerning the regular p-tuples of equi-isoclinic subspaces in Grassmann spaces we have the following results proved in [23] and [25].

Theorem 9.6.6 *The list of regular p-tuples of equi-isoclinic 2-planes is the following:*

1. *All triples.*
2. *For every $p \geq 4$, p-tuples with associated matrices $M_A = J_n \otimes A$, with $A = I, -I$, or S_0, J_p consists of 0's on the diagonal and of 1's elsewhere and $\otimes$ is the tensor product.*
3. *Quadruples with associated matrix $M(R_{\frac{\pi}{2}}, R_{-\frac{\pi}{2}}, R_{\frac{\pi}{2}})$.*
4. *Quintuples with associated matrix $M(S_{-\frac{2\pi}{3}}, S_0, S_{\frac{2\pi}{3}}, S_{\frac{2\pi}{3}}, S_0, S_{-\frac{2\pi}{3}})$.*

Theorem 9.6.7 *The list of all direct regular p-tuples of equi-isoclinic 3-planes is the following*

1. *All direct triples.*
2. *All quadruples described in Theorem 3 of [25].*
3. *For any $p \geq 5$, all p-tuples $(\Gamma_1, \ldots, \Gamma_p)$ which are associated with some Seidel matrix all blocks of which are either I_3 or rotation $R(\pi)$ with angle π.*
4. *The quintuple associated to the Seidel Matrix*

$$\begin{pmatrix} 0 & I_3 & I_3 & I_3 & I_3 \\ I_3 & 0 & A & C & B \\ I_3 & A & 0 & B & C \\ I_3 & C & B & 0 & A \\ I_3 & B & C & A & 0 \end{pmatrix},$$

   where A, B and C are half-turns with coplanar axes.
5. *Let $u_1 = \frac{\sqrt{2}}{2}\left(0\ 1\ -1\right)^T$, $u_2 = \frac{\sqrt{2}}{2}\left(1\ 0\ -1\right)^T$, $u_3 = \frac{\sqrt{2}}{2}\left(1\ 1\ 0\right)^T$, $u_4 = \frac{\sqrt{2}}{2}\left(-1\ 0\ -1\right)^T$, $u_5 = \frac{\sqrt{2}}{2}\left(-1\ 1\ 0\right)^T$, $u_6 = \frac{\sqrt{2}}{2}\left(0\ -1\ -1\right)^T$.*
   The quintuple associated to the Seidel Matrix

$$\begin{pmatrix} 0 & I_3 & I_3 & I_3 & I_3 \\ I_3 & 0 & A & C^T & D^T \\ I_3 & A^T & 0 & B & E \\ I_3 & C & B^T & 0 & F \\ I_3 & D & E^T & F^T & 0 \end{pmatrix},$$

   where A, B, C, D, E and F are half-turns with axes generated respectively by $u_4, u_1, u_3, u_6, u_5, u_2$.
6. *The quintuple associated to the Seidel Matrix*

$$\begin{pmatrix} 0 & I_3 & I_3 & I_3 & I_3 \\ I_3 & 0 & A & C^T & D^T \\ I_3 & A^T & 0 & B & E \\ I_3 & C & B^T & 0 & F \\ I_3 & D & E^T & F^T & 0 \end{pmatrix},$$

   where A, B, C, D, E and F are rotations with angle $\arctan(\frac{5}{\sqrt{2}})$ and with oriented axes generated respectively by $u_1, u_2, u_3, u_4, u_5, u_6$.

Remark The six lines generated respectively by $u_1, \ldots, u_6$ yield a 2-distance set in the real projective plane $\mathbb{RP}^2$. These are lines in $\mathbb{R}^3$ at angles $\frac{\pi}{3}$ and $\frac{\pi}{2}$ [50].

9.7   Complex Conference Matrices

A complex $n \times n$ *conference matrix* C is a matrix with $C_{ii} = 0$ and $\left| C_{ij} \right| = 1, i \neq j$
that satisfies

$$C^* C = (n - 1) I_n.$$

Real conference matrices have been heavily studied in literature in connection with
combinatorial designs in geometry, engineering, statistics, and algebra. Belevitch
[3] initiated the study of these matrices, which he called conference matrices. In
statistics they were treated in connection with weighing designs by Raghavarao [44].
   Complex conference matrices have received considerable attention in the past
few years due to their application in quantum information theory and in geometry.
Here two matrices are equivalent if one can be obtained from the other by
simultaneous permutation of rows and corresponding columns, and or multiplication
of some rows or columns with complex unit numbers. In other words, two complex
conference matrices A and A' are equivalent if there exist unitary diagonal matrices
D_1 and D_2 and a permutation matrix S such that $A' = D_1 S A S^T D_2$.
   The matrix of order 2

$$\begin{pmatrix} 0 & 1 \\ 1 & 0 \end{pmatrix}$$

is symmetric but equivalent to the skew-symmetric matrix

$$\begin{pmatrix} 0 & 1 \\ -1 & 0 \end{pmatrix}.$$

Here is the unique skew-symmetric conference matrix of order 4:

$$\begin{pmatrix} 0 & 1 & 1 & 1 \\ -1 & 0 & 1 & -1 \\ -1 & -1 & 0 & 1 \\ -1 & 1 & -1 & 0 \end{pmatrix}.$$

It is known from Sylvester that if C is a real skew-symmetric conference matrix of
order n then by construction the matrix

$$C_{2n} = \begin{pmatrix} C & C - I_n \\ C + I_n & -C \end{pmatrix},$$

is a real skew-symmetric conference matrix of order $2n$. Here is the unique 6-order real symmetric conference matrix

$$\begin{pmatrix} 0 & 1 & 1 & 1 & 1 & 1 \\ 1 & 0 & -1 & -1 & 1 & 1 \\ 1 & -1 & 0 & 1 & 1 & -1 \\ 1 & -1 & 1 & 0 & -1 & 1 \\ 1 & 1 & 1 & -1 & 0 & -1 \\ 1 & 1 & -1 & 1 & -1 & 0 \end{pmatrix}.$$

Here is the smallest complex conference matrix

$$\begin{pmatrix} 0 & \omega & \omega^2 & \omega^2 & \omega \\ \omega & 0 & \omega & \omega^2 & \omega^2 \\ \omega^2 & \omega & 0 & \omega & \omega^2 \\ \omega^2 & \omega^2 & \omega & 0 & \omega \\ \omega & \omega^2 & \omega^2 & \omega & 0 \end{pmatrix},$$

with $\omega = e^{\frac{i2\pi}{3}}$

Questions in the theory of polytopes, posed by Coxeter [13], led Paley [43] to the construction of real symmetric conference matrices. Paley in 1933 constructed real symmetric and skew-symmetric conference matrices with orders $p^\alpha + 1 \equiv 2$ (mod 4), and $p^\alpha + 1 \equiv 0 \pmod 4$ p odd prime, α non-negative integer.

The following necessary conditions for the existence of a real symmetric (respectively skew-symmetric) conference matrix of order n are known :

$n \equiv 2 \pmod 4$ and $n - 1 = a^2 + b^2$, a and b integers (respectively $n = 2$ or $n \equiv 0 \pmod 4$) [31].

The only real conference matrices that have been constructed so far are symmetric matrices of order $n = p^\alpha + 1 \equiv 2 \pmod 4$, p prime, α non-negative integer (Paley) or $n = (q - 1)^2 + 1$, where q is the order of a conference symmetric or skew symmetric matrix (Goethals and Seidel [31]) or $n = (q + 2)q^2 + 1$, where $q = 4t - 1 = p^\alpha$, p prime and $q + 3$ is the order of a conference symmetric matrix (Mathon [41]), or $n = 5 \cdot 9^{2\alpha+1} + 1$, α non-negative integer (Seberry and Whiteman [48]), and skew symmetric matrices of order $n = 2^s \prod_{i=1}^{r} (p_i^{\alpha_i} + 1)$,

$p_i^{\alpha_i} + 1 \equiv 0 \pmod 4$, p_i primes, s, r and α_i non-negative integers (Williamson [52]).

In fact Delsarte et al. [15] proved that essentially there are no other real conference matrices. Precisely they proved that any real conference matrix of order $n > 2$ is equivalent, under multiplication of rows and columns by -1, to a conference symmetric or to a skew symmetric matrix according as n satisfies $n \equiv 2 \pmod 4$ or $n \equiv 0 \pmod 4$.

In addition we observe that n must be even. This is not the case for complex conference matrices.

There is no complex conference matrix of order 3; however we can find such a matrix of order 5. Some complex conference matrices of even orders can be easily constructed. If C is a real symmetric conference matrix then iC is a complex symmetric conference matrix of even order (but equivalent to the real one). However only one method to construct complex conference matrices of odd orders is known and is reported in Sect. 9.7.

Observe that complex Hermitian or skew-symmetric conference matrices exist only if their orders are even.

9.7.1 Real Symmetric Conference Matrices

Let $GF(q)$ be the Galois field of order $q = p^\alpha$, $p^\alpha \equiv 1 \pmod 4$, p odd prime, α non-negative integer. Let χ denote the Legendre symbol, defined by $\chi(0) = 0$, $\chi(x) = 1$ or -1 according as x is or not a square in $GF(p^\alpha)$. It is well known that $\chi(-1) = 1$, $\chi(xy) = \chi(x)\chi(y)$, $\chi(x^{-1}) = (\chi(x))^{-1}$, $\chi(-x) = \chi(x)$ and $\sum_{x \in GF(q)} \chi(x) = 0$. The Paley matrix P of order $q = 2k - 1$ is defined by

$$P_{\alpha\alpha} = 0, \alpha = 1, \ldots, q, \text{ and } P_{\alpha\beta} = \chi(a_\alpha - a_\beta), \alpha \neq \beta, \alpha, \beta = 1, \ldots, q.$$

Here is an important formula of Jacobstal, which he used in a classification of some quadratic forms.

Theorem 9.7.1 *For any $b \in GF(q)^*$ we have*

$$\sum_{a \in GF(q)} \chi(a)\chi(a + b) = -1.$$

Paley proved that $P^2 = (2k - 1)I_{2k-1} - J_{2k-1}$ and $PJ_{2k-1} = 0$ by use of Jacobsthal's theorem. Paley extended the matrix P to obtain a real symmetric conference matrix of order $2k$ as follows:

$$C = \begin{pmatrix} 0 & j^\top \\ j & P \end{pmatrix},$$

where j is the $(2k - 1) \times 1$ matrix consisting solely of $1's$.

9.7.2 Complex Hermitian Conference Matrices

Goethals and Seidel proved in [31] that any Paley matrix of order $n = 2k = p^\alpha + 1 \equiv 2 \pmod 4$ is equivalent to a matrix of the form

$$
C = \begin{pmatrix} 0 & 1 & j^T & j^T \\ 1 & 0 & -j^T & j^T \\ j & -j & A & B \\ j & j & B^T & -A \end{pmatrix},
$$

where A and B are square matrices of order $k - 1$, j is the $(k - 1) \times 1$ matrix consisting solely of $1's$. The matrices A and B satisfy :

$$A^T = A,\, AJ = J,\, BJ = JB = 0, \tag{9.6}$$

$$AB = BA,\, BB^T = B^T B, \tag{9.7}$$

$$A^2 + BB^T = (2k - 1)I - 2J, \tag{9.8}$$

The following theorem is proved in [20].

Theorem 9.7.2 *Let $k \geq 3$ be an odd integer such that $2k = p^\alpha + 1$. There exists an infinite family of complex Hermitian conference matrices of order $2k$ depending on one complex parameter b of modulus 1.*

It is easy to check that the matrix

$$
C(b) = \begin{pmatrix} 0 & 1 & j^T & j^T \\ 1 & 0 & -j^T & j^T \\ j & -j & A & bB \\ j & j & \bar{b}B^T & -A \end{pmatrix},
$$

is a complex Hermitian conference matrix of order $2k$.

9.8 Complex Equiangular Tight Frames

Important for coding and quantum information theories are real and complex equiangular tight frames. In a Hilbert space $\mathcal{H}$, a subset $F = \{f_i\}_{i \in I} \subset \mathcal{H}$ is called a *frame* for $\mathcal{H}$ provided there are two constants $C, D > 0$ such that

$$
C \|x\|^2 \leq \sum_{i \in I} |\langle x, f_i \rangle|^2 \leq D \|x\|^2
$$

holds for every $x \in \mathcal{H}$. If $C = D = 1$ then the set is called *normalized tight* or a *Parseval frame*. Throughout this paragraph we use the term (n, k) frame to refer to a Parseval frame of n vectors in $\mathbb{C}^k$ equipped with the usual inner product. The ratio $\frac{n}{k}$ is called the *redundancy* of the (n, k) frame. It is well known that any Parseval frame induces an isometric embedding of $\mathbb{C}^k$ into $\mathbb{C}^n$ which maps $x \in \mathbb{C}^k$ to its frame coefficients $(Vx)_j = \langle x, f_j \rangle$, called the analysis operator of the frame. Because V is linear, we may identify V with an $n \times k$ matrix and the vectors $\{f_1, \ldots, f_n\}$ denote the columns of V^*, the Hermitian conjugate of V. From Holmes and Paulsen [36] we know that a (n, k) frame is determined up to a unitary equivalence by its Gram matrix VV^*, which is a self-adjoint projection of rank k. If in addition the frame is uniform and equiangular, that is, $\|f_i\|^2$ and $|\langle f_i, f_j \rangle|$ are constants for all $1 \le i \le n$ and for all $i \ne j$, $1 \le i, j \le n$, respectively, then

$$VV^* = \frac{k}{n} I_n + \sqrt{\frac{k(n-k)}{n^2(n-1)}} Q,$$

where Q is a self-adjoint matrix with diagonal entries all 0 and off-diagonal entries all of modulus 1, and I_n is the identity matrix of order n. The matrix Q is called the *Seidel matrix* or *signature matrix* associated with the (n, k) frame. The existence of an equiangular Parseval frame is known from Holmes and Paulsen to be equivalent to the existence of a Seidel matrix with two eigenvalues. Using the above theorem we arrive to the following result [20].

Theorem 9.8.1 *For any integer $k \ge 3$ such that $2k = p^\alpha + 1 \equiv 2 \pmod 4$ there exists a $(2k, k)$ CETF (complex equiangular tight frame).*

However Zauner constructed in his Phd thesis [53] $(q + 1, (q + 1)/2)$ $CETFs$ for any odd prime power q. But I proved in [20] that the associated Seidel matrices of these frames are real symmetric conference matrices or the product by i of real skew-symmetric conference matrices. First we recall the construction of Zauner. For any odd prime power $q = p^m$ let $GF(q)$ be the Galois field of order q, χ be the Legendre symbol which is a multiplicative character of $GF(q)^*$ and let ψ be the additive character defined by $\psi(a) = e^{2i\pi Tr(a)/p}$ where the Tr is the linear mapping from $GF(q)$ to F_p such that $Tr(a) = a^p + \ldots + a^{p^m}$. Now let $a_1, \ldots, a_q$ be the elements of $GF(q)$, $b_1, \ldots, b_{(q-1)/2}$ be the non-zero squares and $b'_1, \ldots, b'_{(q-1)/2}$ be the non-zero non squares. The following vectors are given in Zauner's thesis.

$$x_1 = (1, 0, \ldots, 0),$$

$$x_2 = (1/\sqrt{q}, \sqrt{2/q}\psi(b_1 a_1), \ldots, \sqrt{2/q}\psi(b_{(q-1)/2} a_1), \ldots,$$

$$x_{q+1} = (1/\sqrt{q}, \sqrt{2/q}\psi(b_1 a_q), \ldots, \sqrt{2/q}\psi(b_{(q-1)/2} a_q)).$$

Based on a formula on additive characters Zauner showed in that his vectors are unit and that the absolute value of any Hermitian product $\langle x_k, x_l \rangle$ with $k \ne l$, is equal to $\frac{1}{\sqrt{q}}$. In the following the Hermitian products $\langle x_k, x_l \rangle$ with $k \neq l$ are computed. For any $2 \leq k < l \leq q + 1$ the Hermitian product $\langle x_k, x_l \rangle$ is equal to

$$\frac{1}{q} + \frac{2}{q} \sum_{s=1}^{\frac{q-1}{2}} \psi(b_s(a_k - a_l)).$$

On the one hand

$$\sum_{s=1}^{\frac{q-1}{2}} \psi(b_s(a_k - a_l)) + \sum_{s=1}^{\frac{q-1}{2}} \psi(b_s'(a_k - a_l)) = \sum_{\alpha \in GF(q)} \psi(\alpha) - 1 = -1,$$

because $\sum_{\alpha \in GF(q)} \psi(\alpha) = 0$.

On the other hand

$$\sum_{s=1}^{\frac{q-1}{2}} \psi(b_s(a_k - a_l)) - \sum_{s=1}^{\frac{q-1}{2}} \psi(b_s'(a_k - a_l)) = \frac{1}{\chi(a_k - a_l)} \sum_{\alpha \in GF(q)} \chi(\alpha)\psi(\alpha).$$

However the last sum is a general Gauss sum which was computed by Berndt and Evans [4]. It turns out that this sum is equal to

$$(-1)^{m-1} \sqrt{q} \text{ if } p \equiv 1 (\mathrm{mod}\, 4) \text{ and}$$

$$- (-i)^m \sqrt{q} \text{ if } p \equiv -1 (\mathrm{mod}\, 4).$$

From this follows clearly that

$$\langle x_k, x_l \rangle = (-1)^{m-1} \frac{1}{\sqrt{q}} \chi(a_k - a_l) \text{ if } p \equiv 1 (\mathrm{mod}\, 4) \text{ and}$$

$$\langle x_k, x_l \rangle = -(-i)^m \frac{1}{\sqrt{q}} \chi(a_k - a_l) \text{ if } p \equiv -1 (\mathrm{mod}\, 4).$$

We see from the Gram matrices that with this construction we find again Paley matrices and no other complex conference matrices. That is Zauner's construction does not lead to new $(2k, k)$ $CETFs$ in comparison with our previous theorem. In the same time it is interesting to see how Zauner obtained again the Paley matrices using an additive character on $GF(q)^*$ instead of the Legendre symbol which is a multiplicative character.

Any real conference skew-symmetric matrix C of order $2k$ leads to a Seidel matrix iC with two eigenvalues and then to a $(2k, k)$ $CETF$. Note that the $2k$ vectors of this frame generate a set of equiangular lines called in my Phd thesis an *F-regular 2k-tuple* in $\mathbb{CP}^{k-1}$ [16]. This is a tuple in which all triples of lines are pairwise congruent. It is seen in [17] that there exists an F-regular 2k-tuple in $\mathbb{CP}^{k-1}$ if and only if there exists a real skew-symmetric conference matrix of order 2k.

9.9 Complex Symmetric Conference Matrices of Odd Orders

Let $q = 2k - 1$, ω be any complex number of modulus one, (a_α), $\alpha = 1, \ldots, q$ be the elements of $GF(q)$, $q = p^\alpha$, $p^\alpha \equiv 1 \pmod{4}$, and define the square matrix $C(\omega)$ of order q by

$$c_{\alpha\alpha} = 0, \alpha = 1, \ldots, q, \text{ and}$$

$$c_{\alpha\beta} = \omega^{\chi(a_\alpha - a_\beta)}, \alpha \neq \beta, \alpha, \beta = 1, \ldots, q.$$

Here is an analogous theorem to Jacobsthal's theorem [21].

Theorem 9.9.1 *For any $b \in GF(q)^*$ we have*

$$\sum_{a \in GF(q)^* \setminus \{-b\}} \omega^{\chi(a) - \chi(a+b)} = k - 2 + (k - 1)\Re(\omega^2).$$

Proof Define $z = 1 + \frac{b}{a}$. As a ranges over all non-zero elements of GF(q) except $-b$, z ranges over all non-zero elements of GF(q) except unity. However,

$$\frac{1 - \chi(z)}{\chi(z - 1)} = \frac{\chi(a) - \chi(a + b)}{\chi(b)}.$$

Thus

$$\sum_{a \in GF(q)^* \setminus \{-b\}} \omega^{\chi(a) - \chi(a+b)} = \sum_{z \in GF(q)^* \setminus \{1\}} \omega^{\chi(b) \frac{1 - \chi(z)}{\chi(z-1)}}.$$

We claim that this last sum is independent of b. Clearly

$$\sum_{z \in GF(q)^* \setminus \{1\}} \omega^{-\frac{1 - \chi(z)}{\chi(z-1)}} = \sum_{z \in GF(q)^* \setminus \{1\}} \omega^{\frac{1 - \chi(z)}{\chi(z-1)}},$$

because replacing z with $\frac{1}{z}$ negates exponents.

Recall that there are $k - 1$ non-zero squares and $k - 1$ non-zero non-squares. Clearly our sum is then of the form

$$r.1 + s\omega^2 + t\omega^{-2}.$$

Whence $r = k - 2$. Now ω^2 and ω^{-2} appear in pairs because if a non-square z yields ω^2 (respectively ω^{-2}) then its inverse is another non-square and yields ω^{-2} (respectively ω^2). According $s = t = \frac{k-1}{2}$. This proves the assertion. $\square$

Theorem 9.9.2 *The complex symmetric matrix $C(\omega)$ of order $2k - 1$ satisfies*

$$C^*C = (2k - 2 - c)I + cJ, \text{ with}$$

$$c = k - 2 + (k - 1)\Re(\omega^2).$$

As a consequence we obtain the following result.

Corollary 9.9.3 *The complex symmetric matrix $C(\omega)$ of order $2k - 1$ such that $\Re(\omega^2) = \frac{2-k}{k-1}$ satisfies*

$$C^*C = (2k - 2)I.$$

Here is the 9-order complex symmetric conference matrix:

$$\begin{pmatrix}
0 & \omega & \omega & \omega & \omega & \omega^{-1} & \omega^{-1} & \omega^{-1} & \omega^{-1} \\
\omega & 0 & \omega^{-1} & \omega^{-1} & \omega & \omega & \omega & \omega^{-1} & \omega^{-1} \\
\omega & \omega^{-1} & 0 & \omega & \omega^{-1} & \omega & \omega^{-1} & \omega & \omega^{-1} \\
\omega & \omega^{-1} & \omega & 0 & \omega^{-1} & \omega^{-1} & \omega & \omega^{-1} & \omega \\
\omega & \omega & \omega^{-1} & \omega^{-1} & 0 & \omega^{-1} & \omega^{-1} & \omega & \omega \\
\omega^{-1} & \omega & \omega & \omega^{-1} & \omega^{-1} & 0 & \omega & \omega & \omega^{-1} \\
\omega^{-1} & \omega & \omega^{-1} & \omega & \omega^{-1} & \omega & 0 & \omega^{-1} & \omega \\
\omega^{-1} & \omega^{-1} & \omega & \omega^{-1} & \omega & \omega & \omega^{-1} & 0 & \omega \\
\omega^{-1} & \omega^{-1} & \omega^{-1} & \omega & \omega & \omega^{-1} & \omega & \omega & 0
\end{pmatrix},$$

where ω is a unit complex number such that $\Re(\omega^2) = -\frac{3}{4}$.

The following operations on the set of complex symmetric conference matrices of the same order q:

1. multiplication by a unit complex number of any row and the corresponding column,
2. interchange of rows and, simultaneously, of the corresponding columns,

generate a relation , called *equivalence*. Concerning this equivalence relation we have the following result [21].

Theorem 9.9.4 *For any $q = 2k - 1 = p^{\alpha}$, p odd prime, $k \geq 3$ there exist four complex symmetric conference matrices of order q and all are equivalent.*

We have seen above that for any integer n for which there exists a real symmetric conference matrix of Paley type there exists a complex symmetric conference matrix of order $n - 1$. Notice that there exist other real symmetric conference matrices which are not of Paley type. The smallest integer for which there exist nonequivalent real symmetric conference matrices is 26. More precisely there are four nonequivalent real symmetric conference matrices. Classification of real both symmetric and skew-symmetric conference matrices is still at its infancy. Recently with Blokhuis and Brehm, we generalised the previous result. We proved [5] that

Theorem 9.9.5 *If there exists a real symmetric conference matrix of order n then there exists a complex symmetric conference matrix of order $n - 1$.*

We also constructed other complex symmetric conference matrices called dihedral and which are not obtained by the previous construction.

More recently, with Makhlouf, we constructed in [24] complex skew-symmetric conference matrices, from which we can construct complex symmetric conference matrices. These matrices can be used for our geometric problem. Let C be a complex conference matrix of order n. We may assume

$$C = \begin{pmatrix} 0 & j^{\top} \\ j & P \end{pmatrix},$$

and C satisfies $C^* C = (n - 1)I_n$. Thus P of order $n - 1$ satisfies

$$P^* P = (n - 1)I_{n-1} - J_{n-1}, \tag{9.9}$$

$$P^* J = JP = O. \tag{9.10}$$

The matrix of order $(n - 1)^2$ defined by

$$T = P \otimes P + I_{n-1} \otimes J_{n-1} - J_{n-1} \otimes I_{n-1}$$

is symmetric (if P is symmetric or skew-symmetric) and satisfies

$$T^* T = (n - 1)^2 I_{(n-1)^2} - J_{(n-1)^2}, \tag{9.11}$$

$$T^* J = JT = O. \tag{9.12}$$

Hence T could be extended to a complex symmetric conference matrix of order $(n - 1)^2 + 1$. We also classified all complex conference matrices up to order 5 and we fully classified 6-order complex symmetric, skew-symmetric and Hermitian complex conference matrices. Among other things we proved that any complex symmetric conference matrix of order 6 is equivalent to the real symmetric conference matrix of order 6 of Paley type.

9.10   Equi-Isoclinic Planes in Euclidean Odd Dimensional Spaces

We know from Lemmens and Seidel, that there exist n equi-isoclinic planes which span $\mathbb{R}^r$ with parameter λ if and only if there exists a real symmetric matrix of order $2n$, partitioned into square blocks of order 2 with zero blocks on the diagonal and orthonormal blocks elsewhere, whose smallest eigenvalue equals $-\lambda^{\frac{-1}{2}}$ and has multiplicity $2n - r$. All the results of this section can be found in [21]. Let us set $\omega = e^{i\theta}$ with $\cos(2\theta) = \frac{2-k}{k-1}$ and define the block matrix S of order $2(2k-1)$ as follows :

$$S_{\alpha\alpha} = \begin{pmatrix} 0 & 0 \\ 0 & 0 \end{pmatrix},$$

$\alpha = 1, \ldots, q$, and

$$S_{\alpha\beta} = \begin{pmatrix} \cos(\theta\chi(a_\alpha - a_\beta)) & \sin(\theta\chi(a_\alpha - a_\beta)) \\ \sin(\theta\chi(a_\alpha - a_\beta)) & -\cos(\theta\chi(a_\alpha - a_\beta)) \end{pmatrix},$$

$\alpha \neq \beta, \alpha, \beta = 1, \ldots, q$. As in the case of complex symmetric conference matrices of odd orders we need the following formula.

Theorem 9.10.1 *Let θ be any real number, r_η the plane rotation with angle η and $b \in GF(q)^*$. Then*

$$\sum_{a \in GF(q)^* \setminus \{-b\}} r_{\theta(\chi(a) - \chi(a+b))} = (k - 2 + (k-1)\cos(2\theta))I_2.$$

This theorem leads to the following results.

Theorem 9.10.2 *The matrix S of order $4k - 2$ is symmetric and satisfies*

$$S^2 = (2k - 2)I_{4k-2}.$$

Corollary 9.10.3 *S has two eigenvalues with equal multiplicities.*

Corollary 9.10.4 *For any integer $k \geq 3$ such that $2k = p^\alpha + 1 \equiv 2 \pmod{4}$, p odd prime, α non-negative integer, $v_{\frac{1}{2k-2}}(2, 2k-1) = 2k - 1$.*

Corollary 9.10.5 *$v_{\frac{1}{4}}(2, 5) = 5$, $v_{\frac{1}{8}}(2, 9) = 9$, $v_{\frac{1}{12}}(2, 13) = 13. \ldots$*

In order to give more examples, we consider all odd $k \geq 3$ and $k \leq 51$ such that $2k = p^\alpha + 1$, p odd prime, α non-negative integer, and we obtain the following corollary.

Corollary 9.10.6 *If k is odd, $3 \leq k \leq 51$, we may construct the $(2k - 1)$-order complex symmetric conference matrix and thus the maximal $(2k - 1)$-tuple of equi-isoclinic planes with parameter $\frac{1}{2k-2}$ in $\mathbb{R}^{2k-1}$, except possibly in the cases $k = 11, 17, 23, 29, 33, 35, 39, 43, 47$.*

A matrix of order n with unimodular entries and satisfying $HH^* = nI_n$ is called *complex Hadamard*. Complex *conference* matrices are also important in the setting of complex Hadamard matrix theory because if C_n is a complex conference matrix of order n then by construction the matrix

$$H_{2n} = \begin{pmatrix} C_n + I_n & C_n^* - I_n \\ C_n - I_n & -C_n^* - I_n \end{pmatrix},$$

is a complex Hadamard matrix of order $2n$.

9.11   Equi-Isoclinic Planes in $\mathbb{C}^4$

Let $\Phi_{n,m}$ be the function from n by m quaternionic matrices to $2n$ by $2m$ complex matrices with the properties

$$\forall X \in \mathbb{H}^{m \times n}, \forall Y \in \mathbb{H}^{n \times p}, \Phi_{m,p}(XY) = \Phi_{m,n}(X)\Phi_{n,p}(Y),$$

$$\forall X, Y \in \mathbb{H}^{m \times n}, \forall s, t \in \mathbb{R}, \Phi_{m,n}(sX + tY) = s\Phi_{m,n}(X) + t\Phi_{m,n}(Y),$$

$$\forall X \in \mathbb{H}^{m \times n}, \Phi_{n,m}(X^*) = (\Phi_{m,n}(X))^*.$$

It is also known that $X \in \mathbb{H}^{m \times n}$ has rank r if and only if $\Phi_{m,n}(X)$ has rank $2r$. These results can be found in [46] (Chapter 3).

$$U = \{z \in \mathbb{C} \mid |z| = 1\}.$$

The matrices of homotheties of $M_2(\mathbb{C})$, the ring of 2×2 matrices over $\mathbb{C}$, are denoted by the corresponding complex number ($1 = I, \ldots$). Hoggar's theorem leads to the following propositions.

Proposition 9.11.1 $v_\lambda(2, 4, \mathbb{C}) = 2$ *if* $0 \leq \lambda < \frac{1}{4}$, $v_\lambda(2, 4, \mathbb{C}) = 3$ *if* $\frac{1}{4} \leq \lambda < \frac{1}{3}$, $v_\lambda(2, 4, \mathbb{C}) = 4$ *if* $\frac{1}{3} \leq \lambda < \frac{3}{8}$, $v_\lambda(2, 4, \mathbb{C}) = 5$ *if* $\frac{3}{8} \leq \lambda < \frac{2}{5}$, $v_{\frac{2}{5}}(2, 4, \mathbb{C}) = 6$ *and* $v_\lambda(2, 4, \mathbb{C}) = 5$ *if* $\frac{2}{5} < \lambda < 1$, *whence* $v(2, 4, \mathbb{C}) = 6$.

Proposition 9.11.2 $v_\lambda(1, 2, \mathbb{H}) = 2$ *if* $0 \leq \lambda < \frac{1}{4}$, $v_\lambda(1, 2, \mathbb{H}) = 3$ *if* $\frac{1}{4} \leq \lambda < \frac{1}{3}$, $v_\lambda(1, 2, \mathbb{H}) = 4$ *if* $\frac{1}{3} \leq \lambda < \frac{3}{8}$, $v_\lambda(1, 2, \mathbb{H}) = 5$ *if* $\frac{3}{8} \leq \lambda < \frac{2}{5}$, $v_{\frac{2}{5}}(1, 2, \mathbb{H}) = 6$ *and* $v_\lambda(1, 2, \mathbb{H}) = 5$ *if* $\frac{2}{5} < \lambda < 1$, *therefore* $v(1, 2, \mathbb{H}) = 6$.

The two propositions hold word by word when replacing planes in $\mathbb{C}^4$ by lines in $\mathbb{H}^2$. However there are some interesting differences which are developed below. For instance we have three congruence classes of quadruples of equi-isoclinic planes in $\mathbb{C}^4$ with parameter $\frac{1}{3}$ but just one congruence class of quadruples of equiangular lines in $\mathbb{H}^2$ with the same parameter. In the last section we characterize all the p-tuples of equi-isoclinic planes in $\mathbb{C}^{2r}$ which are the images by $\Phi_{r,r}$ of some p-tuples of equiangular lines in $\mathbb{H}^r$. The maximum number of equiangular lines in $\mathbb{H}^2$ and the maximum number of equi-isoclinic planes in $\mathbb{C}^4$ are equal by the two propositions, however Hoggar did not give a 6-tuple of equiangular lines of $\mathbb{H}^2$. In this section we give a continuous one parameter family of quadruples of equiangular lines, two continuous one parameter families of quintuples of equiangular lines in $\mathbb{H}^2$. In addition we obtain two isometry classes of 6-tuples of equiangular lines in $\mathbb{H}^2$ with parameter $\frac{2}{5}$.

9.11.1 Triples

Here is given the list of all triples of equi-isoclinic planes in $\mathbb{C}^4$. Clearly a triple of planes in $\mathbb{C}^6$ has the following normal Seidel matrix

$$S = \begin{pmatrix} 0 & 1 & 1 \\ 1 & 0 & M \\ 1 & M^* & 0 \end{pmatrix}, M \in U(2).$$

A computation of its characteristic polynomial leads to the following result proved in [22].

Theorem 9.11.3 *Let* $c \in]0, 1[$ *and* $\omega = \arccos\left(\frac{1}{2c}\left(3 - \frac{1}{c^2}\right)\right)$.

1. *If* $c < 1/2$, *then no triple of planes with parameter* c *exist in* $\mathbb{C}^4$.
2. *If* $c = \frac{1}{2}$, *there exists a unique congruence class of triples of planes with parameter* c. *A normal Seidel matrix is*

$$\begin{pmatrix} 0 & 1 & 1 \\ 1 & 0 & -1 \\ 1 & -1 & 0 \end{pmatrix}.$$

3. *If $c > \frac{1}{2}$, there exists three isometry classes of triples of planes with parameter c. The associated Seidel matrices are:*

$$\begin{pmatrix} 0 & 1 & 1 \\ 1 & 0 & e^{i\omega} \\ 1 & e^{-i\omega} & 0 \end{pmatrix}, \begin{pmatrix} 0 & 1 & 1 \\ 1 & 0 & e^{-i\omega} \\ 1 & e^{i\omega} & 0 \end{pmatrix}, \begin{pmatrix} 0 & 1 & 1 \\ 1 & 0 & \begin{pmatrix} e^{i\omega} & 0 \\ 0 & e^{-i\omega} \end{pmatrix} \\ 1 & \begin{pmatrix} e^{-i\omega} & 0 \\ 0 & e^{i\omega} \end{pmatrix} & 0 \end{pmatrix}.$$

Remark The symmetry group of any triple of equi-isoclinic planes in $\mathbb{C}^2$ with Seidel matrix containing homotheties is isomorphic to the alternating group A_3.

9.11.2 Quadruples

This subsection is devoted to list all of the 4-tuples of equi-isoclinic planes embedded in $\mathbb{C}^4$. Although the usual method, involving the Seidel matrices gives the desired result in the case $\mathbb{F} = \mathbb{R}$, the calculus in the complex case seems too difficult. Hence we choose to investigate the problem from another point of view. We use the fact that very few triples of equi-isoclinic planes imbedded in $\mathbb{C}^4$ exist (see Theorem 9.11.3), and that each sub-triple of any 4-tuple has to be isometric to one of them.

A normal Seidel matrix of a quadruple of equi-isoclinic planes is of the form:

$$S = \begin{pmatrix} 0 & 1 & 1 & 1 \\ 1 & 0 & A & C^* \\ 1 & A^* & 0 & B \\ 1 & C & B^* & 0 \end{pmatrix},$$

where A, B, C are $e^{\pm i\omega}$ or $R \stackrel{def}{=} \begin{pmatrix} e^{i\omega} & 0 \\ 0 & e^{-i\omega} \end{pmatrix}$. In addition ABC is also of the same type. We use the following notation

$$M(\gamma, \alpha, \beta) \stackrel{def}{=} \begin{pmatrix} \cos\gamma\, e^{i\alpha} & -\sin\gamma\, e^{-i\beta} \\ \sin\gamma\, e^{i\beta} & \cos\gamma e^{-i\alpha} \end{pmatrix}.$$

From [22] we have

Theorem 9.11.4

1. If $c < 1/\sqrt{3}$ there is no quadruple of equi-isoclinic planes in $\mathbb{C}^4$.
2. If $c > 1/\sqrt{3}$, there exists a unique congruence class of quadruples in $\mathbb{C}^4$ with parameter c . The normal Seidel matrix is

$$
S(c) \overset{def}{=} \begin{pmatrix} 0 & 1 & 1 & 1 \\ 1 & 0 & R & Z \\ 1 & R^* & 0 & M \\ 1 & Z^* & M^* & 0 \end{pmatrix},
$$

with $M = \frac{c}{1-c^2}(1 + R^*Z - cR^* - cZ)$, $Z = M(\gamma, \alpha, 0)$, $R = M(0, \omega, \ldots)$, $\omega = \arccos\left(\frac{1}{2c}\left(3 - \frac{1}{c^2}\right)\right)$, $\alpha = \arctan\frac{(1-c^2)(2c^2-1)}{(3c^2-1)\sqrt{4c^2-1}}$, and $\gamma = \arccos\frac{\cos\omega}{\cos\alpha}$.

3. If $c = \frac{1}{\sqrt{3}}$, there exist three isometry classes of quadruples of equi-isoclinic planes in $\mathbb{C}^4$ with parameter $\frac{1}{3}$ with normal Seidel matrices

$$
S_+ = \begin{pmatrix} 0 & 1 & 1 & 1 \\ 1 & 0 & i & -i \\ 1 & -i & 0 & i \\ 1 & i & -i & 0 \end{pmatrix}, \; S_- = \begin{pmatrix} 0 & 1 & 1 & 1 \\ 1 & 0 & -i & i \\ 1 & i & 0 & -i \\ 1 & -i & i & 0 \end{pmatrix} \; and
$$

$$
\lim_{c \to 1/\sqrt{3}} S(c) = \begin{pmatrix} 0 & 1 & 1 & 1 \\ 1 & 0 & R & -R \\ 1 & -R & 0 & R \\ 1 & R & -R & 0 \end{pmatrix} \; where \; R = \begin{pmatrix} i & 0 \\ 0 & -i \end{pmatrix}.
$$

Remark

1. The block R of the third matrix $S\left(1/\sqrt{3}\right)$ is conjugate to the plane rotation $R(\pi/2) = \begin{pmatrix} 0 & -1 \\ 1 & 0 \end{pmatrix}$ by the unitary $\begin{pmatrix} \frac{i}{\sqrt{2}} & -\frac{i}{\sqrt{2}} \\ \frac{1}{\sqrt{2}} & \frac{1}{\sqrt{2}} \end{pmatrix}$. Then we obtain the 4 equi-isoclinic planes in $\mathbb{R}^4$.

2. Let $H_4 = (\Gamma_1, \Gamma_2, \Gamma_3, \Gamma_4)$ be four planes spanned respectively by the columns of

$$
T_1 = \begin{pmatrix} 1 \\ 0 \end{pmatrix}, \; T_2 = \begin{pmatrix} c \\ s \end{pmatrix}, \; T_3 = \begin{pmatrix} c \\ \frac{c}{s}(R-c) \end{pmatrix},
$$

$$
T_4 = \begin{pmatrix} c \\ \frac{c}{s}(Z-c) \end{pmatrix}.
$$

Each equi-isoclinic 4-tuple with parameter $c > \frac{1}{\sqrt{3}}$ imbedded in $\mathbb{C}^4$ is isometric to H_4. Applying some permutation $\sigma \in S_4$ to the planes of H_4, we obtain another 4-tuple imbedded in $\mathbb{C}^4$ which is therefore isometric to H_4. It follows that H_4 is regular, that is its symmetry group is isomorphic to the symmetric group S_4.

3. The two quadruples arising from homotheties in the case $c = \frac{1}{\sqrt{3}}$ have both their symmetry group isomorphic to the alternating group A_4.

9.11.3 Quintuples

In the following we give all the quintuples of equi-isoclinic planes imbedded in $\mathbb{C}^4$. The quadruples in $\mathbb{C}^4$ are all of type "all homotheties" or "all in $SU(2)$" thus we cannot have a quintuple with mixed quadruples. A careful investigation of all cases [22] leads to the following theorem.

Theorem 9.11.5

1. If $c < \sqrt{\frac{3}{8}}$ there is no quintuple of equi-isoclinic planes in $\mathbb{C}^4$.

2. If $c > \sqrt{\frac{3}{8}}$ there are two congruence classes of quintuples of equi-isoclinic planes in $\mathbb{C}^4$ with parameter c with Seidel matrices, S_ε $(\varepsilon = \pm 1)$

$$
S_\varepsilon = \begin{pmatrix}
0 & 1 & 1 & 1 & 1 \\
1 & 0 & R & Z & Z_\varepsilon \\
1 & R^* & 0 & M & M_\varepsilon \\
1 & Z^* & M^* & 0 & N_\varepsilon \\
1 & (Z_\varepsilon)^* & (M_\varepsilon)^* & (N_\varepsilon)^* & 0
\end{pmatrix},
$$

$M = \frac{c}{1-c^2}(1 + R^*Z - cR^* - cZ)$, $Z = M(\gamma, \alpha, 0)$, $R = M(0, \omega, \ldots)$,

$\omega = \arccos\left(\frac{1}{2c}\left(3 - \frac{1}{c^2}\right)\right)$, $\alpha = \arctan\frac{(1-c^2)(2c^2-1)}{(3c^2-1)\sqrt{4c^2-1}}$, $\gamma = \arccos\frac{\cos\omega}{\cos\alpha}$,

$M_\varepsilon = \frac{c}{1-c^2}(1 + R^*Z_\varepsilon - cR^* - cZ_\varepsilon)$, $Z_\varepsilon = M(\gamma, \alpha, -2\varepsilon\beta_0)$, $\beta_0 = \frac{1}{2}\arccos\left(h\left(c^2\right)\right)$, $N_\varepsilon = \frac{c}{1-c^2}(1 + Z^*Z_\varepsilon - c(Z^* + Z_\varepsilon))$.

3. If $c = \sqrt{\frac{3}{8}}$ there is one congruence class of quintuples of equi-isoclinic planes in $\mathbb{C}^4$ with parameter c with Seidel matrix $S_1 = S_{-1}$.

9.11.4 Sextuples

The following results are proved in [22].

Theorem 9.11.6 *There exist two congruence classes of 6-tuples of equi-isoclinic planes in* $\mathbb{C}^4$ *with parameter* $\sqrt{2/5}$ *(and then* $\beta_0 = \gamma = \pi/3$*). Their Seidel matrices are* S_ε

$$
S_\varepsilon = \begin{pmatrix}
0 & 1 & 1 & 1 & 1 & 1 \\
1 & 0 & R & Z & Z_\varepsilon & Z_{-\varepsilon} \\
1 & R^* & 0 & M & M_\varepsilon & M_{-\varepsilon} \\
1 & Z^* & M^* & 0 & N_\varepsilon & N_{-\varepsilon} \\
1 & Z_\varepsilon^* & M_\varepsilon^* & N_\varepsilon^* & 0 & W \\
1 & (Z_{-\varepsilon})^* & (M_{-\varepsilon})^* & (N_{-\varepsilon})^* & W^* & 0
\end{pmatrix},
$$

where $R, Z, M, Z_\varepsilon, M_\varepsilon, N_\varepsilon$ *are as in Theorem 9.11.5. W is given in [22].*

Theorem 9.11.7 *No 7-tuple of equi-isoclinic planes can exist in* $\mathbb{C}^4$.

Remark Let Γ_5, Γ_6 be the planes generated by the columns of

$$
T_5 = \begin{pmatrix} c \\ \frac{c}{s}(Z_\epsilon - c) \end{pmatrix}, \quad T_6 = \begin{pmatrix} c \\ \frac{c}{s}(Z_{-\epsilon} - c) \end{pmatrix}.
$$

$H_6 = (\Gamma_1, \Gamma_2, \Gamma_3, \Gamma_4, \Gamma_5, \Gamma_6)$ and $H_6' = (\Gamma_1, \Gamma_2, \Gamma_3, \Gamma_4, \Gamma_6, \Gamma_5)$ are our sextuples of equi-isoclinic planes. They are obviously the only (up to isometry) 6-tuples of equi-isoclinic planes imbedded in $\mathbb{C}^4$, and no such 7-tuple can exist in $\mathbb{C}^4$. Then it is clear that H_6 and H_6' are not isometric. If not, H_5 and H_5' would be isometric which is not possible.

The symmetric group S_p (right) acts on the set of p-tuples of equi-isoclinic planes in a natural way: for any p-tuple $H = (\Gamma_1, \ldots, \Gamma_p)$ and any $\sigma \in S_p$, we denote by $H\sigma$ the p-tuple $(\Gamma_{\sigma(1)}, \ldots, \Gamma_{\sigma(p)})$. It is easy to see that the action induces an action on the quotient and that S_p also acts on the set $\mathcal{H}_p(\mathbb{C}^4)$ of congruence classes of p-tuples of equi-isoclinic planes imbedded in $\mathbb{C}^4$. Let us denote by $\bar{H} \in \mathcal{H}_p(\mathbb{C}^4)$ the congruence class of H. It is easy to see that the stabilizer of $\bar{H}$ is also the symmetry group of H. We denote it by $G(H)$.

The transposition (5 6) obviously does not belong to $G(H_6)$, which is thereby a proper subgroup of S_6. Hence the orbit of $\overline{H_6}$ in $\mathcal{H}_p(\mathbb{C}^4)$ has exactly two elements. It follows that the index $[S_6 : G(H_6)] = 2$. It is a well-known fact that A_p is the only 2-index subgroup of S_p for $p \geq 5$ whence $G(H_6) \cong A_6$. Of course, the above argument also holds for $G(H_6')$.

Since no transposition belongs to $G(H_6)$, H_6 (1 2) is isometric to H_6', and so it is of H_5 (1 2) and H_5'. Hence (1 2) does not belong to $G(H_5)$. Thus the symmetry group of any quintuple is isomorphic to the alternating group A_5.

The method displayed in this section was used efficiently to obtain $v(3, 8, \mathbb{R})[25]$.

9.12 Quaternionic Equiangular Lines

9.12.1 A One to One Correspondence

In this section we prove the following theorem.

Theorem 9.12.1 *There is a one to one correspondence between congruence classes of n-tuples of equi-isoclinic planes in $\mathbb{C}^{2r}$ with parameter λ whose associated Seidel matrices contain zero blocks on the diagonal, and blocks in $SU(2)$ elsewhere and congruence classes of n-tuples of equiangular lines in $\mathbb{H}^r$ with angle $\arccos \sqrt{\lambda}$.*

Proof If $([e_1], \ldots, [e_n])$ is an n-tuple of equiangular lines with angle $\arccos c$ embedded in $\mathbb{H}^r$ then $H = (\langle e_p, e_q \rangle)_{p,q=1,\ldots,n}$ is a Hermitian positive semi-definite and of rank r. Hence $H = A^*A$ with $A = (e_1 \ldots e_n)$, where $e_1, \ldots, e_n$ are the columns of the matrix A. However,

$$\Phi_{n,n}(H) = \Phi_{n,n}(A^*A) = \Phi_{n,r}(A^*)\Phi_{r,n}(A).$$

Now put $\Phi_{r,n}(A) = (e_{11}, e_{12}, \ldots, e_{n1}, e_{n2}) \in \mathbb{C}^{2r \times 2n}$. The entry M_{pq} of $\Phi_{n,n}(H) \in \mathbb{C}^{2n \times 2n}$, $p \neq q$, $p, q = 1, \ldots, n$, is

$$M_{pq} = \begin{pmatrix} \langle e_{p1}, e_{q1} \rangle & \langle e_{p1}, e_{q2} \rangle \\ \langle e_{p2}, e_{q1} \rangle & \langle e_{p2}, e_{q2} \rangle \end{pmatrix}.$$

If we set $h_{pq} = \langle e_p, e_q \rangle = c(z_{pq} + j w_{pq})$, then

$$M_{pq} = \begin{pmatrix} cz_{pq} & -c\overline{w_{pq}} \\ cw_{pq} & c\overline{z_{pq}} \end{pmatrix},$$

hence $\langle e_{p1}, e_{q1} \rangle = cz_{pq}$, $\langle e_{p1}, e_{q2} \rangle = -c\overline{w_{pq}}$. Thus for any $p \neq q$, $p, q = 1, \ldots, n$, $\frac{1}{c}M_{pq} \in SU(2)$ and the planes $(\Gamma_1, \Gamma_2, \ldots, \Gamma_n)$ generated respectively by (e_{11}, e_{12}), $(e_{21}, e_{22}) \ldots$, and (e_{n1}, e_{n2}) are pairwise isoclinic with the same angle $\arccos c$.

Conversely let $(\Gamma_1, \ldots, \Gamma_n)$ be a n-tuple of equi-isoclinic planes in $\mathbb{C}^{2r}$ with parameter λ and with associated matrix containing non-diagonal blocks of the form

$$\begin{pmatrix} z_{pq} & -\overline{w_{pq}} \\ w_{pq} & \overline{z_{pq}} \end{pmatrix},$$

where $z_{pq}, w_{pq} \in \mathbb{C}$ such that $|z_{pq}|^2 + |w_{pq}|^2 = 1$ for every $p < q$, $p, q = 2, \ldots, n$. Its Gram matrix is $G = I + cM$ where the non-diagonal entries of M are all of the form $\begin{pmatrix} z_{pq} & -\overline{w_{pq}} \\ w_{pq} & \overline{z_{pq}} \end{pmatrix}$. Consider $H = I + N$ where N is the matrix containing zeros on the diagonal and $s_{pq} = z_{pq} + j w_{pq}$ elsewhere. We have $\Phi_{n,n}(H) = G$ by construction. But G is Hermitian positive semi-definite of rank $2r$. Thus H is Hermitian positive semi-definite and with rank r. Whence there exists an n-tuple $([e'_1], \ldots, [e'_n])$ of equiangular lines with angle $\arccos c$ imbedded in $\mathbb{H}^r$. This means that if we set $A' = (e'_1 \ldots e'_n)$ then $H = A'^*A'$. Therefore $\Phi_{n,n}(H) = \Phi_{n,r}(A'^*)\Phi_{r,n}(A') = G$. Thus the Gram matrix of the columns of $\Phi_{r,n}(A') = (e'_{11}, e'_{12}, \ldots, e'_{n1}, e'_{n2})$ is equal to G. This implies that there exists an isometry Ψ of $\mathbb{C}^{2r}$ such that $\Psi(e_{pq}) = e'_{pq}$ for all $p = 1, \ldots, n$ and $q = 1, 2$. This completes the proof. □

An analogue of this proof can be used to prove Theorem 9.5.2. We can find in [19] another proof of Theorem 9.5.2.

9.12.2 Quaternionic Equiangular Lines in $\mathbb{H}^2$

Using our one to one correspondence we derive the list of all p-tuples of equiangular lines in $\mathbb{H}^2$. Let $e_1, e_2, \ldots, e_6$ be the following vectors

$$e_1 = (1, 0), \ e_2 = (c, s), \ e_3 = \left(c, \frac{c}{s}(e^{i\omega} - c)\right),$$

$$e_4 = \left(c, \frac{c}{s}(\cos(\gamma)e^{i\alpha} - c + \sin(\gamma)j)\right),$$

$$e_5 = \left(c, \frac{c}{s}(\cos(\gamma)e^{i\alpha} - c + \sin(\gamma)e^{2i\beta_0}j)\right),$$

$$e_6 = \left(c, \frac{c}{s}(\cos(\gamma)e^{i\alpha} - c + \sin(\gamma)e^{-2i\beta_0}j)\right),$$

where $s = \sqrt{1 - c^2}$, $\omega = \arccos\left(\frac{1}{2c}\left(3 - \frac{1}{c^2}\right)\right)$, $\alpha = \arctan\frac{(1-c^2)(2c^2-1)}{(3c^2-1)\sqrt{4c^2-1}}$, $\gamma = \arccos\frac{\cos\omega}{\cos\alpha}$ and $\beta_0 = \frac{1}{2}\arccos\left(\frac{2c^2-1}{2(3c^2-1)}\right)$ for $c > \frac{1}{\sqrt{3}}$,

Using our one to one correspondence and Theorems 9.11.4, 9.11.5 and 9.11.6 we easily arrive at the following.

Theorem 9.12.2

1. For $c < \frac{1}{2}$, there is no 3-tuple of equiangular lines with parameter c embedded in $\mathbb{H}^2$.

2. For any parameter $c \in \left[\frac{1}{2}, \frac{1}{\sqrt{3}}\right[$ there exists a unique (up to a global isometry) 3-tuple $([e_1], [e_2], [e_3])$ of equiangular lines with parameter c embedded in $\mathbb{H}^2$. The triple with parameter $\frac{1}{2}$ degenerates in $\mathbb{R}^2$. All of these triples have their symmetry group isomorphic to the symmetric group S_3.

3. *For any parameter $c \in \left[\frac{1}{\sqrt{3}}, \sqrt{\frac{3}{8}} \right[$ there exists a unique (up to a global isometry) 4-tuple $([e_1], [e_2], [e_3], [e_4])$ of equiangular lines with parameter c embedded in $\mathbb{H}^2$. The quadruple with parameter $\frac{1}{\sqrt{3}}$ degenerates in $\mathbb{C}^2$. All of these quadruples have their symmetry group isomorphic to the symmetric group S_4.*

4. *There exists a unique (up to a global isometry) 5-tuple $([e_1], [e_2], [e_3], [e_4], [e_5])$ of equiangular lines with parameter $\sqrt{\frac{3}{8}}$ embedded in $\mathbb{H}^2$. The symmetry group of this quintuple is isomorphic to the alternating group A_5. The vectors are the following*

$$e_1 = (1,0), \ e_2 = \left(\sqrt{\frac{3}{8}}, \sqrt{\frac{5}{8}}\right), \ e_3 = \left(\sqrt{\frac{3}{8}}, -\frac{1}{6}\sqrt{\frac{5}{2}} + \frac{\sqrt{5}}{3}i\right),$$

$$e_4 = \left(\sqrt{\frac{3}{8}}, -\frac{1}{3}\sqrt{\frac{5}{8}} - \frac{\sqrt{5}}{6}i + \frac{5}{6}\sqrt{\frac{3}{5}}j\right),$$

$$e_5 = \left(\sqrt{\frac{3}{8}}, -\frac{1}{3}\sqrt{\frac{5}{8}} - \frac{\sqrt{5}}{6}i - \frac{5}{6}\sqrt{\frac{3}{5}}j\right).$$

5. *For any parameter $c \in \left] \sqrt{\frac{3}{8}}, \sqrt{\frac{2}{5}} \right[$ there exists two (up to a global isometry) 5-tuples $([e_1], [e_2], [e_3], [e_4], [e_5])$ and $([e_1], [e_2], [e_3], [e_4], [e_6])$ of equiangular lines with parameter c embedded in $\mathbb{H}^2$. Both quintuples have their symmetry group isomorphic to A_5.*

6. *There exists two (up to a global isometry) 6-tuples $([e_1], [e_2], [e_3], [e_4], [e_5], [e_6])$ and $([e_1], [e_2], [e_3], [e_4], [e_6], [e_5])$ of equiangular lines with parameter $\sqrt{\frac{2}{5}}$ embedded in $\mathbb{H}^2$. The symmetry group of both of these sextuples is isomorphic to the alternating group A_6. The vectors are the following*

$$e_1 = (1,0), \ e_2 = \left(\sqrt{\frac{2}{5}}, \sqrt{\frac{3}{5}}\right), \ e_3 = \left(\sqrt{\frac{2}{5}}, -\frac{1}{4}\sqrt{\frac{3}{5}} + \frac{3}{4}i\right),$$

$$e_4 = \left(\sqrt{\frac{2}{5}}, -\frac{1}{4}\sqrt{\frac{3}{5}} - \frac{1}{4}i + \frac{\sqrt{2}}{2}j\right),$$

$$e_5 = \left(\sqrt{\frac{2}{5}}, -\frac{1}{4}\sqrt{\frac{3}{5}} - \frac{1}{4}i - \frac{\sqrt{2}}{4}j + \frac{\sqrt{6}}{4}k\right),$$

$$e_6 = \left(\sqrt{\frac{2}{5}}, -\frac{1}{4}\sqrt{\frac{3}{5}} - \frac{1}{4}i - \frac{\sqrt{2}}{4}j - \frac{\sqrt{6}}{4}k\right).$$

7. *For any parameter* $c \in \left]\sqrt{\frac{2}{5}}, 1\right[$ *there exists two (up to a global isometry) 5-tuples* $([e_1], [e_2], [e_3], [e_4], [e_5])$ *and* $([e_1], [e_2], [e_3], [e_4], [e_6])$ *of equiangular lines with parameter* c *imbedded in* $\mathbb{H}^2$. *Both of these quintuples have their symmetry group isomorphic to* A_5.

Remark Notice that the quadruple of 4 equiangular lines with parameter $\frac{1}{3}$ given in [35] Example 1.5 is isometric to our quadruple of item 2 of Theorem 9.12.2.

The sets of five and six equiangular lines in $\mathbb{H}^2$ were first calculated in [26] using the Hopf map. This technique though does not immediately generalise to other dimensions, like that of [22]. In [11] is proved that 15 equiangular lines in $\mathbb{H}^3$ exist by presenting computer-assisted proofs, rooted in numerical methods. By the way they also proved the existence of 27 equiangular lines in the octonionic space O^3.

References

1. D.M. Appleby, C.A. Fuchs, H. Zhu, Group theoretic, Lie algebraic and Jordan algebraic formulations of the SIC existence problem. Quantum Info. Comput. 15 (2015), no. 1–2, 61–94.
2. I. Balla, F. Draxler, P. Keevash, B. Sudakov, Equiangular lines and spherical codes in Euclidean space. Invent. Math. 211(1), 172–212 (2018)
3. V. Belevitch, Theory of 2n-terminal networks with applications to conference telephony. Elect. Commun. **27**, 231–244 (1950)
4. B.C. Berndt, R.J. Evans, The determination of Gauss sums. Bull. A.M.S. **5**(2), 107–129 (1981)
5. A. Blokhuis, U. Brehm, B. Et-Taoui, Complex conference matrices and equi-isoclinic planes in Euclidean spaces. Beit. Alg. Geom. 491–500 (2017)
6. L.M. Blumenthal, *Theory and Applications of Distance Geometry* (Oxford University Press, Oxford, 1953)
7. L.M. Blumenthal, Congruence and superposability in elliptic space. Trans. Am. Math. Soc. **62**, 431–451 (1947)
8. U. Brehm, B. Et-Taoui, Congruence criteria of finite subsets of complex projective and complex hyperbolic spaces. Manuscripta Math. **96**, 81–95 (1998)
9. U. Brehm, B. Et-Taoui, Congruence criteria of finite subsets of quaternionic projective and hyperbolic spaces. Geom. Dedicata 81–95 (2001)
10. B. Bukh, Bounds on equiangular lines and on related spherical codes. SIAM J. Discrete Math. **30**(1), 549–554 (2016)
11. H. Cohn, A. Kumar, G. Minton, Optimal simplices and codes in projective spaces. Geom. Topol. **20**, 1289–1357 (2016)
12. J.H. Conway, R.H. Hardin, N.J.A. Sloane, Packing lines, Planes, ect.: packings in Grassmannian spaces. Exp. Math. **5**, 139–159 (1996)
13. H.S.M. Coxeter, Regular compound polytopes in more than four dimensions. J. Math. Phys.**12**, 334–345 (1933)
14. D. de Caen, Large equiangular sets of lines in Euclidean space. Electron. J. Combin. **7**, Research Paper 55, 3 pp. (2000)
15. P. Delsarte, J.-M. Goethals, J.J. Seidel, Orthogonal matrices with zero diagonal II. Can. J. Math. XXXIII **5**, 816–832 (1971)
16. B. Et-Taoui, Sur les m-uplets F-réguliers dans les espaces projectifs complexes. Geom. Dedicata **63**, 297–308 (1996)
17. B. Et-Taoui, Equiangular lines in $\mathbb{C}^r$. Indag. Math. N.S. **13**(4), 201–201 (2000)

18. B. Et-Taoui, Equi-isoclinic planes in Euclidean dimensional spaces. Indag. Math. N.S. **17**(2), 205–219 (2006)
19. B. Et-Taoui, Equi-isoclinic planes in Euclidean even dimensional spaces. Adv. Geom. **7**(3), 379–384 (2007)
20. B. Et-Taoui, Complex conference matrices, complex Hadamard matrices and equiangular tight frames, in *Convexity and Discrete Geometry Including Graph Theory*, ed. by K.A. Adiprasito, I. Bárány, C. Vilcu (Springer, Berlin, 2016), pp. 181–191
21. B. Et-Taoui, Infinite family of equi-isoclinic planes in Euclidean odd dimensional spaces and of complex conference matrices of odd orders. Lin. Alg. Appl. 373–380 (2018)
22. B. Et-Taoui, Quaternionic equiangular lines. Adv. Geom. **20**(2), 273–284 (2020)
23. B. Et-Taoui, A. Fruchard, Sous-espaces equi-isoclins des espaces euclidiens. Adv. Geom. **9**, 471–515 (2009)
24. B. Et-Taoui, A. Makhlouf, Complex skew-symmetric conference matrices. Lin. Mult. Alg. 1–16 (2021)
25. B. Et-Taoui, J. Rouyer, Equi-isoclinic 3-planes of Euclidean spaces. Indag. Math. **20**(4), 491–525 (2009)
26. M.K. Fard, Regular structures of lines in complex spaces. ProQuest LLC, Ann Arbor, MI, 2008. Thesis (Ph.D.)–Simon Fraser University, Canada
27. M.C. Fickus, J. Jasper, D.G. Mixon, C.E. Watson, A brief introduction to Equi-chordal and equi-isoclinic tight fusion frames. AFIT Scholar 1–10 (2017)
28. G.R.W. Greaves, P. Yatsyna, On equiangular lines in 17 dimensions and the characteristic polynomial of a Seidel matrix. Math. Comput. **88**(320), 3041–3061 (2019)
29. G. Greaves, J. Koolen, A. Munemasa, F. Szollosi, Equiangular lines in Euclidean spaces. J. Combin. Theory Ser. A **138**, 208–235 (2016)
30. G.R.W. Greaves, J. Syatriadi, P. Yatsyna, Equiangular lines in low dimensional Euclidean spaces. Combinatorica. **41**, 839–872 (2021)
31. J.-M. Goethals, J.J. Seidel, Orthogonal matrices with zero diagonal. Can. J. Math. **19**, 1001–1010 (1967)
32. J. Haantjes, Equilateral point-sets in elliptic two and three-dimensional spaces. Nieuw.Arch. Wisk. **22**, 355–362 (1948)
33. R.W. Heath, T. Strohmer, Grassmannian frames with applications to coding and communication. Appl. Comput. Harmon. Anal. **14**, 257–275 (2003)
34. S.G. Hoggar, Quaternionic equi-isoclinic n-planes . Ars combinatoria **2**, 11–13 (1976)
35. S.G. Hoggar, New sets of equi-isoclinic n-planes from old. Proc. Edin. Math. Soc. **20**, 287–291 (1976)
36. R.B. Holmes, V.I. Paulsen, Optimal frames for erasures. Linear Algebra Appl. **377**, 31–51 (2004)
37. Z. Jiang, J. Tidor, Y. Yao, S. Zhang, Y. Zhao, Equiangular lines with a fixed angle (2019). arXiv:1907.12466
38. P.W.H. Lemmens, J.J. Seidel, Equi-isoclinic subspaces of Euclidean spaces. Nederl. Akad. Wet. Proc. Ser. A **76**, 98–107 (1973)
39. P.W.H. Lemmens, J.J. Seidel, Equiangular lines. J. Algebra **24**, 494–512 (1973)
40. J.H. Van Lint, J.J. Seidel, Equilateral point sets in elliptic geometry. Koninkl. Nederl. Proc. Series A **69**(3) and Indag. Math. **28**(3), 335–348 (1966)
41. R. Mathon, Symmetric conference matrices of order $pq^2 + 1$. Can. J. Math. **30**, 321–331 (1978)
42. K. Menger, Untersuchungen über allgemeine Metrik. Math. Ann. **100**, 113–141 (1928)
43. R.E.A.C. Paley, On orthogonal matrices. J. Math. Phys. **12**, 311–320 (1933)
44. D. Raghavarao, Some aspects of weighing designs. Ann. Math. Stat. **31**, 878–884 (1960)
45. J.M. Renes, R. Blume-Kohout, A.J. Scott, C.M. Caves, Symmetric informationally complete quantum measurements. J. Math. Phys. **45**, 2171–2180 (2004)
46. L. Rodman, *Topics in Quaternion Linear Algebra* (Princeton University Press, Princeton, 2014)
47. A. Scott, M. Grassl, Symmetric informationally complete positive-operator-valued measures: a new computer study. J. Math. Phys. **51**(4), 042203, 16 pp. (2010)

48. J. Seberry, A.L. Whiteman, New Hadamard matrices and conference matrices obtained via Mathon's construction. Graphs Comb. **4**, 355–377 (1988)
49. J.J. Seidel, congruentie-orde van het elliptische vlak. Thesis, Leiden University, Amsterdam, 1948
50. J.J. Seidel, Discrete non-euclidean geometry, in *Handbook of Incidence Geometry*, Chap. 15, ed. by Buekenhout (Elsevier, Amsterdam, 1995), pp. 843–919
51. D.E. Taylor, Regular 2-graphs. Proc. London Math. Soc. **35**, 257–274 (1977)
52. J. Williamson, Hadamard's determinant theorem and the sum of four squares. Duke Math. J. **11**, 65–81 (1944)
53. G. Zauner, Quantum designs: foundations of a non-commutative design theory. Int. J. Quantum Inf. **9**(1), 445–507 (2011)

Chapter 10
On the Geometry of Finite Homogeneous Subsets of Euclidean Spaces

Valeriĭ Nikolaevich Berestovskiĭ and Yuriĭ Gennadievich Nikonorov

Abstract This survey is devoted to results recently obtained on finite homogeneous metric spaces. One of the main subjects of discussion is the classification of regular and semiregular polytopes in Euclidean spaces by whether or not their vertex sets have the normal homogeneity property or the Clifford–Wolf homogeneity property. Every finite homogeneous metric subspace of a Euclidean space represents the vertex set of a compact convex polytope whose isometry group is transitive on the set of vertices and with all these vertices lying on some sphere. Consequently, the study of such subsets is closely related to the theory of convex polytopes in Euclidean spaces. Normal homogeneity and the Clifford–Wolf homogeneity describe stronger properties than homogeneity. Therefore, it is natural to first check the presence of these properties for the vertex sets of regular and semiregular polytopes. The second part of the survey is devoted to the study of the m-point homogeneity property and the point homogeneity degree for finite metric spaces. We discuss some recent results, in particular, the classification of polyhedra with all edges of equal length and with 2-point homogeneous vertex sets. In addition to the classification results, the paper contains a description of the main tools for the study of the relevant objects.

Keywords Archimedean solid · Finite Clifford–Wolf homogeneous metric space · Finite homogeneous metric space · Finite normal homogeneous metric space · Gosset polytope · m-point homogeneous metric space · Platonic solid · Point homogeneous degree · Regular polytope · Semiregular polytope

The work of the first author was carried out within the framework of the state Contract to the IM SD RAS, project FWNF-2022-0006.

V. N. Berestovskiĭ (✉)
Sobolev Institute of Mathematics of the SB RAS, Novosibirsk, Russia
e-mail: vberestov@inbox.ru

Y. G. Nikonorov
Southern Mathematical Institute of VSC RAS, Vladikavkaz, Russia
e-mail: nikonorov2006@mail.ru

2020 Mathematical Subject Classification 54E35, 52B15, 20B05

Introduction

A metric space M is called *homogeneous* if its isometry group acts transitively on it. The main object of our study are homogeneous metric spaces with special properties. For instance, we are interested in *finite* homogeneous metric spaces. A finite homogeneous metric subspace of a Euclidean space represents the vertex set of a compact convex polytope whose isometry group is transitive on the vertex set; in each case, all vertices lie on a sphere. In [7, 9, 10], the authors obtained some structure and classification results on special subclasses of finite homogeneous metric spaces. The description of the classes under consideration was given in terms of graph theory. This description allows to construct some particular examples of finite metric spaces with unusual properties. For instance, the Kneser graphs are fruitful sources of such quite unexpected examples [7].

In [7, 9, 10], the authors obtained a complete description of the metric properties of the vertex sets of regular and semiregular polytopes in Euclidean spaces from the point of view of normal homogeneity and Clifford–Wolf homogeneity, see also the survey [11]. In the first part of this present survey, we discuss the corresponding classification along with other important properties of finite homogeneous metric spaces, in particular, homogeneous polytopes in Euclidean spaces.

The corresponding classes of (generalized) normal homogeneous and Clifford–Wolf homogeneous Riemannian manifolds were studied earlier in [3–6, 8].

In the recent paper [11], the authors started the study of the m-point homogeneity of finite subspaces of Euclidean spaces; this is another special property, which is a strengthening of the notion of homogeneity. The main results of [11] are discussed in the three last sections of this survey.

Since the vertex sets of regular polytopes, as well as of some their generalizations, are homogeneous, we pay much attention to the study of the homogeneity properties of the vertex sets of polytopes in Euclidean spaces. In addition, much attention has been paid to the development of methods and tools for studying the relevant objects.

As a rule, we use standard notation. When we deal with some metric space (M, d), for given $c \in M$ and $r \geq 0$, we consider $S(c, r) = \{x \in M \mid d(x, c) = r\}$, $U(c, r) = \{x \in M \mid d(x, c) < r\}$, and $B(c, r) = \{x \in M \mid d(x, r) \leq r\}$, respectively the sphere, the open ball, and the closed ball with center c and radius r. For $x, y \in \mathbb{R}^n$, the symbol $[x, y]$ means the closed interval in $\mathbb{R}^n$ with ends x and y.

This chapter is organized as follows. In Sect. 10.1 we consider three important subclasses of the class of homogeneous metric spaces: (generalized) normal homogeneous, Clifford–Wolf homogeneous, and m-point homogeneous spaces. In Sect. 10.2 we recall some properties of finite homogeneous metric spaces. Some special properties of finite homogeneous metric subsets of Euclidean spaces are considered in Sect. 10.3. In Sect. 10.4, we give some important information on regular and semiregular polytopes in Euclidean spaces. In Sect. 10.5 we reproduce the classification of regular and semiregular polytopes with normal homogeneous or Clifford–Wolf homogeneous vertex sets. In Sect. 10.6 we study the m-point homogeneity property for some finite homogeneous metric spaces, mainly the vertex sets of regular and semiregular polytopes. In Sect. 10.7, we prove some important results on m-point homogeneous polyhedra in $\mathbb{R}^3$. Finally, in Sect. 10.8 we discuss some results on the point homogeneity degree for some important classes of polytopes. In the concluding section, some information about the prospects for future research is given.

10.1   Homogeneous Metric Spaces and Their Special Subclasses

For a given metric space (M, d), we denote by $\mathrm{Isom}(M, d)$ its isometry group.

Definition 10.1 A metric space (M, d) is called *homogeneous*, if for every $x, y \in M$ there exists an isometry of (M, d), moving x to y, i. e. the isometry group $\mathrm{Isom}(M, d)$ acts transitively on M.

It should be noted that some proper subgroups of $\mathrm{Isom}(M, d)$ could also act transitively on M. For example, the isometry group of the sphere S^{2m-1} with the metric induced by the Euclidean metric of $\mathbb{R}^{2m}$, is the orthogonal group $O(2m)$, which acts transitively. On the other hand, the special orthogonal group $SO(2m)$, the unitary group $U(m)$, and the special unitary group $SU(m)$ also act transitively on S^{2m-1}.

Definition 10.2 Let (M, d) be a metric space and $x \in M$. An isometry $f : M \to M$ is called *a $\delta(x)$-translation* or *a δ-translation at the point x*, if x is a point of maximal displacement of f, i. e. for every $y \in M$ the relation $d(y, f(y)) \leq d(x, f(x))$ holds.

Definition 10.3 Let (M, d) be a metric space. An isometry $f : M \to M$ is called *a Clifford–Wolf translation* (CW-translation), if f moves all points of (M, d) the same distance, i. e. $d(y, f(y)) = d(x, f(x))$ for every $x, y \in M$.

Let us recall one well known fact which gives us a useful technical tool.

Proposition 10.1 *Let (M, d) be a metric space. If $f \in \mathrm{Isom}(M, d)$ is such that the group $G = \{g \in \mathrm{Isom}(M, d) \mid gf = fg\}$, i. e. the centralizer of f in $\mathrm{Isom}(M, d)$, acts transitively on M, then f is a Clifford–Wolf translation on (M, d).*

Proof Let us take $x, y \in M$ and prove that $d(x, f(x)) = d(y, f(y))$. Since G acts transitively on M, there is $g \in G$ such that $g(x) = y$. Further, we have

$$d(y, f(y)) = d\big(g(x), f(g(x))\big) = d\big(g(x), g(f(x))\big) = d(x, f(x)),$$

since g is an isometry of (M, d) and $fg = gf$.   ∎

Now we are ready to define generalized normal homogeneous and Clifford–Wolf homogeneous metric spaces, very remarkable subclasses of the class of homogeneous metric spaces.

Definition 10.4 A metric space (M, d) is called *generalized normal homogeneous* (respectively, *Clifford–Wolf homogeneous*), if for every $x, y \in M$ there exists a $\delta(x)$-translation (respectively, *Clifford-Wolf translation*) of (M, d) moving x to y.

Clearly, any Clifford–Wolf translation is a $\delta(x)$-translation for all $x \in M$, hence, any Clifford–Wolf homogeneous space is generalized normal homogeneous, and the latter one is homogeneous.

Since $\mathbb{R}^n$ is a commutative group (with respect to vector addition), then, by Proposition 10.1, every Euclidean space is Clifford–Wolf homogeneous (of course, it is easy to find all suitable CW-translations explicitly in this special example). Let us consider a more general example.

Example 10.1 Let G be a group, supplied with a metric d, that is bi-invariant, i. e. invariant both with respect to left and right shifts on G (for an element $a \in G$, the maps $l_a : x \mapsto a \cdot x$ and $r_a : x \mapsto x \cdot a$ are called the left shift and the right shift by a respectively). It is clear that the group of left shifts, as well as the group of right shifts, acts transitively on G. Moreover, every left shift centralizes the group of right shifts, hence, it is a Clifford–Wolf translation of (G, d). Analogously, every right shift is a Clifford–Wolf translation of (G, d). This implies that (G, d) is Clifford–Wolf homogeneous.

Note also that every odd-dimensional sphere (with the standard metric of constant curvature) is Clifford–Wolf homogeneous, see details e. g. in [5], [8, Chapter 7], or [48].

The study of generalized normal homogeneous and Clifford–Wolf homogeneous metric spaces (in particular, in the Riemannian and Finsler cases) is the subject of many publications, which we cannot list here. We refer the interested reader especially to the book [8] where one can also find a comprehensive list of references on the given classes of spaces.

We are now ready to define m-point homogeneous metric spaces, which are also natural distinguished homogeneous metric spaces.

Definition 10.5 A metric space (M, d) is called *m-point homogeneous*, $m \in \mathbb{N}$, if for every m-tuples $(A_1, A_2, \ldots, A_m)$ and $(B_1, B_2, \ldots, B_m)$ of elements of M such that $d(A_i, A_j) = d(B_i, B_i)$, $i, j = 1, \ldots, m$, there is an isometry $f \in \mathrm{Isom}(M)$ with the following property: $f(A_i) = B_i, i = 1, \ldots, m$.

From a formal point of view, there can be coinciding points among the points $A_1, \ldots, A_m$ (as well as among the points $B_1, \ldots, B_m$) in the above definition. This observation immediately implies the following result.

Proposition 10.2 *If a metric space (M, d) is m-point homogeneous for some $m \geq 2$, then it is k-point homogeneous for any $k = 1, \ldots, m - 1$. If the set M*

is finite, l is its cardinality, and (M, d) is l-point homogeneous, then (M, d) is m-point homogeneous for all $m \geq 1$.

The notion of n-point homogeneity for a metric space was introduced by G. Birkhoff in [16] by the following sentence: "Any isometry between two (sub)sets of n points or fewer points can be extended to a self-isometry of space". In the same paragraph he noticed that the Urysohn space [45] has n-point homogeneity for every finite order n. This posthumous paper by P. Urysohn contains his construction of a remarkable complete separable metric space U which besides the n-point homogeneity for every finite order n is universal in the sense that every separable metric space is isometric to a subspace of U. Moreover, any space with these properties is isometric to U [45].

H. Busemann considered 2-point homogeneous Busemann G-spaces (geodesic spaces) in 1942 [21]. The latter can be described as complete locally compact metric spaces (M, d) such that any two points in M can be joined by a *shortest arc*, i. e. an isometric embedding $p : [0, d(x, y)] \rightarrow (M, d)$; shortest arcs are locally extendable; any shortest arc has a unique extension to a *geodesic*, i. e. locally isometric mapping $\gamma : \mathbb{R} \rightarrow (M, d)$.

It is well known that every 2-point homogeneous Riemannian manifold is isometric to a Euclidean space or a Riemannian symmetric space of rank 1, see e. g. Theorem 8.12.8 and Corollary 8.12.9 in [48]. In fact, this result holds in a more general situation, namely, for 2-point homogeneous Busemann G-spaces. This can be easily deduced from the papers [46, 47] by H.C. Wang for compact spaces and [44] by J. Tits for the non-compact case. Currently, there are various proofs of the above characterization of 2-point homogeneous Riemannian manifolds. Especially important is the very short paper [42] by Z.I. Szabó, where he showed also that the above 2-point homogeneous spaces are symmetric not using any classification, thereby solving a problem of Wolf from [48].

H. Busemann proved in [22] that if for every two geodesics of a compact Busemann G-space there exists an isometry of the space moving one of them to the other (this condition is much weaker than the 2-point homogeneity of the space) then the space is isometric to a compact Riemannian symmetric space of rank 1. V.N. Berestovskii in [1] proved that non-compact Busemann G-space with isometry group transitive on the family of its geodesics is isometric to a Euclidean space or a non-compact Riemannian symmetric space of rank 1, thus giving a positive answer to a Busemann problem.

As a corollary of the results quoted, every 3-point homogeneous Busemann G-space (in particular, Riemannian manifold) is isometric to a Euclidean space, spherical space, elliptic space, or hyperbolic space with appropriate constant sectional curvature. This implies that in any such space every isometry between every two of their subsets can be extended to an isometry of the entire space, which is extremely different from the case with the vertex sets of polytopes.

It should be noted that it makes no sense to classify (even finite) m-point homogeneous metric spaces up to isometry or similarity without additional restrictions, because any such metric space (M, d) generates a class of m-point homogeneous metric spaces of the following type: $(M, \psi \circ d)$, where ψ is any increasing concave (convex up) function such that $\psi(0) = 0$. For instance, one can take $\psi(t) = \ln(1 + t)$, $t \geq 0$, or $\psi(t) = 2 \sin(t/2)$, $0 \leq t \leq \pi$. A similar remark can be fully applied also to generalized normal homogeneous and Clifford–Wolf homogeneous metric spaces.

10.2 Some Properties of Finite Homogeneous Metric Spaces

Here we recall some important properties of finite homogeneous metric spaces. The following are the following sources for such spaces (see details in [7]):

1. a homogeneous space G/H of a finite group G by some subgroup H, endowed with an invariant metric;
2. a compact convex polytope in Euclidean space whose isometry group acts transitively on the vertex set;
3. a vertex-symmetric (in another terminology, vertex-transitive) connected finite graph with the natural metric;
4. the Cayley graph of a finite group for a minimal generating set.

Definition 10.6 A map of metric spaces $f : (M_1, d_1) \to (M_2, d_2)$ is called a *submetry*, if it maps every closed ball $B(x, s) \subset (M_1, d_1)$ with center x and radius s onto the closed ball $B(f(x), s) \subset (M_2, d_2)$ with center $f(x)$ and radius s [2].

Definition 10.7 A finite homogeneous metric space (M, d) is called *normal homogeneous* if for its isometry group $\mathrm{Isom}(M, d)$ and its stabilizer H at a point $x_0 \in M$, there exists a subgroup Γ of the group $\mathrm{Isom}(M, d)$ which is transitive on M and a bi-invariant metric σ on Γ such that the canonical projection $\pi : (\Gamma, \sigma) \to (\Gamma/(\Gamma \cap H), d) = (M, d)$ is a submetry.

Remark 10.1 It should be noted that there are more restrictive definitions of *normal homogeneity* for some special classes of metric spaces. For instance, this is the case with Riemannian manifolds, see details in [4, 6, 8].

The following result shows that the property to be generalized normal homogeneous (which is a pure metric property) is equivalent to the property to be a normal homogeneous (which is an algebraic property in fact) in the case of finite metric spaces.

Proposition 10.3 ([7]) *A finite metric space (M, d) is generalized normal homogeneous if and only if it is normal homogeneous.*

Proof *First, let us prove that a finite normal homogeneous metric space (M, d) is generalized normal homogeneous.* Let us denote $\Gamma \cap H$ by H' (see Definition 10.7). We identify elements of M with left cosets $\alpha H' = \pi(\alpha)$, $\alpha \in \Gamma$. Since the canonical projection $\pi : (\Gamma, \sigma) \to (\Gamma/H', d) = (M, d)$ is a submetry, then the following statements hold: (I) the map π does not increase distances; (II) for every three points $x, y \in M, \xi \in \pi^{-1}(x)$, there exists a point $\eta \in \pi^{-1}(y)$ such that $\sigma(\xi, \eta) = d(x, y)$.

Let us consider some $x, y \in M$ and $\xi \in \pi^{-1}(x)$. We know (by (II)) that there is $\eta \in \pi^{-1}(y)$ such that $\sigma(\xi, \eta) = d(x, y)$. Let us consider $\gamma = \eta\xi^{-1}$. According to Example 10.1, the left shift by γ is a Clifford–Wolf translation of the space (Γ, σ). On the other hand, γ is an isometry of $(\Gamma/H' = M, d)$ with $\gamma(x) = y$ (since $\gamma\xi H' = \eta H'$). Further, for any $z \in M$ and any $\zeta \in \pi^{-1}(z)$, we have

$$d(x, \gamma(x)) = d(x, y) = \sigma(\xi, \eta) = \sigma(\xi, \gamma\xi) = \sigma(\zeta, \gamma\zeta)) \geq \quad \text{(by (I))}$$

$$\geq d(\pi(\zeta), \pi(\gamma\zeta)) = d(\zeta H', \gamma\zeta H') = d(z, \gamma(z)).$$

Therefore, γ is a $\delta(x)$-translation. Since $x, y \in M$ could be arbitrary, we get that the metric space (M, d) is generalized normal homogeneous.

Let us prove that a finite generalized normal homogeneous metric space (M, d) *is normal homogeneous.* We consider the group $G = \text{Isom}(M, d)$ and the stabilizer H of a certain point $x_0 \in M$ in G. Let us define

$$\sigma(g, h) := \max_{x \in M} d(g(x), h(x)), \quad g, h \in G.$$

It is easy to verify that σ is a bi-invariant metric on G. We state that

$$\pi : (G, \sigma) \to (G/H, d) = (M, d)$$

is a submetry, hence, (M, d) is normal homogeneous. The definition of σ implies

$$d(\pi(g), \pi(h)) = d(g(x_0), h(x_0)) \leq \sigma(g, h)$$

for every $g, h \in G$, i. e. π does not increase distances. Let x be any point in M. Since (M, d) is generalized normal homogeneous, there exists $g \in G$ such that $g(x_0) = x$ and $d(x_0, x) = d(x_0, g(x_0)) \geq d(y, g(y))$ for all $y \in M$ (i. e. g is a $\delta(x_0)$-translation). Therefore,

$$\sigma(e, g) = d(x_0, g(x_0)) = d(\pi(e), \pi(g)).$$

From the above reasoning it follows that $\pi(B(e, r)) = B(x_0, r) = B(\pi(e), r)$ for each number $r \geq 0$. Since the metric σ is left-invariant, we get $\pi(B(g, r)) = \pi(l_g(B(e, r))) = B(\pi(g), r)$ for any $r \geq 0, g \in G$, i. e. π is a submetry. ∎

Let us denote by $FGBM, FGLM, FCWHS, FGNHS, FNHS, FHS$ respectively the classes of finite groups with bi-invariant metrics, finite groups with left-invariant metrics, finite Clifford–Wolf homogeneous spaces, finite generalized normal homogeneous spaces, finite normal homogeneous spaces, and finite homogeneous spaces. Proposition 10.3 implies the equality $FGNHS = FNHS$. It is known also that

$$FGBM \subset FCWHS \subset FGNHS = FNHS \subset FHS,$$

$$FGBM \subset FGLM \subset FHS.$$

Moreover, all the above inclusions are strict, see details in [7].

In what follows, we will focus mainly on the following subclasses of the class of finite homogeneous metric spaces: finite Clifford–Wolf homogeneous metric spaces, finite (generalized) normal homogeneous spaces, and m-point homogeneous finite metric spaces.

10.3   Finite Homogeneous Subspaces of Euclidean Spaces

Starting from this section, we restrict ourselves to the study of finite homogeneous subsets of the Euclidean spaces $\mathbb{R}^n$. We assume that any such set M is supplied with the metric d induced from $\mathbb{R}^n$.

Since the barycenter of a finite system of material points (with one and the same mass) in any Euclidean space is preserved for any bijection (in particular, any isometry) of this system, we immediately get the following result.

Proposition 10.4 ([7]) *Let $M = \{x_1, \ldots, x_m\}$, $m \geq n+1$, be a finite homogeneous metric subspace of a Euclidean space $\mathbb{R}^n$, $n \geq 2$, which does not lie in a hyperplane. Then M is the vertex set of a convex polytope P that is situated on some sphere in $\mathbb{R}^n$ with radius $r > 0$ and center $x_0 = \frac{1}{m} \cdot \sum_{k=1}^m x_k$. In particular, $\mathrm{Isom}(M, d) \subset O(n)$. Up to similarity, any homogeneous finite metric subspace in $\mathbb{R}^n$, $n \geq 2$, is a homogeneous metric subspace of the unit sphere $S^{n-1} \subset \mathbb{R}^n$.*

This result shows that *the theory of convex polytopes* is very important for the study of finite homogeneous subspaces of Euclidean spaces. Now, we recall several important definitions. For a more detailed acquaintance with the theory of convex polytopes, we recommend [13, 27, 28, 36].

Let P be a non-degenerate convex polytope in $\mathbb{R}^n$ with barycenter at the origin $O = (0, 0, \ldots, 0) \in \mathbb{R}^n$ (the property to be non-degenerate means that $\mathrm{Lin}(P) = \mathbb{R}^n$ or, equivalently, O is in the interior of P). The symmetry group $\mathrm{Symm}(P)$ of P is the group of isometries of $\mathbb{R}^n$ that preserve P. It is clear that each $\psi \in \mathrm{Symm}(P)$ is an orthogonal transformation of $\mathbb{R}^n$ (obviously, $\psi(O) = O$ for any symmetry ψ of P).

If M is the vertex set of a polytope P supplied with the metric d induced from the Euclidean one in $\mathbb{R}^n$, then the isometry group $\mathrm{Isom}(M, d)$ of the metric space (M, d) is the same as $\mathrm{Symm}(P)$.

We say that an n-dimensional polytope P in $\mathbb{R}^n$ is *homogeneous* (or *vertex-transitive*) if its isometry group acts transitively on the set of its vertices.

By virtue of Proposition 10.4, the following three definitions are very natural.

Definition 10.8 A convex polytope P in $\mathbb{R}^n$ is called normal homogeneous if its vertex set (with induced metrics d from $\mathbb{R}^n$) is normal homogeneous.

Definition 10.9 A convex polytope P in $\mathbb{R}^n$ is called Clifford–Wolf homogeneous if its vertex set (with the induced metric d from $\mathbb{R}^n$) is Clifford–Wolf homogeneous.

Definition 10.10 A convex polytope P in $\mathbb{R}^n$ is called m-point homogeneous if its vertex set (with induced metrics d from $\mathbb{R}^n$) is m-point homogeneous.

In particular, the property to be 1-point homogeneous means just the property to be homogeneous.

Now, we are going to recall the definitions of *regular, semiregular,* and *uniform* convex polytopes in Euclidean spaces.

A one-dimensional polytope is a closed segment, bounded by two endpoints. It is regular by definition. Two-dimensional regular polytopes are the convex regular polygons on Euclidean plane. For higher dimensions, regular polytopes are defined inductively. A convex n-dimensional polytope for $n \geq 3$ is called *regular*, if it is homogeneous and if all its facets are regular polytopes congruent to each other. This definition is equivalent to other definitions of regular convex polytopes (see [40]).

We also recall the definition of a wider class of semiregular convex polytopes. For $n = 1$ and $n = 2$, semiregular polytopes are defined as regular. A convex n-dimensional polytope for $n \geq 3$ is called *semiregular* if it is homogeneous and all its facets are regular polytopes.

A generalization of the class of semiregular polytopes is the class of uniform polytopes. For $n \leq 2$, uniform polytopes are defined as regular. For other dimensions, uniform polytopes are defined inductively. A convex n-dimensional polytope for $n \geq 3$ is called *uniform* if it is homogeneous and all its facets are uniform polytopes. In particular, for $n = 3$, the classes of uniform and semiregular polytopes coincide, and for $n = 4$ the facets of a uniform polytope must be semiregular three-dimensional polytopes. This class of polytopes is far from completely classified; see the known results in [34, 40].

The following proposition contains some important properties of uniform polytopes.

Proposition 10.5 *All edges of an arbitrary uniform polytope have the same length. All vertices, adjacent to any given vertex, lie in some hyperplane.*

Proof The first assertion is obvious for two-dimensional polytopes (regular polygons in the plane). We can continue this argument by induction. The facets of every uniform polytope are also uniform. Hence (by the induction hypothesis) all edges incident to one facet are of equal length. In addition, if two facets have a common edge then they have edges of equal length. This implies the first assertion of the proposition.

Let us prove the second assertion. For a given vertex u of a uniform polytope P in $\mathbb{R}^n$, the distance to all adjacent vertices is some number $r > 0$. Therefore, all these adjacent vertices are in the sphere $S(u, r)$. Moreover, since P is uniform, and hence, homogeneous, all vertices of P are on one and the same distance from the barycenter of P. This implies that all vertices, adjacent to u, are in the same hyperplane ($\|x - a\| = C$ and $\|x - b\| = D$ for $x, a, b \in \mathbb{R}^n$ implies $2(x, a - b) = \|a\|^2 - \|b\|^2 - C^2 + D^2$).   ∎

The above proposition allows us to introduce the following definition. A *vertex figure* of an n-dimensional uniform polytope, $n \geq 3$, is an $(n - 1)$-dimensional polytope which is the convex hull of the vertices having a common edge with a given vertex and different from it.

10.4   Regular and Semiregular Polytopes

In this section, we recall some important properties of regular and semiregular polytopes in Euclidean spaces. Faces of dimension $n - 1$ (hyperfaces) of a n-dimensional polytope are commonly referred to as *facets*. Note that they are also called *cells* for $n = 4$.

The classification of regular polytopes of arbitrary dimension was first obtained by Ludwig Schläfli and is presented in his book [41], see also Harold Coxeter's book [27]. The list of semiregular polytopes of arbitrary dimension was first presented without proof in Thorold Gosset's paper [35]. Later this list appeared in the work of Emanuel Lodewijk Elte [33]. The proof of the completeness of this list was obtained much later by Gerd Blind and Rosvita Blind, see [17] and the references therein. Semiregular (non-regular) polytopes in $\mathbb{R}^n$ for $n \geq 4$ are called *Gosset polytopes*. A lot of additional information can be found in [29].

We briefly recall the classification of regular and semiregular polytopes in Euclidean spaces.

An one-dimensional polytope (closed segment) is regular. Two-dimensional regular polyhedra (polygons) have equal sides and are inscribed in a circle.

For regular three-dimensional polyhedra, the vertex figure is a regular polygon. It is well known that there are only five regular three-dimensional polyhedra: the tetrahedron, cube, octahedron, dodecahedron, and icosahedron. These polyhedra are traditionally called *Platonic solids*, see Fig. 10.1. Some important properties of these polyhedra may be found in Table 10.1, where V, E, and F mean respectively the

a)   b)   c)   d)   e)

Fig. 10.1 Platonic solids: (**a**) tetrahedron; (**b**) cube; (**c**) octahedron; (**d**) dodecahedron; (**e**) icosahedron

Table 10.1 Regular 3-dimensional polyhedra

Polyhedron	V	E	F	α	Face
Tetrahedron	4	6	4	$2\arcsin(1/\sqrt{3})$	△
Cube (hexahedron)	8	12	6	$\pi/2$	☐
Octahedron	6	12	8	$2\arcsin(\sqrt{2/3})$	△
Dodecahedron	20	30	12	$2\arcsin\left(\sqrt{\varphi}/\sqrt[4]{5}\right)$	⬠
Icosahedron	12	30	20	$2\arcsin(\varphi/\sqrt{3})$	△

numbers of vertices, edges, and faces; α is the dihedral angle; the number $\varphi :=$ $\frac{1+\sqrt{5}}{2}$ is known as *the golden ratio*.

There are 6 regular 4-dimensional polytopes: the hypertetrahedron (5-cell), hypercube (8-cell), hyperoctahedron (16-cell), 24-cell, 120-cell, and 600-cell. A detailed description of the structure of four-dimensional regular polytopes could be found e. g. in [9, Section 3].

In each dimension $n \geq 5$, there exist exactly three regular polytopes: the n-dimensional simplex, the hypercube (n-cube) and the hyperoctahedron (n-orthoplex).

We recall some information on the structure of these polytopes and their symmetry groups. In what follows, S_n denotes the symmetric group of degree n, i. e. the group of all permutations on n symbols. Recall that the order of S_n is $n!$. We will also use the cyclic groups $\mathbb{Z}_k$ of order k.

Clearly, the symmetry group of the regular n-dimensional simplex P is the group S_{n+1}.

The symmetry group of the n-cube P is the group $\mathbb{Z}_2^n \rtimes S_n$. This is easy to check if we represent P as a convex hull of the points $(\pm a, \pm a, \cdots, \pm a) \in \mathbb{R}^n$, where $a > 0$, and the sign could take value $-$ or $+$ on each place independently on other places. Here, S_n is the group of permutation of all coordinates, and the group $\mathbb{Z}_2^n$ is generated by the maps $x \mapsto -x$ for all coordinates.

The symmetry group of the n-orthoplex P is also the group $\mathbb{Z}_2^n \rtimes S_n$. This is easy to check if we represent P as a convex hull of the points $\pm b\,e_i$, $i = 1, 2, \ldots, n$, where $b > 0$ and if e_i is the i-th basic vector in $\mathbb{R}^n$ (it has all zero coordinates except 1 in the i-th place). This result could be obtained more simply. Since the n-orthoplex is dual to the n-cube, they have one and the same symmetry group. The duality means that the centers of all facets of the n-orthoplex (the n-cube) constitute the vertex set of the n-cube (respectively, the n-orthoplex).

We now proceed to a brief description of the semiregular (non-regular) polytopes.

In three-dimensional space (in addition to Platonic solids), there are the following semiregular polyhedra: 13 *Archimedean solids* and two infinite series of regular prisms and right antiprisms. A detailed description of Archimedean solids could be found in Sections 4 and 5 of [9].

A *right prism* is a polyhedron whose two-faces (called bases) are congruent (equal) regular polygons, lying in parallel planes, while the other faces (called lateral ones) are rectangles (perpendicular to the bases). If lateral faces are squares then the prism is said to be *regular*. In this case we get an infinite family of semiregular convex polyhedra.

A *right antiprism* is a semiregular polyhedron, with two parallel faces (bases) being equal regular n-gons, while the other $2n$ (lateral) faces are regular triangles. Note that the octahedron is an antiprism with triangular bases.

For $n = 4$ we have exactly three semiregular polytopes: the rectified 4-simplex, rectified 600-cell, and snub 24-cell. A detailed description of these polytopes could be found in Section 5 of [10].

The unique (up to similarity) semiregular Gosset polytope in $\mathbb{R}^n$ for $n \in \{5, 6, 7, 8\}$ we denote by the symbol Goss $_n$. Detailed descriptions of these polytopes could be found in Sections 6,7,8, and 9 of [10] respectively.

A polytope is called a *polytope with regular faces* (respectively, a *polytope with congruent faces*), if all its facets are regular (respectively, congruent) polytopes.

Theorem 10.1 ([49]) *Besides regular and semiregular (three-dimensional) polyhedra, there exist up to isometry or homothety only 92 convex polyhedra with regular faces.*

This theorem is a confirmation of a conjecture by N.W. Johnson in [38]. It is an easy corollary of the following theorem (which Zalgaller called "the main theorem").

Theorem 10.2 ([49]) *Besides (semi)regular prisms and antiprisms (excluding the regular octahedron) there exist up to an isometry or homothety only 28 simple convex polyhedra $M_1 - M_{28}$ with regular faces.*

Here a convex polyhedron P with regular faces is said to be *simple* if P cannot be dissected by a plane into two others convex polyhedra with regular faces. In the opposite case, P is called *composite*. In [49] are presented pictures of all polyhedra $M_1 - M_{28}$ and the schemes which shows how any of the 92 mentioned and *titled there composite* polyhedra is composed of the mentioned simple polyhedra (see also [14]).

Example 10.2 The five-dimensional Gosset polytope Goss $_5$ is a demihypercube in $\mathbb{R}^5$. Recall that an n-dimensional hypercube with edge length 2 can be represented as a convex hull of points of the form $(\pm 1, \pm 1, \dots, \pm 1) \in \mathbb{R}^n$ with arbitrarily chosen signs of each coordinate. It is clear that the group $(\mathbb{Z}_2)^n$ acts simply transitively on the set of vertices of this hypercube. Let us consider a subset of vertices G of the hypercube, consisting of all vertices with an even number of signs "$-$" in coordinates and identify it with a subset of the group $(\mathbb{Z}_2)^n$, which will also be denoted by G. It is easy to check that the product of elements from G remains in G. Thus G is a subgroup of index 2 in the group $(\mathbb{Z}_2)^n$. The convex hull of all vertices in G is, by definition, a demihypercube in $\mathbb{R}^n$.

Example 10.3 Let us consider a brief explicit description of Goss $_6$. This polytope can be implemented in different ways. Let us set it with the coordinates of the vertices in $\mathbb{R}^6$, as it is done in [33]. Let us put $a = \frac{\sqrt{2}}{4}$ and $b = \frac{\sqrt{6}}{12}$. We define the points $A_i \in \mathbb{R}^6$, $i = 1, \ldots, 27$, as follows:

$A_1 = (0, 0, 0, 0, 0, 4b)$, $\quad$ $A_2 = (a, a, a, a, a, b)$, $\quad$ $A_3 = (-a, -a, a, a, a, b)$,
$A_4 = (-a, a, -a, a, a, b)$, $\quad$ $A_5 = (-a, a, a, -a, a, b)$, $\quad$ $A_6 = (-a, a, a, a, -a, b)$,
$A_7 = (a, -a, -a, a, a, b)$, $\quad$ $A_8 = (a, -a, a, -a, a, b)$, $\quad$ $A_9 = (a, -a, a, a, -a, b)$,
$A_{10} = (a, a, -a, -a, a, b)$, $\quad$ $A_{11} = (a, a, -a, a, -a, b)$, $\quad$ $A_{12} = (a, a, a, -a, -a, b)$,
$A_{13} = (-a, -a, -a, -a, a, b)$, $A_{14} = (-a, -a, -a, a, -a, b)$, $A_{15} = (-a, -a, a, -a, -a, b)$,
$A_{16} = (-a, a, -a, -a, -a, b)$, $A_{17} = (a, -a, -a, -a, -a, b)$, $A_{18} = (2a, 0, 0, 0, 0, -2b)$,
$A_{19} = (0, 2a, 0, 0, 0, -2b)$, $\quad$ $A_{20} = (0, 0, 2a, 0, 0, -2b)$, $\quad$ $A_{21} = (0, 0, 0, 2a, 0, -2b)$,
$A_{22} = (0, 0, 0, 0, 2a, -2b)$, $\quad$ $A_{23} = (-2a, 0, 0, 0, 0, -2b)$, $\quad$ $A_{24} = (0, -2a, 0, 0, 0, -2b)$,
$A_{25} = (0, 0, -2a, 0, 0, -2b)$, $\quad$ $A_{26} = (0, 0, 0, -2a, 0, -2b)$, $\quad$ $A_{27} = (0, 0, 0, 0, -2a, -2b)$.

The Gosset polytope Goss$_6$ is the convex hull of these points. It is easy to check that $d(A_1, A_i) = 1$ for $2 \leq i \leq 17$ and $d(A_1, A_i) = \sqrt{2}$ for $18 \leq i \leq 27$.

It is clear that the points $A_2 - A_{17}$ are vertices of a five-dimensional demihypercube (the corresponding hypercube has 32 vertices of the form $(\pm a, \pm a, \pm a, \pm a, \pm a, b)$), and the points $A_{18} - A_{27}$ are the vertices of the five-dimensional hyperoctahedron (orthoplex), which is a facet of the polytope Goss$_6$ (lying in the hyperplane $x_6 = -2b$). The origin $O = (0, 0, 0, 0, 0, 0) \in \mathbb{R}^6$ is the center of the hypersphere described around Goss$_6$ with radius $4b = \sqrt{2/3}$.

Example 10.4 Let us consider a description of the Gosset polytope Goss$_7$. Let us set it with the coordinates of the vertices in $\mathbb{R}^7$, as it is done in [33]. In what follows, $\overline{a_1, a_2, \ldots, a_m}$ means the set of all permutations of the elements $a_1, a_2, \ldots, a_m$.

Let us put $a = \frac{\sqrt{2}}{4}$, $b = \frac{\sqrt{6}}{12}$, and $c = \frac{\sqrt{3}}{6}$. We define the points $B_i \in \mathbb{R}^7$, $i = 1, \ldots, 56$, as follows:

$B_1 = (0, 0, 0, 0, 0, 0, 3c)$, $\quad$ $B_2 = (0, 0, 0, 0, 0, 4b, c)$, $\quad$ $B_3 = (a, a, a, a, a, b, c)$,

$B_4 - B_{13} = (\overline{-a, -a, a, a, a}, b, c)$, $\quad$ $B_{14} - B_{18} = (\overline{-a, -a, -a, -a, a}, b, c)$,

$B_{19} - B_{23} = (\overline{2a, 0, 0, 0, 0}, -2b, c)$, $\quad$ $B_{24} - B_{28} = (\overline{-2a, 0, 0, 0, 0}, -2b, c)$,

$B_{29} - B_{33} = (\overline{-2a, 0, 0, 0, 0}, 2b, -c)$, $\quad$ $B_{34} - B_{38} = (\overline{2a, 0, 0, 0, 0}, 2b, -c)$,

$B_{39} - B_{48} = (\overline{a, a, -a, -a, -a}, -b, -c)$, $\quad$ $B_{49} - B_{53} = (\overline{a, a, a, a, -a}, -b, -c)$,

$B_{54} = (-a, -a, -a, -a, -a, -b, -c)$, $\quad$ $B_{55} = (0, 0, 0, 0, 0, -4b, -c)$,

$B_{56} = (0, 0, 0, 0, 0, 0, -3c)$.

The Gosset polytope Goss$_7$ is the convex hull of these points. It is clear that the last 28 points could be obtained from the first 28 points via the central symmetry with respect to the origin O. Hence, the circumradius of this polytope is $3c = \frac{\sqrt{3}}{2}$. It is clear that the convex hull of the points B_i for $2 \leq i \leq 28$ is the Gosset polytope Goss$_6$ (the "first" Goss$_6$, that is a vertex figure for the vertex $B_1 \in$ Goss$_7$). The point B_i for $29 \leq i \leq 55$ constitute the polytope that is symmetric to the previous one (the "second" Goss $_6$).

It is easy to check that $d(B_1, B_i) = 1$ for $2 \leq i \leq 28$, $d(B_1, B_i) = \sqrt{2}$ for $29 \leq i \leq 55$, and $d(B_1, B_{56}) = \sqrt{3}$.

It is easy to see that the subgroup of the isometries of Goss $_7$ that preserve all points of the straight line Ox_7, acts transitively on the vertices B_i for $2 \leq i \leq 28$, as well as for $29 \leq i \leq 55$. Indeed, any isometry of Goss $_6$ (that is identified with the convex hull of the points B_i for $2 \leq i \leq 28$) could be uniquely extended to the isometry of Goss $_7$, that fixes B_1 (as well as B_{56}). In particular, the isotropy group $I(B_1)$ acts transitively on all (non-empty) spheres $S(B_1, r)$. It is clear that there are exactly four non-empty spheres $S(B_1, r)$: for $r = 0$ and $r = 6c = \sqrt{3}$ we get one-point sets, whereas for $r = 1$ and $r = \sqrt{2}$ we have two copies of Goss $_6$.

10.5   Regular and Semiregular Polytopes with Normal Homogeneous or Clifford–Wolf homogeneous Vertex Sets

In [7, 9, 10], the authors obtained a complete description of the metric properties of the sets of vertices of regular and semiregular polytopes in Euclidean spaces from the point of view of the normal homogeneity and the Clifford–Wolf homogeneity. Recall that any set M in the Euclidean space $\mathbb{R}^n$ is supposed to be supplied with the metric d induced from $\mathbb{R}^n$. Here we collect all related results in the following theorem.

Theorem 10.3 ([7, 9, 10]) *For a given regular or semiregular polytope P in $\mathbb{R}^n$, the vertex set M of P is normal homogeneous or Clifford–Wolf homogeneous if and only if there is the sign "+" in the suitable place of the intersection of the row corresponding to P with the fifth column of Table 10.2, where NH means the normal homogeneity and CWH means the Clifford–Wolf homogeneity.*

Remark 10.2 In the fourth column of Table 10.2 we clarify the degree of regularity of P: R and SR mean respectively a regular polytope and a semiregular (non-regular) polytope. The last column of Table 10.2 contains the sources for the corresponding results.

We discuss in more detail some tools that were used for obtaining the suitable results.

The most usual way to prove that a given metric space (M, d) is Clifford–Wolf homogeneous, is to supply M with a group structure, such that d is invariant both under left and right shifts, see Example 10.1. For instance, the vertex set M of a regular polygon with m vertices could be identified with the cyclic group C_m (with the generator that is a rotation around center of the polygon with rotation angle $2\pi/m$). Hence the vertices of every regular two-dimensional polyhedron (regular polygon) form Clifford–Wolf homogeneous metric spaces. The vertex sets of the n-dimensional simplex and the hypercube (n-cube) can be considered respectively

Table 10.2 Metric properties of regular and semiregular polytopes

N	Polytope	Dimension	Regularity	(NH, CWH)	Source
1	n-simplex	n	R	$(+, +)$	[7]
2	n-cube	n	R	$(+, +)$	[7]
3	n-orthoplex	n	R	$(+, +)$	[7]
4	Any regular polygon	2	R	$(+, +)$	[7]
5	Dodecahedron	3	R	$(-, -)$	[7]
6	Icosahedron	3	R	$(+, -)$	[7]
7	24-cell	4	R	$(+, +)$	[9]
8	120-cell	4	R	$(-, -)$	[9]
7	600-cell	4	R	$(+, +)$	[9]
8	Any regular prism	3	SR	$(+, +)$	[9]
9	Any right antiprism	3	SR	$(+, +)$	[9]
10	Any Archimedean solid	3	SR	$(-, -)$	[9]
11	Rectified 4-simplex	4	SR	$(+, -)$	[10]
12	Snub 24-cell	4	SR	$(-, -)$	[10]
13	Rectified 600-cell	4	SR	$(-, -)$	[10]
14	Goss $_5$	5	SR	$(+, +)$	[10]
15	Goss $_6$	6	SR	$(+, -)$	[10]
16	Goss $_7$	7	SR	$(+, -)$	[10]
17	Goss $_8$	8	SR	$(+, +)$	[10]

as the cyclic group C_{n+1} and the group $(\mathbb{Z}_2)^n$ with bi-invariant metrics, see [7, Corollary 2] (the same idea works in the case of the n-dimensional demihypercube, see [10, Proposition 19]). For $n = 1$ we get the case of one-dimensional polytope (segment) in this construction. Note that the vertex sets of the 24-cell, dysphenoidal 288-cell, and 600-cell could be identified with some subgroups of the group S^3 (the group of unit quaternions), hence, these sets are also Clifford–Wolf homogeneous metric spaces, see [9, Proposition 3]. We specially note that *the dysphenoidal 288-cell is neither regular, nor semiregular, nor even uniform*, see Section 3 in [9].

Particularly noteworthy is the study of the eight-dimensional Gosset polytope Goss $_8$ [10].

Example 10.5 Recall that the Gosset polytope Goss $_8$ in $\mathbb{R}^8$ can be realized as the convex hull of some multiplication-closed system of unit *octaves* called *Cayley integers*, see e. g. [26], where explicit expressions for Cayley integers and a discussion of their properties can be found. This system is not a group under the multiplication operation, but is a Moufang loop of unit octaves, consisting of 240 elements [24]. We identify it with the vertex set M of Goss $_8$. Furthermore, $|\cdot|$ means the norm in $\mathbb{O} = \mathbb{R}^8$. It should be taken into account that the equality $|ab| = |a| \cdot |b|$ is true for the product of octaves. Of the properties of Moufang loops, we highlight the presence of a two-sided inverse element for each element of the loop (we use the notation x^{-1} for the two-sided inverse of x) and the equality $(yx^{-1})x = y(x^{-1}x) = y$ for any elements x and y. Below we consider only elements from M (all of them have unit norm). Each element of $x \in M$ can be moved into any other element of $y \in M$ by left multiplication by some element from M:

$$(yx^{-1})x = y(x^{-1}x) = y.$$

Since every left multiplication is an isometry, we get:

$$\rho(xy, xz) = |xy - xz| = |x(y - z)| = |x||y - z| = |y - z| = \rho(y, z).$$

Finally, we get that every left multiplication is a Clifford–Wolf translation:

$$\rho(xy, y) = |xy - y| = |(xy - y)y^{-1}| = |(xy)y^{-1} - yy^{-1}| = |x - 1| = \rho(x, 1).$$

Therefore, the vertex set M of the Gosset polytope $Goss_8$ in $\mathbb{R}^8$ is Clifford–Wolf homogeneous.

It should be noted that we used quite special methods for some of Clifford–Wolf homogeneous metric spaces. For instance, the hyperoctahedron (n-ortoplex) is a Clifford–Wolf homogeneous metric spaces in $\mathbb{R}^n$ for any $n \geq 1$ by [7, Corollary 4].

Let us give an idea of how to prove that a vertex set of a given polytope is not normal homogeneous. It is based on the following

Proposition 10.6 ([11]) *Let M be the vertex set of a polytope $P \subset \mathbb{R}^n$. Suppose that there are adjacent points $O, O' \in M$ such that*

(1) *for any vertex $Q' \in M$ adjacent to O' and $Q' \neq O$ we have $\angle OO'Q' \geq \pi/2$;*
(2) *there are two distinct vertices $Q_1, Q_2 \in M$ adjacent to O and distinct from O' such that $\angle Q_i OO' > \pi/2$, $i = 1, 2$.*
Then (M, d) is not normal homogeneous.

Proof Denote $d(O, O')$ by ρ. Suppose that (M, d) is normal homogeneous. Then there is an isometry ψ of the metric space (M, d) shifting all points by a distance at most ρ and such that $\psi(O) = O'$ (ψ is a δ-translation at the point O).

Since ψ is an isometry and the vertex Q_i is adjacent to O, then $\psi(Q_i)$ is adjacent to O', $i = 1, 2$. Since $\psi(Q_2) \neq \psi(Q_1)$, one of these points, say $\psi(Q_1)$, is distinct from O. Then we have $\angle OO'\psi(Q_1) \geq \pi/2$ and $\angle Q_1 OO' > \pi/2$, therefore, $d(Q_1, \psi(Q_1)) > \rho$ which is impossible (even the orthogonal projection of the line segment $[Q_1, \psi(Q_1)]$ to the straight line OO' is longer than ρ). Hence, the map ψ with the desirable properties does not exist. The proposition is proved. ∎

This proposition could be used to prove that the vertex sets of the dodecahedron in $\mathbb{R}^3$, the 120-cell in $\mathbb{R}^4$, and the Archimedian solids are not normal homogeneous.

To study all other regular and semiregular polytopes, we refer the reader to the detailed reasoning in [7, 9], and [10].

10.6 General Results on the m-Point Homogeneity and m-Point Homogeneous Subspaces of Euclidean Spaces

Let us consider an arbitrary homogeneous metric space (M, d) with isometry group Isom(M). For any $x \in M$, the group $I(x) = \{\psi \in \text{Isom}(M) \mid \psi(x) = x\}$ is called *the isotropy subgroup at the point x*. If $\eta \in \text{Isom}(M)$ is such that $y = \eta(x)$, then, obviously, $I(y) = \eta \circ I(x) \circ \eta^{-1}$. Hence, the isotropy subgroups for any two points of a given homogeneous finite metric M space are conjugate to each other in the isometry group. In particular, the cardinality of M is the quotient of the cardinality of Isom(M) by the cardinality of $I(x)$ (for any $x \in M$).

We will discuss some properties of 2-point homogeneous metric spaces. According to Definition 10.5, a metric space (M, d) is *two-point homogeneous* (*2-point homogeneous*), if for every pairs (A_1, A_2) and (B_1, B_2) of elements of M such that $d(A_1, A_2) = d(B_1, B_2)$, there is an isometry $f \in \text{Isom}(M)$ with the following property: $f(A_i) = B_i, i = 1, 2$.

Remark 10.3 There are some publications, where some types of two-point homogeneity for finite sets with additional structures are studied. For instance, we refer to [43], where two-point homogeneous quandles are studied. On the other hand, we are focused on the study of metric spaces.

For a given $x \in M$ and $r > 0$ we consider the sphere $S(x, r) = \{y \in M \mid d(x, y) = r\}$ with center x and radius r.

Remark 10.4 If (M, d) is a finite homogeneous subset of the Euclidean space $\mathbb{R}^n$, then every sphere $S(x, r)$ in (M, d) lies in some $(n - 1)$-dimensional Euclidean subspace, since all points of this sphere are also on one and the same distance from the barycenter of $M \subset \mathbb{R}^n$ (recall that $\|x - a\| = C$ and $\|x - b\| = D$ for $x, a, b \in \mathbb{R}^n$ implies $2(x, a - b) = \|a\|^2 - \|b\|^2 - C^2 + D^2$).

It is easy to prove the following useful result.

Proposition 10.7 *A homogeneous metric space (M, d) is two-point homogeneous if and only if for any point $x \in M$ the following property holds: for every $r > 0$ and every $y, z \in S(x, r)$, there is $f \in I(x)$ such that $f(y) = z$. In other words, a homogeneous metric space (M, d) is two-point homogeneous if and only if for any point $x \in M$, the isotropy subgroup $I(x)$ acts transitively on every sphere $S(x, r)$, $r > 0$.*

Remark 10.5 It should be noted that there could be some isometries of a given sphere $S(x, r)$ that are not generated by the elements of the isotropy subgroup $I(x)$.

The most interesting results on m-homogeneous finite metric spaces, obtained in [12], are related to finite subsets of Euclidean space $\mathbb{R}^n$. Let us recall that any such subset M is supplied with the metric d induced from $\mathbb{R}^n$.

Recall that regular as well as semiregular polytopes in Euclidean spaces are homogeneous.

There are many equivalent definition of regular (convex) polytopes [40]. One of this is as follows: A polytope P is regular if its symmetry group acts transitively on flags of P. Recall that *a flag* in an n-dimensional polytope P is a sequence of faces $F_0, F_1, \ldots, F_{n-1}$, where F_k is a k-dimensional face and $F_i \subset F_{i+1}$, $i = 0, 1, \ldots, n-2$. See the discussion of this property e. g. in [30, Chapter 4].

The above definition leads to the conjecture, that *the vertex set of any n-dimensional regular polytope is n-point homogeneous*. But this conjecture is false (for instance, by Theorem 10.6, the dodecahedron in $\mathbb{R}^3$ is not 3-point homogeneous). Therefore, the following problem is natural:

Problem 10.1 *Classify all convex polytopes P in $\mathbb{R}^n$ whose vertex sets are m-point homogeneous, where $m \geq 2$.*

It is easy to see that Proposition 10.7 implies

Proposition 10.8 *The vertex set of any regular polytope in $\mathbb{R}^3$ is 2-point homogeneous.*

Lemma 10.1 *Let P be an n-dimensional convex centrally symmetric polytope with vertex set M inscribed in the sphere $S^{n-1}(O, r)$ with center in the origin of $\mathbb{R}^n$, M_1, M_2 be some ordered subsets of M, for which there exists an isometry f of the polytope P and the sphere $S^{n-1}(O, r)$ on themselves such that $f(M_1) = M_2$, $v \in M_1$, $w \in M_2$, $f(v) = w$. Then*

$$f(M_1 \cup \{-v\}) = M_2 \cup \{-w\}, \quad f[(M_1 - \{v\}) \cup \{-v\}] = (M_2 - \{w\}) \cup \{-w\}.$$

Proof Clearly, $f(-v) = -f(v) = -w$. This implies the first equality of the lemma. The second equality is a consequence of the first one. ∎

Theorem 10.4 ([12]) *Let P be a polytope in $\mathbb{R}^n$, $n \geq 2$, whose vertex set M is homogeneous, symmetric with respect to the center of P, and such that the distances between distinct vertices of P constitute a 3-element set, say $\{d_1, d_2, d_3\}$, where $0 < d_1 < d_2 < d_3$. If for some vertex $v \in M$ the isotropy subgroup $I(v)$ acts $(m-1)$-point transitively on the sphere $S(v, d_1)$, i. e. $S(v, d_1)$ is $(m-1)$-point homogeneous under the action of $I(v)$ for some $m \geq 2$, then M is m-point homogeneous.*

Proof Let M_1, M_2 be arbitrary ordered subsets of M consisting of m vertices such that the distance between two arbitrary points in M_1 is equal to the distance between corresponding two points in M_2. Due to the homogeneity of M, we may assume that the first vertices in M_1 and M_2 are both equal to v.

If $M_1 \subset S(v, d_1) \cup \{v\}$, then the same is true for M_2 and the conditions of the theorem imply that there is $f \in I(v)$ such that $f(M_1) = M_2$.

Otherwise, $M_1 = M_1' \cup M_1''$ and $M_2 = M_2' \cup M_2''$, where $M_i' \subset S(v, d_1) \cup \{v\}$ and $M_i'' \subset S(v, d_2) \cup \{-v\}$, $i = 1, 2$. It is clear that the ordered sets $M_1' \cup (-M_1'')$ and $M_2' \cup (-M_2'')$ of vertices in M are mutually isometric and both are contained in $S(v, d_1) \cup \{v\}$. The conditions of the theorem imply that there is $f \in I(v)$ such that $f(M_1' \cup (-M_1'')) = M_2' \cup (-M_2'')$. Applying several times the second assertion of

Lemma 10.1 to the subsets $-M_1''$ and $-M_2''$, we obtain $f(M_1) = M_2$. Hence, M is m-point homogeneous.   ∎

Remark 10.6 It is clear that the sphere $S(v, d_1)$ could be replaced by the sphere $S(v, d_2)$ in the statement of Theorem 10.4. Indeed, $S(v, d_1)$ is the vertex figure for the vertex v, while $S(v, d_2)$ is the vertex figure for the vertex $-v$.

Question 10.1 *What is the classification of polytopes P in $\mathbb{R}^n$, $n \geq 2$, such that its vertex set M is homogeneous, symmetric with respect to the center of P, and the distances between distinct vertices of P constitute a 3-element set?*

It is easy to verify the validity of the following result (see e. g. [12]).

Lemma 10.2 *For any subset of the sphere $\{v_1, \ldots, v_n\} \subset S^{n-1}(0, 1) \subset \mathbb{R}^n$, where $n \geq 2$, not lying on any great subsphere of the sphere $S^{n-1}(0, 1)$ of smaller dimension, every point $w \in S^{n-1}(0, 1)$ is uniquely determined by its distances to the points $v_1, \ldots, v_n$.*

This lemma implies the following useful results.

Theorem 10.5 ([12]) *Let $n \geq 3$ and P be an n-dimensional convex $(n-1)$-point homogeneous polytope in $\mathbb{R}^n$ with vertex set M of cardinality $k \geq n+1$. Then P is k-point homogeneous if and only if for any two isometric non-degenerate ordered n-tuples of vertices M_1 and M_2 in M with the first points v, there exists an isometry f from the isotropy subgroup $I(v)$ such that $f(M_1) = M_2$. If P is also centrally symmetric then M in this statement can be replaced by an arbitrary maximal (with respect to inclusion) subset $N \subset M \cap B(v, \pi/2)$, which contains no diametrically opposite vertices.*

Corollary 10.1 *Every n-dimensional convex n-point homogeneous polytope in $\mathbb{R}^n$ which has vertex set M with cardinality $k \geq n+1$, is k-point homogeneous.*

Let us consider some simple examples.

Example 10.6 Any regular n-dimensional simplex S is $(n+1)$-point homogeneous. Indeed, the distances between any vertices S coincide and the symmetry group of S is the group S_{n+1} of all permutations of the vertex set.

Example 10.7 Any regular m-gon P in $\mathbb{R}^2$, $m \geq 3$, is m-point homogeneous. Indeed, it is clear that P is 2-point homogeneous. Now the statement follows from Corollary 10.1.

Example 10.8 The vertex set M of the 3-cube P is (at least) 4-point homogeneous. There are three possible distances between distinct vertices of P, $0 < d_1 < d_2 < d_3$, and P has the center of symmetry. Moreover, for any vertex v we see that $S(v, d_1)$ is the vertex set of a regular triangle. The isotropy subgroup $I(v)$ acts 3-point transitively on $S(v, d_1)$. Therefore, M is 4-point homogeneous by Theorem 10.4.

Example 10.9 The vertex set M of the icosahedron P is (at least) 6-point homogeneous. There are three possible distances between distinct vertices of P, $0 < d_1 < d_2 < d_3$, and P has the center of symmetry. Next, for any vertex v we see that

$S(v, d_1)$ is the vertex set of a regular pentagon. The isotropy subgroup $I(v)$ contains all isometries of this pentagon, hence $I(v)$ acts 5-point transitively on $S(v, d_1)$ (see Example 10.7). Therefore, M is 6-point homogeneous by Theorem 10.4.

Corollary 10.1 together with Examples 10.8 and 10.9 imply

Corollary 10.2 *The 3-dimensional cube is 8-point homogeneous. The icosahedron is 12-point homogeneous.*

Corollary 10.3 *The n-orthoplex in $\mathbb{R}^n$ is 2n-point homogeneous for every $n \in \mathbb{N}$.*

Proof This follows from Theorem 10.4 and Corollary 10.1 applied to the n-orthoplex. ∎

Example 10.10 Let us check that the convex hull of the union of regular n-dimensional simplex and the simplex which is centrally symmetric to it, is $2(n + 1)$-point homogeneous.

As the first simplex S_1, we can take the convex hull of $n + 1$ points $v_i \in \mathbb{R}^{n+1}$, $i = 1, \ldots, n + 1$, where the j-th coordinate of v_i is $v_{i,j} = -1$ if $j \neq i$ and $v_{i,i} = n$.

One can easily see that the convex hull P^n of $S_1 \cup (-S_1)$ is the image of the orthogonal projection (along $(1, \ldots, 1) \in \mathbb{R}^{n+1}$) of the standard $(n + 1)$-orthoplex Ort (in $\mathbb{R}^{n+1}$), multiplied by $n + 1$, onto the n-dimensional hyperplane

$$H^n = \{x = (x_1, \ldots, x_{n+1}) \in \mathbb{R}^{n+1} : x_1 + \ldots x_{n+1} = 0\}.$$

It is easy to see that P^n is homogeneous and satisfies the conditions of Theorem 10.4 with $m - 1 = n$. Then by Theorem 10.4 and Corollary 10.1, P^n is $2(n + 1)$-point homogeneous.

Remark 10.7 P^2 is the regular hexagon and P^3 is the cube.

The following two examples are more involved.

Example 10.11 Let us check that the vertex set M of the polytope Goss_6 (see Example 10.3) is two-point homogeneous.

At first, we note that the spheres $S(A_1, r)$ are non-empty exactly for $r = 1$ and $r = \sqrt{2}$. Indeed, $d(A_1, A_i) = 1$ for $2 \leq i \leq 17$ and $d(A_1, A_i) = \sqrt{2}$ for $18 \leq i \leq 27$. The points $A_2 - A_{17}$ are vertices of a five-dimensional demihypercube (the corresponding hypercube has $32 = 2^5$ vertices of the form $(\pm a, \pm a, \pm a, \pm a, \pm a, b)$), and the points $A_{18} - A_{27}$ are the vertices of the five-dimensional hyperoctahedron (5-orthoplex), which is a facet of the polytope Goss_6 (lying in the hyperplane $x_6 = -2b$).

It is easy tho check that the isotropy subgroup $I(A_1)$ of the group $\text{Isom}(M) = \text{Symm}(\text{Goss}_6)$ contains the orthogonal operators of the following two forms:

(1) $B(x_1, x_2, x_3, x_4, x_5, x_6) = (x_{\sigma(1)}, x_{\sigma(2)}, x_{\sigma(3)}, x_{\sigma(4)}, x_{\sigma(5)}, x_6)$, where $\sigma \in S_5$ is any permutation of $(1, 2, 3, 4, 5)$;

(2) $B(x_1, x_2, x_3, x_4, x_5, x_6) = (\pm x_1, \pm x_2, \pm x_3, \pm x_4, \pm x_5, x_6)$, where the number of signs "$-$" is even.

Now, it is easy to see that a composition of some suitable maps of the above forms moves any given point in the sphere $S(A_1, 1)$ or in the sphere $S(A_1, \sqrt{2})$ to any other given point in the same sphere. By Proposition 10.7, M is two-point homogeneous.

Example 10.12 By Proposition 10.7, the polytope Goss$_7$ (see Example 10.4) is 2-point homogeneous. Moreover, Goss$_7$ is even 3-point homogeneous by Theorem 10.4 and Example 10.11.

10.7   On m-Point Homogeneous Polyhedra in $\mathbb{R}^3$

Recall that any regular polytope in $\mathbb{R}^3$ is 2-point homogeneous (see Proposition 10.8). The main results of this section are Theorems 10.7 and 10.8. The following result deserves special mention.

Theorem 10.6 ([12]) *The point homogeneity degree of the dodecahedron D is 2 (i. e. M is 2-point homogeneous but not 3-point homogeneous).*

Proof In the argument below, we will use the spherical distance between points. Besides that we can assume that the vertices of the dodecahedron are centers of the facets for the icosahedron I with the center O, so that the circumscribed sphere for I is the inscribed sphere for D, while the spherical distance d_1 between adjacent vertices for D is the angle between the outer normals of adjacent faces for I. Let us recall some property of the dodecahedron D.

The dihedral angle under the edge of the icosahedron is equal to $\alpha = 2\arcsin(\varphi/\sqrt{3})$ (see Table 10.1). Therefore

$$d_1 = 2\pi - (2\pi/2 + \alpha) = \pi - \alpha = 2(\pi/2 - \arcsin(\varphi/\sqrt{3})) = 2\arccos(\varphi/\sqrt{3}),$$

$$\cos d_1 = \frac{2\varphi^2}{3} - 1 = \frac{2\varphi^2 - 3}{3} = \frac{3 + \sqrt{5} - 3}{3} = \frac{\sqrt{5}}{3}.$$

By the spherical cosine theorem, for the next spherical distance

$$\cos d_2 = \cos^2 d_1 + \cos\left(\frac{2\pi}{3}\right)\sin^2 d_1 = \cos^2 d_1 - \frac{1}{2}(1 - \cos^2 d_1)$$

$$= \frac{3\cos^2 d_1 - 1}{2} = \frac{1}{3}.$$

All possible spherical distances between vertices of the dodecahedron are equal to

$$0 < \arccos\left(\frac{\sqrt{5}}{3}\right) < \arccos\left(\frac{1}{3}\right) < \arccos\left(\frac{-1}{3}\right) < \arccos\left(\frac{-\sqrt{5}}{3}\right) < d_5 = \pi.$$

The isotropy subgroup $I(v)$ of the vertex $v \in M \subset D$ in the symmetry group of D consists of 6 isometries, three rotations around the axis $v(-v)$ by angles 0, $2\pi/3$, $4\pi/3$ and three mirror reflections relative to every plane, passing through O and one of the edges of the dodecahedron, which includes v. As a corollary, the group $\text{Isom}(D)$ of all isometries of the dodecahedron has order $6 \cdot 20 = 120$.

The spheres $S(v, d_1)$, $S(v, d_2)$, $S(v, d_3)$, $S(v, d_4)$ in $S^2(O, 1)$ contain respectively 3, 6, 6, 3 vertices of the dodecahedron D, while D has also the vertices v and $-v$. The isotropy group $I(v)$ acts twice transitively on the sets $M \cap S(v, d_1)$ and $M \cap S(v, d_4)$ and simply transitively on the sets $M \cap S(v, d_2)$ and $M \cap S(v, d_3)$.

We get from the above argument and the homogeneity of the dodecahedron D that the vertex set M of D is 2-point homogeneous.

The set $M \cap S(v_1, d_2)$, where $v_1 \in M$, has vertices v_2, v_3, v_3' with the spherical distances between any two distinct vertices in $\{v_2, v_3, v_3'\}$ equal to d_3. Consequently, the ordered triples $M_1 = \{v_1, v_2, v_3\}$ and $M_1' = \{v_1, v_2, v_3'\}$ are isometric. At the same time, there is no isometry of M and $S^2(0, 1)$ such that $f(M_1) = M_1'$ because the isotropy subgroup $I(v_1)$ acts simply transitively on $M \cap S(v_1, d_2)$ due to the above observations (see also Fig. 10.2a). ∎

The above results give a description of all regular polyhedra from the point of view of the m-point homogeneity.

Theorem 10.7 ([12]) *The tetrahedron, cube, octahedron and icosahedron are m-point homogeneous polyhedra for every natural m, while the dodecahedron is 2-point homogeneous but is not 3-point homogeneous.*

Proof We know that the tetrahedron, cube, octahedron, icosahedron are at least 3-point homogeneous polyhedra (see Example 10.6, Corollaries 10.3 and 10.2). Hence, it suffices to apply Corollary 10.1 to these four regular polyhedra. The corresponding result for the dodecahedron follows from Theorem 10.6. ∎

Now, we study semiregular polyhedra. Recall that all semiregular polytopes are (1-point) homogeneous.

Lemma 10.3 *If a semiregular polyhedron $P \in \mathbb{R}^3$ is 2-point homogeneous and is not regular, then P is possibly the cuboctahedron or the icosidodecahedron.*

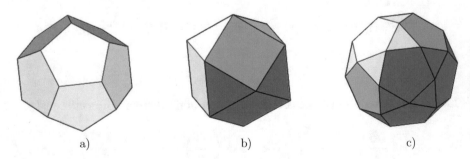

Fig. 10.2 (**a**) dodecahedron; (**b**) cuboctahedron; (**c**) icosidodecahedron

Proof All semiregular (right) prisms different from the cube are not 2-point homogeneous because there is no isometry of the polyhedron, moving the horizontal edge to the vertical one. Analogously, all semiregular antiprisms different from the octahedron are not 2-point homogeneous.

Among the 13 Archimedean polyhedra, possibly only two, the cuboctahedron and the icosidodecahedron (see Fig. 10.2), are two-point homogeneous because all the others have different edges incident to nonisometric pairs of faces containing them.   ∎

The isotropy subgroups of all vertices of the cuboctahedron and the icosahedron coincide as subgroups of the Euclidean isometry group: they are generated by mirror reflections with respect to two mutually perpendicular planes and they include also only the identity mapping and the rotation by the angle π around straight line that is the intersection of these planes. The cuboctahedron and icosidodecahedron are, respectively, the convex hulls of the set of midpoints of all edges of the cube (or octahedron) and of the dodecahedron (or icosahedron) and therefore have 12 and 30 vertices, respectively [9]. Therefore, their isometry groups have orders 48 and 120 which are equal to the orders of the isometry groups of the cube (or octahedron) and of the dodecahedron (or icosahedron). On the other hand, we have the following results.

Proposition 10.9 ([12]) *The (semiregular) cuboctahedron is* 12-*point homogeneous (hence, its point homogeneity degree is* ∞).

Proposition 10.10 ([12]) *The icosidodecahedron is not* 2-*point homogeneous (hence, its point homogeneity degree is* 1).

The following theorem gives a description of all semiregular polyhedra (and even some more general polyhedra) from the point of view of the m-point homogeneity.

Theorem 10.8 ([12]) *If all edges of a given* 2-*point homogeneous polyhedron* $P \subset \mathbb{R}^3$ *have the same length, then* P *is either the cuboctahedron, or a regular polyhedron.*

To conclude this section, we give examples of homogeneous polyhedra in $\mathbb{R}^3$ with different edge lengths.

Example 10.13 The right three-dimensional (anti)prism with regular bases is three-point homogeneous if the length of the lateral edge is longer than the sides and diagonals of the bases.

Example 10.14 Let us consider a simplex S in $\mathbb{R}^3$ with given three edge lengths $0 < a < b < c$, $c^2 < a^2 + b^2$, such that every pair of edges without common vertex have equal Euclidean lengths and every of number a, b, c is a length of edges for some pair. Hence we have a 3-parameter family of simplices in $\mathbb{R}^3$. One can prove that every such simplex is 4-point homogeneous. Hence we have a 3-parameter family of 4-point homogeneous simplices (4-point sets) in $\mathbb{R}^3$.

Example 10.15 If a finite metric space (M, d) is homogeneous and if for one of its points the distances to all other points are pairwise distinct, then (M, d) is k-point homogeneous for every $k \geq 1$. Indeed, it suffices to move only one point of the first k-tuple to the corresponding point of the second k-tuple with a suitable isometry and all other points will automatically be placed at the required places.

Proposition 10.11 ([12]) *Let P be a 3-dimensional polyhedron isomorphic to the octahedron, with diametrically opposite pairs of vertices $A, B; C, D; E, F$. Let d_1, d_2 be positive numbers such that $d_1 < d_2$, $d_1 + d_2 = \pi$. Let us set distances*

$$d(A, B) = d(C, D) = d(E, F) = \pi;$$

$$d(A, C) = d(A, E) = d(B, D) = d(B, F) = d(C, E) = d(D, F) = d_1;$$

while the remaining 6 distances to be equal to d_2. Then (P, d) is realized as the three-dimensional convex centrally symmetric polyhedron in $\mathbb{R}^3$ with vertices on $S^2(0, 1)$ and with indicated spherical distances between the vertices. Moreover, (P, d) is a 6-point homogeneous polyhedron.

10.8 Some Results on the Point Homogeneity Degree

Definition 10.11 For a given convex polytope P, let q be a natural number such that P is q-point homogeneous, but is not $(q + 1)$-point homogeneous. Then q is called the point homogeneity degree of the polytope P. If P is m-homogeneous for all natural m then we define its point homogeneity degree as ∞.

It should be noted that there are polytopes with point homogeneity degree ∞ (compare with Corollary 10.1). For instance, this property is satisfied by the following polytopes: any regular m-gon P in $\mathbb{R}^2$ for $m \geq 3$ (Example 10.7), the n-dimensional regular simplex (Example 10.6) and the n-orthoplex for all $n \geq 1$ (Corollary 10.3); the n-cube for $n = 1, 2, 3$, the icosahedron (Corollary 10.2).

If the vertex set of a polytope P is homogeneous then the point homogeneity degree of P is ≥ 1. Otherwise, it is equal to 0. We will show that the n-cube has point homogeneity degree 3 for $n \geq 4$ (see Proposition 10.12 below), the dodecahedron has point homogeneity degree 2 (Theorem 10.6 below). Note also that the cuboctahedron has the point homogeneity degree 2, while all other Archimedean polyhedra have point homogeneity degree 1 (Theorems 10.7 and 10.6).

On the other hand, we do not know what is the point homogeneity degree for the 24-cell, 120-cell, 600-cell, as well as for Gosset polytopes (with the exception of the four-dimensional truncated simplex and Goss_5, whose point homogeneity degree are respectively 2 and 3 due to Propositions 10.13 and 10.14).

Remark 10.8 It should be noted that the 24-cell, 120-cell, and 600-cell contains a 4-cube (some 16 vertices of any of these polytope are the vertices of a 4-cube). We know that the 4-cube is not 4-point homogeneous. It is possible that the 24-cell, 120-cell, and 600-cell have the same property.

Question 10.2 *What is the point homogeneity degree of a given regular polytope P?*

Question 10.3 *What is the point homogeneity degree of a given semiregular polytope P?*

We know that the 6-dimensional Gosset polytope Goss $_6$ has the point homogeneity degree ≥ 2 (Example 10.11) and the 7-dimensional Gosset polytope Goss $_7$ has the point homogeneity degree ≥ 3 (Example 10.12).

We are going to find the point homogeneity degree of *the n-cube for $n \geq 4$.*

Proposition 10.12 ([12]) *The vertex set of the n-cube, $n \geq 4$, is 3-point homogeneous but is not 4-point homogeneous. Consequently, the point homogeneity degree of the n-cube is 3 for all $n \geq 4$.*

Let us consider some argument for the 4-point homogeneity in the above proposition. Let us consider the following two 4-tuples of vertices of the standard 4-cube:

$$\Big((1, 1, 1, 1),\ (-1, -1, 1, 1),\ (-1, 1, -1, 1),\ (-1, 1, 1, -1)\Big)$$

$$\Big((1, 1, 1, 1),\ (-1, -1, 1, 1),\ (-1, 1, -1, 1),\ (1, -1, -1, 1)\Big).$$

Let us suppose that there is an isometry f of the standard 4-cube that maps the first 4-tuple onto the second one. It is easy to see that the distances between the corresponding pairs of the points in these 4-tuples coincide. We see also that the first three vertices in these 4-tuples coincide. Now, any isometry that preserves these first three vertices (in particular, f) should be the identical map that is impossible.

If $n > 4$, we can consider two 4-tuple of vertices of the standard n-cube, such that the first four coordinates are the same as the above 4-tuples for the standard 4-cube, but all other coordinates are equal to 1. The same argument shows that the standard n-cube, $n > 4$, is not 4-homogeneous.

Remark 10.9 It should be noted that the vectors in both 4-tuples in the above example generate the following Hadamard matrices:

$$\begin{pmatrix} 1 & 1 & 1 & 1 \\ -1 & -1 & 1 & 1 \\ -1 & 1 & -1 & 1 \\ -1 & 1 & 1 & -1 \end{pmatrix} \quad \text{and} \quad \begin{pmatrix} 1 & 1 & 1 & 1 \\ -1 & -1 & 1 & 1 \\ -1 & 1 & -1 & 1 \\ 1 & -1 & -1 & 1 \end{pmatrix}.$$

The necessary information about Hadamard matrices and related results can be found, for example, in [37].

We are going to find the point homogeneity degree of *the n-dimensional demihypercube.* We consider the standard n-dimensional demihypercube DC_n, $n \geq 1$, as the convex hull of the points $(\pm 1, \pm 1, \ldots, \pm 1) \in \mathbb{R}^n$, where the quantity of the signs "$-$" is even (if we consider odd quantity of such signs, then we get another one n-dimensional demihypercube). It is clear that DC_n has a center of symmetry if and only if n is even. The symmetry group of DC_n consists of those symmetries of the standard n-cube that preserved DC_n (it contains all permutation of coordinates and all inversions of signs for even quantity of coordinates).

It is easy to see that DC_2 (which is a segment) is two-point homogeneous and DC_3 (which is a regular simplex) is 4-point homogeneous. Therefore, the point homogeneity degree of DC_n is ∞ for $n = 1, 2, 3$.

Proposition 10.13 ([12]) *The vertex set of DC_n, $n \geq 4$, is* 3-*point homogeneous but is not* 4-*point homogeneous. Consequently, the point homogeneity degree of the n-dimensional demihypercube is* 3 *for all $n \geq 4$.*

We shall find the point homogeneity degree of *the truncated n-dimensional simplex*.

The standard truncated n-dimensional simplex TS_n can be represented as the convex hull of the points $B_{i,j} \in \mathbb{R}^{n+1}$, $i, j = 1, \ldots, n+1$, $i < j$, such that all coordinates of the point $B_{i,j}$ are zero except for its i-th and j-th coordinates, which are equal to 1. It is clear that there are exactly $C_n^2 = n(n-1)/2$ such points. The distance between (different) points $B_{i,j}$ and $B_{s,t}$ is $\sqrt{2}$ for $i = s$ or $j = t$ and 2 for $\{i, j\} \cap \{s, t\} = \emptyset$. Isometry group of the corresponding simplex (the convex hull of points $A_i \in \mathbb{R}^{n+1}$, $i = 1, \ldots, n+1$, such that all coordinates of A_i are zero except for its i-th coordinate, which is 2) acts transitively on the vertex set of the truncated simplex under consideration. Let M be the vertex set of TS_n. The above argument shows that M is a finite homogeneous two-distance set. Moreover, TS_n is uniform in every dimension [25].

Note that TS_1 is a one-point set, TS_2 is a regular triangle, TS_3 is an orthoplex (octahedron), TS_4 is a semiregular polytope.

The isometry group of TS_n is the symmetric group S_{n+1} for $n = 2$ and $n \geq 4$ (the group of permutations of the symbols $1, 2, \ldots, n, n+1$), see Theorem 1 in [15]. The isometry group of TS_3 is the isometry group of the octahedron $\mathbb{Z}_2^3 \rtimes S_3 = \mathbb{Z}_2 \rtimes S_4$.

We will denote the vertex $B_{i,j}$ also by the symbol $\{i, j\}$ (the set of two indices). Let us fix the vertex $B_{1,2} = \{1, 2\}$. We see that the sphere $S(\{1, 2\}, \sqrt{2})$ (respectively, the sphere $S(\{1, 2\}, 2)$) consists of vertices $\{i, j\}$ such that the set $\{1, 2\} \cap \{i, j\}$ contains exactly one element (respectively, $\{1, 2\} \cap \{i, j\} = \emptyset$). The isotropy group $I(\{1, 2\})$ is the direct product $S_2 \times S_{n-1}$ (independent permutations of the elements 1, 2 as well as the elements $3, 4, \ldots, n+1$). Obviously, $I(\{1, 2\})$ acts transitively on the spheres $S(\{1, 2\}, \sqrt{2})$ and $S(\{1, 2\}, 2)$. Therefore M is 2-point homogeneous.

Let us show that M is not 3-point homogeneous. For this we consider the following triples of vertices:

$$(\{1, 2\}, \{1, 3\}, \{2, 3\}), \qquad (\{1, 2\}, \{1, 3\}, \{1, 4\}).$$

If the isometry f maps the first triple onto the second one, then from $f(\{1, 2\}) = \{1, 2\}$ we see that f either interchanges the coordinates with numbers 1 and 2 or preserves both coordinates. Next, from $f(\{1, 3\}) = \{1, 3\}$ we see that f preserves the coordinates with numbers 1, 2 and 3. Therefore, the equality $f(\{2, 3\}) = \{1, 4\}$ is impossible. In this construction, we used four distinct coordinates of the vertices, that is important.

Hence, we have proved the following result.

Proposition 10.14 ([12]) *The point homogeneity degree of the truncated n-dimensional simplex is 2 for $n \geq 4$ and is ∞ for $n = 1, 2, 3$.*

10.9   Conclusion

Here we would like to outline further options for continuing the study of finite homogeneous subsets of Euclidean spaces. It should be noted that an arbitrary orbit of an arbitrary finite subgroup G in the group $O(n)$ of orthogonal transformations of Euclidean space $\mathbb{R}^n$ is a finite homogeneous subset of $\mathbb{R}^n$. Therefore, it would be natural to single out some of the most interesting and promising special cases.

Two-distance subsets of Euclidean spaces can become a remarkable object of study. A set M in Euclidean space $\mathbb{R}^n$, is called *a two-distance set* if the distance between distinct points of M assumes only two values. The maximum size of such a set is 5 in $\mathbb{R}^2$ (Kelly), and 6 in $\mathbb{R}^3$ (Croft). Moreover, we know that $\text{card}(M) \leq \frac{1}{2}(n+1)(n+2)$ for the cardinality of any two-distance set M in the Euclidean space $\mathbb{R}^n$ [18].

A discussion of maximal two-distance sets in $\mathbb{R}^n$ for $n \leq 10$ could be found in [39]. It is clear that (maximal) one-distance sets in $\mathbb{R}^n$ are exactly the vertex sets of (n-dimensional) regular simplexes (simplices).

It is helpful also to consider *spherically two-distance sets*, i. e., two-distance sets in $\mathbb{R}^n$ that are on the unit round sphere. Recall that every finite homogeneous two-distance subset of $\mathbb{R}^n$ is a spherically two-distance set, up to a homothety.

The classical examples of two-distance homogeneous subsets of Euclidean spaces are the vertex sets of truncated simplices (see the discussion before Proposition 10.14). Another very notable example is the vertex set of the six-dimensional Gosset polytope Goss_6 (see Example 10.3)

Problem 10.2 *Classify all normal homogeneous (respectively, Clifford–Wolf homogeneous) subspaces of Euclidean spaces among two-distant finite homogeneous subspaces.*

Problem 10.3 *For a given $m \geq 2$, classify all m-point homogeneous subspaces of Euclidean spaces among two-distant finite homogeneous subspaces.*

An important source of two-distance homogeneous subspaces in Euclidean spaces are special realizations of *strongly regular graphs* in Euclidean spaces, see details e. g. in [20, 23].

A strongly regular graph Γ with parameters (v, k, λ, μ) is an undirected regular graph on v vertices of valency k such that each pair of adjacent vertices (notation

$x \sim y$) has λ common neighbors, and each pair of nonadjacent vertices (notation $x \not\sim y$) has μ common neighbors. The adjacency matrix A (under some numeration of the vertex set) of Γ is symmetric and has the following properties:

$$AJ = JA = kJ, \qquad A^2 + (\mu - \lambda)A + (\mu - k)I = \mu J, \qquad (10.9.1)$$

where I is the identity matrix and J is the $(v \times v)$-matrix with all entries equal to 1. It is clear that the vector $x_0 = (1, 1, \dots, 1) \in \mathbb{R}^v$ is an eigenvector of A with eigenvalue k. Any other eigenvector of A is orthogonal to x_0, so the corresponding eigenvalue t satisfies the quadratic equation $t^2 + (\mu - \lambda)t + \mu - k = 0$.

It follows that the eigenvalues of A are k, r, s, with $k > r > 0 > s$, where r and s are the two solutions of $t^2 + (\mu - \lambda)t + \mu - k = 0$, so that $(A - rI)(A - sI) = \mu J$.

The multiplicities of k, r, s are $1, f, g$ (respectively), where $1 + f + g = v$ and $k + fr + gs = 0$ (this follows from $\mathrm{trace}(A) = 0$), i.e.

$$f, g = \frac{1}{2}\left(v - 1 \pm \frac{(v-1)(\mu - \lambda) - 2k}{\sqrt{(\lambda - \mu)^2 + 4(k - \mu)}} \right).$$

If we multiply the second equality in (10.9.1) by J and take the trace (using the first equality), then we get

$$(v - k - 1)\mu = k(k - \lambda - 1).$$

Let us consider the orthogonal decomposition $\mathbb{R}^v = V_0 \oplus V_1 \oplus V_2$, where V_0, V_1, V_2 are the eigenspaces of A with the eigenvalues k, r, s respectively and $\dim(V_0) = 1, \dim(V_1) = f, \dim(V_2) = g$.

The parameters may be conveniently expressed in terms of the eigenvalues as follows:

$$\lambda = k + r + s + rs, \qquad \mu = k + rs.$$

Let us prove that the matrix

$$M := A - sI - \frac{k - s}{v} J$$

has rank f. Indeed, since $[A, J] = 0$ by (10.9.1), we have $[A, M] = 0$, and the subspaces V_0, V_1, V_2 are invariant under the action of M. In particular, $x_0 = (1, 1, \dots, 1) \in \mathbb{R}^v$ is the eigenvector of M with eigenvalue 0, i. e. M map V_0 to the trivial subspace. The same is true about V_2. On the other hand, M maps V_1 onto V_1. Hence, M has rank f. Therefore, the map $x \mapsto \bar{x}$ which sends each vertex x to row $\bar{x}$ of M with the same index is a representation of Γ in $\mathbb{R}^f$. It is clear that the inner product $(\bar{x}, \bar{y})$ depends only on whether $x = y$, $x \sim y$ or $x \not\sim y$.

Let us denote by P_Γ the convex hull of images of all vertices of Γ under the map $x \mapsto \bar{x}$. Since the sum of all elements in every rows and column of M is 0, we get that the barycenter of P_Γ is 0. It is also clear from the above reasoning that the vertex set of P_Γ is a homogeneous and two-distance subset of the Euclidean space $\mathbb{R}^f$. The two-distance sets constructed in this way play an important role in various branches of mathematics, see details in [20, 23].

It should be noted that even homogeneous simplices in Euclidean spaces are not classified (compare with the 3-dimensional Example 10.14). It is proved in [31] (see Theorem 2.1) that the isometry group of an equifacetal n-simplex acts transitively on the vertices and, equivalently, on the facets. The description of all homogeneous simplices in $\mathbb{R}^n$ is very hard. It is related to the problem of classification of all transitive subgroups of the symmetric groups S_{n+1}, see details in [31, 32].

Problem 10.4 *Classify all two-distant homogeneous simplices in Euclidean spaces.*

Problem 10.5 *Classify all two-distant homogeneous simplices in Euclidean spaces with normal homogeneous (respectively Clifford–Wolf homogeneous) vertex sets.*

Problem 10.6 *For a given $m \geq 2$, classify all two-distant homogeneous simplices in Euclidean spaces with m-point homogeneous vertex sets.*

A convex n-dimensional polytope is said to be *simple* if any of its vertices is incident to exactly n edges. A convex polytope is said to be *simplicial* if any of its facets is a simplex. A convex polytope is simple (respectively, simplicial) if and only if its *dual polytope* is simplicial (respectively, simple) [19].

Problem 10.7 *Study convex simple and simplicial polytopes that are additionally m-point homogeneous, Clifford–Wolf homogeneous, or normal homogeneous.*

We hope that the solution to these problems will lead to a better understanding of finite homogeneous subsets of Euclidean spaces that have additional properties.

References

1. V.N. Berestovskii, Solution of a Busemann problem (in Russian). Sib. Mat. Zh. **51**(6), 1215–1227 (2010). English translation: Siberian Math. J., 2010, V. 51, N 6. P. 962–970. https://doi.org/10.1007/s11202-010-0095-3
2. V.N. Berestovskii, L. Guijarro, A metric characterization of Riemannian submersions. Ann. Global Anal. Geom. **18**(6). 577–588 (2000). https://doi.org/10.1023/A:1006683922481
3. V.N. Berestovskii, Y.G. Nikonorov, Killing vector fields of constant length on Riemannian manifolds (in Russian). Sib. Mat. Zh. **49**(3), 497–514 (2008). English translation: Siberian Math. J., 2008, V. 49, N 3. P. 395–407. https://doi.org/10.1007/s11202-008-0039-3
4. V.N. Berestovskii, Y.G. Nikonorov, On δ-homogeneous Riemannian manifolds. Diff. Geom. Appl. **26**(5), 514–535 (2008). https://doi.org/10.1016/j.difgeo.2008.04.003
5. V.N. Berestovskii, Y.G. Nikonorov, Clifford–Wolf homogeneous Riemannian manifolds. J. Differ. Geom. **82**(3), 467–500 (2009). https://doi.org/10.4310/jdg/1251122544
6. V.N. Berestovskii, Y.G. Nikonorov, Generalized normal homogeneous Riemannian metrics on spheres and projective spaces. Ann. Global Anal. Geom. **45**(3), 167–196 (2014), https://doi.org/10.1007/s10455-013-9393-x
7. V.N. Berestovskii, Y.G. Nikonorov, Finite homogeneous metric spaces (in Russian). Sib. Mat. Zh. **60**(5), 973–995 (2019). English translation: Siberian Math. J., 2019, V. 60, N 5. P. 757–773. https://doi.org/10.1134/S0037446619050021

8. V.N. Berestovskii, Y.G. Nikonorov, *Riemannian Manifolds and Homogeneous Geodesics*. Springer Monographs in Mathematics (Springer, Cham, 2020). https://doi.org/10.1007/978-3-030-56658-6

9. V.N. Berestovskii, Y.G. Nikonorov, Finite homogeneous subspaces of Euclidean spaces (in Russian). Mat. Trudy **24**(1), 3—34 (2021). English translation: Siberian Advances in Mathematics, 2021, V. 31, N 3, P. 155–176. https://doi.org/10.1134/S1055134421030019

10. V.N. Berestovskii, Y.G. Nikonorov, Semiregular Gosset polytopes (in Russian). Izv. Ross. Akad. Nauk, Ser. Mat. **86**(4), 51–84 (2022). English translation: Izv. Math., 2022, 86 (in press). https://doi.org/10.1070/IM9169

11. V.N. Berestovskii, Y.G. Nikonorov, On finite homogeneous metric spaces. Vladikavkaz. Mat. Zh. **24**(2), 51–61 (2022). https://doi.org/10.46698/h7670-4977-9928-z

12. V.N. Berestovskii, Y.G. Nikonorov, On m-point homogeneous polytopes in Euclidean spaces. Filomat, **37**(25), 8405–8424 (2023). https://www.pmf.ni.ac.rs/filomat-content/2023/37-25/37-25-1-18943.pdf

13. M. Berger, *Geometry I*. Universitext (Springer, Berlin, 2009)

14. M. Berman, Regular-faced convex polyhedra. J. Franklin Inst. **291**(5), 329–352 (1971)

15. L.W. Berman, B. Monson, D. Oliveros, G.I. Williams, Fully truncated simplices and their monodromy groups. Adv. Geom. **18**(2), 193–206 (2018). https://doi.org/10.1515/advgeom-2017-0047

16. G. Birkhoff, Metric foundations of geometry I. Trans. Am. Math. Soc. **55**(3), 465–492 (1944)

17. G. Blind, R. Blind, The semiregular polytopes. Comment. Math. Helv. **66**(1), 150–154 (1991). https://doi.org/10.1007/BF02566640

18. A. Blokhuis, A new upper bound for the cardinality of 2-distance sets in Euclidean space. Ann. Discrete Math. **20**, 65–66 (1984). https://doi.org/10.1016/S0304-0208(08)72809-3

19. A. Bronsted, *An Introduction to Convex Polytopes*. Graduate Texts in Mathematics, vol. 90 (Springer, New York, 1983)

20. A.E. Brouwer, W.H. Haemers, *Spectra of Graphs*. Universitext (Springer, Berlin, 2012). **Zbl.**1231.05001

21. H. Busemann, *Metric Methods in Finsler Spaces and in the Foundation of Geometry*. Annals of Mathematics Studies, vol. 8 (Princeton University Press, Princeton, 1942)

22. H. Busemann, Groups of motions transitive on sets of geodesics. Duke Math. J. **24**, 539–544 (1956). https://doi.org/10.1215/S0012-7094-56-02352-3

23. P.J. Cameron, Strongly regular graphs, in *Topics in Algebraic Graph Theory*, ed. by L.W. Beineke et al. Encyclopedia of Mathematics and Its Applications, vol. 102 (Cambridge University Press, Cambridge, 2004), pp. 203–221. **Zbl.**1067.05079

24. J.H. Conway, D.A. Smith, *On Quaternions and Octonions: Their Geometry, Arithmetic, and Symmetry* (A K Peters, Natick, 2003)

25. H.S.M. Coxeter, Wythoff's construction for uniform polytopes. Proc. London Math. Soc. **38**, 327–339 (1934). https://doi.org/10.1112/plms/s2-38.1.327

26. H.S.M. Coxeter, Integral Cayley numbers. Duke Math. J. **13**, 561–578 (1946)

27. H.S.M. Coxeter, *Regular Polytopes*. 3rd edn. (Dover, New York, 1973)

28. P.R. Cromwell, *Polyhedra* (Cambridge University Press, Cambridge, 1997)

29. M. Dutour Sikirić, http://mathieudutour.altervista.org/Regular/ (2022)

30. P. Du Val, *Homographies, Quaternions and Rotations*. Oxford Mathematical Monographs (Clarendon Press, Oxford, 1964)

31. A.L. Edmonds, The geometry of an equifacetal simplex. Mathematika **52**, 31—45 (2005). https://doi.org/10.1112/S0025579300000310

32. A.L. Edmonds, The partition problem for equifacetal simplices. Beiträge Algebra Geom. **50**(1), 195–213 (2009). **Zbl.**1159.52015, **MR**2499788

33. E.L. Elte, *The Semiregular Polytopes of the Hyperspaces*, vol. 1912 (University of Groningen, Groningen, 2022). https://quod.lib.umich.edu/u/umhistmath/ABR2632.0001.001

34. Four-Dimensional Euclidean Space (2022). http://eusebeia.dyndns.org/4d/index

35. T. Gosset, On the regular and semi-regular figures in space of *n* dimensions. Messenger Math. **29**, 43–48 (1900)

36. B. Grünbaum, *Convex Polytopes*. Graduate Texts in Mathematics, vol. 221, 2nd edn. (Springer, New York, 2003)
37. A. Hedayat, W.D. Wallis, Hadamard matrices and their applications. Ann. Stat. **6**(6), 1184–1238 (1978). https://doi.org/10.1214/aos/1176344370
38. N.W. Johnson, Convex polyhedra with regular faces. Can. J. Math. **18**(1), 169–200 (1966). https://doi.org/10.4153/CJM-1966-021-8
39. P. Lisoněk, New maximal two-distance sets. J. Combin. Theory Ser. A **77**(2), 318–338 (1997). https://doi.org/10.1006/jcta.1997.2749
40. H. Martini, A hierarchical classification of Euclidean polytopes with regularity properties. *Polytopes: Abstract, Convex and Computational*, ed. by T. Bisztriczky et al., Proceedings of the NATO Advanced Study Institute, Scarborough, Ontario, August 20–September 3, 1993 (Kluwer, Dordrecht, 1994), pp. 71–96. NATO ASI Ser., Ser. C, Math. Phys. Sci. 440
41. L. Schläfli, *Theorie der vielfachen Kontinuität*. Hrsg. im Auftrage der Denkschriften-Kommission der schweizerischen naturforschenden Gesellschaft von J.H. Graf (Georg, Zürich, 1901)
42. Z.I. Szabó, A short topological proof for the symmetry of 2 point homogeneous spaces. Invent. Math. **106**(1), 61–64 (1991). https://doi.org/10.1007/BF01243903
43. H. Tamaru, Two-point homogeneous quandles with prime cardinality. J. Math. Soc. Japan **65**(1), 1117–1134 (2013). https://doi.org/10.2969/jmsj/06541117
44. J. Tits, Sur certaines classes d'espaces homogenes de groupes de Lie. Acad. Roy. Belg. Cl. Sci. Mém Coll. **29**(3), 268pp. (1955)
45. P. Urysohn, Sur un espace métrique universel. Bull. Sci. Math. **51**, 43–64, 74–90 (1927)
46. H.C. Wang, Two theorems on metric spaces. Pac. J. Math. **1**, 473–480 (1951)
47. H.C. Wang, Two-point homogeneous spaces. Ann. Math. **55**(1), 177–191 (1952). https://doi.org/10.2307/1969427
48. J.A. Wolf, *Spaces of Constant Curvature*, 6th edn. (AMS Chelsea Publishing, Providence, 2011)
49. V.A. Zalgaller, *Convex Polyhedra with Regular Faces*, Zapiski Nauchnykh Seminarov LOMI (in Russian), vol. 2 (1967), pp. 5–221. English translation: New York: Consultants Bureau, 1969

Chapter 11
Discrete Coxeter Groups

Gye-Seon Lee and Ludovic Marquis

Abstract Coxeter groups are a special class of groups generated by involutions. They play important roles in the various areas of mathematics. This survey particularly focuses on how one uses Coxeter groups to construct interesting examples of discrete subgroups of Lie groups.

Keywords Coxeter groups · Discrete subgroups of Lie groups · Reflection groups · Convex cocompact subgroups · Anosov representations

AMS Codes 20F55, 51F15

11.1 Introduction

It is a fundamental problem in geometry and topology to understand discrete subgroups of Lie groups G. For example, when G is the isometry group $\mathrm{Isom}(\mathbb{H}^d)$ of the hyperbolic d-space $\mathbb{H}^d$, the study of discrete subgroups of $\mathrm{Isom}(\mathbb{H}^d)$ is closely related to that of complete hyperbolic d-manifolds. More precisely, there is a one-to-one correspondence between torsion-free discrete subgroups Γ of $\mathrm{Isom}(\mathbb{H}^d)$ (up to conjugation) and complete hyperbolic d-manifolds $\mathbb{H}^d / \Gamma$ (up to isometry).

Convex cocompact subgroups of rank-one Lie groups G are specially an important class of discrete subgroups of G. In particular, given a finitely generated group Γ, the space of representations $\rho : \Gamma \to G$ whose image is convex cocompact is *open* in the representation space $\mathrm{Hom}(\Gamma, G)$, i.e., the space of all representations

G.-S. Lee (✉)
Department of Mathematical Sciences and Research Institute of Mathematics, Seoul National University, Gwanak-gu, Seoul, South Korea
e-mail: gyeseonlee@snu.ac.kr

L. Marquis
Université Rennes, CNRS Rennes, France
e-mail: ludovic.marquis@univ-rennes1.fr

$\rho : \Gamma \to G$. So, if ρ is convex cocompact and is not isolated, then all the nearby representations of ρ are again convex cocompact, and hence discrete.

Recently, new notions of representations were introduced to generalize convex cocompact subgroups of rank-one Lie groups: Anosov representations in real semisimple Lie groups (see [38, 46]) and convex cocompact subgroups in real projective spaces (see [26]). Such representations also have the property of openness. As new theories are developed, it is also important to have many examples to support them. From this perspective, the role of Coxeter groups is crucial.

The aim of this survey is to illustrate how one can build interesting examples of discrete Coxeter groups.

11.2 Coxeter Groups

11.2.1 What Is a Coxeter Group?

A *Coxeter matrix* $M = (m_{s,t})_{s,t \in S}$ on a finite set S is a symmetric matrix with entries $m_{s,t} \in \{1, 2, \ldots, m, \ldots, \infty\}$ such that the diagonal entries $m_{s,s} = 1$ and off-diagonal entries $m_{s,t} \neq 1$. From any Coxeter matrix $M = (m_{s,t})_{s,t \in S}$, one may obtain the *Coxeter group* W of M given by generators and relations:

$$W = \langle s \in S \mid (st)^{m_{s,t}} = 1, \ \forall s, t \in S \rangle.$$

Here, $(st)^\infty = 1$ means that there is no relation between s and t. Since $m_{s,s} = 1$, each generator s is an involution, i.e., $s^2 = 1$. We shall use the notation W, W_S or $W_{S,M}$ for a Coxeter group, depending on what is important to stress. One should remember that a Coxeter group is a group with a preferred generating set, namely S. The *rank* of W_S is the cardinality $\#S$ of S.

The *Coxeter diagram* of the Coxeter group W_S is the labeled graph $\mathscr{G}_W$ such that:

(i) the set of nodes[1] of $\mathscr{G}_W$ is the set S;
(ii) two nodes $s, t \in S$ are connected by an edge $\overline{st}$ of $\mathscr{G}_W$ if $m_{s,t} \in \{3, 4, \ldots, \infty\}$;
(iii) the label of the edge $\overline{st}$ is $m_{s,t}$ if $m_{s,t} \in \{4, 5, \ldots, \infty\}$.

It is well-known that for any subset T of S, the subgroup of W_S generated by T is the Coxeter group $W_{T,M'}$ with generating set T and exponents $m'_{s,t} = m_{s,t}$ for every $s, t \in T$ (see [14, Chap. IV, Th. 2]). Such a subgroup W_T is called a *standard subgroup* of W_S.

The connected components of the graph $\mathscr{G}_{W_S}$ are graphs of the form $\mathscr{G}_{W_{S_i}}$, $i \in I$, where the $(S_i)_{i \in I}$ form a partition of S. The standard subgroups W_{S_i} are called the

---

[1] A Coxeter group often comes with a Coxeter polytope in such a way that the nodes of the Coxeter diagram are in bijection with the facets of the Coxeter polytope. We shall use the word *node* of the Coxeter diagram rather than *vertex* to make a distinction between the vertices of the Coxeter polytope and the vertices of the Coxeter diagram.

irreducible components of W_S. Since $m_{s,t} = 2$ if and only if $st = ts$, we see that the group W_S is the direct product of the subgroups W_{S_i} for $i \in I$. A Coxeter group W_S is *irreducible* when the Coxeter diagram $\mathscr{G}_{W_S}$ is connected, i.e., $\#I = 1$. A subset T of S is said to be "*something*" if the Coxeter group W_T is "something". For example, the word "something" can be replaced by "irreducible", and so on. Two subsets $T, U \subset S$ are *orthogonal* if $m_{t,u} = 2$ for every $t \in T$ and every $u \in U$.

The *Cosine matrix* of $W_{S,M}$ is the $S \times S$ symmetric matrix C_W whose entries are:

$$\left(C_W \right)_{s,t} = -2 \cos \left(\frac{\pi}{m_{s,t}} \right) \qquad \text{for every } s, t \in S$$

An irreducible Coxeter group W is said to be *spherical* (resp. *affine*) when the Cosine matrix C_W is positive definite (resp. positive semi-definite but not definite).

Theorem 11.1 (Coxeter [22, 23] and Margulis–Vinberg [54]) *Let W_S be an irreducible Coxeter group. Then exactly one of the following is true:*

(i) *If W_S is spherical, then W_S is a finite group.*

(ii) *If W_S is affine, then $\#S \geqslant 2$ and W_S is virtually[2] $\mathbb{Z}^{\#S-1}$.*

(iii) *Otherwise, W_S is* large, *i.e., there exists a surjective homomorphism of a finite index subgroup of W_S onto a free group on two generators.*

Remark 11.1 These three cases are clearly exclusive. Consequently, if an irreducible Coxeter group W_S is finite (resp. infinite and virtually abelian), then W_S is spherical (resp. affine).

Remark 11.2 The irreducible spherical and irreducible affine Coxeter groups were classified by Coxeter [22, 23]; see also Witt [71]. The complete list can be found in Fig. 11.1.

A Coxeter group (not necessarily irreducible) is *spherical* (resp. *affine*) when all its irreducible components are spherical (resp. affine).

11.3   Hyperbolic Reflection Groups

11.3.1   Hyperbolic Polytopes

Let $\mathbb{R}^{d,1}$ be the vector space $\mathbb{R}^{d+1}$ endowed with a non-degenerate symmetric bilinear form $\langle \cdot, \cdot \rangle$ of signature[3] $(d, 1)$, and let q be the associated quadratic form.

---

[2] A group G is *virtually "something"* if there is a finite index subgroup $H \leqslant G$ such that H is "something".

[3] The *signature* of a symmetric matrix B is the triple (p, q, r) of the positive, negative, and zero indices of inertia of B. In the case $r = 0$, we simply say that B is of signature (p, q).

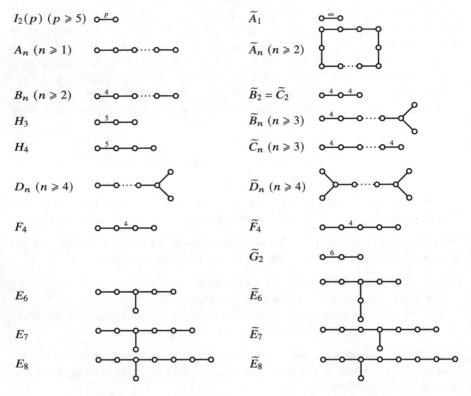

Fig. 11.1 The irreducible spherical Coxeter diagrams on the left and irreducible affine Coxeter diagrams on the right

A coordinate representation of q with respect to some basis of $\mathbb{R}^{d,1}$ is:

$$q(x) = x_1^2 + \cdots + x_d^2 - x_{d+1}^2.$$

A *hyperbolic d-space* $\mathbb{H}^d$ is a connected component of a hyperquadric:

$$\mathbb{H}^d = \{x \in \mathbb{R}^{d,1} \mid q(x) = -1 \text{ and } x_{d+1} > 0\}.$$

The isometry group of $\mathbb{H}^d$ is $O_{d,1}^+(\mathbb{R})$, which consists of the elements of $O_{d,1}(\mathbb{R})$ that preserve $\mathbb{H}^d$. We often work with the projective model of $\mathbb{H}^d$:

$$\mathbb{H}^d = \{x \in \mathbb{R}^{d,1} \mid q(x) < 0 \text{ and } x_{d+1} > 0\}/\mathbb{R}_+,$$

where the set $\mathbb{R}_+$ of positive scalars acts on $\mathbb{R}^{d,1} \smallsetminus \{0\}$ by multiplication. If we set

$$\mathbb{S}(\mathbb{R}^{d,1}) := (\mathbb{R}^{d,1} \smallsetminus \{0\})/\mathbb{R}_+,$$

then $\mathbb{H}^d$ is an open subset of the *projective sphere* $\mathbb{S}(\mathbb{R}^{d,1})$. The closure $\overline{\mathbb{H}^d}$ of $\mathbb{H}^d$ in $\mathbb{S}(\mathbb{R}^{d,1})$ is the *compactification* of $\mathbb{H}^d$.

A subset of $\mathbb{H}^d$ is a *hyperbolic d-polytope* if it is the intersection of a finite family of closed half-spaces of $\mathbb{H}^d$ and if it has non-empty interior. A *hyperbolic Coxeter d-polytope* (or simply a $\mathbb{H}^d$-Coxeter polytope) is a hyperbolic d-polytope P such that all its dihedral angles are sub-multiples of π. In other words, if two facets[4] s, t of P are adjacent,[5] then the dihedral angle $\theta(s, t)$ between s and t is equal to π/m for some integer $m \geqslant 2$. When two facets s, t are parallel, it is common to say that $\theta(s, t) = 0$. A hyperbolic Coxeter polytope is *right-angled* if all its dihedral angles are $\pi/2$.

Associated with a hyperbolic Coxeter polytope P is a Coxeter matrix $M = (m_{s,t})_{s,t \in S}$ on the set S of facets of P: if $s, t \in S$ are adjacent, then $m_{s,t} = \pi/\theta(s,t)$; otherwise, $m_{s,t} = \infty$. We denote by W_P the Coxeter group of the Coxeter matrix M, and call it the *Coxeter group of P*. If s is a facet of P, then σ_s denotes the reflection in the hyperplane containing s.

Theorem 11.2 (Poincaré [57]) *Let P be a $\mathbb{H}^d$-Coxeter polytope, and W_P the Coxeter group of P. Then the homomorphism $\sigma : W_P \rightarrow \mathrm{Isom}(\mathbb{H}^d)$ defined by*

$$\sigma(s) = \sigma_s \quad \text{for each } s \in S$$

is injective and the image $\Gamma_P := \sigma(W_P)$ is discrete. Moreover, P is a fundamental domain for the action of Γ_P on $\mathbb{H}^d$. In particular, if P has finite volume (resp. is compact), then Γ_P is a lattice (resp. uniform lattice) of $\mathrm{Isom}(\mathbb{H}^d)$.

A subgroup H of $\mathrm{Isom}(\mathbb{H}^d)$ is called a *hyperbolic reflection group* if $H = \Gamma_P$ for some $\mathbb{H}^d$-Coxeter polytope P. In this case, we call Γ_P the *reflection group of P*. Theorem 11.2 provides a nice way to construct discrete subgroups of $\mathrm{Isom}(\mathbb{H}^d)$, even lattices. We shall review in the next section the classification of $\mathbb{H}^d$-Coxeter polytopes of finite volume in small dimensions and their non-existence in large dimensions.

11.3.2 Classical Results in Dimensions 2 and 3

11.3.2.1 Dimension 2

A necessary and sufficient condition for the existence of hyperbolic Coxeter 2-polytopes of finite volume follows immediately from:

---

[4] A face of P of codimension 1 (resp. 2) is called a *facet* (resp. *ridge*) of P.

[5] Two facets s and t of P are *adjacent* if $s \cap t$ is a ridge of P.

Theorem 11.3 *Let* $\theta_1, \ldots, \theta_n$ *be real numbers such that* $0 \leqslant \theta_i < \pi$ *for each* $i = 1, \ldots, n$. *Then there exists a hyperbolic polygon of finite volume with dihedral angles* $\theta_1, \ldots, \theta_n$ *if and only if* $\sum_{i=1}^{n} \theta_i < (n-2)\pi$.

11.3.2.2 Dimension 3

Hyperbolic Coxeter 3-polytopes of finite volume are well understood, notably thanks to the classification of hyperbolic 3-polytopes with dihedral angles $\leqslant \pi/2$ due to Andreev [6, 7]. During his PhD, Roeder found and fixed a gap in the original proof of Andreev (see [61]). Hodgson and Rivin [39] gave a characterization of hyperbolic 3-polytopes, which generalizes Andreev's theorem, in terms of a generalized Gauss map.

To express properly Andreev's theorem, one needs some definitions. Two compact polytopes $\mathcal{P}, \mathcal{P}'$ of the Euclidean space $\mathbb{R}^d$ are *combinatorially equivalent* if there is a bijection between their faces that preserves the inclusion relation. A combinatorial equivalence class is called a *combinatorial polytope* . Note that if a hyperbolic polytope $P \subset \mathbb{H}^d$ is of finite volume, then the closure $\overline{P}$ of P in $\overline{\mathbb{H}^d}$ is combinatorially equivalent to a compact polytope of $\mathbb{R}^d$. A *labeled polytope* is a combinatorial polytope $\mathcal{P}$ with a labeling θ, that is, a function from the ridges of $\mathcal{P}$ to $[0, \pi/2]$. A hyperbolic polytope P of finite volume *realizes* a labeled polytope $\mathcal{P}$ if there exists a combinatorial equivalence ϕ between the faces of $\overline{P}$ and $\mathcal{P}$ such that the dihedral angle at the ridge e of $\overline{P}$ is the label $\theta(\phi(e))$ of $\mathcal{P}$.

Let $\mathcal{P}$ be a labeled 3-polytope with labeling θ. A *k-circuit* of $\mathcal{P}$ is a sequence of distinct facets $s_1, \ldots, s_k$ such that $e_i := s_i \cap s_{i+1}$ (indices are modulo k) is an edge of $\mathcal{P}$. A k-circuit is *prismatic* if all the (closed) edges e_i are disjoint. The *angle sum* of a k-circuit is the real number $\sum_{i=1}^{k} \theta(e_i)$. A k-circuit is *spherical* (resp. *Euclidean*, resp. *hyperbolic*) if its angle sum is bigger than (resp. equal to, resp. less than) $(k-2)\pi$. A vertex v of $\mathcal{P}$ is *spherical* (resp. *Euclidean*) if the circuit consisting of the facets that contain v is spherical (resp. Euclidean). The *graph* $W_{\mathcal{P}}$ of $\mathcal{P}$ is the graph whose nodes are the facets of $\mathcal{P}$ and such that two nodes s, t are connected if and only if $s \cap t$ is not an edge, or $s \cap t$ is an edge and $\theta(s \cap t) < \pi/2$.

Theorem 11.4 (Andreev [6, 7]; see also Roeder–Hubbard–Dunbar [61]) *Let* $\mathcal{P}$ *be a labeled 3-polytope whose underlying polytope is not a tetrahedron. Then there exists a compact (resp. finite volume) hyperbolic 3-polytope P that realizes $\mathcal{P}$ if and only if:*

- *(i) all the vertices of $\mathcal{P}$ are spherical (resp. spherical or Euclidean);*
- *(ii) all the prismatic 3- and 4-circuits of $\mathcal{P}$ are hyperbolic;*
- *(iii) the graph $W_{\mathcal{P}}$ of $\mathcal{P}$ is connected.*

In that case, the polytope P is unique up to an isometry of $\mathbb{H}^3$.

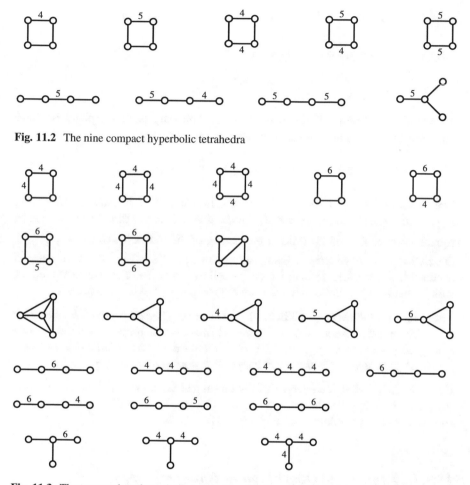

Fig. 11.2 The nine compact hyperbolic tetrahedra

Fig. 11.3 The twenty-three hyperbolic tetrahedra of finite volume which are not compact

Remark 11.3 A careful reader probably wonders what happen when the underlying polytope of $\mathcal{P}$ is a tetrahedron. This case is explained in [60]. The list of all Coxeter tetrahedra of finite volume can be found in Figs. 11.2 and 11.3.

Remark 11.4 The condition on the connectivity of $W_{\mathcal{P}}$ is often expressed by inequalities. One may notice that if the conditions (i) and (ii) are satisfied, then the disconnectedness of $W_{\mathcal{P}}$ implies that (see [61, Prop 1.5]):

- $\mathcal{P}$ is a right triangular prism, i.e., all the labels between the base facets and the joining facets are $\pi/2$, or
- $\mathcal{P}$ is a quadrilateral pyramid with Euclidean apex such that the labels of two opposite edges in the base are $\pi/2$.

11.3.3 The Gram Matrix of a Hyperbolic Polytope

A hyperbolic d-polytope P is of the form

$$P = \cap_{i=1}^{N} H_i^-,$$

where H_i^- is a closed half-space of $\mathbb{H}^d$ whose boundary is a hyperplane H_i. Each H_i^- corresponds to a unique vector u_i with the property that:

$$H_i^- = \mathbb{H}^d \cap \{x \in \mathbb{R}^{d,1} \mid \langle x, u_i \rangle \leqslant 0\} \quad \text{and} \quad \langle u_i, u_i \rangle = 1.$$

The *Gram matrix* $G = (g_{i,j})$ of P is a symmetric $N \times N$ matrix with entries $g_{i,j} = \langle u_i, u_j \rangle$. A square matrix B is *reducible* if B is the direct sum of smaller square matrices B_1 and B_2 (after a reordering of the indices), i.e., $B = \begin{pmatrix} B_1 & 0 \\ 0 & B_2 \end{pmatrix}$. Otherwise, B is *irreducible* . Every square matrix B shall be the direct sum of irreducible submatrices, each of which we call a *component* of B. The next theorem tells us that the hyperbolic polytopes are determined by the Gram matrices.

Theorem 11.5 (Vinberg [70, Th. 2.1]) *Let $G = (g_{i,j})$ be an irreducible symmetric $N \times N$ matrix of signature $(d, 1, N - d - 1)$ such that the diagonal entries $g_{i,i} = 1$ and off-diagonal entries $g_{i,j}$ are $\leqslant 0$. Then there exists a hyperbolic d-polytope P whose Gram matrix is G, and the polytope P is unique up to an isometry of $\mathbb{H}^d$.*

Remark 11.5 A detailed analysis of the Gram matrix can reveal the combinatorial structure of the hyperbolic polytope P (see [70, Th. 3.1 & 3.2]) and whether P is compact or of finite volume (see [70, Th. 4.1]).

11.3.4 Lannér and Quasi-Lannér Coxeter Groups

A Coxeter group W_S is *Lannér* (resp. *quasi-Lannér*) if $\det(\mathbf{C}_W) < 0$ and if for every proper subset $T \subset S$, the Coxeter group W_T is spherical (resp. spherical or irreducible affine). If P is a compact (resp. finite volume) Coxeter simplex, then the Coxeter group W_P is Lannér (resp. quasi-Lannér). Conversely, if W is a Lannér (resp. quasi-Lannér) Coxeter group of rank $d + 1$, then its Cosine matrix $\mathbf{C}_W$ has signature $(d, 1)$ and there exists a compact (resp. finite volume) $\mathbb{H}^d$-Coxeter polytope whose Gram matrix is $\frac{1}{2}\mathbf{C}_W$.

In short, Lannér (resp. quasi-Lannér) Coxeter groups correspond to compact (resp. finite-volume) $\mathbb{H}^d$-Coxeter simplices. Such Coxeter simplices exist only in small dimensions.

Theorem 11.6 (Lannér [47], Koszul [45] and Chein [20]) *If W_S is a Lannér (resp. quasi-Lannér) Coxeter group, then $\#S \leqslant 5$ (resp. $\#S \leqslant 10$). Table 11.1 indicates the number of Lannér (resp. quasi-Lannér) Coxeter groups of a given rank.*

Table 11.1 The numbers of quasi-Lannér or Lannér Coxeter groups

Dimension $d = \#S - 1$	♯ of quasi-Lannér not Lannér	♯ of Lannér Coxeter groups
2	∞	∞
3	23	9
4	9	5
5	12	0
6	3	0
7	4	0
8	4	0
9	3	0

The list of Lannér and quasi-Lannér Coxeter groups of rank $\geqslant 4$ can be found in [20].

The non-existence of hyperbolic Coxeter simplices in large dimensions is in fact the first side of a more general non-existence theorem in Sect. 11.3.5.

Remark 11.6 Figures 11.2 and 11.3 give us the list of all the Lannér or quasi-Lannér Coxeter groups of rank 4, which correspond to compact or finite volume hyperbolic tetrahedra, respectively.

11.3.5 Absence in Large Dimension

Theorem 11.7 (Vinberg [69] and Prokhorov [59]) *If Γ_P is a discrete reflection group of* $\mathrm{Isom}(\mathbb{H}^d)$ *with compact (resp. finite volume) fundamental domain P, then $d \leqslant 29$ (resp. $d \leqslant 995$).*

The proof of Vinberg (resp. Prokhorov) uses Nikulin's inequality for simple polytopes (resp. edge-simple polytopes) established in [56] (resp. [42]). The upper bounds in the right-angled case are better:

Theorem 11.8 (Potyagailo–Vinberg [58] and Dufour [33]) *If Γ_P is a discrete reflection group of* $\mathrm{Isom}(\mathbb{H}^d)$ *with compact (resp. finite volume) right-angled fundamental domain P, then $d \leqslant 4$ (resp. $d \leqslant 12$).*

Except for compact right-angled polytopes, the upper bounds are far from being sharp.

- There exists a compact right-angled hyperbolic 4-polytope: 120-cell.
- Examples of finite volume right-angled d-polytopes are known in dimension $d \leqslant 8$ (see [58]).
- Examples of compact d-polytopes are known in dimension $d \leqslant 8$ (see [15, 16] for $d = 7, 8$).
- Examples of finite volume d-polytopes are known in dimension $d \leqslant 21$ and $d \neq 20$ (see [5] and the references therein for $d \leqslant 19$ and [13] for $d = 21$).

Remark 11.7 Moussong observed that the argument of Vinberg [69] may be extended to show that if a Coxeter group W is word hyperbolic and the nerve of W is a generalized homology $(d - 1)$-sphere, then $d \leqslant 29$ (see [24, Prop. 12.6.7]). Here, the *nerve* of a Coxeter group W_S is the poset of all nonempty spherical subsets of S partially ordered by inclusion, which is an abstract simplicial complex, and a *generalized homology d-sphere* is a homology d-manifold with the same homology as the d-sphere.

11.3.6 Hyperbolic Coxeter Polytopes with Few Facets

The complete classification of compact or finite-volume hyperbolic polytopes is not an easy task. Only compact d-polytopes with N facets ($N \leqslant d + 3$) and finite-volume d-polytopes with N facets ($N \leqslant d + 2$) were classified. For more details, we refer the reader to the web page maintained by Felikson and Tumarkin:

   https://www.maths.dur.ac.uk/users/anna.felikson/Polytopes/polytopes.html

This webpage contains all the known examples of hyperbolic Coxeter polytopes of dimension $\geqslant 4$.

11.3.7 Convex Cocompact Hyperbolic Reflection Groups

A subgroup Γ of $\mathrm{Isom}(\mathbb{H}^d)$ is *convex cocompact* if there exists a nonempty Γ-invariant convex subset $\mathscr{C}$ of $\mathbb{H}^d$ such that Γ acts properly discontinuously on $\mathscr{C}$ with compact quotient. Using the following theorems, one can easily check when a reflection group Γ_P of $\mathrm{Isom}(\mathbb{H}^d)$ is convex cocompact.

Theorem 11.9 (Desgroseilliers–Haglund [31, Th. 4.12]) *Let P be an $\mathbb{H}^d$-Coxeter polytope and let Γ_P be its reflection group. Then Γ_P is convex cocompact if and only if*

 (i) Γ_P *is word-hyperbolic, and*
 (ii) P *has no pair of asymptotic facets.*

Remark 11.8 The condition (ii) may be replaced by (ii') there is no pair of facets s, t of P such that $g_{s,t} = -1$, where $G = (g_{s,t})$ is the Gram matrix of P.

Theorem 11.10 (Moussong's Hyperbolicity Criterion [55]) *Let W_S be a Coxeter group. Then W_S is word hyperbolic if and only if S does not contain two orthogonal non-spherical subsets, nor any affine subset of rank $\geqslant 3$.*

Remark 11.9 Desgroseilliers and Haglund [31, Th. 1.1] found a class of Coxeter groups which can be realized as a convex cocompact subgroup of $\mathrm{Isom}(\mathbb{H}^d)$, which is not a reflection group, and they conjectured that there exists a word-hyperbolic Coxeter group which admits a convex cocompact representation into $\mathrm{Isom}(\mathbb{H}^d)$ but which cannot be realized as a convex cocompact *reflection* group of $\mathrm{Isom}(\mathbb{H}^n)$ for any $n \in \mathbb{N}$.

11.4   Projective Reflection Groups

11.4.1   Tits–Vinberg's Theorem

Let V be a vector space over $\mathbb{R}$, and let $\mathbb{S}(V)$ be the projective sphere. We denote by $\mathrm{SL}^{\pm}(V)$ the group of automorphism of $\mathbb{S}(V)$, i.e.,

$$\mathrm{SL}^{\pm}(V) = \{g \in \mathrm{GL}(V) \mid \det(g) = \pm 1\}.$$

We denote by $\hat{\mathbb{S}}$ the natural projection of $V \smallsetminus \{0\}$ to $\mathbb{S}(V)$, and let $\mathbb{S}(W) := \hat{\mathbb{S}}(W \smallsetminus \{0\})$ for any subset W of V. The complement of a projective hyperplane in $\mathbb{S}(V)$ consists of two connected components, each of which we call an *affine chart* of $\mathbb{S}(V)$. A *cone* is a subset of V which is invariant under multiplication by positive scalars. A subset $\mathscr{C}$ of $\mathbb{S}(V)$ is *convex* if there exists a convex cone U of V such that $\mathscr{C} = \mathbb{S}(U)$, *properly convex* if it is convex and its closure lies in some affine chart, and *strictly convex* if in addition its boundary does not contain any nontrivial projective line segment. Hyperbolic spaces are special examples of strictly convex open subsets of $\mathbb{S}(V)$.

A *projective polytope* is a properly convex subset P of $\mathbb{S}(V)$ such that P has a non-empty interior and $P = \cap_{i=1}^{N} \mathbb{S}(\{x \in V \mid \alpha_i(x) \leqslant 0\})$, where α_i, $i = 1, \ldots, N$, are linear forms on V. We always assume that P has N facets, i.e., to define P, we need all the N linear forms $(\alpha_i)_{i=1}^{N}$. A *projective reflection* is an element of $\mathrm{SL}^{\pm}(V)$ of order 2 which is the identity on a hyperplane. Every projective reflection σ can be written as:

$$\sigma = \mathrm{Id} - \alpha \otimes v, \quad \text{i.e.,} \quad \sigma(x) = x - \alpha(x)v \quad \forall x \in V,$$

where α is a linear form on V and v is a vector of V such that $\alpha(v) = 2$.

Let P be a projective polytope and S the set of facets of P. A *reflection in a facet* $s \in S$ is a projective reflection σ_s which fixes each point of s. A *pre-mirror polytope* is a projective polytope P together with one reflection σ_s in each facet s of P. So, one may choose $\sigma_s = \mathrm{Id} - \alpha_s \otimes v_s$ with $\alpha_s(v_s) = 2$ such that $P = \cap_{s \in S} \mathbb{S}(\{x \in V \mid \alpha_s(x) \leqslant 0\})$. Note that the pairs (α_s, v_s) are uniquely determined only up to multiplication by a positive real number.

If P is a pre-mirror polytope, then Γ_P denotes the group generated by the reflections in the facets of P. We say that Γ_P is a *projective reflection group* if for any $\gamma \in \Gamma_P$,

$$\gamma(\mathring{P}) \cap \mathring{P} \neq \varnothing \quad \Rightarrow \quad \gamma = \mathrm{Id},$$

where $\mathring{P}$ denotes the interior of P.

In the next paragraph, we introduce a relevant tool to formulate Proposition 11.1 and Theorem 11.11 which express necessary and sufficient conditions for Γ_P of a pre-mirror polytope P to be a projective reflection group. A key notion is that of Cartan matrix of mirror polytope which generalizes the twice of the Gram matrix of hyperbolic polytope.

Definition 11.1 A *Cartan matrix* on a set S is a $S \times S$ matrix $\mathscr{A}_S = (a_{s,t})_{s,t \in S}$ which satisfies the conditions: (i) $a_{s,s} = 2$, $\forall s \in S$; (ii) $a_{s,t} \leqslant 0$, $\forall s \neq t \in S$; (iii) $a_{s,t} = 0 \Leftrightarrow a_{t,s} = 0$, $\forall s \neq t \in S$.

A *mirror polytope* is a pre-mirror polytope P such that the matrix $\mathscr{A}_P := (\alpha_s(v_t))_{s,t \in S}$ is a Cartan matrix. In this case, we call $\mathscr{A}_P$ the *Cartan matrix of P*. A Cartan matrix $\mathscr{A}$ on S is of *Coxeter type* if for any $s \neq t \in S$,

$$a_{s,t} a_{t,s} < 4 \quad \Rightarrow \quad \frac{\pi}{\arccos\left(\frac{1}{2}\sqrt{a_{s,t} a_{t,s}}\right)} \in \mathbb{N}.$$

A *projective Coxeter polytope* is a mirror polytope P whose Cartan matrix $\mathscr{A}_P$ is of Coxeter type. For each pair of adjacent facets s, t of P, the *dihedral angle* of the ridge $s \cap t$ is said to be $\pi/m_{s,t}$ if $a_{s,t} a_{t,s} = 4 \cos^2(\pi/m_{s,t})$.

Proposition 11.1 (Vinberg [68, Prop. 17]) *Let P be a pre-mirror polytope. If the group Γ_P is a projective reflection group, then P is a projective Coxeter polytope.*

A Cartan matrix $\mathscr{A}_S$ and a Coxeter group W_S are *compatible* when:

(i) $\forall s, t \in S$, $\quad m_{s,t} = 2 \Leftrightarrow a_{s,t} = 0$;
(ii) $\forall s, t \in S$, $\quad m_{s,t} < \infty \Leftrightarrow a_{s,t} a_{t,s} = 4 \cos^2(\pi/m_{s,t})$;
(iii) $\forall s, t \in S$, $\quad m_{s,t} = \infty \Leftrightarrow a_{s,t} a_{t,s} \geqslant 4$.

It is clear that there is at most one Coxeter group compatible with a given Cartan matrix, and that a Cartan matrix $\mathscr{A}_S$ is compatible with some Coxeter group W_S if and only if $\mathscr{A}_S$ is of Coxeter type. If P is a projective Coxeter polytope, then W_P denotes the unique Coxeter group compatible with $\mathscr{A}_P$. The following is a generalization of Theorem 11.2 to the projective setting.

Theorem 11.11 (Bourbaki [14, Chap. V] and Vinberg [68, Th. 2])[6] *Let P be a projective Coxeter polytope of $\mathbb{S}(V)$ with Coxeter group W_P, and let Γ_P be the group generated by the projective reflections $(\sigma_s)_{s \in S}$ in the facets of P. Then the following hold:*

(i) the homomorphism $\sigma : W_P \to \mathrm{SL}^{\pm}(V)$ defined by $\sigma(s) = \sigma_s$ is an isomorphism onto Γ_P;
(ii) the group Γ_P is a discrete projective reflection group;

---

[6] Theorem 11.11 was proved by Tits for Δ_W, which we define in Sect. 11.4.2, and by Vinberg for the general case.

(iii) the union $\mathscr{C}_P$ of the Γ_P-translates of P is a convex subset of $\mathbb{S}(V)$;

(iv) if Ω_P is the interior of $\mathscr{C}_P$, then Γ_P acts properly discontinuously on Ω_P.

11.4.2 From Cartan Matrices to Mirror Polytopes

Given a Cartan matrix $\mathscr{A}_S$, there is a simple process to build a canonical mirror polytope $\Delta_{\mathscr{A}}$ such that $\mathscr{A}_{\Delta_{\mathscr{A}}} = \mathscr{A}_S$. In this construction, $\Delta_{\mathscr{A}}$ will be a simplex of dimension $\#S - 1$.

Let $V = \mathbb{R}^S$. We denote by $(e_s)_{s \in S}$ the canonical basis of V and $(e_s^*)_{s \in S}$ its dual basis. We set $\alpha_s := e_s^*$ and $v_s := \sum_t \mathscr{A}_{t,s} e_t$, i.e., v_s is the s-column vector of $\mathscr{A}_S$. Hence, by taking Δ to be (the projectivization of) the negative quadrant in $\mathbb{S}(V)$ and $\sigma_s = \mathrm{Id} - \alpha_s \otimes v_s$ to be the reflection in the facet $\Delta \cap \mathbb{S}(\mathrm{Ker}\,\alpha_s)$, we obtain a mirror polytope $\Delta_{\mathscr{A}}$ whose underlying polytope is a simplex of dimension $\#S - 1$. We call $\Delta_{\mathscr{A}}$ the *mirror simplex associated with $\mathscr{A}_S$*.

In the case where W_S is any Coxeter group and $\mathscr{A}_S = \mathbf{C}_W$, the mirror simplex associated with $\mathscr{A}_S$ is called the *Tits simplex associated with W_S* and denoted by Δ_W. The corresponding representation $\sigma : W_S \to \mathbb{S}(\mathbb{R}^S)$ is dual to Tits geometric representation described in [14].

For example, if W is spherical (resp. irreducible affine), then Δ_W gives rise to the classical tiling of $\mathbb{S}(V)$ (resp. of an affine chart) with Γ_{Δ_W} in the isometry group of the sphere (resp. Euclidean space). If W is Lannér (resp. quasi-Lannér), then Ω_{Δ_W} is the projective model of the hyperbolic space and Δ_W (resp. $\Delta_W \cap \Omega_{\Delta_W}$) is the hyperbolic Coxeter polytope whose Coxeter group is W.

By the Perron–Frobenius theorem, an irreducible Cartan matrix $\mathscr{A}$ admits an eigenvector with positive entries corresponding to the lowest eigenvalue $\lambda_{\mathscr{A}}$ of $\mathscr{A}$. We say that $\mathscr{A}$ is of *positive, zero* or *negative type* when $\lambda_{\mathscr{A}}$ is positive, zero or negative, respectively. For example, the Gram matrix of a hyperbolic polytope of finite volume is always of negative type. Now, the following is a generalization of Theorem 11.5 to the projective setting.

Theorem 11.12 (Vinberg [68, Cor. 1]) *Let $\mathscr{A}$ be a Cartan matrix of size $N \times N$. Assume that $\mathscr{A}$ is irreducible, of negative type and of rank $d + 1$. Then there exists a unique mirror d-polytope P, up to automorphism of $\mathbb{S}(\mathbb{R}^{d+1})$, such that $\mathscr{A}_P = \mathscr{A}$.*

Remark 11.10 Theorem 11.12 is not explicitly stated in [68, Cor. 1] for non-Coxeter polytopes, but may be proved from [68, Prop. 13 & 15].

11.4.3 Anosov Reflection Groups

Anosov representations are discrete representations of word-hyperbolic groups into semisimple Lie group with good dynamical properties. They have received a lot of attention and have been much studied recently (see e.g. [38, 46] for the definition of Anosov representation). But examples of Anosov representations of word hyperbolic groups, which are more complicated than free groups and surface groups, into Lie group of higher rank are less known. The following theorem, which generalizes Theorem 11.9, tells us that any infinite, word hyperbolic, irreducible Coxeter group admits Anosov representations.

Theorem 11.13 ([30, Cor. 1.18]) *Let P be a projective Coxeter polytope of $\mathbb{S}(V)$ with Coxeter group W_S. Suppose that W_S is word-hyperbolic. Then the following are equivalent:*

- *the representation $\sigma : W_S \to \mathrm{SL}^{\pm}(V)$ defined by $\sigma(s) = \sigma_s$ is P_1-Anosov (i.e., Anosov with respect to the stabilizer of a line in V);*
- *$\mathscr{A}_{s,t}\mathscr{A}_{t,s} > 4$ for all $s \neq t$ with $m_{s,t} = \infty$.*

Remark 11.11 Anosov reflection groups in $O(p, q)$ can be used to give a new proof of Theorem 11.10 (Moussong's hyperbolicity criterion); see [27, 48].

Projective reflection groups may also be used to construct interesting Anosov representations in $O(d, 2)$, as demonstrated in the following:

Theorem 11.14 ([48, Th. A]) *In dimension $d = 4, \ldots, 8$, there exists a projective Coxeter polytope of $\mathbb{S}(\mathbb{R}^{d+2})$ with Coxeter group W_S such that:*

- *the group W_S is word-hyperbolic and its boundary is a $(d-1)$-sphere;*
- *the image of the representation $\sigma : W_S \to \mathrm{SL}^{\pm}(\mathbb{R}^{d+2})$ defined by $\sigma(s) = \sigma_s$ lies in $O_{d,2}(\mathbb{R})$;*
- *the representation $\sigma : W_S \to O_{d,2}(\mathbb{R})$ is P-Anosov, where P is the stabilizer of an isotropic line;*
- *the group W_S is not quasi-isometric to $\mathbb{H}^d$.*

11.4.4 Convex Cocompact Projective Reflection Groups

An infinite discrete subgroup Γ of $\mathrm{SL}^{\pm}(V)$ is *convex cocompact in* $\mathbb{S}(V)$ if it acts properly discontinuously on some properly convex open subset Ω of $\mathbb{S}(V)$ and cocompactly on a nonempty Γ-invariant closed convex subset $\mathscr{C}$ of Ω whose closure in $\mathbb{S}(V)$ contains all accumulation points of all possible Γ-orbits $\Gamma \cdot y$ with $y \in \Omega$.

The notion of convex cocompactness in $\mathbb{S}(V)$ introduced in [26], in some sense, generalizes that of Anosov representation, but it does not require that the group Γ is word hyperbolic. There is also a simple characterization of convex cocompactness for projective reflection groups:

Theorem 11.15 ([30, Th. 1.3]) *Let P be a projective Coxeter polytope of $\mathbb{S}(V)$ with infinite irreducible Coxeter group W_S, and $\sigma : W_S \to \mathrm{SL}^{\pm}(V)$ the representation defined by $\sigma(s) = \sigma_s$. If $\sigma(W_S)$ is convex cocompact in $\mathbb{S}(V)$, then W_S satisfies the following two conditions:*

(i) S does not contain two orthogonal non-spherical subsets;

(ii) *if S contains an irreducible affine subset T of rank $\geqslant 3$, then W_T is of type $\widetilde{A}_k$ where $k = \#T - 1$.*

Theorem 11.16 ([30, Th. 1.8]) *Let P be a projective Coxeter polytope of $\mathbb{S}(V)$ with infinite irreducible Coxeter group W_S, $\mathscr{A}_S = (\mathscr{A}_{s,t})_{s,t \in S}$ the Cartan matrix of P, and $\sigma : W_S \to \mathrm{SL}^{\pm}(V)$ the representation defined by $\sigma(s) = \sigma_s$. If W_S satisfies the conditions (i) and (ii) of Theorem 11.15, then the following are equivalent:*

- *$\sigma(W)$ is convex cocompact in $\mathbb{S}(V)$;*
- *for any irreducible standard subgroup W_T of W_S with $\varnothing \neq T \subset S$, the Cartan submatrix $\mathscr{A}_T := (\mathscr{A}_{s,t})_{s,t \in T}$ is not of zero type;*
- *$\det(\mathscr{A}_T) \neq 0$ for all $T \subset S$ with W_T of type $\widetilde{A}_k$, $k \geqslant 1$.*

As a result, any infinite, irreducible Coxeter group W_S satisfying the conditions (i) and (ii) of Theorem 11.15 admits projective reflection groups, which are convex cocompact in $\mathbb{S}(\mathbb{R}^N)$ with $N = \#S$ (see [30, Th. 1.3]).

11.4.5 Divisible and Quasi-Divisible Domains

Every properly convex open subset Ω of $\mathbb{S}(V)$ admits a Hilbert metric d_Ω on Ω so that the group $\mathrm{Aut}(\Omega)$ of automorphisms of $\mathbb{S}(V)$ preserving Ω acts on Ω by isometries for d_Ω. A properly convex domain Ω is *divisible* (resp. *quasi-divisible*) *by* Γ if there exists a discrete subgroup Γ of $\mathrm{Aut}(\Omega)$ such that Ω / Γ is compact (resp. of finite volume with respect to the Hausdorff measure induced by d_Ω). (see e.g. [51] for more details for the Hilbert metric and the Hausdorff measure).

In general, it is difficult to construct divisible or quasi-divisible domains with various properties. But, in small dimensions, one can use *perfect* or *quasi-perfect* projective Coxeter polytopes to build such domains. We first introduce the definition of 2-perfect polytopes, which is slightly more general than that of perfect or quasi-perfect polytopes, and in Sect. 11.5 we give some interesting examples of divisible domains.

Let P be a projective Coxeter polytope of $\mathbb{S}(V)$ and S the set of facets of P. Given a vertex v of P, we denote by S_v the set of facets that contain v. For any $s \in S_v$, the projective reflection σ_s induces a projective reflection $\overline{\sigma}_s$ of the projective space $\mathbb{S}(V/\langle v \rangle)$, where $\langle v \rangle$ is the subspace spanned by v and $V/\langle v \rangle$ is the quotient vector space. The projection of P to $\mathbb{S}(V/\langle v \rangle)$ with the reflections $(\overline{\sigma}_s)_{s \in S_v}$ define a projective Coxeter polytope P_v of $\mathbb{S}(V/\langle v \rangle)$, called the *link of P at v*.

Definition 11.2 A projective Coxeter d-polytope P is *elliptic* (resp. *parabolic* , resp. *loxodromic*) when each component of $\mathscr{A}_P$ is of positive type (resp. zero type, resp. negative type) and the rank of $\mathscr{A}_P$ is $d + 1$ (resp. d, resp. $d + 1$).

Remark 11.12 If P is elliptic, then W_P is a spherical Coxeter group and P is the Tits simplex associated with W_P. If P is parabolic, then W_P is an affine

Coxeter group, P is the Cartesian product of the Tits simplices associated with the irreducible components of W_P, and Ω_P is an affine chart of $\mathbb{S}(V)$.

A projective Coxeter polytope P is *perfect* (resp. *quasi-perfect* , resp. *2-perfect*) when all its vertex links are elliptic (resp. elliptic or parabolic, resp. perfect). For example, quasi-perfect Coxeter polytopes should be 2-perfect.

Remark 11.13 By Vinberg [68, Prop. 26], a perfect Coxeter polytope is either elliptic, parabolic or irreducible loxodromic.

Let P be an irreducible loxodromic Coxeter polytope and Γ_P the projective reflection group of P. Then Ω_P is a properly convex domain, hence it admits a Hilbert metric d_{Ω_P}. By Vinberg [68, Th. 2], a projective Coxeter polytope P is perfect if and only if the action of Γ_P on Ω_P is cocompact. So, in this case, the domain Ω_P is divisible by Γ_P.

The action of Γ_P on Ω_P is said to be of *finite covolume* if $P \cap \Omega_P$ has finite volume with respect the Hausdorff measure μ_{Ω_P} induced by d_{Ω_P}, and *geometrically finite* if $\mu_{\Omega_P}(P \cap C(\Lambda_{\Omega_P})) < \infty$, where Λ_P is the limit set of Γ_P and $C(\Lambda_P)$ is the convex hull of Λ_P of Ω_P (see [52] for more details).

Theorem 11.17 ([52, Th. A]) *Let P be an irreducible, loxodromic, 2-perfect Coxeter polytope of $\mathbb{S}(V)$. Then the action of Γ_P on Ω_P is always geometrically finite, and*

- Γ_P *is of finite covolume if and only if P is quasi-perfect;*
- Γ_P *is convex cocompact in $\mathbb{S}(V)$ if and only if all the vertex links of P are elliptic or loxodromic.*

11.4.6 Cocompact Action of Coxeter Groups

There are many examples of discrete Coxeter subgroups of $\mathrm{SL}^{\pm}(V)$ other than projective reflection groups. However, if a Coxeter group Γ divides a properly convex domain, then $\Gamma = \Gamma_P$ for some projective Coxeter polytope P:

Theorem 11.18 (Davis [24, Prop. 10.9.7] and Charney–Davis [17]; see [48, Lem. 5.4]) *Let W be a Coxeter group, and let $\rho : W \to \mathrm{SL}^{\pm}(V)$ be a faithful representation. Suppose that there exists a properly convex domain Ω divisible by $\rho(W)$. Then the following hold:*

(i) for each $s \in S$, the image $\rho(s)$ of s is a projective reflection of $\mathbb{S}(V)$;
(ii) $\rho(W)$ is a projective reflection group generated by $(\rho(s))_{s \in S}$.

Remark 11.14 It is an open question whether Theorem 11.18 still holds when the word "divisible" is replaced by "quasi-divisible".

11.5 Examples of Projective Reflection Groups

The construction of projective reflection groups had led to several existence theorems in convex projective geometry.

A properly convex domain Ω of $\mathbb{S}(V)$ is *decomposable* if a cone of V lifting Ω is a non-trivial direct product of two smaller cones, and *homogeneous* if the group $\mathrm{Aut}(\Omega)$ acts transitively on Ω. Since the homogeneous quasi-divisible domains are well-understood by [43, 67], only inhomogeneous ones are of interest to us. So, all properly convex domains in this section are assumed to be inhomogeneous and indecomposable.

11.5.1 Kac–Vinberg's Example

The first example of divisible 2-domain which is not a hyperbolic plane was found by Kac and Vinberg [40]. They used perfect projective Coxeter triangles P with Cartan matrix $\mathscr{A}_P = (\mathscr{A}_{i,j})_{i,j=1,2,3}$ such that (i) each entry $\mathscr{A}_{i,j}$ is an integer, (ii) $\det(\mathscr{A}_P) < 0$, and (iii) $\mathscr{A}_{1,2}\mathscr{A}_{2,3}\mathscr{A}_{3,1} \neq \mathscr{A}_{1,3}\mathscr{A}_{3,2}\mathscr{A}_{2,1}$. Here, the condition (i) implies that the projective reflection group Γ_P of P is a subgroup of $\mathrm{SL}^{\pm}(3, \mathbb{Z})$, (ii) implies that $\mathscr{A}_P$ is of negative type, and finally (iii) implies that Ω_P is not a hyperbolic plane (see Fig. 11.4).

Remark 11.15 Let $\hat{\Gamma}_P$ be any finite-index torsion-free subgroup of Γ_P. Then $\hat{\Gamma}_P$ is an infinite index subgroup of $\mathrm{SL}(3, \mathbb{Z})$ and is Zariski dense in $\mathrm{SL}(3, \mathbb{R})$. In other words, $\hat{\Gamma}_P$ is a *thin* surface group (see [44] for an introduction to thin groups).

Fig. 11.4 Triangles with dihedral angles $\pi/3$, $\pi/3$ and $\pi/6$ on the left, with dihedral angles $\pi/3$, $\pi/4$ and $\pi/6$ on the center, and with dihedral angles $\pi/6$, $\pi/6$ and $\pi/6$ on the right

Fig. 11.5 A collection of properly embedded triangles in the non-strictly convex divisible 3-domains is colored. Each triangle is preserved by a subgroup of Γ, which is virtually $\mathbb{Z}^2$

11.5.2 Benoist's Examples and More

The first known examples of divisible d-domains Ω which are not strictly convex were introduced by Benoist [11] in dimension $d = 3, \ldots, 7$ (see Fig. 11.5). In such examples, the discrete group Γ which divides Ω is relatively hyperbolic with respect to virtual $\mathbb{Z}^{d-1}$. Later, different examples of non-strictly convex divisible d-domains were found in [21] in dimension $d = 4, \ldots, 8$, and the group Γ dividing such d-domain is relatively hyperbolic with respect to a collection of virtually free abelian subgroup of rank $< d - 1$. Except in dimension 3 (see [9]), all the known examples were built from projective reflection groups.

Remark 11.16 A generalization of Thurston's hyperbolic Dehn filling theorem to the projective setting led to the examples in [21].

In [12], Benoist found the first example of word-hyperbolic group Γ, not quasi-isometric to the hyperbolic space, that divides a properly convex 4-domain Ω, again using projective reflection groups. Since Γ is word hyperbolic, Ω should be strictly convex by [10]. Shortly after, Kapovich [41] found examples in any dimension $d \geqslant 4$, using Gromov–Thurston manifolds [37].

11.6 Hitchin Component of Polygon Groups

Let P be a compact hyperbolic polygon with dihedral angles $\pi/m_1, \ldots, \pi/m_k$, and let W be the Coxeter group of P. The conjugacy classes of discrete and faithful representations of W to $\mathrm{PGL}(2, \mathbb{R})$ form a connected component $\mathcal{T}$ of $\chi(W, \mathrm{PGL}(2, \mathbb{R})) := \mathrm{Hom}(W, \mathrm{PGL}(2, \mathbb{R}))/\mathrm{PGL}(2, \mathbb{R})$, i.e., the space of conjugacy classes of representations of W to $\mathrm{PGL}(2, \mathbb{R})$.

For any $n \geqslant 2$, there is a unique irreducible representation $\kappa : \mathrm{PGL}(2, \mathbb{R}) \rightarrow \mathrm{PGL}(n, \mathbb{R})$ up to conjugation. This gives rise to an embedding:

$$\mathcal{T} \rightarrow \chi(W, \mathrm{PGL}(n, \mathbb{R})) := \mathrm{Hom}(W, \mathrm{PGL}(n, \mathbb{R}))/\mathrm{PGL}(n, \mathbb{R})$$

The image of this embedding is called the *Fuchsian locus* and the component of $\chi(W, \mathrm{PGL}(n, \mathbb{R}))$ containing the Fuchsian locus is the *Hitchin component* $\mathrm{Hit}(W, \mathrm{PGL}(n, \mathbb{R}))$. A representation $\rho : W \rightarrow \mathrm{PGL}(n, \mathbb{R})$ is called a *Hitchin representation* if its $\mathrm{PGL}(n, \mathbb{R})$-conjugacy class is an element of $\mathrm{Hit}(W, \mathrm{PGL}(n, \mathbb{R}))$.

Theorem 11.19 ([4, Th. 1.1 & 1.2]) *Let P be a compact hyperbolic polygon with dihedral angles $\pi/m_1, \ldots, \pi/m_k$, and W its Coxeter group. Then each Hitchin representation in $\mathrm{Hit}(W, \mathrm{PGL}(n, \mathbb{R}))$ is discrete and faithfull, and $\mathrm{Hit}(W, \mathrm{PGL}(n, \mathbb{R}))$ is an open cell of dimension*

$$- (n^2 - 1) + \sum_{\ell=2}^{n} \sum_{i=1}^{k} \left\lfloor \ell \left(1 - \frac{1}{m_i}\right) \right\rfloor,$$

where $\lfloor x \rfloor$ denotes the biggest integer not bigger than x.

For example, the $\mathrm{PGL}(2m, \mathbb{R})$ (resp. $\mathrm{PGL}(2m + 1, \mathbb{R})$) Hitchin component of the Coxeter group associated with a right-angled hyperbolic k-gon ($k \geqslant 5$) is an open cell of dimension $(k - 4)m^2 + 1$ (resp. $(k - 4)(m^2 + m)$).

Remark 11.17 In the case of $n = 2$ (resp. $n = 3$), Theorem 11.19 was proved by Thurston [64] (resp. Choi–Goldman [19]).

Remark 11.18 Let ρ be any Hitchin representation in $\mathrm{Hit}(W, \mathrm{PGL}(n, \mathbb{R}))$. In the case $n \geqslant 4$, the image of each generator of W should be an involution but not a projective reflection, hence ρ is not a projective reflection group.

11.7 Properly Discontinuous Affine Groups

11.7.1 Auslander's Conjecture and Milnor's Question

In the 1960s, Auslander raised the following conjecture:

Conjecture 11.1 (Auslander [8]) *Every discrete subgroup Γ of the affine group $\mathrm{Aff}(\mathbb{R}^d)$ which acts properly discontinuously and cocompactly on $\mathbb{R}^d$ is virtually solvable.* □

In the 1970s, Milnor asked if Auslander's conjecture still holds without the condition that the action is cocompact:

Question 11.2 (Milnor [53]) Is every discrete subgroup Γ of $\mathrm{Aff}(\mathbb{R}^d)$ which acts properly discontinuously on $\mathbb{R}^d$ virtually solvable?   □

In 1983, Fried and Goldman [34] showed that Auslander's conjecture is true in dimension 3, and Margulis answered Milnor's question negatively:

Theorem 11.20 (Margulis [49, 50]) *There exists a properly discontinuous affine action of the free group on two generators on $\mathbb{R}^3$.*

Even if some progress have been made over the years towards Auslander's conjecture (see e.g. [3, 66] for a proof assuming $d \leqslant 6$ and [2, 35, 65] for a proof assuming the linear part is contained in a particular class of semisimple Lie subgroups), Auslander's conjecture is still open.

Back to Milnor's question, the existence and property of properly discontinuous affine action of free groups on $\mathbb{R}^n$ have been actively studied (see e.g. [1, 18, 25, 32, 36, 62, 63] or the survey [28]).

11.7.2 Properly Discontinuous Affine Coxeter Groups

Before the following theorem, properly discontinuous affine actions by non-virtually solvable non-free groups were unknown.

Theorem 11.21 (Danciger–Guéritaud–Kassel [29, Th. 1.1]) *Any right-angled Coxeter group of rank k admits a properly discontinuous affine action on $\mathbb{R}^{k(k-1)/2}$.*

Remark 11.19 The action preserves a bilinear form and in some particular cases, one can find much smaller affine space on which the Coxeter groups act (see [29, Prop. 1.6]).

Acknowledgments G.-S. Lee was supported by the National Research Foundation of Korea(NRF) grant funded by the Korea government(MSIT) (No. 2020R1C1C1A01013667). L. Marquis acknowledges support by the Centre Henri Lebesgue (ANR-11-LABX-0020 LEBESGUE).

We would like to thank Athanase Papadopoulos for carefully reading this paper and suggesting several improvements.

References

1. H. Abels, G.A. Margulis, G.A. Soifer, On the Zariski closure of the linear part of a properly discontinuous group of affine transformations. J. Differ. Geom. **60**(2), 315–344 (2002)
2. H. Abels, G.A. Margulis, G.A. Soifer, The linear part of an affine group acting properly discontinuously and leaving a quadratic form invariant. Geom. Dedicata **153**, 1–46 (2011)
3. H. Abels, G.A. Margulis, G.A. Soifer, The Auslander conjecture for dimension less then 7 (2020). arXiv:2011.12788
4. D. Alessandrini, G.-S. Lee, F. Schaffhauser, Hitchin components for orbifolds. J. Eur. Math. Soc. (JEMS) **25**(4), 1285–1347 (2023)

5. D. Allcock, Infinitely many hyperbolic Coxeter groups through dimension 19. Geom. Topol. **10**, 737–758 (2006)
6. E.M. Andreev, On convex polyhedra in Lobachevskiĭ spaces. Math. USSR, Sb. **10**(3), 413–440 (1971)
7. E.M. Andreev, On convex polyhedra of finite volume in Lobachevskiĭ space. Math. USSR, Sb. **12**(2), 255–259 (1971)
8. L. Auslander, The structure of complete locally affine manifolds. Topology **3**, 131–139 (1964)
9. S.A. Ballas, J. Danciger, G.S. Lee, Convex projective structures on nonhyperbolic three-manifolds. Geom. Topol. **22**(3), 1593–1646 (2020)
10. Y. Benoist, Convexes divisibles. I, in *Algebraic Groups and Arithmetic* (Tata Institute of Fundamental Research, Mumbai, 2004), pp. 339–374
11. Y. Benoist, Convexes divisibles. IV. Structure du bord en dimension 3. Invent. Math. **164**(2), 249–278 (2006)
12. Y. Benoist, Convexes hyperboliques et quasiisométries. Geom. Dedicata **122**, 109–134 (2006)
13. R. Borcherds, Automorphism groups of Lorentzian lattices. J. Algebra **111**(1), 133–153 (1987)
14. N. Bourbaki, *Groupes et algèbres de Lie, Chapitre IV, V, VI* (Hermann, Paris, 1968)
15. V.O. Bugaenko, Groups of automorphisms of unimodular hyperbolic quadratic forms over the ring $\mathbf{Z}[(\sqrt{5}+1)/2]$. Vestnik Moskov. Univ. Ser. I Mat. Mekh. **5**, 6–12 (1984)
16. V.O. Bugaenko, Arithmetic crystallographic groups generated by reflections, and reflective hyperbolic lattices, in *Lie Groups, Their Discrete Subgroups, and Invariant Theory*. Advances in Soviet Mathematics, vol. 8 (American Mathematical Society, Providence, 1992), pp. 33–55
17. R. Charney, M. Davis, When is a Coxeter system determined by its Coxeter group? J. London Math. Soc. **61**(2), 441–461 (2000)
18. V. Charette, T.A. Drumm, W.M. Goldman, Proper affine deformations of the one-holed torus. Transform. Groups **21**(4), 953–1002 (2016)
19. S. Choi, W.M. Goldman, The deformation spaces of convex $\mathbb{RP}^2$-structures on 2-orbifolds. Amer. J. Math. **127**(5), 1019–1102 (2005)
20. M. Chein, Recherche des graphes des matrices de Coxeter hyperboliques d'ordre $\leqslant$ 10. Rev. Française Informat. Recherche Opérationnelle **3**, 3–16 (1969)
21. S. Choi, G.-S. Lee, L. Marquis, Convex projective generalized Dehn filling. Ann. Sci. Éc. Norm. Supér. **53**(1), 217–266 (2020)
22. H.S.M. Coxeter, The polytopes with regular-prismatic vertex figures. Proc. London Math. Soc. **34**(2), 126–189 (1932)
23. H.S.M. Coxeter, Discrete groups generated by reflections. Ann. Math. **35**(3), 588–621 (1934)
24. M.W. Davis, *The Geometry and Topology of Coxeter Groups*. London Mathematical Society Monographs Series, vol. 32 (Princeton University Press, Princeton, 2008)
25. J. Danciger, F. Guéritaud, F. Kassel, Margulis spacetimes via the arc complex. Invent. Math. **204**(1), 133–193 (2016)
26. J. Danciger, F. Guéritaud, F. Kassel, Convex cocompact actions in real projective geometry (2017). arXiv:1704.08711
27. J. Danciger, F. Guéritaud, F. Kassel, Convex cocompactness in pseudo-Riemannian hyperbolic spaces. Geom. Dedicata **192**, 87–126 (2018)
28. J. Danciger, T.A. Drumm, W.M. Goldman, I. Smilga, Proper actions of discrete groups of affine transformations, in Dynamics, geometry, number theory–the impact of Margulis on modern mathematics (University of Chicago Press, Chicago, 2022), pp. 95–168
29. J. Danciger, F. Guéritaud, F. Kassel, Proper affine actions for right-angled Coxeter groups. Duke Math. J. **169**(12), 2231–2280 (2020)
30. J. Danciger, F. Guéritaud, F. Kassel, G.-S. Lee, L. Marquis, Convex cocompactness for Coxeter groups (2021). arXiv:2102.02757
31. M. Desgroseilliers, F. Haglund, On some convex cocompact groups in real hyperbolic space. Geom. Topol. **17**(4), 2431–2484 (2013)
32. T.A. Drumm, Fundamental polyhedra for Margulis space-times. Topology **31**(4), 677–683 (1992)

33. G. Dufour, Notes on right-angled Coxeter polyhedra in hyperbolic spaces. Geom. Dedicata **147**, 277–282 (2010)
34. D. Fried, W.M. Goldman, Three-dimensional affine crystallographic groups. Adv. Math. **47**(1), 1–49 (1983)
35. W.M. Goldman, Y. Kamishima, The fundamental group of a compact flat Lorentz space form is virtually polycyclic. J. Differ. Geom. **19**(1), 233–240 (1984)
36. W.M. Goldman, F. Labourie, G.M. Margulis, Proper affine actions and geodesic flows of hyperbolic surfaces. Ann. Math. **170**(3), 1051–1083 (2009)
37. M. Gromov, W. Thurston, Pinching constants for hyperbolic manifolds. Invent. Math. **89**(1), 1–12 (1987)
38. O. Guichard, A. Wienhard, Anosov representations: domains of discontinuity and applications. Invent. Math. **190**(2), 357–438 (2012)
39. C.D. Hodgson, I. Rivin, A characterization of compact convex polyhedra in hyperbolic 3-space. Invent. Math. **111**(1), 77–111 (1993)
40. V.G. Kac, É.B. Vinberg, Quasi-homogeneous cones. Math. Not. **1**, 231–235 (1967)
41. M. Kapovich, Convex projective structures on Gromov-Thurston manifolds. Geom. Topol. **11**, 1777–1830 (2007)
42. A.G. Khovanskiĭ, Hyperplane sections of polyhedra, toric varieties and discrete groups in Lobachevskiĭ space. Funktsional. Anal. i Prilozhen. **20**(1), 50–61 (1986)
43. M. Koecher, *The Minnesota Notes on Jordan Algebras and Their Applications*. Lecture Notes in Mathematics, vol. 1710 (Springer, Berlin, 1999)
44. A. Kontorovich, D.D. Long, A. Lubotzky, A.W. Reid, What is . . . a thin group? Notices Am. Math. Soc. **66**(6), 905–910 (2019)
45. J.-L. Koszul, *Lectures on Hyperbolic Coxeter Groups* (University of Notre Dame, Dame, 1967)
46. F. Labourie, Anosov flows, surface groups and curves in projective space. Invent. Math. **165**(1), 51–114 (2006)
47. F. Lannér, On complexes with transitive groups of automorphisms. Commun. Sém. Math. Univ. Lund [Medd. Lunds Univ. Mat. Sem.] **11**, 71 (1950)
48. G.-S. Lee, L. Marquis, Anti-de Sitter strictly GHC-regular groups which are not lattices. Trans. Am. Math. Soc. **372**(1), 153–186 (2019)
49. G.A. Margulis, Free completely discontinuous groups of affine transformations. Dokl. Akad. Nauk SSSR **272**(4), 785–788 (1983)
50. G.A. Margulis, Complete affine locally flat manifolds with a free fundamental group. J. Soviet Math. **36**(1), 129–139 (1987)
51. L. Marquis, Around groups in Hilbert geometry, in *Handbook of Hilbert Geometry*. IRMA Lectures in Mathematics and Theoretical Physics, vol. 22 (European Mathematical Society, Zürich, 2014), pp. 207–261
52. L. Marquis, Coxeter group in Hilbert geometry. Groups Geom. Dyn. **11**(3), 819–877 (2017)
53. J. Milnor, On fundamental groups of complete affinely flat manifolds. Adv. Math. **25**(2), 178–187 (1977)
54. G.A. Margulis, É.B. Vinberg, Some linear groups virtually having a free quotient. J. Lie Theory **10**(1), 171–180 (2000)
55. G. Moussong, *Hyperbolic Coxeter Groups* (ProQuest LLC, Ann Arbor, 1988). Thesis (Ph.D.)– The Ohio State University
56. V.V. Nikulin, Quotient-groups of groups of automorphisms of hyperbolic forms by subgroups generated by 2-reflections. Algebro-geometric applications, in *Current Problems in Mathematics, Vol. 18* Akad. Nauk SSSR, Vsesoyuz (Inst. Nauchn. i Tekhn. Informatsii, Moscow, 1981), pp. 3–114
57. H. Poincaré, Mémoire: Les groupes kleinéens. Acta Math. **3**, 49–92 (1883)
58. L. Potyagailo, É.B. Vinberg, On right-angled reflection groups in hyperbolic spaces. Comment. Math. Helv. **80**(1), 63–73 (2005)
59. M.N. Prokhorov, Absence of discrete groups of reflections with a noncompact fundamental polyhedron of finite volume in a Lobachevskiĭ space of high dimension. Izv. Akad. Nauk SSSR Ser. Mat. **50**(2), 413–424 (1986)

60. R.K.W. Roeder, Compact hyperbolic tetrahedra with non-obtuse dihedral angles. Publ. Mat. Barc. **50**(1), 211–227 (2006)

61. R.K.W. Roeder, J.H. Hubbard, W.D. Dunbar, Andreev's theorem on hyperbolic polyhedra. Ann. Inst. Fourier **57**(3), 825–882 (2007)

62. I. Smilga, Proper affine actions: a sufficient criterion. Math. Ann. **382**(1-2), 513–605 (2022)

63. I. Smilga, Proper affine actions on semisimple Lie algebras. Ann. Inst. Fourier **66**(2), 785–831 (2016)

64. W. Thurston, *The Geometry and Topology of Three-Manifolds*. Lecture Notes (American Mathematical Society, Providence, 1979)

65. G. Tomanov, The virtual solvability of the fundamental group of a generalized Lorentz space form. J. Differ. Geom. **32**(2), 539–547 (1990)

66. G. Tomanov, Properly discontinuous group actions on affine homogeneous spaces. Tr. Mat. Inst. Steklova **292**, 268–279 (2016)

67. É.B. Vinberg, The theory of homogeneous convex cones. Trudy Moskov. Mat. Obv č. **12**, 303–358 (1963)

68. É.B. Vinberg, Discrete linear groups that are generated by reflections. Izv. Akad. Nauk SSSR Ser. Mat. **35**, 1072–1112 (1971)

69. É.B. Vinberg, Absence of crystallographic groups of reflections in Lobachevskiĭ spaces of large dimension. Trudy Moskov. Mat. Obshch. **47**, 68–102, 246 (1984)

70. É.B. Vinberg, Hyperbolic groups of reflections. Uspekhi Mat. Nauk **40**(1), 29–66, 255 (1985)

71. E. Witt, Spiegelungsgruppen und Aufzählung halbeinfacher Liescher Ringe. Abh. Math. Sem. Hansischen Univ. **14**(1), 289–322 (1941)

Chapter 12
Isoperimetry in Finitely Generated Groups

Bruno Luiz Santos Correia and Marc Troyanov

Abstract We revisit the isoperimetric inequalities for finitely generated groups introduced and studied by N. Varopoulos, T. Coulhon and L. Saloff-Coste. Namely we show that a lower bound on the isoperimetric quotient of finite subsets in a finitely generated group is given by the $\mathcal{U}-$transform of its growth function, which is a variant of the Legendre transform. From this lower bound, we obtain some asymptotic estimates for the Følner function of the group. The paper also includes a discussion of some basic definitions from Geometric Group Theory and some basic properties of the $\mathcal{U}$-transform, including some computational techniques and its relation with the Legendre transform.

Keywords Finitely generated groups · Growth · Isoperimetric inequality · Følner function · Lambert W-function

AMS Subject Classification 20F65, 20F69

12.1 Introduction

In the early 1990s, T. Coulhon and L. Saloff-Coste proved the following general isoperimetric inequality relating the size of an arbitrary finite set D in an infinite, finitely generated group Γ to the size of its boundary:

$$\frac{|\partial_S D|}{|D|} \geq \frac{1}{4|S|} \cdot \frac{1}{\phi_S(2|D|)}. \tag{12.1.1}$$

Here S is a finite symmetric set of generators for Γ and ϕ_S is the *inverse growth function* associated with the word metric in Γ defined by S. Precise definitions are given in the next section. The remarkable feature of this inequality is that, in some

B. L. Santos Correia · M. Troyanov (✉)
Institut de Mathématiques EPFL, Lausanne, Switzerland
e-mail: bruno.santoscorreia@epfl.ch; marc.troyanov@epfl.ch

sense, the size of metric balls controls the boundary size of arbitrary finite subsets of
the group Γ. The proof of this inequality appeared in [7, Theorem 1], see also [18,
Theorem 3.2]; however, a version of this result for groups of polynomial growth has
been proved earlier by Varopoulos in [22]. M. Gromov later improved the constant
in the above inequality; he proved:

$$\frac{|\partial_S D|}{|D|} \geq \frac{1}{2} \cdot \frac{1}{\phi_S(2|D|)}, \tag{12.1.2}$$

see Chapter 6, p. 346, in [11]. The proof also applies to finite groups, provided one
assumes $|D| < \frac{1}{2}|\Gamma|$. This version of the isoperimetric inequality is now standard
and appears in textbooks such as [6, Theorem 14.95] and [16, Theorem 6.29].

The constant 2 appearing in the previous inequality is suboptimal; optimizing
the argument in Gromov's proof, leads us to a new formulation of the isoperimetric
inequality in which the inverse growth ϕ_S is replaced by another transformation
of the growth function γ, that we call the $\mathcal{U}$-transform of that function. In
Theorem 12.3.4 we obtain the following inequality

$$\frac{|\partial_S D|}{|D|} \geq \mathcal{U}_\gamma(|D|),$$

where $\mathcal{U}_\gamma(t)$ is defined in (12.3.4). As explained in Sect. 12.4.4, the $\mathcal{U}$-transform
$\gamma \mapsto \mathcal{U}_\gamma$ is equivalent to the Legendre transform after some change of variables. The
main results of this Chapter are Theorems 12.3.4 and 12.3.6, and various corollaries
discussed in Sect. 12.3.

The rest of the chapter is organized as follows: In Sect. 12.2 we recall some basic
definitions from Geometric Group Theory. In Sect. 12.3 we formulate and prove the
main result of the chapter and discuss some of its consequences. The basic properties
of the $\mathcal{U}$−transform, including some computational techniques and its relation with
the Legendre transform are postponed till the last section of the chapter.

12.2   Preliminaries from Geometric Group Theory

We recall here some basic notions of geometric group theory and refer to the books
[6] and [12] for a thorough introduction to the subject. Starting with a group Γ
generated by the finite symmetric set $S = S^{-1} \subset \Gamma$, we define:

(i) **The word metric and the Cayley graph.** For any $r \in \mathbb{N}$, we denote by
    $B_S(r) \subset \Gamma$ the subset of those elements in Γ that can be written as a product
    of at most r generators in S. The right invariant *word metric* on Γ with respect
    to the generating set S is the function $d_S : \Gamma \times \Gamma \to \mathbb{N} \cup \{0\}$ defined as

$$d_S(x, y) = \min\{r \in \mathbb{N} \cup \{0\} \mid xy^{-1} \in B_S(r)\}.$$

With this definition, $B_S(r)$ is the ball in the metric space (Γ, d_S) centered at the identity element $e \in \Gamma$. We shall also denote by $\|x\|_S = d_S(e, x)$ the distance from e to $x \in \Gamma$.

The *Cayley graph* of Γ with respect to S is the undirected graph $X(\Gamma, S)$ whose vertex set is the group Γ itself and the edges are the pairs $\{x, y\} \subset \Gamma$ such that $d_S(x, y) = 1$ (i.e. such that $y = sx$ for some $s \in S$). Note that this is a regular graph of degree $d = |S|$. By construction, the distance d_S between two elements in the group Γ is the length of the shortest edge path joining them.

As a simple example, consider the Dihedral group $D_4 = \langle r, s \mid r^4, s^2, (rs)^2 \rangle$. Note that $r^k s = s r^{4-k}$ for $0 \le k \le 4$, and the corresponding Cayley graph is

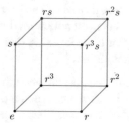

Simple examples of infinite Cayley graphs are the infinite regular tree of degree $2q$ for the free group F_q on q generators and the standard integer lattice in $\mathbb{R}^d$ for the free abelian group on d generators.

(ii) **The growth function.** The *growth function* of Γ with respect to the generating set S is the function $\gamma_s : \mathbb{N} \to \mathbb{N}$ counting the number of elements in the ball $B(n)$. That is,

$$\gamma_S(n) = |B_S(n)|,$$

where we denote by $|D|$ the cardinality of a finite set D. Some examples of finitely generated groups and their growth are:

o A group has bounded growth if and only if it is finite.
o In particular, for the dihedral group D_4 with the above presentation we have $\gamma(0) = 1$, $\gamma(1) = 4$, $\gamma(2) = 7$ and $\gamma(n) = 8$ for any $n \ge 3$.
o The infinite cyclic group $\mathbb{Z}$, generated by $S = \{+1, -1\}$, has linear grow $\gamma(n) = 2n + 1$. Conversely, any finitely generated group Γ whose growth satisfies $\limsup_{n \to \infty} \gamma(n)/n < \infty$ is either finite or contains an infinite cyclic group of finite index. This was proved by J. Justin in 1971, see [17, Theorem 3.1] or [6, Theorem 7.26].
o A finitely generated abelian group of rank d has polynomial growth of degree d for any system of generators, that is $c_1 n^d \le \gamma(n) \le c_2 n^d$ for some constants c_1, c_2. The precise growth function of the abelian group $\mathbb{Z}^d$, with respect to its standard set of generators, is given in [13, page 9].

o The free group F_q on q generators has exponential growth if $q \geq 2$. More precisely, we have:

$$\gamma(n) = \frac{q(2q-1)^n - 1}{q-1}$$

This is easily proved by induction from the fact that $\gamma(0) = 1$ and $\gamma(n) - \gamma(n-1) = 2q(2q-1)^{n-1}$ is the number of elements in F_q at distance exactly n from e.

o The Heisenberg group $\mathbb{H}_3 \subset GL_3(\mathbb{Z})$ of 3×3 upper-triangular matrices with 1 on the diagonal has polynomial growth of degree 4. This follows from [2, Theorem 2], which in fact computes the polynomial growth degree of an arbitrary finitely generated nilpotent group.

o A celebrated Theorem, proved in 1981 by M. Gromov, states that a finitely generated group has at most polynomial growth if and only if it contains a nilpotent subgroup of finite index. A detailed, self-contained reference for that result is Chapter 12 in [6].

In general, the growth $\gamma_s(n)$ of an infinite, finitely generated group, is at least a linear and at most an exponential function of n. An important development of Geometric Group Theory has been the discovery in 1983 by R. Grigorchuk of a class of groups with "intermediate growth", that is finitely generated groups whose growth γ_s satisfies

$$\lim_{n\to\infty} \frac{\log(\gamma_s(n))}{n} = 0 \quad \text{and} \quad \liminf_{n\to\infty} \frac{\gamma_s(n)}{n^d} = \infty, \quad \forall d \in \mathbb{N}.$$

For an updated presentation on the subject of intermediate-growth groups, we refer to [9] and [17], and the recent papers [8] and [4].

(iii) The inverse growth function. The *inverse growth function* is the function $\phi_S : \mathbb{R}_+ \to \mathbb{N} \cup \{\infty\}$ defined as

$$\phi_S(t) = \min\{n \in \mathbb{N} \cup \{0\} \mid \gamma_s(n) \geq t\}. \tag{12.2.1}$$

This is the smallest function such that $\gamma_s(\phi_S(t)) \geq t$ for any $t \in \mathbb{R}_+$, equivalently $\phi_S(t)$ is the smallest integer n such that $B_S(n)$ contains at least t elements. Note that $\phi_S(t) = \infty$ if and only if $t > |\Gamma|$, in particular $\phi_s(t)$ is always finite if Γ is an infinite group. Observe also that ϕ_S is a left inverse of γ_S, that is $\phi_S(\gamma_S(n)) = n$ for any integer $n \in \mathbb{N}$.

(iv) Various definitions of the boundary of a subset in Γ. There are several ways to define a notion of boundary of a non empty subset $D \subset \Gamma$ with respect to a symmetric generating set $S \subset \Gamma$. We first define the *inner boundary* as

$$\partial_S D = \{x \in D \mid \text{dist}_S(x, \Gamma \setminus D) = 1\} = \{x \in D \mid \exists s \in S \text{ such that } sx \notin D\}.$$

We also define the *outer boundary* of $D \subset \Gamma$ as

$$\partial'_S D = \{x \in \Gamma \setminus D \mid \mathrm{dist}_S(x, D) = 1\}$$
$$= \{x \in \Gamma \setminus D \mid \exists s \in S \text{ such that } sx \in D\}.$$

A third boundary that is sometimes considered is the *edge boundary*, which is the set of edges in the Cayley graph joining a point in D to a point in $\Gamma \setminus D$. We denote it by

$$E_S(D) = \big\{\{x, y\} \mid x \in D, \ y \in \Gamma \setminus D, \ d_S(x, y) = 1\big\}.$$

Observe that $(\partial_S D \cup \partial'_S D, E_S(D))$ is a bipartite subgraph of $X(\Gamma, S)$ such that each vertex in $\partial_S D$ is related to at least one and at most $|S|$ vertices in $\partial_S D'$, and vice versa. In particular we have

$$\max\{|\partial_S D|, |\partial'_S D|\} \leq |E_S(D)| \leq |S| \cdot \min\{|\partial_S D|, |\partial'_S D|\}.$$

(v) **The isoperimetric profile.** The *isoperimetric profile* of (Γ, S), introduced by Gromov in [10], is the smallest function $I_S : \mathbb{N} \to \mathbb{N}$ such that $|\partial_S D| \geq I_s(|D|)$ for any subset $D \subset \Gamma$. More explicitly, it is the function

$$I_S(m) = \min\{|\partial_S D| \mid D \subset \Gamma, \ |D| = m\}.$$

(vi) **Følner sequences and the Følner function.** A *Følner sequence* in the group Γ is a sequence of finite nonempty sets $D_j \subset \Gamma$ such that

$$\lim_{j \to \infty} \frac{|D_i \setminus x D_i|}{|D_i|} = 0$$

for any $x \in \Gamma$. Equivalently, if $S \subset \Gamma$ is a finite generating set, then $\{D_j\}$ is a Følner sequence if and only if

$$\lim_{j \to \infty} \frac{|\partial_S D_i|}{|D_i|} = 0.$$

It has been proved by E. Følner in 1955 that a finitely generated group is amenable if and only if there exists a Følner sequence in that group, see e.g. [6, Corollary 14.24].

A related notion is the *Følner function* of (Γ, S), introduced by A. Vershik in [23]. It is the function $\mathrm{Føl} : \mathbb{N} \to \mathbb{N} \cup \{\infty\}$ defined as

$$\mathrm{Føl}(n) = \min \big\{k \in \mathbb{N} \mid \exists D \subset \Gamma \text{ s.t. } |D| = k \text{ and } k \geq n|\partial_S D|\big\},$$

with the convention that $\min(\emptyset) = +\infty$. In other words, $\text{Føl}(n)$ is the cardinality of the smallest set $D \subset \Gamma$ such that

$$\frac{|\partial_S D|}{|D|} \leq \frac{1}{n}.$$

Observe also the following relation between the isoperimetric profile and the Følner function:

$$I_S(\text{Føl}(n)) \leq \frac{\text{Føl}(n)}{n},$$

for all $n \in \mathbb{N}$ such that $\text{Føl}(n) < \infty$.

We also mention that the Følner function is sometimes defined in a slightly different way:

$$\Phi(n) = \min \left\{ k \in \mathbb{N} \mid \exists D \subset \Gamma \text{ s.t. } |D| = k \text{ and } |s^{-1} D \triangle D| \cdot n \leq k, \; \forall s \in S \right\},$$

where

$$s^{-1} D \triangle D = (s^{-1} D \cup D) \setminus (s^{-1} D \cap D)$$

$$= \{x \in D \mid sx \notin D\} \cup \{x \in s^{-1} D \mid x \notin D\}$$

is the symmetric difference. Observe that

$$\frac{1}{|S|} |\partial_S D| \leq \max_{s \in S} |s D \triangle D| \leq 2 |\partial_S D|,$$

therefore

$$\Phi\left(\tfrac{n}{2}\right) \leq \text{Føl}(n) \leq \Phi(|S|n).$$

Some simple examples of Følner functions are:

○ It always holds that $\text{Føl}(1) = 1$ and $\text{Føl}(n) \geq n$ if $n \leq |\Gamma|$.
○ For a finite group, we have $\text{Føl}(n) = |\Gamma|$ if $n \geq |\Gamma|$ (because $\partial_S \Gamma = \emptyset$).
○ Returning to the Dihedral group D_4, we have $\text{Føl}(1) = 1$, $\text{Føl}(2) = 7$ and $\text{Føl}(n) = 8$ for $n \geq 3$.
○ The Følner function of $\mathbb{Z}$ is $\text{Føl}(1) = 1$ and $\text{Føl}(n) = 2n$ if $n \geq 2$.
○ For the free group on $q \geq 2$ generators F_q, we have $\text{Føl}(n) = \infty$ if $n \geq 2$, this follows from Proposition 12.3.1 below.
○ The group Γ is amenable if and only if $\text{Føl}(n) < \infty$ for all integers n.

The sequence of balls $\{B_S(n)\}$ in an amenable finitely generated group Γ is not always a Følner sequences, but the following facts are known:

(a) The sequence of balls $\{B_S(n)\} \subset \Gamma$ is Følner for any group of polynomial growth, see [17, prop. 12.7].
(b) Conversely, if the sequence of balls $\{B_S(n)\}$ is Følner, then Γ has subexponential growth, see [17, prop. 12.7].
(c) If $|\partial B_S(n)| \geq c|B_S(n)|$ for any $n \in \mathbb{N}$ and some $c > 0$, then Γ has exponential growth. In fact we clearly have $0 < c < 1$ and $\gamma(n) - \gamma(n-1) \geq c\gamma(n)$. Therefore $\gamma(n)(1-c) \geq \gamma(n-1)$ for any integer $n \geq 1$ and since $\gamma(0) = 1$ we have $\gamma(n) \geq \left(\frac{1}{1-c}\right)^n$.
(d) If Γ has subexponential growth, then there exists a sequence $\{n_j\} \subset \mathbb{N}$ such that $\{B_S(n_j)\}$ is Følner, see [12, §3.4].

(vii) **The Cheeger constant.** We finally define the *Cheeger constant* for a finite group Γ with respect to S as follows:

$$h_S(\Gamma) = \min\left\{ \frac{|E_S(D)|}{|D|} \;\middle|\; D \subset \Gamma,\; 0 < |D| \leq \frac{1}{2}|\Gamma| \right\}.$$

The Buser–Cheeger inequality for a finite group tells us that

$$\frac{1}{2}\lambda_1 \leq h_S(\Gamma) \leq \sqrt{2|S|\lambda_1},$$

where λ_1 is the smallest eigenvalue of the combinatorial Laplace operator $L : \mathbb{R}^\Gamma \to \mathbb{R}^\Gamma$, which is defined as

$$Lf(x) = \sum_{d_S(x,y)=1} (f(x) - f(y)).$$

See Propositions 4.2.4 and 4.2.5 in [15].

12.3 The Isoperimetric Inequality in Finitely Generated Groups and Some Consequences

12.3.1 Isoperimetric Inequalities in Free Groups and Free Abelian Groups

The general isoperimetric inequality for general finitely generated groups is discussed in the next subsection. Here we first consider the special case of free groups and free abelian groups. The following elementary result is probably well known to the experts, but we did not find it in the literature:

Proposition 12.3.1 *Let F_q be the free group on q generators $\{s_1, \ldots, s_q\} \subset F_q$, and consider the symmetric generating set $S = \{s_1, \ldots, s_q, s_1^{-1}, \ldots, s_q^{-1}\}$. Then for any finite, non empty subset $D \subset F_q$, we have*

$$\frac{|\partial D|}{|D|} \geq \frac{q-1}{q} + \frac{m}{q|D|}, \tag{12.3.1}$$

where m is the number of connected components of D. Furthermore, if the set D is connected, then the outer boundary satisfies the following identity:

$$|\partial' D| = (2q - 2)|D| + 2. \tag{12.3.2}$$

By definition, two vertices in a subset $D \subset F_q$ belong to the same connected component if and only if they can be joined by an edge path contained in D.

Proof Recall that the Cayley graph of F_q is the infinite regular tree of degree $2q$. We first prove the second statement by induction on the cardinality of D. If $|D| = 1$, then D contains exactly one point and the outer boundary of D is the set of all neighbors of that point. There are $2q$ such neighbors, parameterized by S, therefore Eq. (12.3.2) is trivially satisfied. Assume now that D is connected and contains at least two points and chose a base point x_0 in X and a point $x \in D$ at maximum distance from x_0 (such a point is sometimes called a *leaf* in D). Then x has one neighbor in D and $2q - 1$ neighbors outside D, denote them by $\{y_1, \ldots, y_{2q-1}\}$. Clearly, the outer boundary of $D^- = D \setminus \{x\}$ is given by

$$\partial' D^- = \left(\partial' D \setminus \{y_1, \ldots, y_{2q-1}\}\right) \cup \{x\}.$$

Therefore we have

$$|D| = |D^-| + 1 \quad \text{and} \quad |\partial' D| = |\partial' D^-| + (2q - 2),$$

and we conclude the proof of (12.3.2) by induction.

The proof of the first statement follows immediately for a connected set $D \subset F_q$ from (12.3.2) and the fact that $|\partial D| \geq \frac{1}{2q}|\partial' D|$. If D has several connected components $D_1, \ldots, D_m$, then the inner boundary ∂D is the disjoint union of the $\partial D_j = D_j \cap \partial D$ for any $j = 1, \ldots, m$ and we have

$$(2q - 2)|D| + 2m = \sum_{j=1}^{m} \left((2q - 2)|D_j| + 2\right)$$

$$= \sum_{j=1}^{m} |\partial' D_j| \leq 2q \sum_{j=1}^{m} |\partial D_j| = 2q|\partial D|,$$

which implies the inequality (12.3.1). $\qquad\qquad\square$

Note that the reason why (12.3.2) may fail for a disconnected set is that the outer boundaries of two different connected components of that set may not be disjoint. Regarding free abelian groups, we have:

Proposition 12.3.2 *For any finite, non empty, subset $D \subset \mathbb{Z}^d$ we have*

$$\frac{|\partial D|}{|D|} \geq \frac{1}{|D|^{1/d}}. \tag{12.3.3}$$

Proof The results follows from the *Loomis-Whitney Inequality* [14], which states that for any finite subset $D \subset \mathbb{Z}^d$ we have

$$|D|^{d-1} \leq \prod_{j=1}^{d} |\pi_j(D)|,$$

where $\pi_j : \mathbb{Z}^d \to \mathbb{Z}^{d-1}$ is the projection in direction of the jth coordinate. We thus have

$$|\partial D| \geq \max_{1 \leq j \leq d} |\pi_j(D)| \geq \left(\prod_{j=1}^{d} |\pi_j(D)| \right)^{\frac{1}{d}} \geq |D|^{\frac{d-1}{d}}.$$

$\square$

Inequality (12.3.3) can be improved for the outer boundary, see [16, Theorem 6.22].

12.3.2   Statement of the Main Result

The heart of the argument in Gromov's proof of Inequality (12.1.2) involves an estimate of the number of elements in a finite set $D \subset \Gamma$ that are moved outside D by the action of a ball $B(r) \subset \Gamma$ whose radius r is well chosen. To quantitatively express this estimate in terms of the growth function of the group, it is convenient to introduce the following notion:

Definition 12.3.3 Given an arbitrary subset $E \subset \mathbb{R}_+ = [0, +\infty)$ and a function $g : E \to \mathbb{R}_+$, we define a new function $\mathcal{U}_{E,g} : \mathbb{R}_+ \to \mathbb{R} \cup \{\infty\}$ by

$$\mathcal{U}_{E,g}(t) = \sup \left\{ \frac{1}{r} \left(1 - \frac{t}{g(r)} \right) \mid r \in E \setminus \{0\} \right\}. \tag{12.3.4}$$

We will call this function the *$\mathcal{U}$-transform* of g. When the domain $E \subset \mathbb{R}_+$ is fixed, we usually write $\mathcal{U}_g(t)$ instead of $\mathcal{U}_{E,g}(t)$.

As a first example, we mention that the $\mathcal{U}$-transform of the polynomial function $g(r) = (d + 1)r^d$ is the function $\mathcal{U}_g(t) = \frac{d}{d+1}t^{-1/d}$. More examples are given in Sect. 12.4, where some computational techniques and basic properties of the $\mathcal{U}$-transform will be given. We will in particular explain in Sect. 12.4.4 that the $\mathcal{U}$-transform is nothing else than a variant of the classical Legendre transform obtained by some change of variables.

Using the $\mathcal{U}$-transform, we now state the main result of the present chapter, which is the following version of isoperimetric inequality in finitely generated groups:

Theorem 12.3.4 *In an arbitrary group Γ, generated by the finite symmetric set $S = S^{-1} \subset \Gamma$, the following isoperimetric inequality holds for any non empty finite subset $D \subset \Gamma$:*

$$\frac{|\partial_S D|}{|D|} \geq \mathcal{U}_{\gamma_S}(|D|), \tag{12.3.5}$$

where the growth function $\gamma_S : \mathbb{N} \to \mathbb{N}$ and the boundary $\partial_S D$ are defined with respect to the generating set S. Furthermore, the following holds for the Følner function of (Γ, S):

$$\mathcal{U}_{\gamma_S}(\text{Føl}(n)) \leq \frac{1}{n}. \tag{12.3.6}$$

If Γ is a finite group, then its Cheeger constant satisfies

$$h_S(\Gamma) \geq \mathcal{U}_{\gamma_S}\left(\frac{1}{2}|\Gamma|\right). \tag{12.3.7}$$

We will prove this Theorem in the next subsection. We first derive the following result, which is stated in [19] and [21], and implies the inequality of Coulhon and Saloff-Coste:

Corollary 12.3.5 *Let Γ be a finitely generated group. For any non empty finite subset $D \subset \Gamma$ and any real number $\lambda > 0$, we have*

$$\frac{|\partial_S D|}{|D|} \geq \left(1 - \frac{1}{\lambda}\right)\frac{1}{\phi_S(\lambda|D|)}. \tag{12.3.8}$$

Choosing $\lambda = 2$ in this inequality gives us (12.1.2).

Proof If $\lambda|D| > |\Gamma|$, then $\phi_S(\lambda|D|) = \infty$ and there is nothing to prove. We thus assume that $\lambda|D| \leq |\Gamma|$ and set $r = \phi_S(\lambda|D|)$. Then $\lambda|D| \leq \gamma_S(r)$ and Theorem 12.3.4 implies that

$$\frac{|\partial_S D|}{|D|} \geq \mathcal{U}_{\gamma_S}(|D|) \geq \frac{1}{r}\left(1 - \frac{|D|}{\gamma_S(r)}\right) \geq \left(1 - \frac{1}{\lambda}\right)\frac{1}{\phi_S(\lambda|D|)}.$$

$\square$

12.3.3 Proof of the Main Theorem

Theorem 12.3.4 will be a direct consequence of the following stronger result:

Theorem 12.3.6 *Given an arbitrary group Γ generated by the finite symmetric set $S = S^{-1} \subset \Gamma$, the following isoperimetric inequality holds for any finite, non empty subset $D \subset \Gamma$:*

$$\frac{|\partial_S D|}{|D|} \geq \sup_{r \in \mathbb{N}} \left(\frac{\gamma_S(r) - |D|}{r \gamma_S(r) - \sum_{k=0}^{r-1} \gamma_S(k)} \right), \qquad (12.3.9)$$

where the growth function $\gamma_S : \mathbb{N} \to \mathbb{N}$ and the boundary $\partial_S D$ are defined with respect to the generating set S.

We will use the following notation: For any $k \in \mathbb{N}$, we denote by $S_S(k) = \{x \in \Gamma \mid \|x\|_S = k\}$ the sphere of radius k in Γ, and we set $\sigma_S(k) = |S_S(k)|$. Note that the ball $B_S(r)$ is the disjoint union of the spheres $S_S(k)$ for $0 \leq k \leq n$. In particular we have

$$\gamma_S(r) = \sum_{k=0}^{r} \sigma_S(k) \quad \text{and} \quad \sigma_S(k) = (\gamma_S(k) - \gamma_S(k-1)). \qquad (12.3.10)$$

Proof of Theorem 12.3.6 The proof follows a strategy similar to that in Gromov's book [11, §33], see also [16, §6.7]. We first claim that for any $y \in S(k)$, we have

$$|\{x \in D \mid yx \notin D\}| \leq k |\partial D|. \qquad (12.3.11)$$

Indeed, if $k = 0$ then $y = e$ and the claim is trivial. Moreover, the following inclusion holds for any $y \in \Gamma$ and any $s \in S$:

$$\{x \in D \mid syx \notin D\} \subset \{x \in D \mid yx \notin D\} \cup \{x \in D \mid yx \in \partial_S D\}, \qquad (12.3.12)$$

thus (12.3.11) follows by induction on k. Let us now set $P_r(D) = \{(x, y) \in D \times B_S(r) \mid yx \notin D\}$, we then have from (12.3.11):

$$|P_r(D)| = \sum_{k=0}^{r} |\{(x, y) \in D \times S_S(k) \mid yx \notin D\}| \leq |\partial D| \sum_{k=0}^{r} k \sigma(k). \qquad (12.3.13)$$

On the other hand, the following inequality is obvious for every $x \in D$ and any $r \in \mathbb{N}$:

$$|\{y \in B_S(r) \mid yx \notin D\}| = |\{y \in B_S(r) \mid y \notin Dx^{-1}\}| \geq |B_S(r)| - |D|.$$

This can be written as

$$\gamma_S(r) - |D| \leq |\{y \in B_S(r) \mid yx \notin D\}|. \tag{12.3.14}$$

From (12.3.13) and (12.3.14), one obtains the following inequalities for any $r \in \mathbb{N}$:

$$|D|\,(\gamma_S(r) - |D|) \leq |P_r(D)| \leq |\partial D| \sum_{k=0}^{r} k\,\sigma_S(k),$$

from which the inequality

$$\frac{|\partial_S D|}{|D|} \geq \sup_{r \in \mathbb{N}} \left(\frac{\gamma_S(r) - |D|}{\sum_{k=1}^{r} k\sigma(k)} \right) \tag{12.3.15}$$

follows immediately. Inequality (12.3.9) follows now from (12.3.15) and the obvious identity

$$\sum_{k=1}^{r} k\sigma(k) = r\gamma_S(r) - \sum_{k=0}^{r-1} \gamma_S(k).$$

□

Proof of Theorem 12.3.4 The inequality (12.3.5) follows now immediately from (12.3.9) and the definition (12.3.4) of the $\mathcal{U}$-transform:

$$\frac{|\partial_S D|}{|D|} \geq \sup_{r \in \mathbb{N}} \left(\frac{\gamma_S(r) - |D|}{r\gamma_S(r) - \sum_{k=0}^{r-1} \gamma_S(k)} \right) \geq \sup_{r \in \mathbb{N}} \left(\frac{\gamma_S(r) - |D|}{r\gamma_S(r)} \right) = \mathcal{U}_S(|D|).$$

Inequalities (12.3.6) and (12.3.7) are immediate consequences of (12.3.5) and the definitions of the Følner function and the Cheeger constant.

□

12.3.4 Some Consequences of the Main Result

In this section and the next one, we derive some consequences of Theorem 12.3.4. The proofs use the basic properties of the $\mathcal{U}$-transform developed in Sect. 12.4. We begin with the following statement on groups with polynomial growth:

Corollary 12.3.7 *Let* Γ *be a finitely generated group whose growth function* γ_S *satisfies* $\gamma_S(n-1) \geq Cn^d$ *for some constants* $C > 0$ *and* $d \geq 1$ *and any integer*

$n \geq 1$. *Then the following isoperimetric inequality*

$$\frac{|\partial_S D|}{|D|} \geq \frac{C^{\frac{1}{d}} d}{(d+1)^{1+\frac{1}{d}}} |D|^{-\frac{1}{d}} \tag{12.3.16}$$

holds for any finite, non empty subset $D \subset \Gamma$.

Proof The inequality (12.3.16) follows from Theorem 12.3.4 together with Lemma 12.4.2 and the computation (12.4.4) of the polynomial $\mathcal{U}$-transform given in Example 12.4.4 below.

□

By comparison, for the same growth function, the inequality (12.1.2) gives us the estimate:

$$\frac{|\partial_S D|}{|D|} \geq \frac{C^{\frac{1}{d}}}{2^{1+\frac{1}{d}} \left(1 - \left(\frac{C}{2|D|}\right)^{\frac{1}{d}}\right)} |D|^{-\frac{1}{d}}.$$

Note that for large d, the constant in the latter inequality is about one half that in (12.3.16).

For group with exponential growth, we have the following

Corollary 12.3.8 *Let Γ be a finitely generated group whose growth function γ_S satisfies $\gamma_S(n-1) \geq C e^{bn^\alpha}$ for some constants $0 < \alpha \leq 1$, $C, b > 0$, and any integer $n \geq 1$. Then we have*

$$\frac{|\partial_S D|}{|D|} \geq \left(\frac{b}{\log(|D|) + o(\log(|D|))}\right)^{1/\alpha} \tag{12.3.17}$$

for any finite subset $D \subset \Gamma$ with at least two elements.

Proof The result follows from Theorem 12.3.4 combined with the inequality (12.4.8) in Example 12.4.5 below.

□

Note that for the same growth function, (12.1.2) gives us

$$\frac{|\partial_S D|}{|D|} \geq \frac{1}{2^{1/\alpha}} \left(\frac{b}{\log(|D|) + \log(2/C)}\right)^{1/\alpha}.$$

Our next result gives lower bounds for the Følner function:

Corollary 12.3.9 *Let Γ be a finitely generated group with growth function γ_s.*

(a) If $\gamma_s(n-1) \geq Cn^d$ for some $d \geq 1$, then $\mathrm{Føl}(n) \geq \dfrac{Cd^d}{(1+d)^{1+d}} \cdot n^d$,

(b) If $\gamma_s(n-1) \geq C\,e^{bn}$ for some $b > 0$, then $\mathrm{Føl}(n) \geq \dfrac{C}{\alpha e b} \cdot \dfrac{\exp(bn^\alpha)}{n^\alpha}$.

Proof Using (12.3.6) and the calculation in Example 12.4.4, we see that if $\gamma_s(n-1) \geq Cn^d$, then

$$\frac{1}{n} \geq \mathcal{U}_{\gamma_S}(\mathrm{Føl}(n)) = \frac{dC^{\frac{1}{d}}}{(d+1)^{1+\frac{1}{d}}} \cdot \frac{1}{(\mathrm{Føl}(n))^{\frac{1}{d}}},$$

which proves (a).

To prove (b), we use the calculation in Example 12.4.5. In particular if $\gamma_s(n-1) \geq g(n) = C\,e^{bn^\alpha}$, then Eq. (12.4.9) with $u = 1/n$ and $t = \mathrm{Føl}(n)$ implies that

$$\mathcal{U}_{\gamma_S}(\mathrm{Føl}(n)) \leq \frac{1}{n} \quad \Rightarrow \quad \mathrm{Føl}(n) \geq \frac{C}{\alpha e b} \cdot \frac{\exp(bn^\alpha)}{n^\alpha}$$

$\square$

12.3.5 Asymptotic Estimates

In this section we formulate some asymptotic estimates on the isopermietric ratio and the Følner function for some groups with non polynomial growth. We will need the following somewhat technical definition:

Definition 12.3.10 We will say that a function $g : \mathbb{R}_+ \to \mathbb{R}_+$ has *Tame Superpolynomial Growth*, abbreviated as (TSPG), if it is everywhere differentiable, with $g' > 0$ and

$$\lim_{r \to \infty} \left(\frac{rg'(r)g(\lambda r)}{g(r)^2} \right) = \begin{cases} 0, & \text{if } 0 < \lambda < 1, \\ \infty, & \text{if } \lambda = 1. \end{cases} \tag{TSPG}$$

This condition can also be written for $f(r) = \log(g(r))$ as follows:

$$\lim_{r \to \infty} rf'(r) = \infty \quad \text{and} \quad \lim_{r \to \infty} \frac{rf'(r)}{\exp(f(r) - f(\lambda r))} = 0, \text{ (for any } 0 < \lambda < 1).$$

An obvious example is the exponential function. Other examples are the functions $g_1(r) = \exp(ar^\beta)$ with $a > 0$ and $0 < \beta \le 1$, $g_2(r) = \exp\left(\frac{ar}{\log(r+1)^\alpha}\right)$ for some $a > 0$ and any $\alpha \in \mathbb{R}$ and $g_3(r) = r^{\sqrt{r}}$.

This notion is interesting because the non polynomial group growths that are described in the literature are bounded below by functions satisfying (TSGP), see e.g. [9, 17]. Our next results are formulated under this hypothesis:

Corollary 12.3.11 *Let Γ be a finitely generated group whose growth function γ_s satisfies $\gamma_s(r-1) \ge g(r)$, for any $r \ge 1$, where $g : \mathbb{R}_+ \to \mathbb{R}_+$ satisfies the growth condition (TSPG). Then for any $\varepsilon > 0$, there exists $N = N(\varepsilon) \in \mathbb{N}$ such that the following inequality holds*

$$\frac{|\partial_S D|}{|D|} \ge \frac{1 - \varepsilon}{g^{-1}(|D|)}, \tag{12.3.18}$$

for any finite subset $D \subset \Gamma$ such that $|D| \ge N(\varepsilon)$.

Proof Inequality (12.3.18) follows from Theorem 12.3.4, together with Lemma 12.4.2 and Proposition 12.4.8 below.

$\square$

Using the previous result, we obtain the following asymptotic estimate for the Følner function of groups of intermediate growth, thus completing Corollary 12.3.9.

Corollary 12.3.12 *Let Γ be an infinite amenable group satisfying the hypothesis of Corollary 12.3.11. Then for any $\varepsilon > 0$, there exists $N = N(\varepsilon) \in \mathbb{N}$ such that for any $n \ge N$,*

$$\text{Føl}(n) \ge g((1 - \varepsilon)n). \tag{12.3.19}$$

In other words, we have

$$\liminf_{n \to \infty} \frac{g^{-1}(\text{Føl}(n))}{n} \ge 1. \tag{12.3.20}$$

Proof Suppose (12.3.19) does not hold, then there exists $\eta > 0$ and a sequence $n_j \subset \mathbb{N}$ such that $n_j \to \infty$ and $g^{-1}(\text{Føl}(n_j))/n_j \le (1 - \eta)$ for any integer j. From the definition of the Følner function, one can then find finite subsets $D_j \subset \Gamma$ such that $|D_j| = \text{Føl}(n_j)$ and

$$\frac{|\partial_S D_j|}{|D_j|} \le \frac{1}{n_j} \le \frac{1 - \eta}{g^{-1}(\text{Føl}(n_j))} = \frac{1 - \eta}{g^{-1}(|D_j|)}.$$

This inequality contradicts Corollary 12.3.11 since $\lim_{j \to \infty} |D_j| = \infty$.

$\square$

Remark 12.3.13 *Using the other version* Φ *of the Følner function, L. Bartholdi gave a direct proof of the following inequality:*

$$\Phi(n) \geq \frac{1}{2}g(n),$$

for any $n \in \mathbb{N}$, *see [1, page 455]. Because* $F\o l(n) \geq \Phi\left(\frac{n}{2}\right)$, *the above inequality implies*

$$F\o l(n) \geq \frac{1}{2}g\left(\frac{n}{2}\right),$$

which is slightly better than Proposition 14.100 in [6]. Note that Corollaries 12.3.9 and 12.3.12 improve this inequality.

The following question is inspired by the recent work of C. Pittet and B. Stankov in [19, 20], and the previous Corollary.

Question 12.3.14 Given a smooth, unbounded monotone increasing function g : $\mathbb{R} \to \mathbb{R}$, we ask for the asymptotically smallest possible Følner function among all groups Γ with a finite, symmetric, generating set S whose growth function satisfies $\gamma_S(n-1) \geq g(n)$. Specifically, given the function g, we ask for the value of

$$K_g = \inf\left\{ \liminf_{n\to\infty} \frac{g^{-1}(F\o l(n))}{n} \;\middle|\; \exists(\Gamma, S) \text{ with growth } \gamma_S(n-1) \geq g(n)\right\}.$$
$$(12.3.21)$$

For the polynomial case $g(r) = Cr^d$, Corollary 12.3.9 (a) gives us the following lower bound:

$$K_{Cr^d} \geq \frac{d}{(1+d)^{1+1/d}}.$$

On the other hand Corollary 12.3.12 implies that if for any function g satisfying (TSPG) we have

$$K_g \geq 1.$$

See also the more explicit Corollary 12.3.9 (b) for the exponential case.

To find an upper bound for K_g, one needs to estimate the size of a Følner sequence in an amenable group with growth $\geq g$. In [20], B. Stankov gave an example where $1 \leq K_g \leq 2$. More precisely, he considers the lamplighter group, that is the wreath product $\Gamma = \mathbb{Z} \wr (\mathbb{Z}/2\mathbb{Z})$, with a well chosen generating set S and he computes in this case that $\gamma_s(n-1) \geq 2^{n-1}$ and $\text{Føl}(n) \leq 4^n$. It follows that $g^{-1}(t) = \log_2(2t) = \log_2(t) + 1$ and thus

$$\frac{g^{-1}(\text{Føl}(n))}{n} \leq 2 + \frac{1}{n}.$$

We refer to [19] and [20] for additional results and related questions.

12.3.6 Final Remarks

We conclude this Section with a few specific remarks:

o The proof of Theorem 12.3.6, and therefore all the estimates in the chapter, also holds if we replace the inner boundary $\partial_S D$ with the outer boundary $\partial'_S D$. We only need to change the inclusion (12.3.12) by

$$\{x \in D \mid syx \notin D\} \subset \{x \in D \mid yx \notin D\} \cup \{x \in D \mid syx \in \partial' D\}, \quad (12.3.22)$$

and follow the same argument.
o The inverse growth function ϕ_S defined in (12.2.1) is a variant of the function defined in [18] and [7]. These authors use instead the function

$$\widetilde{\phi}_S(t) = \min\{n \in \mathbb{N} \mid \gamma_s(n) > t\}. \quad (12.3.23)$$

Observe that $\widetilde{\phi}_S(t) \geq \phi_S(t)$, therefore the isoperimetric inequality (12.3.8) still holds if one uses $\widetilde{\phi}_S$ instead of ϕ_S, and it is in fact a slightly weaker statement in that case.
o For further results and updated references on the Følner function and the isoperimetric profile, we refer to Chapter 14 of the book [6], especially Sections 14.10–14.13. In particular an upper bound for the Følner function of a finitely generated nilpotent group in terms of its growth function can be deduced from the proof of Theorem 14.102 in that book.

12.4 Some Properties and Calculations of the $\mathcal{U}$-Transform

In this final Section, we present some basic facts on the $\mathcal{U}$-transform and perform some computations that are used in Sect. 12.3.4. As explained in Sect. 12.4.4 below, the $\mathcal{U}$-transform is equivalent to the Legendre transform after some change of variables. However we find it convenient to give direct proofs of the properties of the $\mathcal{U}$-transform we shall use, rather than obtaining them as consequences of properties of the Legendre transform.

Recall that the $\mathcal{U}-$transform of the function $g : E \subset \mathbb{R}_+ \to \mathbb{R}_+$ is defined as

$$\mathcal{U}_{E,g}(t) = \sup \left\{ \frac{1}{r} \left(1 - \frac{t}{g(r)} \right) \mid r \in E \setminus \{0\} \right\}.$$

12.4.1 Basic Properties

We first gather some elementary facts:

Lemma 12.4.1 *The $\mathcal{U}$-transform satisfies the following properties:*

(i) *The function $t \mapsto \mathcal{U}_{E,g}(t)$ is non-increasing.*

(i) *If $g : E \to \mathbb{R}_+$ and $E' \subset E$, then $\mathcal{U}_{E',g}(t) \leq \mathcal{U}_{E,g}(t)$ for any $t \in \mathbb{R}_+$.*

(iii) *If $g_1, g_2 : E \to \mathbb{R}_+$ and $g_1(r) \leq g_2(r)$ for any $r \in E$, then $\mathcal{U}_{E,g_1}(t) \leq \mathcal{U}_{E,g_2}(t)$ for any $t \in \mathbb{R}_+$.*

(iv) *If $t < \sup_E g(r)$, then $\mathcal{U}_{E,g}(t) > 0$.*

(v) *If g is increasing and $t > \inf_E g(r)$, then $\mathcal{U}_{E,g}(t) < \infty$.*

(vi) *If $g, h : \mathbb{R}_+ \to \mathbb{R}_+$ are any functions such that $h(r) = cg(br)$ for some positive constants b, c, then*

$$\mathcal{U}_{\mathbb{R},h}(t) = b \cdot \mathcal{U}_{\mathbb{R},g}(t/c).$$

Proof The first four statements are easy consequences of the definition. Statement (v) follows from the fact that if $\mathcal{U}_{E,g}(t) = \infty$, then there exists a sequence $\{r_i\} \subset E \setminus \{0\}$ such that $r_i \to 0$ and $g(r_i) > t$, contradicting the hypothesis $\lim g(r_i) = \inf g(r_i) < t$. To prove the last statement, set $s = br$ and start with the identity

$$\frac{1}{r} \left(1 - \frac{t}{h(r)} \right) = \frac{1}{r} \left(1 - \frac{t}{cg(br)} \right) = \frac{b}{s} \left(1 - \frac{t/c}{g(s)} \right).$$

Taking the sup over r on the left hand side and over s on the right hand side proves (vi).

$\square$

The following Lemma is useful when comparing the $\mathcal{U}$-transform of a function defined on the natural numbers to that of a function defined on the whole set of positive real numbers.

Lemma 12.4.2 *Suppose that $g : \mathbb{R}_+ \to \mathbb{R}_+$ and $\gamma : \mathbb{N} \to \mathbb{N}$ are two functions such that g is non decreasing and $\gamma(k-1) \geq g(k)$ for any $k \in \mathbb{N}$, then $\mathcal{U}_{\mathbb{N},\gamma}(t) \geq \mathcal{U}_{\mathbb{R},g}(t)$ for any $t \in \mathbb{R}_+$.*

Proof The statement follows from the fact that the integer part $k = \lfloor r \rfloor$ of any real number r satisfies $k \leq r < k + 1$. From our hypothesis we then have $g(r) \leq g(k+1) \leq \gamma(k)$. Therefore we have

$$\frac{1}{r}\left(1 - \frac{t}{g(r)}\right) \leq \frac{1}{k}\left(1 - \frac{t}{\gamma(k)}\right),$$

and the lemma follows immediately.

$\square$

12.4.2 Computing the $\mathcal{U}$-Transform

To compute the $\mathcal{U}$-transform of a differentiable function $g : [0, \infty) \to \mathbb{R}_+$, one fixes $t > g(0)$ and sets $f(r) = \frac{1}{r}\left(1 - \frac{t}{g(r)}\right)$. We observe that f achieves its maximum on $\mathbb{R}_+$ since $f(r) \leq 0$ if $g(r) \leq t$, $f(r) > 0$ if $g(r) > t$ and $\lim_{r \to \infty} f(r) = 0$. Define now the function $\rho : (g(0), \infty) \to \mathbb{R}$ by

$$\rho(t) = \max\{r \in \mathbb{R}_+ \mid f(r) = \mathcal{U}_g(t)\}, \tag{12.4.1}$$

and observe that

$$\mathcal{U}_g(t) = \frac{1}{\rho(t)}\left(1 - \frac{t}{g(\rho(t))}\right). \tag{12.4.2}$$

We have thus reduced the computation of the $\mathcal{U}$-transform of g to that of $\rho(t)$.

Lemma 12.4.3 *The function $\tau : \mathbb{R}_+ \to \mathbb{R}_+$ defined by*

$$\tau(r) = \frac{g(r)}{1 + r\frac{g'(r)}{g(r)}} \tag{12.4.3}$$

is a left inverse of ρ.

Proof At any point r where the function f achieves its maximum we have

$$f'(r) = \frac{t}{r^2 g(r)}\left(1 + \frac{rg'(r)}{g(r)} - \frac{g(r)}{t}\right) = 0.$$

This condition can be written as $g(r) = t\left(1 + r\frac{g'(r)}{g(r)}\right)$, or equivalently as $t = \tau(r)$. In particular we have $\tau(\rho(t)) = t$.

$\square$

The procedure to compute the $\mathcal{U}$−transform of a differentiable function g is then as follows:

(i) Compute the function $\tau(r)$ from (12.4.3).
(ii) Compute $r = \rho(t)$ by solving $\tau(r) = t$.
(iii) $\mathcal{U}_g(t)$ is then given by (12.4.2).

The procedure works fine provided the function $\tau(r)$ is injective, which will be the case in all the examples we consider.

Example 12.4.4 As a first example, we consider the function $g(r) = cr^d$, then $\tau(r) = \dfrac{c}{d+1}r^d$, therefore $\rho(t) = \left(\frac{(d+1)}{c}t\right)^{1/d}$ and we have

$$\mathcal{U}_g(t) = \frac{1}{\rho(t)}\left(1 - \frac{t}{g(\rho(t))}\right) = \left(\frac{dc^{\frac{1}{d}}}{(d+1)^{1+\frac{1}{d}}}\right) \cdot t^{-\frac{1}{d}}. \qquad (12.4.4)$$

Our next example will be to compute the $\mathcal{U}$-transform of exponential functions. The computation is tricky and it will be convenient to introduce the auxiliary function $f : [e, \infty) \to [1, \infty)$ defined as

$$f(x) = \log(x) + \log(f(x)). \qquad (12.4.5)$$

The function f is well-defined; it is the inverse function of $y \mapsto e^y/y$ (for $y \geq 1$ and $x \geq e$). Moreover it is monotone increasing with $f(e) = 1$ and $\lim_{x \to \infty} f(x) = \infty$. It is clear that $f(x) \geq \log(x)$, but more precisely we claim that

$$f(x) = \log(x) + \log(\log(x)) + o(1). \qquad (12.4.6)$$

To see this, let us write

$$f(x) = a(x)\log(x),$$

then

$$a(x) = \frac{f(x)}{\log(x)} = \frac{f(x)}{f(x) - \log(f(x))}.$$

This function is positive, decreasing, and $\lim_{x \to \infty} a(x) = 1$. Therefore $a(x) = 1 + o(1)$ and we have

$$f(x) = \log(x) + \log(f(x)) = \log(x) + \log\left(a(x)\log(x)\right)$$
$$= \log(x) + \log\left(\log(x)\right) + o(1).$$

as claimed above.

Remark We can also express f in terms of the *Lambert W-function*:

$$f(x) = -W(-1/x).$$

The function W is defined to be the inverse of the function $h(y) = y e^y$. Since h is not injective, W is multivaluated, and here we appeal to the branch (sometimes denoted as W_{-1}) corresponding to the ranges $-1/e < x < 0$ and $-\infty < y < -1$. We refer to [3, 5] for more on the Lambert Function.

Example 12.4.5 We now compute the $\mathcal{U}$-transform of the function $g(r) = ce^{br^\alpha}$. We have $\tau(r) = \frac{ce^{br^\alpha}}{1 + \alpha br^\alpha}$, and the relation $\tau(\rho(t)) = t$ is thus equivalent to

$$b\rho(t)^\alpha = \log(t/c) + \log\left(1 + \alpha b\rho(t)^\alpha\right).$$

This identity can be written as

$$\left(b\rho(t)^\alpha + \frac{1}{\alpha}\right) = \log(\lambda t) + \log\left(b\rho(t)^\alpha + \frac{1}{\alpha}\right),$$

with $\lambda = \frac{\alpha}{c} e^{1/\alpha}$. The solution is given by

$$b\rho(t)^\alpha + \frac{1}{\alpha} = f(\lambda t),$$

where $f(x)$ is the function defined in (12.4.5). From (12.4.2), one then obtains

$$\mathcal{U}_g(t) = \frac{1}{\rho(t)}\left(1 - \frac{t}{ce^{b\rho(t)^\alpha}}\right) = \frac{\alpha b\rho(t)^{\alpha-1}}{1 + \alpha b\rho(t)^\alpha} = \frac{b^{1/\alpha}\left(f(\lambda t) - \frac{1}{\alpha}\right)^{1-\frac{1}{\alpha}}}{f(\lambda t)},$$

which be written as

$$\mathcal{U}_g(t) = \left(\frac{b}{f(\lambda t)}\right)^{\frac{1}{\alpha}}\left(1 - \frac{1}{\alpha f(\lambda t)}\right)^{1-\frac{1}{\alpha}}. \tag{12.4.7}$$

Note that if $0 < \alpha \leq 1$, then $\left(1 - \frac{1}{\alpha f(\lambda t)}\right)^{1-\frac{1}{\alpha}} \geq 1$ and we have therefore

$$\mathcal{U}_g(t) \geq \left(\frac{b}{f(\lambda t)}\right)^{\frac{1}{\alpha}} = \left(\frac{b}{\log(t) + o(\log(t))}\right)^{\frac{1}{\alpha}}. \tag{12.4.8}$$

It will also be useful to consider the inverse of the function $\mathcal{U}_g(t)$. For $g(r) = ce^{br^\alpha}$ we claim that for any $0 < \alpha \leq 1$

$$u = \mathcal{U}_g(t) \quad \Rightarrow \quad t \geq \frac{cu^\alpha}{\alpha b} \cdot \exp\left(\frac{b}{u^\alpha} - 1\right), \tag{12.4.9}$$

with equality if $\alpha = 1$. Indeed, we have

$$u = \mathcal{U}_g(t) \geq \frac{b}{f(\lambda t))} \quad \Leftrightarrow \quad f(\lambda t) \geq \frac{b}{u^\alpha}.$$

Using the relation $x = \exp(f(x))/f(x)$, one gets

$$\lambda t = \frac{\exp(f(\lambda t))}{f(\lambda t)} \geq \frac{u^\alpha}{b} \exp\left(\frac{b}{u^\alpha}\right),$$

which is equivalent to (12.4.9) since $\lambda = \frac{\alpha}{c} e^{1/\alpha}$.

12.4.3 Asymptotic Behavior

The $\mathcal{U}$-transform of a general function is generally not computable in a closed form, but (12.4.2) gives us some useful estimates. The first result in this direction is the following

Lemma 12.4.6 *If $g(r)$ is monotone increasing, then $\mathcal{U}_g(t) \leq \dfrac{1}{g^{-1}(t)}$.*

Proof Since $\tau(r) \leq g(r)$ for any $r > 0$, we have $t = \tau(\rho(t)) \leq g(\rho(t))$ for any t. Because g is monotone increasing, so is its inverse, therefore $g^{-1}(t) \leq \rho(t)$ for any t and we conclude that

$$\mathcal{U}_g(t) = \frac{1}{\rho(t)}\left(1 - \frac{t}{g(\rho(t))}\right) \leq \frac{1}{\rho(t)} \leq \frac{1}{g^{-1}(t)}.$$

$\square$

Lemma 12.4.7 *If $g(r)$ is unbounded and monotone increasing, then $\lim\limits_{t\to\infty} \rho(t) = \infty$.*

Proof Use again that $\tau(r) \leq g(r)$ for any r. This implies that $t = \tau(\rho(t)) \leq g(\rho(t))$ and thus $\rho(t) \geq g^{-1}(t)$ and therefore $\lim_{t \to \infty} \rho(t) = \lim_{t \to \infty} g^{-1}(t) = \infty$.

$\square$

The next result says that the $\mathcal{U}$-transform of any function $g : \mathbb{R}_+ \to \mathbb{R}_+$ satisfying (TSPG) (see Definition 12.3.10) is asymptotically given by $1/g^{-1}(t)$.

Proposition 12.4.8 *Let* $g : \mathbb{R}_+ \to \mathbb{R}_+$ *be a strictly increasing differentiable function satisfying* (TSPG), *then for any* $\varepsilon > 0$, *the following inequalities hold for t large enough:*

$$\frac{1 - \varepsilon}{g^{-1}(t)} \leq \mathcal{U}_g(t) \leq \frac{1}{g^{-1}(t)}. \tag{12.4.10}$$

Proof The inequality $\mathcal{U}_g(t) \leq 1/g^{-1}(t)$ has already been proved in the previous Lemma. To prove the other inequality, we first observe that (TSPG) has the following consequences:

(i) $\lim_{r \to \infty} \dfrac{rg'(r)}{g(r)} = \infty$.

(ii) $\lim_{r \to \infty} \dfrac{g(\lambda r)}{g(r)} = 0$ for any $\lambda \in [0, 1)$.

(iii) $\lim_{r \to \infty} \dfrac{g(\lambda r)}{\tau(r)} = 0$ for any $\lambda \in [0, 1)$.

Using (i) and the definition of $\tau(r)$, we obtain

$$\lim_{t \to \infty} \frac{t}{g(\rho(t))} = \lim_{t \to \infty} \frac{\tau(\rho(t))}{g(\rho(t))} = \lim_{r \to \infty} \frac{\tau(r)}{g(r)} = \lim_{r \to \infty} \frac{1}{1 + r\frac{g'(r)}{g(r)}} = 0$$

(we use Lemma 12.4.7 in the second equality). It follows that for any $\varepsilon > 0$, one can find $t(\varepsilon)$ large enough so that

$$\left(1 - \frac{t}{g(\rho(t))}\right) \geq \left(1 - \frac{\varepsilon}{2}\right)$$

for any $t \geq t(\varepsilon)$.

On the other hand, (iii) implies that for any $0 < \varepsilon < 1$, there exists $r(\varepsilon)$ such that $g((1 - \varepsilon/2)r) \leq \tau(r)$ for any $r > r(\varepsilon)$. Together with the monotonicity of g^{-1}, this implies

$$\left(1 - \frac{\varepsilon}{2}\right)r \leq g^{-1}(\tau(r))$$

for any $r > r(\varepsilon)$. Enlarging $t(\varepsilon)$ if necessary, we may also assume that $\rho(t) \geq r(\varepsilon)$, for any $t \geq t(\varepsilon)$ (we use again Lemma 12.4.7 here), which then implies

$$\left(1 - \frac{\varepsilon}{2}\right) \rho(t) \leq g^{-1}\left(\tau(\rho(t))\right) = g^{-1}(t),$$

or equivalently

$$\left(1 - \frac{\varepsilon}{2}\right) \frac{1}{g^{-1}(t)} \leq \frac{1}{\rho(t)}.$$

We conclude that for any $t \geq t(\varepsilon)$:

$$\mathcal{U}_g(t) = \frac{1}{\rho(t)}\left(1 - \frac{t}{g(\rho(t))}\right) \geq \frac{1}{g^{-1}(t)}\left(1 - \frac{\varepsilon}{2}\right)^2 \geq \frac{1}{g^{-1}(t)}(1 - \varepsilon).$$

$\square$

Remark Some of the above proofs are slightly delicate because a priori τ is not invertible. The next lemma gives us a sufficient condition for this to be the case:

Lemma 12.4.9 *Let* $g : [0, \infty) \to \mathbb{R}_+$ *be a monotone increasing function of class* C^2 *such that* $h(r) = \log(g(r))$ *is concave, then the function* τ *is strictly increasing.*

Proof By concavity of $h(r) = \log(g(r))$ we have

$$\tau'(r) = \tau(r) \cdot \frac{r(h'(r)^2 - h''(r))}{1 + rh'(r)} \geq \tau(r) \cdot \frac{r(h'(r)^2)}{1 + rh'(r)} > 0.$$

$\square$

12.4.4 Comparison with the Legendre Transform

The $\mathcal{U}$-transform defined in (12.3.4) is equivalent to the Legendre transform, but is in some way more adapted to studying the asymptotic behavior of a function at infinity. Recall that the *Legendre transform* (also called the *Legendre–Fenchel transform*) of an arbitrary function $f : \mathbb{R}_+ \to \mathbb{R}$ is the function $\mathcal{L}_f : \mathbb{R}_+ \to \mathbb{R}_+ \cup \{+\infty\}$ defined as

$$\mathcal{L}_f(y) = \sup\{yx - f(x) \mid x \in \mathbb{R}_+\}.$$

The relationship between the $\mathcal{U}$-transform and the Legendre transform is the following

Lemma 12.4.10 *If $f, g : \mathbb{R}_+ \to \mathbb{R}$ are two arbitrary functions such that* $f(x)g\left(\frac{1}{x}\right) \equiv x$, *then*

$$\mathcal{U}_g(t) = t \, \mathcal{L}_f\left(\frac{1}{t}\right). \tag{12.4.11}$$

The proof is elementary: we have with $x = 1/r$ and $y = 1/t$

$$\frac{1}{r}\left(1 - \frac{t}{g(r)}\right) = t\left(\frac{1}{tr} - \frac{1}{rg(r)}\right) = t\,(yx - f(x))\,.$$

Taking the supremum in the above identity with respect to r on the left hand side and x on the right hand side yields (12.4.11).

Acknowledgments The authors are thankful to L. Bartholdi, J. Brieussel, N. Monod, A. Papadopoulos and C. Pittet for their comments and suggestions.

References

1. L. Bartholdi, Amenability of groups and g-sets, in *Sequences, Groups, and Number Theory. Trends in Mathematics*, ed. by V. Berthé, M. Rigo (Birkhäuser, Cham, 2018), pp. 433–544
2. H. Bass, The degree of polynomial growth of finitely generated nilpotent groups. Proc. Lond. Math. Soc. (3) **25**, 603–614 (1972)
3. A.F. Beardon, The principal branch of the Lambert W function. Comput. Methods Funct. Theory **21**(2), 307–316 (2021)
4. J. Brieussel, Growth behaviors in the range $\exp(r^\alpha)$. Afr. Mat. **25**(4), 1143–1163 (2014)
5. R.M. Corless, G.H. Gonnet, D.E.G. Hare, D.J. Jeffrey, D.E.Knuth, On the Lambert W function. Adv. Comput. Math. **5**(4), 329–359 (1996)
6. T. Ceccherini-Silberstein, M. D'Adderio, *Topics in Groups and Geometry*. Springer Monographs in Mathematics (Springer, Cham, 2021)
7. T. Coulhon, L. Saloff-Coste, Isopérimétrie pour les groupes et les variétés. Revista Matemática Iberoamericana **9**(9), 293–314 (1993)
8. A. Erschler, T. Zheng, Growth of periodic Grigorchuk groups. Invent. Math. **219**(3), 1069–1155 (2020)
9. R. Grigorchuk Milnor's problem on the growth of groups and its consequences, in *Frontiers in Complex Dynamics*. Princeton Mathematical Series, vol. 51 (Princeton University Press, Princeton, 2014), pp. 705–773
10. M. Gromov, Asymptotic invariants of infinite groups, in *Geometric Group Theory (Sussex, 1991)*, vol. 2. London Mathematical Society Lecture Note series, vol. 182 (Cambridge University Press, Cambridge, 1993), pp. 1–295
11. M. Gromov, Metric Structures for Riemannian and Non-Riemannian Spaces. Progress in Mathematics, vol. 152 (Birkhäuser Boston, Inc., Boston, 1999)
12. P. de la Harpe, *Topics in Geometric Group Theory* (The University of Chicago Press, Chicago, 2000)
13. P. de la Harpe, On the prehistory of growth of groups. ArXiv:2106.02499

14. L. Loomis, H. Whitney, An inequality related to the isoperimetric inequality. Bull. Am. Math. Soc **55**, 961–962 (1949)
15. A, Lubotzky, *Discrete Groups, Expanding Graphs and Invariant Measures*. With an appendix by Jonathan D. Rogawski. Progress in Mathematics, vol. 125 (Birkhäuser Verlag, Basel, 1994)
16. R. Lyons, Y. Peres, *Probability on Trees and Networks* (Cambridge University Press, Cambridge, 2016)
17. A. Mann, *How Groups Grow*. London Mathematical Society Lecture Note Series, vol. 395 (Cambridge University Press, Cambridge, 2012)
18. C. Pittet, L. Saloff-Coste, Amenable groups, isoperimetric profiles and random walks, in *Geometric Group Theory Down Under*. Proceedings of a Special Year in Geometric Group Theory (Canberra, 1996) (de Gruyter, Berlin, 1999), pp. 293–316
19. C. Pittet, B. Stankov, Coulhon Saloff-Coste isoperimetric inequalities for finitely generated groups. arXiv:2211.03227v1
20. B. Stankov, Exact descriptions of Følner functions and sets on wreath products and Baumslag-Solitar groups Bogdan Stankov. arXiv:2111.09158v2
21. B.-L. Santos Correia, M. Troyanov, On the isoperimetric inequality in finitely generated groups. arXiv:2110.15798
22. N. Varopoulos Analysis on Lie groups. J. Funct. Anal. **76**, 346–410 (1988)
23. A. Vershik, Amenability and approximation of infinite groups. Selecta Math. Soviet. **2**(4), 311–330 (1982)

Index

Printed in the United States
by Baker & Taylor Publisher Services